D1918533

Signaling and Communication in Plants

Series Editors

František Baluška
Department of Plant Cell Biology, IZMB, University of Bonn, Kirschallee 1, D-53115 Bonn, Germany

Jorge Vivanco
Center for Rhizosphere Biology, Colorado State University, 217 Shepardson Building, Fort Collins, CO 80523-1173, USA

For further volumes:
http://www.springer.com/series/8094

Preface

(...) the quest for an answer to the riddle, "What is Life?" is one of the grand themes that resonate through the scientific conversation of this century (...). That riddle embraces and transcends the subject matter of all the biological sciences, and much of physical science as well. A physics that has no place for life is as impoverished as would be a biology not informed by chemistry. The study of life as a natural phenomenon, a fundamental feature of the universe, must not be allowed to slip into the black hole of departmental tribalism.

Franklin M. Harold (2001)

The great successes of science in the last one and a half century built a strong conviction that chemical reactions and interactions between molecules lie at the basis of life. Starting with physiological chemistry, through biochemistry and physiology, up to molecular biology, -omics, systems biology, and now also synthetic biology, they all provided a very detailed picture of the chemical nature of cells and organisms. Only in some areas of natural sciences, the emerging data were suggesting that biology means more than chemistry itself. Electrophysiology, bioenergetics, the phenomenon of photosynthesis on one side, and the properties of wood, cotton fibers, silk, or spiderweb as construction and engineering materials on the other, are only a handful of such cases. Research of recent years, however, is more and more evidently indicating that physical forces are profoundly affecting the functioning of life at all levels of its organization. To detect and to respond to such forces, cells and organisms, among them plants, need to be organized physically, and mechanically in particular (Wang et al. 2009). Although the structure–function relationship is studied for decades at all levels of hierarchical organization, the knowledge about its physical aspects is still in the making. Macromolecular crowding, the importance of electrical forces, regulation of molecular machineries via structural organization of cellular compartments, direct mechanical connections between the interior and peripheries of the cell, cellular adhesion, and mechanical integration of cells within multicellular organism are examples of the recently investigated processes with strong physical side. Due to historical reasons, and also significant medical potential, the importance of mechanical environment on the functioning/fate of the cells is becoming very well recognized in animal cell biology, and also for bacterial cells. The least documented are the mechanobiological phenomena in plant cells and plants itself. The interplay between cell walls and the turgor pressure, the basis for creation of the biggest organisms on Earth, is on the other hand the major barrier in revealing the mysteries of the structural and functional integrity of the cells.

The mechanical aspects of plant life could be analyzed at many different levels of hierarchical organization. Here, we are trying to demonstrate how the awareness of the physical side of life is affecting the interpretation of biological phenomena. This book gathers contributions from many authors describing the importance of mechanical forces/stimuli for or mechanical organization of (1) supramolecular structures, like the cytoskeleton or cell walls; (2) cellular integrity, like cytoplasmic streaming and movement of organelles; (3) supracellular coordination in the processes of plant organ growth and development; (4) integration of plant functioning, e.g., in long-range water transport or plant responses to physical forces or environmental stimuli. The chapters are organized in a way to give the reader the possibility to travel along the ladder of hierarchical levels in a bottom-up approach, i.e., from molecules through cells and organs, and up to plants interacting with their immediate neighborhood or responding to stresses. Thus, the book covers all the major aspects of mechanobiological phenomena, providing also direct or indirect evidence for the organismal nature of plants – a feature which could only very rarely be seen in multicellular animals.

Immanuel Kant, in his *Metaphysische Anfangsgründe der Naturwissenschaft* (1786), noted "(...) in any special doctrine of nature there can be only as much *proper* science as there is *mathematics* therein". Using this saying, one can observe that in any biological process there might be only as much freedom as the physical laws would allow. Mechanical integration of living cells and organisms constitutes a visible expression of the unity and intertwining connections between the living and inanimate parts of nature.

Poznań, January 2011 Przemysław Wojtaszek

References

Harold FM (2001) The way of the cell. Molecules, organisms and the order of life. Oxford University Press, Oxford, New York

Kant I (1786) Metaphysische Anfangsgründe der Naturwissenschaft (Metaphysical Foundations of Natural Science)

Wang N, Tytell JD, Ingber DE (2009) Mechanotransduction at a distance: mechanically coupling the extracellular matrix with the nucleus. Nat Rev Mol Cell Biol 10:75–82

Contents

Introduction: Tensegral World of Plants

Anna Kasprowicz, Michalina Smolarkiewicz, Magdalena Wierzchowiecka, Michał Michalak, and Przemysław Wojtaszek

Abstract In this chapter, we are providing a brief overview of the tensegral concept as applied to plants. Starting with a short introduction to the history of the idea of mechanical integration of the cell and the organism, we then discuss the mechanical design of the plant body. The importance of the mechanical properties of cells, tissues, organs, and their domains is indicated, and the systems of detection of mechanical stimuli are briefly discussed. Finally, the mechanical integration of plant cells is presented based on the various aspects of the functioning of the cell wall–plasma membrane–cytoskeleton continuum spanning the whole cell. The initial stage of knowledge within this area is indicated with special attention paid to different modes of inter- and intracellular communication as well as the utilization of the continuum to functional organization and integration of the whole cell.

1 Introduction

The significant role of mechanical forces in growth and development has been studied for over a century in a wide range of plant and animal species (Darwin and Darwin 1880; Thomson 1992). In the last decade, noticeable progress has been made in understanding the molecular background of these mechanical responses (Kasprowicz et al. 2009; Monshausen and Gilroy 2009). Mechanical forces influence living organisms at all levels of organization, from organismal through tissue and organ down to the cellular level. The specific reactions can be evoked either by macroenvironmental stimuli, such as wind, or microenvironmental stimuli resulting from, for example, differential pressure exerted by neighboring cells. These responses

A. Kasprowicz, M. Smolarkiewicz, M. Wierzchowiecka, M. Michalak, and P. Wojtaszek (✉)
Department of Molecular and Cellular Biology, Faculty of Biology, Adam Mickiewicz University, Umultowska 89, 61-614 Poznań, Poland
e-mail: przemow@amu.edu.pl

P. Wojtaszek (ed.), *Mechanical Integration of Plant Cells and Plants*,
Signaling and Communication in Plants 9, DOI 10.1007/978-3-642-19091-9_1,
© Springer-Verlag Berlin Heidelberg 2011

are thought to regulate not only cell growth and development, but also affect their shape or fate. Although significant knowledge has been gathered about such mechanisms in animals, the full picture of plant reactions is only starting to be built (Vogel and Sheetz 2006; Ingber 2008; Monshausen and Gilroy 2009).

The idea that changes in local mechanical environment can regulate growth, fate, shape, and pattern formation is not a new one. Even before the famous D'Arcy Thompson's book *On Growth and Form* (1992), first published in 1917, such suggestions could be found in literature. Many of the biophysical considerations, for example, linking the properties of the cell walls to plant morphogenesis, were later elaborated in the papers by Paul Green (1962, 1996, 1999), and elegantly summarized by Karl J. Niklas' book *Plant Biomechanics. An Engineering Approach to Plant Form and Function* (1992). However, it was Donald Ingber who first paid particular attention to the role of the cytoskeleton in the responses to mechanical stimuli and the maintenance of structural integrity of animal cells (reviewed in Ingber 2003a, b). Using several diverse approaches to study the spreading of animal cells on patterned substrata, equivalent to extracellular matrix (ECM) islands, his group demonstrated that such differentiated microcontacts affected not only the shape of individual cells, but also cell fate. The mechanical factor was sufficient to direct the cells to proliferation, differentiation, or death (Singhvi et al. 1994; Chen et al. 1997; Parker et al. 2002; Brock et al. 2003). It means that mechanical stress must have been somehow perceived by cell surface and then translated into biochemical message, which altered cell metabolism. To explain this phenomenon, a new conceptual framework was needed. Instead of elastic balloon (plasma membrane) filled with viscous cytoplasm, Ingber visualized cell as a tent (membrane) outstretched on the cytoskeletal backbone (Ingber 1993, 1998). The idea was generally based on a tensegral architecture model. The term "tensegrity" was introduced by American architect Buckminster Fuller, and referred to structures relying on the balance between tensile and compressive components, which results in a dynamic state of constant prestress that mechanically stabilizes all constituents (Fuller 1961). In the animal cell, contractile actomyosin filaments are responsible for generating tension forces which are resisted by ECM, neighboring cells, and other cytoskeletal components, such as microtubules which are considered to be resistant to compression. All elements are in a state of isometric tension and disturbance of one building block immediately induces alterations in other elements. The whole system is well balanced and becomes highly responsive to external mechanical stimuli (reviewed in Ingber et al. 1994; Ingber 2003b). In the tensegral model, a special emphasis is put on the internal, but not a cortical, actin network. This is especially important when the transduction of mechanical signal is considered, and distinguishes the tensegral model among others. In a more classical view, the assumption is that stress is equally distributed to the whole surface of plasma membrane and the primary load-bearing structure is a dense network of cortical actin filaments (Schmid-Schönbein et al. 1995; Ingber 2003a). To the opposite, the tensegral model assumes that mechanical stress is perceived by specialized receptor proteins spanning the plasma membrane (integrins in animals), and then transmitted through cortical and internal cytoskeleton networks to the interior of the cell, most often directly to nucleus (Ingber 2008, 2009; Wang et al. 2009). Thus, a whole living organism can be perceived as not only a hierarchical but also tensegral structure (Ingber 2006). If so, then the mechanical disturbance at the organismal level can be simply transduced to tissue, cellular and finally molecular levels. These general theses of the tensegrity model were formulated based on the experiments done on animal cells. However, after some reconsideration, they can also be applied to plants and plant cells.

2 Mechanical Design of Plant Body

Living organisms must follow physical laws limiting their size, form, and structure. The shape of a plant is defined by growth rate and growth direction (Hamant and Traas 2010), although orientation of cellular divisions is also an important variable. Plant mechanical design relies on cell wall properties, existence of supporting tissues as well as spatial organization of plant body (Niklas 1992; Nachtigal 1994). Biochemical/mechanical relationships influence the morphological/anatomical features that scale up with respect to body size, geometry and shape rather than with achievement of an organized body plan. In Plant Kingdom, one noticeable observation is that different clades have adopted body architectures allowing for survival in a particular ecological niche, due to diversification within and convergence among different lineages (Niklas 2000). In contrast to animals, depending on various nutrient sources (e.g., carnivores and herbivores), all plants require the same resources for growth and development (light, water, atmospheric gases, minerals, space) and are thus more influenced by the abiotic factors than by the biotic ones. This is also the reason for a much higher level of developmental plasticity, as elegantly summarized by E.J.H. Corner (2002): “A plant is a living thing that absorbs in microscopic amounts over its surface what it needs for growth. It spreads therefore an exterior whereas the animal develops, through its mouth, an interior.”

Plant form arises from the multitude of molecular/biochemical and physical interactions that occur within a growing assembly of cells. Although molecular techniques are now providing large amounts of information about the behavior of cells and their functioning, little is known about how those processes integrate within whole plant tissues and organs. Plant organogenesis as well as patterning is dependent on mechanical forces acting inside and outside of the plant body. In complex plant organisms, the position and shapes of cells, tissues, meristems, and organs bear repeated and regular relationships with one another. The spatial organization and regularity of patterns are apparent not only at the macroscopic level, but also in meristems and tissues, which contain repeated and predictable arrangements of various types of cells. Patterning in plants is highly ordered and the mechanisms of interaction and communication between cells are crucial to understanding how cellular activities are coordinated during development (Dupuy et al. 2008). Specific patterns of plant organ emergence and formation are dependent on a complex network of hormonal (mainly auxin) and mechanical signaling (Kuhlemeier 2007). Manual bending of roots can induce lateral root formation (Ditengou et al. 2008). Stress-driven buckling is a model proposed for primordium initiation in sunflower capitulum and phyllotaxial events on shoot apical meristem (Dumais 2007; Newell et al. 2008; Yin et al. 2008; see also Chapter 6).

Because plants germinate only with simple embryonic root and cotyledons, they have the potential to respond to changing environmental conditions during their development and formation of plant body. Roots, stem, branches, leaves, and flowers emerge during plant life and, although genetically encoded, are characterized by developmental plasticity optimized to fit into particular ecological niche (Deak and Malamy 2005). For example, even genetically identical plants grown in slightly different conditions of nutrient availability will develop different root architectures (López-Bucio et al. 2003). Organ plasticity is also important for effective relations between neighboring plants. As plants develop larger branches on less shaded side, the branches of neighboring trees can influence the growth direction of other tree branches (Novoplansky et al. 1989; Henriksson 2001). For sessile organisms, such adaptive strategies are crucial to survive in ever-changing climatic and environmental conditions (Deak and Malamy 2005; Malamy 2005).

Plants consist of a number of tissues characterized by different stiffness and mechanical properties. Sclerified tissues, e.g., sclerenchyma and wood fibers, are composed mainly of dead cell fibers with much stiffer cell walls (Niklas 1992). Hydrostatic and sclerified tissues play the supporting role. In living hydrostatic tissues the flexural and torsional rigidities of cells and tissues are a consequence of active interplay between turgor pressure and mechanical properties of cell walls (Niklas 1992). The hydrostatic pressure across the plasma membrane exceeding 2 MPa could be used for mechanical stabilization of plant bodies (Peters et al. 2000). Small herbs are mostly hydrostatic, whereas trees rely mostly on sclerified tissues. Intermediate situations could be found in grass stems or midribs (Niklas 1992; Moulia and Fournier 1997), where stiff sclerenchymous hypodermic exoskeleton and internal spongy parenchymous tissues are necessary for bending (Moulia et al. 2006). In aqueous environment, hydrostatic, buoyant plant cells do not exert significant pressure on themselves and neighboring cells, and different design of cell walls is required. On the other hand, the terrestrial environment requires adaptive solutions allowing the cell, that is approximately 1,000 times more dense than surrounding air, to withstand internal compressive forces (Niklas 2000). Compression-resistant turgid protoplast is surrounded by and pressed against tension-resistant and mechanically stable cell walls (Wojtaszek 2000; Zonia and Munnik 2007). Lignified and thick cell walls provide tissues with higher resistance to changing water conditions or elastic deformation, and this is indispensible when coping with herbivoral or microbial attack.

To resist exposure to external bending and twisting forces resulting from water flow in aqueous environment or wind in terrestrial one, a cylindrical body shape with thick boundary layer is best suited (Niklas 2000). Compared with the cell walls of the internal cell types, the outer epidermal walls are up to ten times thicker and more resistant to mechanical stress. Other cells are tightly packed inside this rigid cylinder of epidermis. Growth of elongated organs such as cylindrical stem or coleoptiles composed of different tissues with different mechanical properties causes longitudinal tissue tensions, leading to transfer of wall stress from inner to external cell layers that take control over organ growth (Schopfer 2006). This creates an additional supracellular pressure allowing the cells of the organ's interior to sense and adjust their mechanical balance to direct physical environment as well as to transduce mechanical stimuli throughout the organ (Kutschera 1995, 2008). A very good example illustrating this phenomenon is the reorientation of the cell fate. Laser ablation of selected cells from the root meristem of *Arabidopsis thaliana* leads to reorganization of the cell division planes in remaining cells, enabling effective and rapid filling of the empty space. Interestingly, daughter cells change their original fate and adopt a new one according to their new position in the root (van den Berg et al. 1995, 1997).

Plants possess very sensitive detectors of mechanical stimuli. Even subtle, short touch can induce not only mechanisms of signal transduction, but also immediate wall remodeling as evidenced by an induced expression of *TCH4* gene coding for xyloglucan endotransglycosidase (XTH; Braam and Davis 1990; Braam 2005). Genome-wide analysis of expression patterns in touch-stimulated *Arabidopsis* plants revealed that expression of 589 genes was upregulated within 30 min of touch stimulation, while 171 genes were downregulated. Importantly, relatively high proportion of upregulated genes coded for proteins involved in cell wall synthesis and modification (Lee et al. 2005). It was also demonstrated that even slight change in the growth pattern with respect to gravity vector (plants growing on normal or on a slope conditions) is immediately reflected by changes in the root proteome (Di Michele et al. 2006). Upon mechanical stresses a tree modifies its growth to minimize the risk of failure by an increase in secondary growth at the side of the prevailing wind direction (Mattheck and Breloer 1995). As for mechanical stability of plant body, a balance between belowground and aboveground parts is essential, counterbalancing mechanisms exist to ensure simultaneous adaptation of root system to wind (Fourcaud et al. 2008). Tree growth response to wind can be controlled by a reorientation of secondary axes, which is possible through the formation of reaction wood (Fournier et al. 2006; Moulia et al. 2006; Sellier and Fourcaud 2009). When the angle between the primary branches and the stem is small, the resulting wind speed near the distal parts of these branches is higher. Therefore trees with plagiotropic branches appear better suited to withstand high, dynamic winds than trees with orthotropic ones (Sellier and Fourcaud 2009). Trees exposed to changing wind conditions can even reversibly rearrange their crowns. Brittle branches break easily in high winds, decreasing the crown exposure, and preventing damage in more critical zones, such as in the stem or the root system (Niklas 2000; Sellier and Fourcaud 2009).

3 Mechanical Properties of Plant Cells

3.1 *Mechanical Properties of Plant Cell Walls*

The turgor pressure is an isodiametric force, and thus protoplasts, devoid of cell walls, as well as the cells equipped with ideally homogenous walls develop an energetically optimal spherical shape (Baluška et al. 2003b; Mathur 2006). However, plant cells are usually surrounded by the cell walls composed of diverse wall domains exhibiting different mechanical properties (Wojtaszek et al. 2007), and an interplay with the turgor pressure leads to an establishment of anisotropic cell growth (Wojtaszek 2000). The micromechanical design of cell walls relies mostly on their biochemical composition (see also chapter "Micromechanics of Cell Walls"). Although cellulose is the strongest biopolymer in terms of tension resistance, it is rather weak when compressed along its backbone. Therefore, in algal cells surrounded by cellulosic walls, when the turgor pressure is lost, protoplasts deflate, the cell wall stiffness is lost, and, in consequence, the collapse of the cell can be observed (Niklas 2000). Primary cell walls exhibit physical properties enabling for plastic expansion (Niklas 1992). Upon extension, primary cell walls resemble viscoelastic composite that undergoes stress relaxation in a time-dependent manner after stretching (Schopfer 2006). Enrichment of the walls with lignin helps to slide cellulose microfibrils against lateral bending, and, due to its hydrophobic properties, protects cellulose from moisture and thus increases wall stiffness (Niklas 2000).

It is commonly agreed that in the elongating cells, the cellulose microfibrils are located perpendicularly to the growth axis, determining the direction of growth. However, the mechanism of their orientation remains elusive. The classical point of view is that the deposition of cellulose microfibrils is affected by the alignment of cortical microtubules (Wymer and Lloyd 1996). The geometrical model assumes that new microfibrils are oriented by the cell geometry together with existing wall components, while orientation of microtubules is a simple reflection of the directed delivery of cellulose synthase complexes to the plasma membrane (reviewed by Emons and Mulder 2000; Emons et al. 2007). The orientation of the cellulose microfibrils during deposition might be therefore controlled by the number of cellulose synthase complexes and their distance (Emons and Mulder 1998). However, according to recent biochemical and genetic data, bidirectional flow of information between cortical microtubules and cellulose microfibrils exists. In tobacco suspension-cultured cells, biophysical forces are responsible for the spatial organization of microtubules and microtubules themselves can respond to vectorial changes of such forces (Wymer et al. 1996). Moreover, cellulose microfibrils through localization of the cellulose synthesis machineries provide spatial cues for the internal organization of microtubules (Fisher and Cyr 1998; Paredez et al. 2006, 2008). In effect, microtubules change dynamically their orientation in response to internal and/or external stimuli, such as directional mechanical stress (Fischer and Schopfer 1998). It should be mentioned, however, that filamentous actin is also essential for cell elongation (Baluška et al. 2001) and for the directed delivery of cellulose synthase complexes to the sites of wall synthesis (Wightman and Turner 2008).

In maturing plant cells, the processes of specific orientation of cellulose microfibrils during wall formation enable plant to control the mechanical properties of the apoplast at the tissue/organ/plant levels (Reiterer et al. 1999; see also chapter "Micromechanics of Cell Walls"). The deposition of cellulose in the cell wall can be also adjusted to prestressed tissues or to actuate movement of the organ upon swelling or shrinking of the cell wall (reviewed in Burgert and Fratzl 2009). For the establishment and maintenance of mechanical properties of plants, the angle of microfibrils in the wood is crucial. Tissues with lower microfibril angle reveal higher modulus of elasticity (Cave 1969; Reiterer et al. 1999; Burgert et al. 2002; Groom et al. 2002). The microfibril angle is also specific to the age of the tree. In young trees high microfibril angles can be found, allowing for plastic deformations after yielding and leading to higher flexibility and streamlining of the stem, whereas in mature trees the wood containing the walls built of small angle microfibrills is stiffer and able to withstand the wind (Lindström et al. 1998; Bonham and Barnett 2001). For the reaction wood in leaning stems and branches on their upper side (tension wood) and lower side (compression wood), respectively, extremely low or high cellulose microfibril angles are specific (Côté and Day 1965; Wardrop 1965). The variations between microfibril angles are crucial for generating stresses causing bending movements of plant organs (Burgert et al. 2007; Goswami et al. 2008; Burgert and Fratzl 2009). The orientation of cellulose microfibrils is crucial during and even after senescence, allowing for passive actuation of organs under changes of humidity. This mechanics is based upon properties of cellulose microfibrils that do not swell axially, thus allowing for swelling/shrinking only in the directions perpendicular to the fibrils. The composition of tissues and cells with cell walls of different orientations of cellulose fibrils allows for complex movements at the organ level. This mechanism is used to control seed dispersal including the spore capsules of mosses that show a moisture-dependent seed dispersal mechanism with hygrosensitive openings and closures of the capsule (Ingold 1959), the release of ripe seeds from conifer cones (Dawson et al. 1997), and enables the motility of seed dispersal units of wheat (Elbaum et al. 2007).

3.2 Linking Cell Walls and Cell Interior

Due to the tensegral organization of the plant cell, the main mechanical force influencing cellular properties is the turgor pressure acting on the plasma membrane and the nonextensible cell walls. Looking from the opposite direction, many external perturbation acts on cell walls, and because of internal pressure caused by turgor, local deformations of the wall domains are conveyed to the plasma membrane. This could trigger specific biochemical responses. During cell growth, localized loosening of the wall structure is a common phenomenon, and it could also be sensed by potential receptors (Monshausen and Gilroy 2009). Opposite to the animal model where tensed elements are pulling against compressed ones (Ingber 2009), in plant cells compression elements push outward the cell wall introducing tension. The interplay between turgor pressure and cell wall mechanics is crucial in determining growth and development patterns in plant. The idea that in plants mechanical stimuli are converted to biochemical information in a similar way as in the animal systems seems very promising. In animal cells, the molecular bridge between ECM and cytoskeleton is known as focal adhesion. The central point in this bridge is formed by transmembrane proteins – integrins – mechanoreceptors converting mechanical forces to biochemical signals such as protein phosphorylation or activation of calcium influx. At the cytoplasmic side, integrins are physically linked with the cytoskeleton through a macromolecular complex, which includes, among others, actin-associated proteins, such as talin, zynxin, and vinculin. The generated biochemical signal is then transduced into changes in gene expression. Moreover, because the cell elements are prestressed, mechanical forces loaded on integrins can easily and quickly move through the cytoskeleton network all over the cell (Geiger et al. 2001; Ingber 2006). This kind of structural and functional continuum between ECM, plasma membrane, and cytoskeleton in animal cells is well documented. The existence of analogous functional network connecting cell *w*all, plasma *m*embrane and *c*ytoskeleton (WMC) in plant cells has also been postulated (Wyatt and Carpita 1993), but knowledge about individual elements of this continuum is rather poor (Wojtaszek et al. 2004). For a long time cell wall was recognized as a dead structure surrounding living protoplast. This view of the walls is systematically evolving to a highly responsive, dynamic structure responsible for many fundamental events including perception of environmental stimuli. Accordingly, identification of specific or general molecular linkers connecting cell walls with plasma membrane and cytoskeleton is crucial in our understanding of an interplay between external and internal environments of the cell. Available experimental data point to several proteins which could function as a potential linker candidates (Baluška et al. 2003a; Gouget et al. 2006; Humphrey et al. 2007).

Despite intensive research efforts, none of the characterized higher plant genomes seems to contain true integrin homologues. One of the possible reasons might lie in the different chemical composition of exocellular matrices: plant cell wall is composed mainly of carbohydrates, while animal ECM is of proteinaceous nature. This chemical diversity might lead to an adaptation of completely different molecules to perform the same receptor functions (Baluška et al. 2003a; Monshausen and Gilroy 2009). Interestingly, however, some experimental data show the presence of integrin-like proteins in plants. In animal cells, the Arg–Gly–Asp (RGD) motif can be found in the ECM proteins responsible for adhesion, and this motif is normally recognized by integrins. Senchou et al. (2004) identified the protein in *A. thaliana* that specifically bound peptides containing RGD sequence. Moreover, addition of such peptides to plasmolysed *Arabidopsis* cells disrupted adhesion sides between cell wall and plasma membrane. Phage display approach showed that among the 12 specific RGD-binding proteins identified in this experiment, eight belong to the receptor-like kinases (RLK) superfamily. Plant RLKs are transmembrane proteins with extracellular domains located at amino-terminal part of protein, and intracellular kinase domains forming the C terminus. This protein architecture shows some similarity to domain organization in the animal receptor tyrosine kinases, such as the receptor of epidermal growth factor. In *Arabidopsis* RLKs belong to a large gene family with over 600 members, but the function of individual proteins is still poorly understood (Shiu and Bleecker 2001). Four of the previously mentioned RLKs identified by phage display technique (Senchou et al. 2004) contained specific lectin-like extracellular domains which were responsible for RGD binding as well as interactions with carbohydrates (Gouget et al. 2006; Humphrey et al. 2007; Bouwmeester and Govers 2009). Although lectin receptor kinases (LecRKs) are relatively well characterized at the molecular level, very little is known of their exact function in plants. They are thought to be involved in processes such as hormone responses, disease resistance, and stress adaptation (Wan et al. 2008). It seems, however, that the ability to bind RGD-containing proteins, combined with kinase activity makes them good candidates for mechanical signal receptors as well.

Another example of RLK which may potentially be involved in mechanical signal transduction between cell wall and the cellular interior are wall-associated kinases (WAK). WAKs are till date the best characterized potential receptor proteins. Five direct WAK isoforms and 22 WAK-like genes have been identified in *A. thaliana* (Anderson et al. 2001; Verica and He 2002). WAKs are defined by the presence of a highly conserved C-terminal kinase domain, transmembrane region, and relatively variable N-terminal sequence responsible for cell wall interactions (Wagner and Kohorn 2001). This amino-terminal region resembles the vertebrate epidermal growth factor motifs, which, in all cases studied, are involved in protein–protein interactions. The presence of other motifs, specific for metazoan proteins, e.g., collagen-like or neurexine-like sequences has also been documented, but their function remains unclear (He et al. 1999; Kohorn 2000; Anderson et al. 2001). Initial data suggested covalent interactions of WAKs with cell walls (Kohorn 2001). However, it was later shown that they interact through the N-terminal domain via noncovalent binding to pectins, and this interaction was shown to be calcium dependent (Decreux and Messiaen 2005; Humphrey et al. 2007). Expression profiling also indicates the role of WAK family proteins in different environmentally induced processes, such as responses to aluminum or pathogens. Interestingly, the expression of WAKs is most noticeably induced in organs undergoing expansion, such as meristems (Humphrey et al. 2007). It was also shown that in protoplasts WAK1 fusion with GFP tended to accumulate in specific cytoplasmic compartments that also contained pectin. Moreover, migration of WAK1 complex to the cell surface was shown to be related to cellulose synthesis (Kohorn et al. 2006). Combined with its pectin-binding properties, the data strongly suggest the possible role of WAKs in the emerging cell wall–plasma membrane functional integrity.

In recent years, very interesting data are emerging from studies on the family of CrRLK1 proteins. The name of the family originates from RLK1 from

Catharanthus roseus, a novel protein kinase of unknown function (Schulze-Muth et al. 1996). In Arabidopsis there are 17 proteins which share similarity with CrRLK1, and few of them have been studied intensively (Hématy and Höfte 2008). THESEUS 1 (THE1) is thought to act as a sensor of the cell wall integrity and mediator of signaling induced by the cell wall damage. The protein was shown to inhibit cell elongation in case of impaired cellulose synthesis (Hématy et al. 2007). FERONIA (FER), another member of this family, was shown to be involved in elongation, and its possible function as a growth arrest factor in the elongating pollen tube was indicated. This growth cessation is crucial at the last step of fertilization process (Hématy and Höfte 2008). Similarly, the HERCULES1 (HERK1) was shown to be involved in cell elongation during vegetative growth (Guo et al. 2009a). More importantly, all three kinases are transcriptionally induced by brassinosteroids (Guo et al. 2009b).

Plant proteins belonging to the formin family seem to have a potential to act as putative linker elements within the WMC continuum. Formins are actin-binding proteins responsible for nucleation and elongation of actin microfilaments. They are defined by the presence of a conserved FH2 domain (formin homology 2), as well as other domains characteristic for distinct types of proteins (Paul and Pollard 2009). There are over 20 formin-coding genes in the *A. thaliana* genome. According to predicted domain architecture they are divided into two classes. Class I formins in plants share the presence of N-terminal transmembrane domain, and a short proline-rich region located at the extracellular side of plasma membrane. The FH2 domain, responsible for interactions with the cytoskeleton is located at the C-terminus of protein, and protrudes to cytoplasm. Class II formins are located in the cytoplasm and, apart from FH2 domain, they usually contain PTEN-like domain. However, due to mutation, this latter domain probably lacks conventional phosphatase activity (Deeks et al. 2002; Cvrcková et al. 2004; Blanchoin and Staiger 2008; Grunt et al. 2008). Class I formins are involved in the tip growth, particularly of pollen tubes and root hairs (Cheung and Wu 2004; Deeks et al. 2005), as well as in cytokinesis (Ingouff et al. 2005). An intriguing observation is that some formins tend to localize at the cross-walls which, in axial organs such as roots, are thought to be involved in signal transduction between neighboring cells (Deeks et al. 2005; Wojtaszek et al. 2007). Because of experimentally confirmed localization at the plasma membrane, and specific domain architecture with one end potentially interacting with the walls and the other with the cytoskeleton, class I formins are often indicated as potential linkers within the WMC continuum. Unfortunately, probably because of the high redundancy, clear phenotypes of individual mutants are unavailable at this moment, and the estimation of the impact of a particular protein for cellular functioning is still a challenge. Recent data on AtFH1 protein indicate that this is, however, possible (Martinière et al. 2011).

Among the variety of cell wall proteins that could also mediate wall–membrane interactions, *a*rabinogalactan *p*roteins (AGP) deserve a special attention. A category of classical AGPs groups hydroxyproline-rich glycoproteins (HRGPs), mostly highly glycosylated, that could be localized predominantly at the cell surfaces. Carbohydrates constitute up to 90% of single arabinogalactan protein mass. Polysaccharide units vary in size between 30 and 150 monomers and are attached to multiple sides on core of the protein (Showalter 2001; Seifert and Roberts 2007). Patterns of sugar moieties on AGPs are thought to be information-bearing structures, and the information could be conveyed via direct interactions with putative membranous receptors. Small carbohydrate fragments, released by an enzymatic cleavage from large oligosaccharide side-chains, could bind to putative receptors and trigger generation of biochemical signaling (Showalter 2001). In fact, such mechanism is suggested as a possible explanation for the regulatory role of chitinases in somatic embryogenesis (van Hengel et al. 2001). At least some AGPs contain an additional *g*lycosyl*p*hosphatidyl*i*nositol (GPI) anchor at the C-terminus that allows them to stay attached to the extracellular side of the plasma membrane (Kohorn 2000; Majewska-Sawka and Nothnagel 2000). This GPI anchor can be removed in a controlled manner by phospholipase C. Such action releases AGPs from the plasma membrane to the cell wall enabling signaling of another type (Borner et al. 2002). In addition to the signaling properties, AGPs are also considered as adhesive molecules connecting plasma membrane with the cell wall. AGPs can bind to pectins (Nothnagel 1997), and potentially interact with WAKs (Gens et al. 2000). A family of fasciclin-like arabinogalactan proteins (FLA) contain fasciclin-like domain which, in animal cells, is responsible for promoting cell adhesion (Johnson et al. 2003). Very importantly, it was recently shown that application of Yariv reagent, which specifically binds to AGPs, leads to disorganization of cortical microtubules in *A. thaliana* roots (Nguema-Ona et al. 2007). The connection between AGPs and cortical microtubules and F-actin was also proved in BY-2 suspension cells (Sardar et al. 2006).

3.3 Actin Cytoskeleton as Mechanical Integrator of Plant Cell

Cytoskeleton plays a very important role in the formation of cellular shape, cell polarity, and functional organization by, among others, the perception of physical forces and different mechanical stimuli. Numerous examples confirm that cytoskeleton is implicated in the maintenance of mechanical integrity of the cell, driving its growth, differentiation, and cell-to-cell communication. In the bacterial model, the actin homolog, MreB, contributes nearly as much to the stiffness of a cell as the peptidoglycan cell wall. MreB is rigidly linked to the cell wall, increasing the mechanical stiffness of the whole cell. These data provide the first evidence that in walled cells the cytoskeleton contributes to the mechanical integrity in similar way as it does in naked animal cells (Wang et al. 2010). It still remains to be demonstrated if the same is true for plant cells.

From studies on animals and some bacteria it is known that cytoskeleton, mainly actin, is involved in the generation of forces necessary for cell movement or shape reorganization (van der Honing et al. 2007). Interestingly, plant cells are equipped with most of the homologues of the actin-binding proteins typical for animals, which are responsible for force generation. Although such force generation by the actin polymerization in plant cells has not been studied yet, it is possible that it plays a role in the organization of their cytoplasm. Evidence for that comes from several experiments on plant mutants. One of the most interesting observations derives from analysis of *Arabidopsis* lines devoid of the subunits of Arp2/3 complex. Such mutation caused disturbances in trichomes development and in root hairs. Trichomes were twisted or with short branches, and the cytoplasmic streaming was largely limited, while root hairs were wavy and had a variable diameter. Epidermal cells of leaves and dark-grown hypocotyls were also affected and displayed abnormal organization (Le et al. 2003; Li et al. 2003; Mathur et al. 2003a, b; El-Din El-Assal et al. 2004). These data suggest that Arp2/3 complex might be involved in the organization of the subapical fine F-actin and, in consequence, in the determination of the architecture of expanding cells (van der Honing et al. 2007). On the other hand, *a*ctin *d*epolymerization *f*actors (ADF) are known to be responsible for the turnover of actin filaments by increasing depolymerization at the pointed end. This provides actin monomers for the elongating barbed end, which, in animal cells, is usually pointing towards the direction of the cellular movement. When ADFs are differentially expressed in plant cells, the amounts of the available G-actin are also changed, and, in effect, the rates of cells and organs development are also affected. Overexpression of ADFs resulted in disappearance of actin bundles, causing reduction of cell expansion and organ growth. Interestingly, inhibition of ADFs expression stimulated cell expansion and organ growth (Dong et al. 2001). These and other analyses of plants lacking or overexpressing actin-binding proteins turn attention to their possible role in force generation. In animal cells, actin filaments are necessary for cell shape changes. The question is, however, how plants may utilize these proteins in cells with the determined cell shape and surrounded by the rigid cell walls? It is possible that the role of actin polymerization-based system is limited just to the directional delivery of exocytic vesicles to cellular peripheries. Differentiation of the rates of exocytosis in various directions within plant cell will provide diverse amounts of cell wall material to different wall domains thus providing the way for creation of the cellular shape. Search for other functions of actin-based force generation in plant cells is now under way (Emons and Mulder 2000; Hussey et al. 2006; van der Honing et al. 2007). One of the possibilities has been just indicated. In plant cells, microfilaments span the whole cell and enable effective intracellular communication through the formation of actin-based transvacuolar strands (van der Honing et al. 2010).

Actomyosin system is involved in the regulation of protoplast volume during plasmolysis (Komis et al. 2003; Wojtaszek et al. 2005). Cells subjected to hyperosmotic conditions reorganize their actin filaments network, and thin cortical F-actin is replaced by cortical, subcortical, and endoplasmic well-organized and thick actin bundles (Komis et al. 2002). The amount of F-actin is generally higher than in control cells, and some of the filaments traverse the Hechtian strands connecting the retracted protoplast with cellular peripheries. Application of anti-actin drug leads to dramatic changes in the pattern of plasmolysis, with resulting greater decrease of the protoplast volume. In some experiments protoplasts adopted amoeboid form or were subdivided into subprotoplasts (Komis et al. 2002). What is even more important is the reorganization of actin is to some extent cell wall dependent as the organization and composition of wall domains surrounding individual cells affect the anchorage of actin bundles (Wojtaszek et al. 2005, 2007). The dynamic reorganization of actin filaments is also essential for the transduction of gravitropic stimuli in root cells (Kordyum 2003; Volkmann and Baluška 2006). Actin mediates positioning, transport, and sedimentation of statoliths, which are the specialized form of amyloplasts, involved in gravity perception by plants (Kordyum 2003). Statocytes, which contain statoliths, have to polarize their protoplast to function as graviperceptive cells. In that way, statoliths sediment in the distal part of the cell in the direction of a gravitational vector, and the nucleus is positioned in the proximal part. It is believed that structures responsible for this positioning and polar arrangement of organelles are actin filaments (Kordyum 2003). Changes in the cytoskeleton architecture were also observed upon localized mechanical stimuli such as pressing the cells with microcapillary. Such stress conditions activate the avoidance response of the chloroplasts, i.e. movement of organelles away from the site of stimulation. As treatment with cytochalasin B (the actin inhibitor) or 2,3-butanedione monoxime (the myosin inhibitor) stopped movements upon stimulation, it is thought that actomyosin motile system plays a role in plant response to touch (Sato et al. 1999). On the other hand, localized mechanical stimulation, which could also be treated as a mimic of fungal pathogen attack, induces very rapid focusing of actin microfilaments beneath the contact site (Hardham et al. 2008).

3.4 Mechanics of Intercellular Communication

The cellular distribution of organelles seems to be essential for proper functioning of living cells and has a great role in maintaining many activities of plants (see chapter "Intracellular Movement: Integration at the Cellular Level as Reflected in the Organization of Organelle Movements"). In animal cells, microtubules rather than actin filaments are considered to be responsible for organelle movements; in contrast, actin filaments are believed to play mostly such a function in plant cells (Muthugapatti et al. 1999; Wada et al. 2003; Kadota et al. 2009). Thus, they are believed to be not only major players in vesicle trafficking between endomembranes compartments, but also responsible for the spatial distribution and movements of most organelles (Chuong et al. 2006; Boutté et al. 2007). Microfilaments congregate in densely packed actin cables, and form the tracks for intracellular organelle trafficking (Schmidt and Panstruga 2007; van der Honing et al. 2007).

In animal and yeast cells, two types of actin-based organelle movements have been identified. The first one is based on myosins, which bind tail domain of organelle cargos and transport them by sliding on actin cables. The second mechanism refers to ARP2/3 complex, which can nucleate actin filaments at the organelle edge, forming "comet tails" generating the motive force to push the organelle (Kadota et al. 2009). It is still unclear how these processes occur in plant cells, but it is likely that they involve the role of myosins, and with usage of the energy of ATP hydrolysis, maintain movement along microfilaments (Vidali et al. 2001; Holweg and Nick 2004; Schmidt and Panstruga 2007). Various organelle movements in plants have been shown to be dependent on the cytoskeleton. Some of the actin-dependent movements of, e.g., peroxisomes (Collings et al. 2002; Jedd and Chua 2002) and mitochondria (Van Gestel et al. 2002) are the part of an active and continuous mass movement called the cytoplasmic streaming (Shimmen and Yokota 2004; Schmidt and Panstruga 2007). Actin filaments are responsible for the movements of both ER tubules and individual Golgi stacks (Knebel et al. 1990; Boevink et al. 1998; Boutté et al. 2007). It was observed that actin inhibitor cytochalasin D causes accumulation of the ER into patches, a fusion of tubules into cisternae and changes in the ER overall shape as well as disruption of Golgi stacks (Knebel et al. 1990; Satiat-Jeunemaitre et al. 1996). In the absence of actin, Golgi bodies clump together and stop moving. Similarly, the movement of the ER tubules were stopped. However, depolymerization of the microtubules had no effect on the ER and Golgi movements (Brandizzi et al. 2003). Interestingly, the organization of ER–Golgi complexes seems to be differentiated: the transport of cargo between ER and Golgi stacks is not dependent on cytoskeleton, while the transport of secretory vesicles from the Golgi stacks to different compartments is the sole responsibility of actin (Boutté et al. 2007). Internalization of small particles from the cell surface is in most cases performed by membrane carrier proteins. Specific uptake of these molecules together with polymers (i.e. pectins) is crucial for cell growth, metabolism, and signaling (Müller et al. 2007). In contrast to animal or fungal endocytosis, where vesicles movement is organized by dyneins and kinesins along microtubules, in plant cells, the pivotal role is played by F-actin and its inherent interactions with plant cell-specific class VIII myosins (Baluška et al. 2002). Latrunculin B treatment revealed a crucial role of F-actin in vesicle docking, fusion as well as endocytic vesicle formation. Inhibition of actin cytoskeleton in pollen tubes affects organelle motility, the vesicle, and small endosome movement pathways in the clear zone and their ability to fuse with the plasma membrane. Moreover, endosomes could play a role as actin nucleation hot spots. Therefore, actin polymerization could underlie directed vesicle movement. In addition, F-actin depolymerization changed tubular and dynamic vacuole into stationary round structure in the subapical region. Hence, motility of vacuole is also actin dependent (Ovecka et al. 2005).

Intracellular distribution of chloroplasts, controlled by actin cytoskeleton, depends mainly on the intensity and spectral quality of light. Chloroplasts move away from strong light irradiation to avoid photodamage, and move towards weak light to maximize photosynthesis (Muthugapatti et al. 1999; Kasahara et al. 2002; Wada et al. 2003; Suetsugu and Wada 2007; Kadota et al. 2009). Application of actin inhibitors caused aberrant distribution of these organelles. On the other hand, microtubules were very rarely observed in connection with chloroplasts, and, accordingly, oryzalin treatment did not affect chloroplasts distribution (Muthugapatti et al. 1999; Chuong et al. 2006). It was also observed that short actin filaments, called cp-actin filaments, form connections between chloroplasts' peripheries and the plasma membrane. The cp-actin filaments appear immediately after inducing signal in the form of light irradiation, and their formation depends on *ch*loroplast *u*nusual *p*ositioning *1* (CHUP1) proteins, which are localized at the chloroplast envelope. Mutation of CHUP1 resulted in disappearance of cp-actin filaments and normal cytoplasmic actin filaments were not influenced (Kadota et al. 2009). Other experiments showed that CHUP1 have not only F-actin binding motif, but it can also bind G-actin or profilin, suggesting that CHUP1 may regulate cp-actin filament dynamics at the chloroplasts envelope (Oikawa et al. 2003, 2008; Schmidt von Braun and Schleiff 2008). Moreover, basket-like structures of actin filaments were noticed around the chloroplasts, anchoring organelles during streaming and allowing for control over proper three-dimensional orientation of chloroplasts with respect to light (for details see Chapter "Intracellular Movement: Integration at the Cellular Level as Reflected in the Organization of Organelle Movements").

The movements of mitochondria are mediated by two components of the cytoskeleton – actin filaments and microtubules, and the involvement of each element varies, depending on a specific cell type and organism (Zheng et al. 2009). For example, it was reported that plant mitochondria move on actin filaments, but their positioning in cortical part of the cell is the responsibility of both actin and myosin (Van Gestel et al. 2002). Myosin inhibitor reduced mitochondrial velocity in a manner similar to that observed in *Arabidopsis* myosin knock-out mutant, confirming that the actomyosin system is the main driving force of mitochondrial movement (Peremyslov et al. 2008; Prokhnevsky et al. 2008; Zheng et al. 2009). Recent analysis also showed that microtubules affect mitochondrial velocity, trajectory, and positioning because they direct the positioning of actin polymerization events (Zheng et al. 2009).

Plant cytoskeleton, especially actin filaments, is responsible for both positioning and movements of nuclei in normal conditions and in response to external stimuli as well as during cell division. It was demonstrated in animal cells that this actin-dependent repositioning of nuclei had an effect on chromatin organization, but affected nuclear movement to a lesser extent (Maniotis et al. 1997a, b; Yang et al. 2008). Interestingly, in plant cells, in some processes, e.g., during cytokinesis, actin cytoskeleton is absolutely required for proper partitioning of organelles (Sheahan et al. 2004). However, microtubules were also observed to function in organelle positioning, and this suggested that both actin and microtubules might function cooperatively during organelle movement. It was thus proposed that the actin-based system provides the mechanism for moving organelles during the early stage of cellular partitioning, whereas microtubules are implicated in proper spatial relationship of organelles at subcellular location (Chuong et al. 2006).

3.5 Maintenance of Hydromechanical Integrity of Plant Cells Due to Balanced Endo- and Exocytosis

Because of the high turgor pressure endocytic processes have been disputed in plant cells on an assumption that the amount of energy needed to aggregate clathrin triskelions is very low. However, it was shown that the energy is inadequate to form a vesicle even in the absence of turgor pressure (Meckel et al. 2005). Adequate energy for pit and consequently vesicle formation is delivered by generating a molecular imbalance in plasma membrane bilayer. Adding phospholipids to the inner leaflet generates enough force for vesicle formation even in unilamellar giant vesicles containing no proteins. Therefore phospholipids asymmetry within bilayer generates enough energy for vesiculation during endocytosis (Meckel et al. 2005). It was proposed that plant cells possess standard, constantly maintained plasma membrane tension. Any changes of tension are detected and followed by adequate response. Therefore, increasing membrane tension triggers exocytic secretion of membrane material, subsequently decreasing tension until standard membrane tension is re-established. Osmotic or pressure stress prompts swelling or shrinking of guard cells protoplasts. Patch clamp experiments revealed that surface changes of these cells are accompanied by removal of membrane material from or its incorporation to plasma membrane. Moreover, in most cases fusion and fission events are not associated with the visible vesicle movement. Therefore, it was proposed that the surface change in guard cell protoplasts is reached mostly via small vesicles of size below diffraction limit (ca. 300 nm) (Meckel et al. 2005). Intact guard cells in hyperosmotic conditions showed uptake of membrane styryl dye FM4-64 in objects whose size varied between 1 μm and diffraction-limited size under 270 nm. Also invaginated tubular structures in the plasma membrane were found (Meckel et al. 2005). Indeed, precise measurements of the size of endocytic vesicles in guard cells with the fluorescent dye Alexa 488 hydrazide provided estimation of the minimal size to be at least 87 nm (Gall et al. 2010).

Endocytic processes are operative in turgid plant cells. They were shown to function during wall remodeling, enabling utilization of cell surface material for the construction of a new cell plate (Dhonukshe et al. 2006). Clathrin-mediated endocytosis constitutes the basis for recycling of PIN auxin efflux carriers (Dhonukshe et al. 2007), and this process itself is regulated by auxin (Paciorek et al. 2005). Finally, endocytic vesicles in intact guard cells are able to carry GFP-tagged plasma membrane K^+ channel KAT1, which shows that endocytosis is also operating in cells with the highest turgor pressure (Meckel et al. 2004). Investigation of the rates of membrane material addition in guard cells showed that the new material is present in the plasma membrane instantly after hydrostatic pressure stressing. Hence, it was proposed that guard cells possess the reservoir of membrane material disposable for fusion with swelling plasma membrane. Moreover, membrane material internalized during cell shrinking could be secreted while cell swells. This shall enable the guard cell protoplasts to swell and shrink several times. Unfortunately, there is still no data showing the origin and quality of membrane material transported to and from the plasma membrane in tension-induced surface changes (Meckel et al. 2005).

The elongation of the pollen tube seems to be orchestrated by transcellular hydrodynamic flow coordinating exocytosis and endocytosis (Zonia and Munnik 2008). Hyperosmotic treatment of pollen tube cells leads to cell shrinking, which stimulates endocytosis at the apex and also arrests exocytic processes. Hyperosmotic conditions induce the reduction of cellular volume and, in consequence, halt cell elongation. On the contrary, hypoosmosis increases the exocytic membrane flow, and decreases endocytosis (Zonia and Munnik 2007). It was proposed that because of the exposure to mechanical agents and therefore possibility of structural defects in expansion, the pollen tube apex is not the place where the secretion of new cell wall compounds occurs and that cell wall synthesis is localized distal from the apex. It was shown that exocytosis takes place adjacent to the apical dome (Zonia and Munnik 2008). Endocytosis was observed in the pollen tube apex in the form of small vesicles that underwent retrograde transport. Respectively, exocytic activity was shown next to the cell apex. During cell elongation excess plasma membrane was moved to the apex and taken up through endocytosis. Therefore predominantly unorganized pollen tube growth seems to be directed by vectorial hydrodynamic flow coupled with exo- and endocytic processes occurring adjacent and at the apex, respectively (Zonia and Munnik 2007). Newly mounted cell wall is more viscoplastic than mature cell wall; therefore it is likely to undergo directed expansion driven by hydrodynamic flux and cooperative exo- and endocytosis. It was assumed that exocytic vesicles are very appropriate for cell wall material transport and showed that they belong to a large class of vesicles with intermediate size (Zonia and Munnik 2008). It should also be added that the cooperative exo- and endocytosis needs to be tightly regulated. Recent estimates of the rates of both processes in root hairs and pollen tubes of *Arabidopsis* showed that 9,204 and 2,686 exocytic vesicles are consumed per minute at 20°C during growth of root hairs and pollen tubes, respectively. More importantly, however, the recycling process involves 86.7% of newly inserted membranes in root hairs, and 79.0% in pollen tubes (Ketelaar et al. 2008).

During plant cell development there are constant changes in cell wall composition realized by directed transport of polysaccharides and proteins in transport vesicles. It is known that internalized membrane arabinogalactan proteins are trafficked to MVB and consequently to vacuoles. Cellulose syntase complex activity is regulated throughout orchestrated recycling of its subunits between plasma membrane and internal membrane compartments. Moreover, ingredients of cell plate during plant cell cytokinesis are partially delivered in endosomes containing components of existing cell walls (Dhonukshe et al. 2006). Endocytic uptake of cell wall pectins and their redirection to the cell plate is important for building new cell wall during cytokinesis (Müller et al. 2007). Pectins cross-linked with boron and calcium are crucial for mechanical strength and spatial organization of plant cell walls. Therefore, actin-dependent endocytosis and secretion of these molecules are pivotal for modulating mechanical properties of plant cells. In consequence, endocytosis might regulate plant growth and morphogenesis.

References

Anderson CM, Wagner TA, Perret M, He ZH, He D, Kohorn BD (2001) WAKs: cell wall-associated kinases linking the cytoplasm to the extracellular matrix. Plant Mol Biol 47:197–206

Baluška F, Jasik J, Edelmann HG, Salajova T, Volkmann D (2001) Latrunculin B-induced plant dwarfism: plant cell elongation is F-actin-dependent. Dev Biol 231:113–124

Baluška F, Hlavacka A, Šamaj J, Palme K, Robinson DG, Matoh T, McCurdy DW, Menzel D, Volkmann D (2002) F-actin-dependent endocytosis of cell wall pectins in meristematic root cells. Insights from brefeldin A-induced compartments. Plant Physiol 130:422–431

Baluška F, Šamaj J, Wojtaszek P, Volkmann D, Menzel D (2003a) Cytoskeleton-plasma membrane-cell wall continuum in plants. Emerging links revisited. Plant Physiol 133:482–491

Baluška F, Wojtaszek P, Volkmann D, Barlow P (2003b) The architecture of polarized cell growth: the unique status of elongating plant cells. BioEssays 25:569–576

Blanchoin L, Staiger CJ (2008) Plant formins: diverse isoforms and unique molecular mechanism. Biochim Biophys Acta 1803:201–206

Boevink P, Oparka K, Santa Cruz S, Martin B, Betteridge A, Hawes C (1998) Stacks on tracks: the plant Golgi apparatus traffics on an actin/ER network. Plant J 15:441–447

Bonham VA, Barnett JR (2001) Fiber length and microfibril angle in Silver Birch (*Betula pendula* Roth). Holzforschung 55:159–162

Borner GH, Sherrier DJ, Stevens TJ, Arkin IT, Dupree P (2002) Prediction of glycosylphosphatidylinositol-anchored proteins in Arabidopsis. A genomic analysis. Plant Physiol 129: 486–499

Boutté Y, Vernhettes S, Satiat-Jeunemaitre B (2007) Involvement of the cytoskeleton in the secretory pathway and plasma membrane organisation of higher plant cells. Cell Biol Int 31:649–654

Bouwmeester K, Govers F (2009) Arabidopsis L-type lectin receptor kinases: phylogeny, classification, and expression profiles. J Exp Bot 60:4383–4396

Braam J (2005) In touch: plant responses to mechanical stimuli. New Phytol 165:373–389

Braam J, Davis R (1990) Rain-, wind-, and touch-induced expression of calmodulin and calmodulin-related genes in Arabidopsis. Cell 60:357–364

Brandizzi F, Saint-Jore C, Moore I, Hawes C (2003) The relationship between endomembranes and the plant cytoskeleton. Cell Biol Int 27:177–179

Brock A, Chang E, Ho C-C, LeDuc P, Jiang X, Whitesides GM, Ingber DE (2003) Geometric determinants of directional cell motility revealed using microcontact printing. Langmuir 19:1611–1617

Burgert I, Fratzl P (2009) Plants control the properties and actuation of their organs through the orientation of cellulose fibrils in their cell walls. Int Comp Biol 49:69–79

Burgert I, Keckes J, Frühmann K, Fratzl P, Tschegg SE (2002) A comparison of two techniques for wood fiber isolation-evaluation by tensile tests on single fibers with different microfibril angle. Plant Biol 4:9–12

Burgert I, Eder M, Gierlinger N, Fratzl P (2007) Tensile and compressive stresses in tracheids are induced by swelling based on geometrical constraints of the wood cell. Planta 226:981–987

Cave ID (1969) The longitudinal Young's modulus of Pinus radiata. Wood Sci Technol 3:40–48

Chen CS, Mrksich M, Huang S, Whitesides GM, Ingber DE (1997) Geometric control of cell life and death. Science 276:1425–1428

Cheung AY, Wu HM (2004) Overexpression of an Arabidopsis formin stimulates supernumerary actin cable formation from pollen tube cell membrane. Plant Cell 16:257–269

Chuong SDX, Franceschi VR, Edwards GE (2006) The cytoskeleton maintains organelle partitioning required for single-cell C4 photosynthesis in chenopodiaceae species. Plant Cell 18:2207–2223

Collings D, Harper J, Marc J, Overall R, Mullen R, Jedd G, Chua NH (2002) Life in the fast lane: actin-based motility of plant peroxisomes. Plant Cell Physiol 43:430–441

Corner EJH (2002) The life of plants. The University of Chicago Press, Chicago, London

Côté WA, Day AC (1965) Anatomy and ultrastructure of reaction wood. In: Côté WA (ed) Cellular ultrastructure of woody plants. Syracuse University Press, New York, pp 391–418

Cvrcková F, Novotný M, Pícková D, Žárský V (2004) Formin homology 2 domains occur in multiple contexts in angiosperms. BMC Genomics 5:44

Darwin CA, Darwin F (1880) The power of movement in plants. John Murray, London

Dawson C, Vincent JFV, Rocca AM (1997) How pine cones open. Nature 390:668

Deak KI, Malamy J (2005) Osmotic regulation of root system architecture. Plant J 43:17–28

Decreux A, Messiaen J (2005) Wall-associated kinase WAK1 interacts with cell wall pectins in a calcium-induced conformation. Plant Cell Phys 46:268–278

Deeks MJ, Hussey PJ, Davies B (2002) Formins: intermediates in signal-transduction cascades that affect cytoskeletal reorganization. Trends Plant Sci 7:492–498

Deeks MJ, Cvrcková F, Machesky LM, Mikitová V, Ketelaar T, Zársky V, Davies B, Hussey PJ (2005) Arabidopsis group Ie formins localize to specific cell membrane domains, interact with actin-binding proteins and cause defects in cell expansion upon aberrant expression. New Phytol 168:529–540

Dhonukshe P, Baluška F, Schlicht M, Hlavacka A, Šamaj J, Friml J, Gadella TWJ (2006) Endocytosis of cell surface material mediates cell plate formation during plant cytokinesis. Dev Cell 10:137–150

Dhonukshe P, Aniento F, Hwang I, Robinson DG, Mravec J, Stierhof YD, Friml J (2007) Clathrin-mediated constitutive endocytosis of PIN auxin efflux carriers in Arabidopsis. Curr Biol 17:520–527

Di Michele M, Chiatante D, Plomion C, Scippa GS (2006) A proteomic analysis of Spanish broom (*Spartium junceum* L.) root growing on a slope condition. Plant Sci 170:926–935

Ditengou FA, Teale WD, Kochersperger P, Flittner KA, Kneuper I, van der Graaff E, Nziengui H, Pinosa F, Li X, Nitschke R, Laux T, Palme K (2008) Mechanical induction of lateral root initiation in Arabidopsis thaliana. Proc Natl Acad Sci USA 105:18818–18823

Dong CH, Xia GX, Hong Y, Ramachandran S, Kost B, Chua NH (2001) ADF proteins are involved in the control of flowering and regulate F-actin organization, cell expansion, and organ growth in Arabidopsis. Plant Cell 13:1333–1346

Dumais J (2007) Can mechanics control pattern formation in plants? Curr Opin Plant Biol 10:58–62

Dupuy L, Mackenzie J, Rudge T, Haseloff J (2008) A system for modelling cell-cell interactions during plant morphogenesis. Ann Bot 101:1255–1265

Elbaum R, Zaltzman L, Burgert I, Fratzl P (2007) The role of wheat awns in the seed dispersal unit. Science 316:884–886

El-Din El-Assal S, Le J, Basu D, Mallery EL, Szymanski DB (2004) DISTORTED2 encodes an ARPC2 subunit of the putative Arabidopsis ARP2/3 complex. Plant J 38:526–538

Emons AM, Mulder BM (1998) The making of the architecture of the plant cell wall: how cells exploit geometry. Proc Natl Acad Sci USA 95:7215–7219

Emons AM, Mulder BM (2000) How the deposition of cellulose microfibrils builds cell wall architecture. Trends Plant Sci 5:35–40

Emons AM, Höfte H, Mulder BM (2007) Microtubules and cellulose microfibrils: how intimate is their relationship? Trends Plant Sci 12:279–281

Fischer K, Schopfer P (1998) Physical strain-mediated microtubule reorientation in the epidermis of gravitropically or phototropically stimulated maize coleoptiles. Plant J 15:119–123

Fisher DD, Cyr RJ (1998) Extending the microtubule/microfibril paradigm. Cellulose synthesis is required for normal cortical microtubule alignment in elongating cells. Plant Physiol 116:1043–1051

Fourcaud T, Zhang X, Stokes A, Lambers H, Korner C (2008) Plant growth modelling and applications: the increasing importance of plant architecture in growth models. Ann Bot 101:1053–1063

Fournier M, Stokes A, Coutand C, Fourcaud T, Moulia B (2006) Tree biomechanics and growth strategies in the context of forest functional ecology. In: Herrel A, Speck T, Rowe N (eds) Ecology and biomechanics – a mechanical approach to the ecology of animals and plants. CRC/Taylor & Francis, Boca Raton, pp 1–33

Fuller B (1961) Tensegrity. Portfolio Artnews Annu 4:112–127

Gall L, Stan RC, Kress A, Hertel B, Thiel G, Meckel T (2010) Fluorescent detection of fluid phase endocytosis allows for in vivo estimation of endocytic vesicle sizes in plant cells with sub-diffraction accuracy. Traffic 11:548–559

Geiger B, Bershadsky A, Pankov R, Yamada KM (2001) Transmembrane extracellular matrix-cytoskeleton crosstalk. Nat Rev Mol Cell Biol 2:793–805

Gens JS, Fujiki M, Pickard BG (2000) Arabinogalactan protein and wall-associated kinase in a plasmalemmal reticulum with specialized vertices. Protoplasma 212:115–134

Goswami L, Dunlop JWC, Jungnikl K, Eder M, Gierlinger N, Coutand C, Jeronimidis G, Fratzl P, Burgert I (2008) Stress generation in tension wood of poplar is based on the lateral swelling power of the G-layer. Plant J 56:531–538

Gouget A, Senchou V, Govers F, Sanson A, Barre A, Rougé P, Pont-Lezica R, Canut H (2006) Lectin receptor kinases participate in protein-protein interactions to mediate plasma membrane-cell wall adhesions in Arabidopsis. Plant Physiol 140:81–90

Green PB (1962) Mechanisms for plant cellular morphogenesis. Science 138:1404–1405

Green PB (1996) Transductions to generate plant form and pattern: an essay on cause and effect. Ann Bot 78:269–281

Green PB (1999) Expression of pattern in plants: combining molecular and calculus-based biophysical paradigms. Am J Bot 86:1059–1076

Groom L, Mott L, Shaler S (2002) Mechanical properties of individual southern pine fibers. Part I. Determination and variability of stress-strain curves with respect to tree height and juvenility. Wood Fiber Sci 34:14–27

Grunt M, Zárský V, Cvrcková F (2008) Roots of angiosperm formins: the evolutionary history of plant FH2 domain-containing proteins. BMC Evol Biol 8:115

Guo H, Li L, Ye H, Yu X, Algreen A, Yin Y (2009a) Three related receptor-like kinases are required for optimal cell elongation in Arabidopsis thaliana. Proc Natl Acad Sci USA 106:7648–7653

Guo H, Ye H, Li L, Yin Y (2009b) A family of receptor-like kinases are regulated by BES1 and involved in plant growth in Arabidopsis thaliana. Plant Signal Behav 4:784–786

Hamant O, Traas J (2010) The mechanics behind plant development. New Phytol 185:369–385

Hardham AR, Takemoto D, White RG (2008) Rapid and dynamic subcellular reorganization following mechanical stimulation of Arabidopsis epidermal cells mimics responses to fungal and oomycete attack. BMC Plant Biol 8:63

He ZH, Cheeseman I, He D, Kohorn BD (1999) A cluster of five cell wall-associated receptor kinase genes, Wak1-5, are expressed in specific organs of Arabidopsis. Plant Mol Biol 39:1189–1196

Hématy K, Höfte H (2008) Novel receptor kinases involved in growth regulation. Curr Opin Plant Biol 11:321–328

Hématy K, Sado PE, Van Tuinen A, Rochange S, Desnos T, Balzergue S, Pelletier S, Renou JP, Höfte H (2007) A receptor-like kinase mediates the response of Arabidopsis cells to the inhibition of cellulose synthesis. Curr Biol 17:922–931

Henriksson J (2001) Differential shading of branches or whole trees: survival, growth, and reproduction. Oecologia 126:482–486

Holweg C, Nick P (2004) Arabidopsis myosin XI mutant is defective in organelle movement and polar auxin transport. Proc Natl Acad Sci USA 101:10488–10493

Humphrey TV, Bonetta DT, Goring DR (2007) Sentinels at the wall: cell wall receptors and sensors. New Phytol 176:7–21

Hussey PJ, Ketelaar T, Deeks MJ (2006) Control of the actin cytoskeleton in plant cell growth. Annu Rev Plant Biol 57:109–125

Ingber DE (1993) Cellular tensegrity: defining new rules of biological design that govern the cytoskeleton. J Cell Sci 104:613–627

Ingber DE (1998) The architecture of life. Sci Am 278:48–57

Ingber DE (2003a) Tensegrity I. Cell structure and hierarchical systems biology. J Cell Sci 116:1157–1173

Ingber DE (2003b) Tensegrity II. How structural networks influence cellular information processing networks. J Cell Sci 116:1397–1408

Ingber DE (2006) Cellular mechanotransduction: putting all the pieces together again. FASEB J 20:811–827

Ingber DE (2008) Tensegrity and mechanotransduction. J Bodyw Mov Ther 12:198–200
Ingber DE (2009) From cellular mechanotransduction to biologically inspired engineering: 2009 Pritzker Award Lecture, BMES Annual Meeting. Ann Biomed Eng 38:1148–1161
Ingber DE, Dike L, Hansen L, Karp S, Liley H, Maniotis A, McNamee H, Mooney D, Plopper G, Sims J, Wang N (1994) Cellular tensegrity: exploring how mechanical changes in the cytoskeleton regulate cell growth, migration, and tissue pattern during morphogenesis. Int Rev Cytol 150:173–224
Ingold CT (1959) Peristome teeth and spore discharge in mosses. Trans Proc Bot Soc 38:76–88
Ingouff M, Fitz Gerald JN, Guérin C, Robert H, Sørensen MB, Van Damme D, Geelen D, Blanchoin L, Berger F (2005) Plant formin AtFH5 is an evolutionarily conserved actin nucleator involved in cytokinesis. Nat Cell Biol 7:374–380
Jedd G, Chua NH (2002) Visualization of peroxisomes in living plant cells reveals actomyosin-dependent cytoplasmic streaming and peroxisome budding. Plant Cell Physiol 43: 384–392
Johnson KL, Jones BJ, Bacic A, Schultz CJ (2003) The fasciclin-like arabinogalactan proteins of Arabidopsis. A multigene family of putative cell adhesion molecules. Plant Physiol 133:1911–1925
Kadota A, Yamada N, Suetsugu N, Hirose M, Saito C, Shoda K, Ichikawa S, Kagawa T, Nakano A, Wada M (2009) Short actin-based mechanism for light-directed chloroplast movement in Arabidopsis. Proc Natl Acad Sci USA 106:13106–13111
Kasahara M, Kagawa T, Oikawa K, Suetsugu N, Miyao M, Wada M (2002) Chloroplast avoidance movement reduces photodamage in plants. Nature 420:829–832
Kasprowicz A, Kierzkowski D, Maruniewicz M, Derba-Maceluch M, Rodakowska E, Zawadzki P, Szuba A, Wojtaszek P (2009) Mechanical integration of plant cell. In: Baluška F (ed) Plant-environment interactions. Springer, Berlin, pp 1–20
Ketelaar T, Galway ME, Mulder BM, Emons AMC (2008) Rates of exocytosis and endocytosis in *Arabidopsis* root hairs and pollen tubes. J Microsc 231:265–273
Knebel W, Quader H, Schnepf E (1990) Mobile and immobile endoplasmic reticulum in onion bulb epidermis cells: short- and long-term observations with a confocal laser scanning microscope. Eur J Cell Biol 52:328–340
Kohorn BD (2000) Plasma membrane-cell wall contacts. Plant Physiol 124:31–38
Kohorn BD (2001) WAKs; cell wall associated kinases. Curr Opin Cell Biol 3:529–533
Kohorn BD, Kobayashi M, Johansen S, Friedman HP, Fischer A, Byers N (2006) Wall-associated kinase 1 (WAK1) is crosslinked in endomembranes, and transport to the cell surface requires correct cell-wall synthesis. J Cell Sci 119:2282–2290
Komis G, Apostolakos P, Galatis B (2002) Hyperosmotic stress-induced actin filament reorganization in leaf cells of *Chlorophyton comosum*. J Exp Bot 53:1699–1710
Komis G, Apostolakos P, Galatis B (2003) Actomyosin is involved in the plasmolytic cycle: gliding movement of the deplasmolyzing protoplast. Protoplasma 221:245–256
Kordyum EL (2003) A role for the cytoskeleton in plant cell gravisensitivity and Ca^{2+} signaling in microgravity. Cell Biol Int 27:219–221
Kuhlemeier C (2007) Phyllotaxis. Trends Plant Sci 12:143–150
Kutschera U (1995) Tissue pressure and cell turgor in axial plant organs: implications for the organism theory of multicellularity. J Plant Physiol 146:126–132
Kutschera U (2008) The growing outer epidermal wall: design and physiological role of a composite structure. Ann Bot 101:615–621
Le J, El-Assal Sel D, Basu D, Saad ME, Szymanski DB (2003) Requirements for Arabidopsis ATARP2 and ATARP3 during epidermal development. Curr Biol 13:1341–1347
Lee D, Polisensky D, Braam J (2005) Genome-wide identification of touch- and darkness-regulated Arabidopsis gene: a focus on calmodulin-like and XTH genes. New Phytol 165: 429–444
Li S, Blanchoin L, Yang Z, Lord EM (2003) The putative Arabidopsis arp2/3 complex controls leaf cell morphogenesis. Plant Physiol 132:2034–2044

Lindström H, Evans JW, Verrill SP (1998) Influence of cambial age and growth conditions on microfibril angle in young Norway spruce (Picea abies [L.] Karst.). Holzforschung 52: 573–581
López-Bucio J, Cruz-Ramírez L, Herrera-Estrella L (2003) The role of nutrient availability in regulating root architecture. Curr Opin Plant Biol 6:280–287
Majewska-Sawka A, Nothnagel EA (2000) The multiple roles of arabinogalactan proteins in plant development. Plant Physiol 122:3–10
Malamy J (2005) Intrinsic and environmental factors regulating root system growth. Plant Cell Environ 28:67–77
Maniotis AJ, Bojanowski K, Ingber DE (1997a) Mechanical continuity and reversible chromosome disassembly within intact genomes removed from living cells. J Cell Biochem 65:114–130
Maniotis AJ, Chen CS, Ingber DE (1997b) Demonstration of mechanical connections between integrins, cytoskeletal filaments, and nucleoplasm that stabilize nuclear structure. Proc Natl Acad Sci USA 94:849–854
Martinière A, Gayral P, Hawes C, Runions J (2011) Building bridges: formin1 of Arabidopsis forms a connection between the cell wall and the actin cytoskeleton. Plant J 66:354–365
Mathur J (2006) Local interactions shape plant cells. Curr Opin Cell Biol 18:40–46
Mathur J, Mathur N, Kernebeck B, Hulskamp M (2003a) Mutations in actin-related proteins 2 and 3 affect cell shape development in Arabidopsis. Plant Cell 15:1632–1645
Mathur J, Mathur N, Kirik V, Kernebeck B, Srinivas BP, Hulskamp M (2003b) Arabidopsis CROOKED encodes for the smallest subunit of the ARP2/3 complex and controls cell shape by region specific fine F-actin formation. Development 130:3137–3146
Mattheck C, Breloer H (1995) The body language of trees: a handbook for failure analysis. HMSO, London
Meckel T, Hurst AC, Thiel G, Homann U (2004) Endocytosis against high turgor: intact guard cells of *Vicia faba* constitutively endocytose fluorescently labelled plasma membrane and GFP-tagged K-channel KAT1. Plant J 39:182–193
Meckel T, Hurst AC, Thiel G, Homann U (2005) Guard cells undergo constitutive and pressure-driven membrane turnover. Protoplasma 226:23–29
Monshausen GB, Gilroy S (2009) Feeling green: mechanosensing in plants. Trends Cell Biol 19:228–235
Moulia B, Fournier M (1997) Mechanics of the maize leaf: a composite beam model of the midrib. J Mater Sci 32:2771–2780
Moulia B, Coutand C, Lenne C (2006) Posture control and skeletal mechanical acclimation in terrestrial plants: implications for mechanical modeling of plant architecture. Am J Bot 93:1477–1489
Müller J, Mettbach U, Menzel D, Šamaj J (2007) Molecular dissection of endosomal compartments in plants. Plant Physiol 145:293–304
Muthugapatti K, Kandasamy MK, Meagher RB (1999) Actin-organelle interaction: association with chloroplast in Arabidopsis leaf mesophyl cell. Cell Motil Cytoskelet 44:110–118
Nachtigal W (1994) On the research history of plant biomechanics. Biomimetics 2:87–108
Newell AC, Shipman PD, Sun Z (2008) Phyllotaxis as an example of the symbiosis of mechanical forces and biochemical processes in living tissue. Plant Signal Behav 3:586–589
Nguema-Ona E, Bannigan A, Chevalier L, Baskin TI, Driouich A (2007) Disruption of arabinogalactan proteins disorganizes cortical microtubules in the root of Arabidopsis thaliana. Plant J 52:240–251
Niklas KJ (1992) Plant biomechanics. An engineering approach to plant form and function. University of Chicago Press, Chicago
Niklas KJ (2000) The evolution of plant body plans. A biomechanical perspective. Ann Bot 85:411–438
Nothnagel EA (1997) Proteoglycans and related components in plant cells. Int Rev Cytol 174:195–291
Novoplansky A, Cohen D, Sachs T (1989) Ecological implications of correlative inhibition between plant shoots. Physiol Plant 77:136–140

Oikawa K, Kasahara M, Kiyosue T, Kagawa T, Suetsugu N, Takahashi F, Kanegae T, Niwa Y, Kadota A, Wada M (2003) Chloroplast unusual positioning1 is essential for proper chloroplast positioning. Plant Cell 15:2805–2815

Oikawa K, Yamasato A, Kong S, Kasahara M, Nakai M, Takahashi F, Ogura Y, Kagawa T, Wada M (2008) Chloroplast outer envelope protein CHUP1 is essential for chloroplast anchorage to the plasma membrane and chloroplast movement. Plant Physiol 148:829–842

Ovecka M, Lang I, Baluška F, Ismail A, Illes P, Lichtscheidl IK (2005) Endocytosis and vesicle trafficking during tip growth of root hairs. Protoplasma 226:39–54

Paciorek T, Zazímalová E, Ruthardt N, Petrásek J, Stierhof YD, Kleine-Vehn J, Morris DA, Emans N, Jürgens G, Geldner N, Friml J (2005) Auxin inhibits endocytosis and promotes its own efflux from cells. Nature 435:1251–1256

Paredez AR, Somerville CR, Ehrhardt DW (2006) Visualization of cellulose synthase demonstrates functional association with microtubules. Science 312:1491–1495

Paredez AR, Persson S, Ehrhardt DW, Somerville CR (2008) Genetic evidence that cellulose synthase activity influences microtubule cortical array organization. Plant Physiol 147: 1723–1734

Parker KK, Brock AL, Brangwynne C, Mannix RJ, Wang N, Ostuni E, Geisse NA, Adams JC, Whitesides GM, Ingber DE (2002) Directional control of lamellipodia extension by constraining cell shape and orienting cell tractional forces. FASEB J 16:1195–1204

Paul AS, Pollard TD (2009) Review of the mechanism of processive actin filament elongation by formins. Cell Motil Cytoskelet 66:606–617

Peremyslov VV, Prokhnevsky AI, Avisar D, Dolja VV (2008) Two class XI myosins function in organelle trafficking and root hair development in *Arabidopsis*. Plant Physiol 146:1109–1116

Peters WS, Hagemann W, Tomos DA (2000) What makes plants different? Principles of extracellular matrix function in "soft" plant tissues. Comp Biochem Physiol A 125:151–167

Prokhnevsky AI, Peremyslov VV, Dolja VV (2008) Overlapping functions of the four class XI myosins in *Arabidopsis* growth, root hair elongation, and organelle motility. Proc Natl Acad Sci USA 105:19744–19749

Reiterer A, Lichtenegger H, Tschegg S, Fratzl P (1999) Experimental evidence for a mechanical function of the cellulose microfibril angle in wood cell walls. Philos Mag A 79:2173–2184

Sardar HS, Yang J, Showalter AM (2006) Molecular interactions of arabinogalactan proteins with cortical microtubules and F-actin in Bright Yellow-2 tobacco cultured cells. Plant Physiol 142:1469–1479

Satiat-Jeunemaitre B, Steele C, Hawes C (1996) Golgi-membrane dynamics are cytoskeleton dependent: a study on Golgi stack movement induced by brefeldin A. Protoplasma 191:21–33

Sato Y, Kadota A, Wada M (1999) Mechanically induced avoidance response of chloroplasts in fern protonemal cells. Plant Physiol 121:37–44

Schmid-Schönbein GW, Kosawada T, Skalak R, Chien S, Stamenović P (1995) Membrane model of endothelial cells and leukocytes. A proposal for the origin of a cortical stress. J Biomech Eng 117:171–178

Schmidt von Braun S, Schleiff E (2008) The chloroplast outer membrane protein CHUP1 interacts with actin and profilin. Planta 227:1151–1159

Schmidt SM, Panstruga R (2007) Cytoskeleton functions in plant–microbe interactions. Physiol Mol Plant Pathol 71:135–148

Schopfer P (2006) Biomechanics of plant growth. Am J Bot 93:1415–1425

Schulze-Muth P, Irmler S, Schröder G, Schröder J (1996) Novel type of receptor-like protein kinase from a higher plant (Catharanthus roseus). cDNA, gene, intramolecular autophosphorylation, and identification of a threonine important for auto- and substrate phosphorylation. J Biol Chem 271:26684–26689

Seifert GJ, Roberts K (2007) The biology of arabinogalactan proteins. Annu Rev Plant Biol 58:137–161

Sellier D, Fourcaud T (2009) Crown structure and wood properties: influence on tree sway and response to high winds. Am J Bot 96:885–896

Senchou V, Weide R, Carrasco A, Bouyssou H, Pont-Lezica R, Govers F, Canut H (2004) High affinity recognition of a Phytophthora protein by Arabidopsis via an RGD motif. Cell Mol Life Sci 61:502–509
Sheahan MB, Rose RJ, McCurdy DW (2004) Organelle inheritance in plant cell division: the actin cytoskeleton is required for unbiased inheritance of chloroplasts, mitochondria and endoplasmic reticulum in dividing protoplasts. Plant J 37:379–390
Shimmen T, Yokota E (2004) Cytoplasmic streaming in plants. Curr Opin Cell Biol 16:68–72
Shiu SH, Bleecker AB (2001) Receptor-like kinases from Arabidopsis form a monophyletic gene family related to animal receptor kinases. Proc Natl Acad Sci USA 98:10763–10768
Showalter AM (2001) Arabinogalactan-proteins: structure, expression and function. Cell Mol Life Sci 58:1399–1417
Singhvi R, Kumar A, Lopez GP, Stephanopoulos GN, Wang DI, Whitesides GM, Ingber DE (1994) Engineering cell shape and function. Science 264:696–698
Suetsugu N, Wada M (2007) Chloroplast photorelocation movement mediated by phototropin family proteins in green plants. Biol Chem 388:927–935
Thomson DW (1992) On growth and form (Canto edition). Cambridge University Press, Cambridge
van den Berg C, Willemsen V, Hage W, Weisbeek P, Scheres B (1995) Cell fate in the Arabidopsis root meristem determined by directional signalling. Nature 378:62–65
van den Berg C, Willemsen V, Hendriks G, Weisbeek P, Scheres B (1997) Short-range control of cell differentiation in the Arabidopsis root meristem. Nature 390:287–289
van der Honing HS, Emons AMC, Ketelaar T (2007) Actin based processes that could determine the cytoplasmic architecture of plant cells. Biochim Biophys Acta 1773:604–614
van der Honing HS, de Ruijter NC, Emons AM, Ketelaar T (2010) Actin and myosin regulate cytoplasm stiffness in plant cells: a study using optical tweezers. New Phytol 185:90–102
Van Gestel K, Kohler RH, Verbelen J (2002) Plant mitochondria move on F-actin, but their positioning in the cortical cytoplasm depends on both F-actin and microtubules. J Exp Bot 369:659–667
van Hengel AJ, Tadesse Z, Immerzeel P, Schols H, van Kammen A, de Vries SC (2001) N-acetylglucosamine and glucosamine-containing arabinogalactan proteins control somatic embryogenesis. Plant Physiol 125:1880–1890
Verica JA, He ZH (2002) The cell wall-associated kinase (WAK) and WAK-like kinase gene family. Plant Physiol 129:455–459
Vidali L, McKenna ST, Hepler PK (2001) Actin polymerization is essential for pollen tube growth. Mol Biol Cell 12:2534–2545
Vogel V, Sheetz M (2006) Local force and geometry sensing regulate cell functions. Nat Rev Mol Cell Biol 7:265–275
Volkmann D, Baluška F (2006) Gravity: one of the driving forces for evolution. Protoplasma 229:143–148
Wada M, Kagawa T, Sato Y (2003) Chloroplast movement. Annu Rev Plant Biol 54:455–468
Wagner TA, Kohorn BD (2001) Wall-associated kinases are expressed throughout plant development and are required for cell expansion. Plant Cell 13:303–318
Wan J, Patel A, Mathieu M, Kim SY, Xu D, Stacey G (2008) A lectin receptor-like kinase is required for pollen development in Arabidopsis. Plant Mol Biol 67:469–474
Wang N, Tytell JD, Ingber DE (2009) Mechanotransduction at a distance: mechanically coupling the extracellular matrix with the nucleus. Nat Rev Mol Cell Biol 10:75–82
Wang S, Arellano-Santoyo H, Combs PA, Shaevitz JW (2010) Actin-like cytoskeleton filaments contribute to cell mechanics in bacteria. Proc Natl Acad Sci USA 107:9182–9185
Wardrop AB (1965) The formation and function of reaction wood. In: Côté WA (ed) Cellular ultrastructure of woody plants. Syracuse University Press, New York, pp 371–390
Wightman R, Turner S (2008) Trafficking of the plant cellulose synthase complex. Plant Physiol 153:427–432
Wojtaszek P (2000) Genes and plant cell walls: a difficult relationship. Biol Rev 75:437–475

Wojtaszek P, Volkmann D, Baluška F (2004) Polarity and cell walls. In: Lindsey K (ed) Polarity in plants. Blackwell, Sheffield, pp 72–121

Wojtaszek P, Anielska-Mazur A, Gabryś H, Baluška F, Volkmann D (2005) Rapid relocations of myosin VIII between cell periphery and plastid surfaces are root-specific and provide the evidence for actomyosin involvement in osmosensing. Funct Plant Biol 32:721–736

Wojtaszek P, Baluška F, Kasprowicz A, Łuczak M, Volkmann D (2007) Domain-specific mechanosensory transmission of osmotic and enzymatic cell wall disturbances to the actin cytoskeleton. Protoplasma 230:217–230

Wyatt SE, Carpita NC (1993) The plant cytoskeleton-cell-wall continuum. Trends Cell Biol 3:413–417

Wymer C, Lloyd C (1996) Dynamic microtubules: implications for cell wall patterns. Trends Plant Sci 1:222–228

Wymer CL, Wymer SA, Cosgrove DJ, Cyr RJ (1996) Plant cell growth responds to external forces and the response requires intact microtubules. Plant Physiol 110:425–430

Yang L, Reece JM, Cho J, Bortner CD, Shears SB (2008) The nucleolus exhibits an osmotically regulated gatekeeping activity that controls the spatial dynamics and functions of nucleolin. J Biol Chem 283:11823–11831

Yin J, Cao Z, Li C, Sheinman I, Chen X (2008) Stress-driven buckling patterns in spheroidal core/shell structures. Proc Natl Acad Sci USA 105:19132–19135

Zheng M, Beck M, Müller J, Chen T, Wang X, Wang F, Wang O, Wang Y, Baluška F, Logan DC, Šamaj J, Lin J (2009) Actin turnover is required for myosin-dependent mitochondrial movements in *Arabidopsis* root hairs. PLoS ONE 4:e5961

Zonia L, Munnik T (2007) Life under pressure: hydrostatic pressure in cell growth and function. Trends Plant Sci 12:90–97

Zonia L, Munnik T (2008) Vesicle trafficking dynamics and visualization of zones of exocytosis and endocytosis in tobacco pollen tubes. J Exp Bot 59:861–873

Micromechanics of Cell Walls

Ingo Burgert and John W.C. Dunlop

Abstract In this chapter, we discuss the mechanical properties of primary and secondary cell walls with regard to cell wall structure and composition. The first sections of the chapter are devoted to defining the mechanical terms used and to giving a general introduction into the mechanical behaviour of composites. The specific structure–property relationships of primary and secondary cell walls are then discussed with a focus on unravelling the mechanical role of individual cell wall components. In these terms, the mechanical characterization of genetically modified plants is highlighted as it allows for the targeted alteration of both cell wall polymers as well as their cross-linking capacities, in turn resulting in a distinct influence on entire cell wall properties.

1 Introduction

A rigid cell wall which supplies mechanical support to the cell is a characteristic feature of the plant kingdom (see, however, Chapter 14). Its high rigidity allows trees to grow more than 100 m tall and build stems with a mass of more than 1,000 tonnes. However, to facilitate growth the cell wall cannot start off by being this rigid after initial deposition by the cell. In contrast, the cell wall during its primary phase must be flexible and plastically deformable to allow growth to a designated shape and size of the cell. This contradiction has fascinated researchers for decades, as it highlights the astonishing ability of plants to both precisely control and adapt the mechanical properties of cell walls on demand. This is all the more remarkable in view of the apoplast nature of the cell walls, meaning that the plant cells need to be able to adjust the performance of extracellular material, both during cell and organ ontogeny as well as upon external stimuli.

I. Burgert (✉) and J.W.C. Dunlop
Department of Biomaterials, Max Planck Institute of Colloids and Interfaces, 14424 Potsdam, Germany
e-mail: ingo.burgert@mpikg.mpg.de; john.dunlop@mpikg.mpg.de

P. Wojtaszek (ed.), *Mechanical Integration of Plant Cells and Plants*, Signaling and Communication in Plants 9, DOI 10.1007/978-3-642-19091-9_2,
© Springer-Verlag Berlin Heidelberg 2011

With this chapter we intend to discuss the mechanics of plant cell walls, highlighting the interrelation with their structural and chemical organization. The first section defines mechanical terms and introduces the basic description of mechanical properties. Here, it is also important to briefly discuss the basic concepts in material testing and the limitations in transferring these concepts from technical materials to biological materials. The following sections deal with general composite theory and explain the analogy between technical fibre composites and the organization of the plant cell wall. The cell wall components of primary and secondary cell walls are briefly introduced along these lines. Thereafter, the structure–function relationships of primary and secondary cell wall are discussed with a focus on the interaction of cell wall components and related approaches to elucidate the underlying principles of cell wall design. Here the crucial role of cellulose fibril orientation is discussed and recent work in the field of mechanical characterization of genetically modified plants with alterations of the cell wall matrix organization is reported. In conclusion, an outlook is given reflecting scientific challenges and potential applications of plant cell wall research.

2 Mechanical Terms and Definitions

This section defines the various parameters used to describe the mechanical behaviour of a material, the experimental methods used to measure them and how this can be applied to the mechanical response of the plant cell wall. For more detailed introductory texts on the mechanics of materials see (e.g. Gere 2002; Callister 2003; Ashby and Jones 2005), for the mechanics of composite materials see (e.g. Chou 1992; Hull and Clyne 1996) and for the mechanics of biological tissues see (e.g. Wainwright et al. 1982; Vincent 1990; Niklas 1992).

The "mechanical behaviour" of a material, refers to how a piece of this material responds to applied forces or displacements, in terms of the amount and rate of deformation, the resistance (force) to deformation and the likelihood of failure. A typical mechanical test involves applying a known displacement (or force) at a given rate to a sample whilst simultaneously measuring the force (or displacement). Forces and displacements, although being the quantities actually measured in an experiment, are inappropriate to describe the material response as they are dependent on sample geometry and size. In order to compare the mechanical response of different samples, force and displacement need to be normalized, giving rise to two important mechanical quantities, stress and strain. Stress, given the symbol σ, is defined as the force acting perpendicular to a surface divided by the cross-sectional area over which this force is applied and has the units of pressure (Newtons per metre square often given in Pascals) (Fig. 1bi). Strain, with the symbol ε, is the change in length divided by a reference length and is unitless (Fig. 1biii).

Loading, of course, can be applied in different directions, meaning stress and strain are defined as tensile (resp. compressive) when the loading results in a lengthening (resp. shortening) of the sample or in a combination of both in bending.

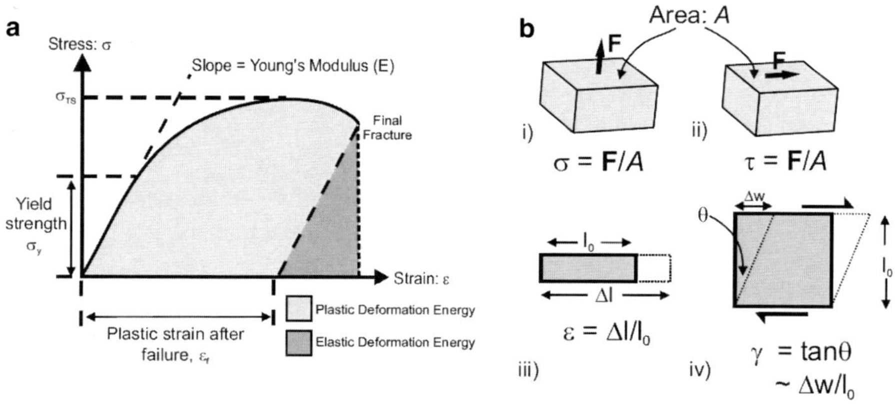

Fig. 1 (**a**) Example of a typical stress–strain curve for a ductile metal. The first portion of the curve is linear with a slope, E, corresponding to the Young's modulus. Below the yield strength, σ_f, the sample deforms completely reversibly, or is linear elastic. Above this stress, the sample deforms irreversibly or plastically, until failure. The areas under the curve give the energies of deformation. The maximum stress, σ_{TS}, or the tensile strength is the maximum stress that the material can support before failure. (**b**) A schematic of how (i) normal and (ii) shear stresses and (iii) normal strain and (iv) shear strains are defined

Loading under shear implies the force is applied parallel to the loaded surface. Both shear and tensile loading cases are sketched in Fig. 1b. Shear stress, τ, is defined as the force acting parallel to a surface divided by the cross-sectional area over which this force is applied and has the same units as tensile stress (Fig. 1bii). Shear strain, γ, is given by the angle of shear, approximated by the sideways shear divided by a perpendicular reference length and is also unitless (Fig. 1biv). As both cross-sectional area and length change due to deformation, different definitions of stress and strain can be found in the literature. For example, engineering stress and strain (often used to present experimental data) are defined using initial area and length, whereas true stress and strain (often used in continuum mechanics modelling) are defined using the current deformed area and length which changes during deformation. The two definitions are readily converted between each other (e.g. Ashby and Jones 2005), however, in the following engineering stress and strain will be used.

The majority of mechanical testing techniques and analysis methods were originally developed by materials scientists and engineers to analyse engineering materials, and are increasingly applied to the mechanical characterization of relatively fragile biological tissues. The use of these methods on natural tissues therefore requires some caution, as many techniques are suitable only for samples with specific shapes and sizes, under the assumption of simple homogeneous microstructures. However, if these limitations are kept in mind or techniques are used in a comparative sense (i.e. two tissue types are compared using the same method), useful information about the mechanics of biological tissue can be obtained. There are several mechanical testing techniques that can be readily performed on plant tissue, such as nano-indentation, and acoustic microscopy which gives localized mechanical information at the submicron scale, whereas micro-indentation and constant strain rate tensile testing give mechanical data at a more microscopic scale.

Micro-indentation and nano-indentation are methods in which a small sharp tip (micro- or nano-scale), with a well-defined shape, is pushed into the surface of the material being analysed while both the force and displacement are recorded (e.g. Hiller et al. 1996; Ebenstein and Pruitt 2006). From the force–displacement curve and a knowledge of the shape of the tip, mechanical parameters, such as elastic modulus and hardness, can be determined at the location of the indent (Oliver and Pharr 1992). Micro-indentation has been used in plant science to measure tissue and cell wall properties of relatively large objects such as pollen tubes (Geitmann et al. 2004; Parre and Geitmann 2005a, b), whereas nano-indentation allows for hardness and stiffness to be directly measured at the submicron level (Wimmer et al. 1997; Gindl et al. 2004). The interpretation of the results requires some care, as the theoretical models used to interpret instrumented indentation often assume that the materials are isotropic and homogeneous, a situation that is rarely true for biological tissues. Contrary, the orthotropic nature of cell walls needs to be considered when determining stiffness properties of the wall (Gindl and Schoberl 2004). Another technique operating at a similar resolution as nano-indentation is scanning acoustic microscopy (SAM). SAM uses focused acoustic waves to map acoustic impedance, a parameter combining stiffness and density, over the surface of a sample (see e.g. Clair et al. 2000; Raum et al. 2006). If the density distribution of the material being measured is known, or at least is constant over the surface, then the acoustic impedance map can be interpreted in terms of elastic modulus. Such measurements are particularly useful in highlighting areas of interest for further structural characterization within complex tissues containing regions of similar chemistry but different nanostructural architectures (e.g. Elbaum et al. 2007).

Perhaps the simplest mechanical test to perform on tissue is the constant strain rate tensile test. The advantage of tensile tests is that there are fewer constraints with respect to sample geometry than in other load cases such as compression or bending. Mechanical properties of small tissue specimens and individual fibrous cells can be determined in uniaxial microtensile tests (Page and El-Hosseiny 1983; Burgert et al. 2002, 2003; Groom et al. 2002; Köhler and Spatz 2002; Ryden et al. 2003; Abasolo et al. 2009). To calculate cell wall properties from these tests, further sample parameters such as structural/geometrical features, density or turgor pressure have to be considered.

In tensile testing, a sample is affixed by glueing or clamping to grips at both ends and is pulled at a given rate of displacement, whilst the force is measured. Such a force–displacement curve can be readily converted into the so-called stress–strain curve, characteristic for a given material. An example stress–strain curve for a ductile metal is given in Fig. 1a. Although the stress–strain curves for plant cell wall material are often more complex (see Fig. 7), the parameters used to characterize the materials response are still the same, and as such the following discussion will focus on simplified stress–strain curves.

The initial portion of a general stress–strain curve is often linear and is called the linear-elastic regime, where the deformation is completely reversible. In this regime, the path of loading corresponds exactly to that of unloading and the sample will return back to the same shape it had before deformation (Fig. 2a). In some materials, such as rubbers, however, the reversible portion of the stress–strain curve is non-linear. The slope of a linear-elastic portion of the stress–strain curve is the

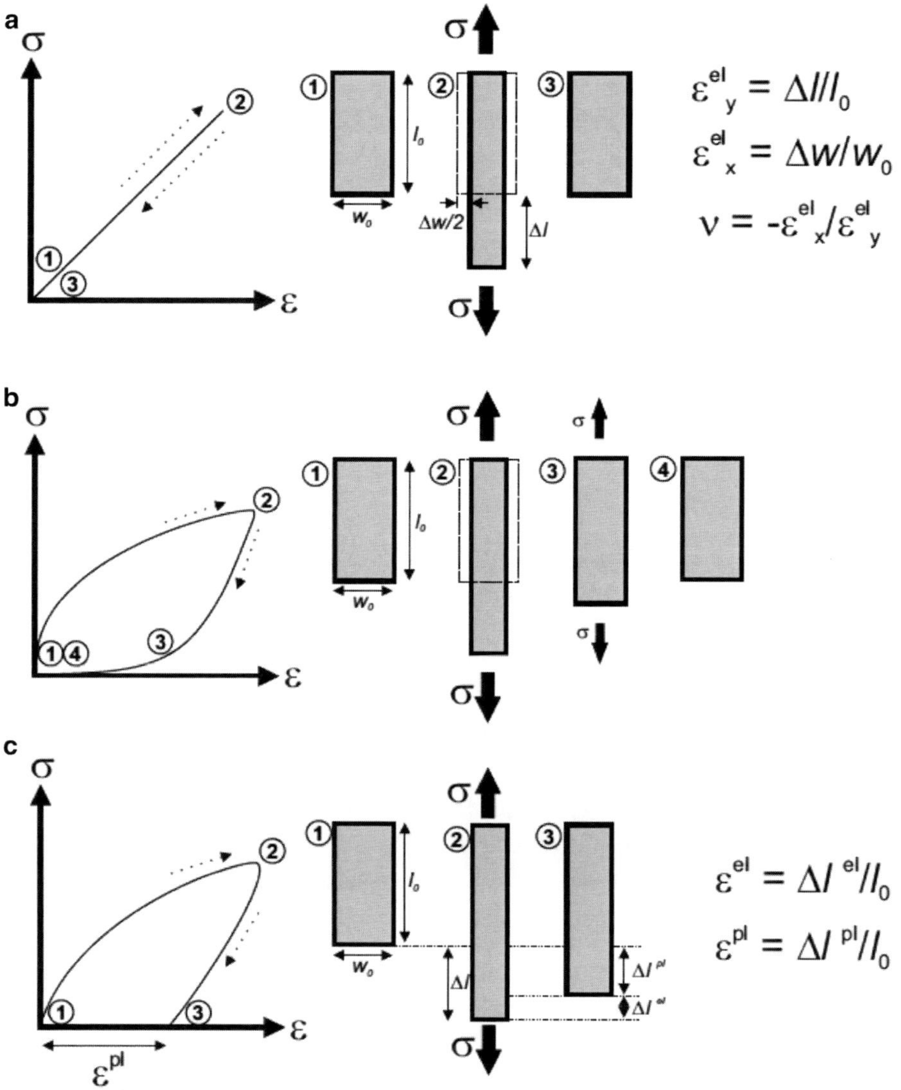

Fig. 2 Examples of stress–strain curves, for (**a**) linear elastic, (**b**) viscoelastic and (**c**) plastic materials with sketches of the changes in sample associated with such deformation behaviours

Young's or elastic modulus and is often given the symbol, E, the constant in Hooke's Law. The Young's modulus gives the stiffness of a material, in other words the resistance of a material to elastic deformation and has the same units as stress (Pascals). The higher E is, the more stress is required to deform to a given strain. Typical values of the Young's modulus range between 70 GPa for aluminium and 200 GPa for steels, normal spruce wood has an elastic modulus around 10 GPa and polymeric materials such as elastomers can have moduli in the 10–100 MPa range. An elastic modulus for shear deformation (Fig. 1bii, biv) can also be defined and is often given the symbol μ (or G). It has the same units as stress and for many materials has about 3/8 of the value of E. A third elastic modulus is the bulk modulus, K, which gives the volume change of a material as a function of pressure.

In some materials after a critical stress, or the yield strength σ_y, is reached (Fig. 1a) the deformation is no longer completely reversible upon unloading. For stresses above this yield stress, the material is said to deform plastically (Fig. 2c). From the yield point onwards the strain consists of two independent contributions, of elastic and plastic strain. By this it is meant that if the sample is unloaded or fails (Figs. 1a and 2c) then the sample will regain partially its initial length by an amount given by ε^{el}, at the point of unloading, but will have a permanent deformation given by ε^{pl}. The highest stress achieved during a tensile test is known as the ultimate tensile stress, σ_{TS} and depending on the material can also ooccur at the point of failure. The integrated area under the stress–strain curve has units of energy per unit volume and corresponds to the energy dissipated per unit volume of material during deformation (Fig. 1a). This area can be separated into elastic energy, which is regained upon unloading, and plastic deformation energy which is lost as heat. The deformation energy dissipated per unit volume is a useful quantity often used as a proxy for the toughness of a material. The higher toughness (area below curve) the more energy is required to fracture a given material.

Many materials display a viscoelastic response in that the elastic behaviour is time dependent with a modulus that is dependent on the rate of loading. Such a viscoelastic response is illustrated in Fig. 2b for a cyclic loading experiment. Although the loading and unloading paths are different, the stress–strain curve returns to the origin after unloading indicating fully reversible deformation. Another method to measure the viscous response of a material is to perform a relaxation test during a tensile test (Fig. 3a, b).

In a relaxation test, the sample is deformed to a given stress, σ_0, the sample displacement is then stopped and fixed at a given strain, ε_0, and the evolution of stress as a function of time is then followed. The rate at which the stress relaxes gives an idea of the contribution of viscous terms to the overall mechanical response. Exponential decay functions are commonly used to fit the relaxation curve, although it is often difficult to get a physical interpretation of the parameters extracted, as relaxation comes from a combination of viscoelastic and plastic deformation. A similar experiment to measure the stress–strain response is the creep test in which a fixed stress is applied to a sample and the resultant deformation observed as a function of time (Fig. 3c, d). In general, creep curves have a linear (elastic) portion corresponding to the loading, a portion of decreasing rate of creep called primary creep,

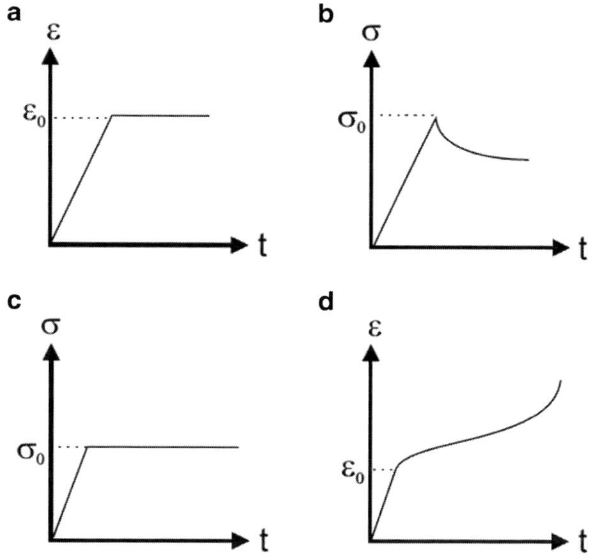

Fig. 3 Other time-dependent mechanical tests (**a**) and (**b**) Stress relaxation test in which the sample is deformed to a given strain (**a**) and then held at this strain whilst the stress (**b**) is measured with time; (**c**) and (**d**) Creep test in which a given load is applied to the sample (**c**) and the strain (**d**) measured as a function of time

followed by a linear portion of steady-state or secondary creep. During tertiary creep, the final portion of the curve displays an accelerating rate until failure.

Up until now in this introductory section, nothing has been mentioned about the fact that specimens are three-dimensional and in principle can deform in all directions. It has been assumed that both stress and strain are measured along the longitudinal tensile direction of the specimen. However, materials can deform also in directions perpendicular to the applied stress. When stretched, they often decrease in cross-section as illustrated in Fig. 2a where the longitudinal strain ε_y^{el} is accompanied by a decrease in sample width ε_x^{el}. This effect is described by the Poisson's ratio, v, which is the negative ratio of the lateral strain to the tensile strain. Typical values of the Poisson's ratio are 0.3 for many metals, 0.5 for rubbers and 0 for materials such as cork. The Poisson's ratio, the elastic, bulk and shear moduli are in fact the four elastic constants required to describe the elastic response of an isotropic material (only three of these are independent). An isotropic elastic response is independent of the loading orientation. In general, however the behaviour of a material is direction dependent, i.e. the material is anisotropic which can, for instance, be easily seen in the directional properties in wood. This means that in order to completely describe the elastic response of anisotropic materials, elastic constants need to be defined for each independent direction in the material. In addition to the difficulty in measuring all of the constants, the fact that the materials properties are direction dependent means that a full characterization of a sample also requires a knowledge of the material orientation (the fibre orientation for example). Fibrous materials are by nature anisotropic and are typically stiffer along the fibres than perpendicular to them. Due to the symmetry of the fibre reinforcement a good approximation for their elastic response is to assume that they display an isotropic response perpendicular to the fibre. This behaviour is known as transverse isotropy and only requires five independent elastic constants in order to describe the elastic response.

3 The Cell Wall as a Composite

Composites are a class of materials made from more than one constituent phase or material. This section outlines some simple models for the mechanical response of composites, and describes how effective materials properties can be calculated as a function of the arrangement and the materials properties of the constituents. For more detailed models and references (see e.g. Chou 1992; Hull and Clyne 1996).

Consider the simplest of composites, a two-phase system under uniaxial loading, comprised of a soft material, M, and a hard material, F. Of the infinite number of possible ways to combine the two materials, there are two limiting architectures with respect to the load direction. In the first of these, the Voigt model, the two materials are oriented in layers parallel to the load, and in the second, the Reuss model, the material layers are oriented perpendicular, or in series, with respect to the load (see insets in Fig. 4a). For both these cases, it is straightforward to estimate effective Young's moduli. For the Voigt model, the effective elastic modulus of the composite, E_C, is simply the sum of the moduli of the components (E_F and E_M) weighted by the volume fractions of each phase. This gives $E_c = fE_F + (1 - f)E_M$ where f is the volume fraction of the hard material F. This model is also known in the literature as the iso-strain model, as it is derived by assuming each phase deforms the same amount. For the Reuss model, which is also known as the iso-stress model, the composite modulus is given by the weighted harmonic sum of the

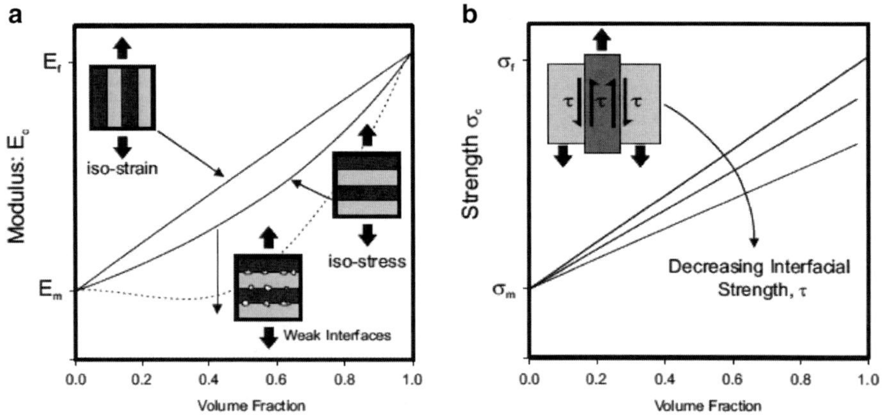

Fig. 4 (**a**) Schematic depiction of elastic modulus of a fibre composite, E_c, as a function of the fibre volume fraction. E_m and E_f are the matrix and fibre stiffnesses, respectively. The *two solid lines* indicate the upper and lower bounds of the stiffness for iso-strain and iso-stress composite models, respectively. The presence of imperfect or weak interfaces results in a reduction of the effective stiffness in the iso-stress as schematized by the *dotted line*. (**b**) Strength of a fibre composite: Strength σ_c, of a composite as a function of fibre volume fraction for an iso-strain composite with different values of interfacial shear strength. σ_m and σ_f are the respective matrix and fibre strengths

moduli, $E_c = (f/E_F + (1-f)/E_M)^{-1}$. This model, however, assumes perfect bonding at the interfaces between the two phases. A sketch of the solutions to these models is given as a function of volume fraction of the hard phase in Fig. 4a. These solutions give strict upper and lower bounds to the elastic modulus of a composite meaning the elastic moduli of real composites lie somewhere between these bounds (Hashin and Shtrikman 1963). The lower bound, however, assumes a perfect bonding between the two phases, and weaker interfaces would lead to a decrease in effective stiffness as indicated by the dotted line in Fig. 4a. Similar models can be developed to estimate other elastic constants necessary to describe the complete elastic response, for more details see (e.g. Chou 1992; Hull and Clyne 1996). The calculation of the failure strength of composite materials is more complex as it depends strongly on how easily stress is transferred via shear through the interface between the two phases. A rough sketch of these effects is made in Fig. 4b, which illustrates how a decreasing interfacial shear strength for a parallel aligned fibre composite results in a decreasing composite strength. With these simple models in mind, it can be seen that by controlling interfacial strength as well as the fibre architecture, by means of fibre's aspect ratio, a large range of material properties can be achieved.

Common technical examples include: glass or carbon fibre-reinforced epoxy resins and concrete and ceramic particle-reinforced metal matrix composites. One of the principal advantages of composite materials is that by combining different materials, a new set of physical properties can be attained that are not available in the constituents. Another important feature is that their properties can be controlled and optimized through the arrangement or architecture of the constituents (Ashby and Brechet 2003). Such an approach is used widely by nature, for example, in mineralized tissues such as bone and shell. By combining stiff and brittle ceramic particles within a soft and tough protein matrix, these mineralized tissues are made to be both tough and strong (see e.g. Fratzl and Weinkamer 2007; Dunlop and Fratzl 2010).

The analogy between technical fibre composites and plant cell walls is due to the organization of stiff cellulose microfibrils of a high aspect ratio, embedded in a matrix of various compliant polymers (Fengel and Wegener 1984; Kerstens et al. 2001; Fratzl et al. 2004; Cosgrove 2005; Fahlen and Salmen 2005). Furthermore, the high contrast in stiffness between both phases further supports this analogy. Cellulose possesses a modulus of elasticity in the axial direction under moist conditions of ~134 GPa, whereas matrix polymers such as hemicelluloses and lignin have much lower moduli of 40 MPa and 2 GPa, respectively (Salmén 2001). Figure 5 illustrates the fibre composite character of primary and secondary cell walls. The cellulose fibrils have a diameter of ~2.5–5 nm and a length of several micrometres. While the diameter of the fibrils can be measured with relatively high precision by small angle X-ray scattering or in TEM studies (Jakob et al. 1995), the length of the fibrils is unknown because of current limitations in experimentally determining this parameter. One approach is to calculate cellulose fibril length from the degree of polymerization. This degree reflects how many β-D-glucose units are linked together by β-1,4-glycosidic

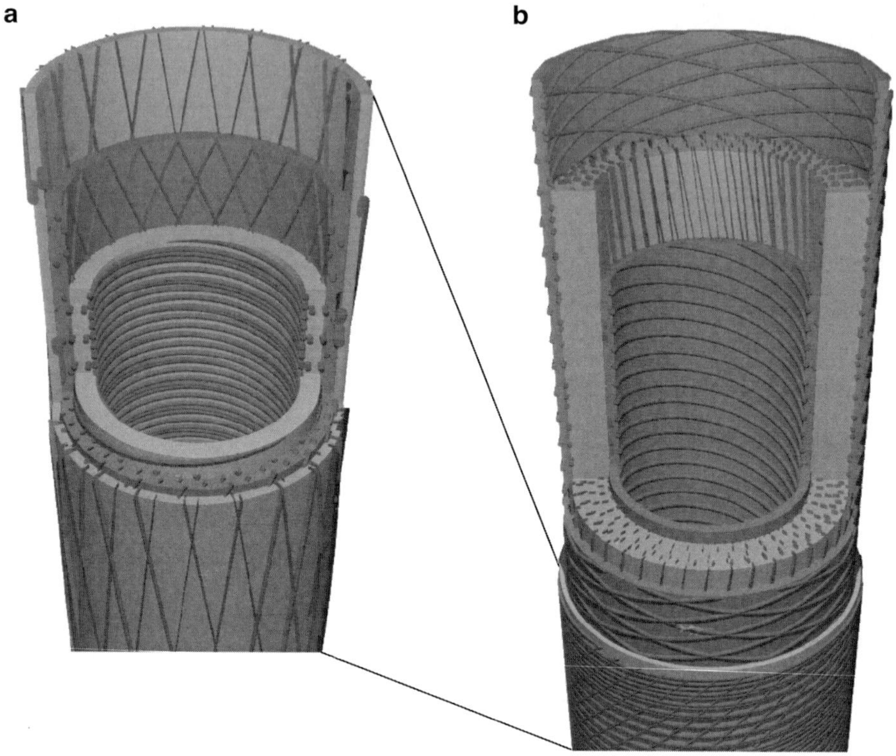

Fig. 5 View of the cell wall assembly as a composite structure (**a**) primary cell wall during growth with cellulose fibrils tilted toward the cell axis in the outer cell wall layers (**b**) secondary cell wall using the example of the wood cell structure

bonds in a linear polymer chain. Depending on the species between 7,000 and 15,000 units have been found for higher plants (Fengel and Wegener 1984). Fourteen thousand units would correspond to a cellulose fibril length of ~7 μm (Somerville et al. 2004). This estimate, however, lacks precision because it requires the cell wall to be deconstructed which can lead to cellulose chain breakage. An indirect visualization in the living cell could be another principle, highlighting cellulose synthase complexes with a fluorescent marker and following their movement by spinning disc confocal microscopy (Paredez et al. 2006). However, this method can only give a lower bound to fibril length because it is only possible to view the cellulose synthase complexes in one plane of observation.

Another characteristic of the cellulose microfibrils is their para-crystalline nature which means that they consist of both crystalline as well as non-crystalline parts. Possible organization patterns of these parts could be either (1) an amorphous shell surrounding a crystalline core, (2) crystalline and amorphous segments alternating along the length axis of the fibril, or (3) a combination of both (Salmen and Bergstrom 2009). The amorphous cellulose units are quite similar to highly ordered hemicelluloses containing either no or short side chains which probably leads to a gradual nanostructural transition between the phases.

Primary and secondary cell walls differ quite substantially in the composition of their matrices which reflects their different functions and emphasizes the crucial role of the matrix for the mechanical performance of the cell wall fibre composite. Primary walls mainly have to facilitate cell growth, which is achieved by a matrix consisting of hemicelluloses, pectin, structural proteins and aromatic substances. To make matters more complicated, the matrix assembly of primary cell walls varies substantially between grasses and other flowering plants, both by means of chemical composition and cross-linking of the macromolecules (Carpita 1996). Secondary cell walls provide mechanical stability to the plant which is supported by a more rigid matrix made up of hemicelluloses and lignin.

Hemicelluloses, of which a variety of different types can be found in all cell wall types, are heteropolymers, consisting of a large variety of neutral sugars (e.g. glucose, mannose, galactose, xylose, fucose, arabinose) and may contain uronic acids. The hemicellulose molecules typically have a backbone accompanied by side branches. Pectins are mainly found in the primary cell walls and in the middle lamella which glues the individual cells together. They are highly heterogeneous, acidic polysaccharides (Caffall and Mohnen 2009). The backbones of the side-branched molecules can be either made up of one polysaccharide unit (e.g. homogalacturonan) or of repetitive building blocks (e.g. rhamnogalacturonan) which consists of repeating disaccharides. Lignin is typically assigned to secondary cell walls, but also the primary walls become lignified after cell growth has been terminated and the cell has reached its final size and shape. Lignin polymerizes inside the cell wall from three different phenylpropane units filling voids in the basic cell wall assembly of cellulose, hemicelluloses and pectins. In secondary cell walls, lignin functions as a further reinforcing agent mainly reducing the risk of cell wall buckling under compressive loading.

In addition to the individual properties of the cell wall macromolecules, their interactions and bonding patterns are of crucial relevance for the performance of the entire cell wall. The most abundant bonding type between the macromolecules in the cell wall is hydrogen bonding. It is the exclusive bonding type found between the surface of cellulose fibrils and the surrounding matrix polymers (supposedly hemicelluloses). However, it has been reported that a fraction of xyloglucan chains can be also incorporated in cellulose fibrils (Pauly et al. 1999). A high number of hydrogen bonds is ideal to establish a tight but flexible connection between macromolecules as the bonds can be easily opened and reformed due to their low bonding strength. Hydrogen bonding is also the most abundant type between the matrix macromolecules, but also several covalent cross-links have been observed. Covalent bonds have been reported between pectin and hemicelluloses (Thompson and Fry 2000) as well as hemicelluloses and lignin (Lawoko et al. 2005; Westbye et al. 2007). Interesting features of pectins are intra-molecular ion-mediated cross-links, e.g. calcium ions in homogalacturonan and borate diesters in rhamnogalacturonan II (Carpita and Gibeaut 1993; O'Neill et al. 1996, 2004; Cosgrove 2005).

In terms of the fibre composite nature of plant cell walls, the mechanical role of cellulose is rather well understood but it is less clear for the matrix polymers. There is also a lack of knowledge on the mechanical interaction of the polymers and possible ways in which they are controlled by the plant. Despite this there are several promising approaches for a better understanding of plant cell wall mechanics which will be introduced in the following sections.

4 Elucidating the Mechanical Design of Plant Cell Walls

One possible way of determining the mechanical role of a certain mechanical component is to alter its structure or binding capacity either by genetic, enzymatic or chemical treatments and then to mechanically characterize the modified cell wall structure. Since the entire cell wall structure is retained, this method resembles a "top-down approach" which is utilized for a better understanding of both primary and secondary cell walls (Köhler and Spatz 2002; Ryden et al. 2003; Goswami et al. 2008; Abasolo et al. 2009). However, one has to be aware that all these treatments are either not highly specific or may cause compensation reactions by the plant. Therefore, further structural, (bio)chemical or genetic characterization is needed to identify clear structure–property relationships of individual cell wall components.

Another way of identifying the mechanical role of certain cell wall components is to reassemble cell wall analogues in a "bottom-up process" and then determine the mechanical performance of the entire composite structure. This approach has been mainly utilized to mimic the primary cell wall structure and study the cross-linking of the polymers (Whitney et al. 1995; Chanliaud et al. 2002). In terms of secondary cell walls, this approach is less prominent as one characteristic feature of secondary cell walls is the strictly parallel orientation of the cellulose microfibrils which can hardly be obtained in a bottom-up process based on cell wall reconstruction.

A further approach which can shed light on the load-bearing elements and the polymer interactions in plant cell walls are so-called in situ experiments which simultaneously combine external mechanical loading with nanostructural observations. For this purpose, mainly spectroscopic analysis (FT-IR, Raman) and X-ray analysis have been utilized (Eichhorn et al. 2000; Akerholm and Salmén 2001; Köhler and Spatz 2002; Keckes et al. 2003; Gierlinger et al. 2006; Martinschitz et al. 2008). For instance, applying an external strain to a plant segment can result in shifts of characteristic Raman bands which are indicative for the load-bearing function of this component. This has been impressively shown in plant tissue and fibre tests for the deformation of the cellulose chains (Eichhorn et al. 2000; Gierlinger et al. 2006; Peetla et al. 2006). In situ X-ray scattering methods can visualize the passive reorientation of cellulose fibrils toward the cell axis during tensile straining which gives insight into cellulose fibre–matrix interactions during deformation. Such a response has been shown for compression wood of conifers (Keckes et al. 2003), coir fibre bundles (Martinschitz et al. 2008) and supporting tissue of *Aristolochia* (Köhler and Spatz 2002). One limitation of both Raman microspectroscopy and X-ray diffraction is that they are highly sensitive to changes in the molecular structure of cellulose fibrils and their orientation but changes of the matrix polymers and their interplay either cannot or can only hardly be detected. Here in situ FT-IR measurements are of advantage which have a lower resolution but are more sensitive, for instance, towards hemicelluloses (Akerholm and Salmén 2001).

5 Structure and Mechanics of Primary Cell Walls

5.1 Elastic Versus Plastic Deformation

In view of the composite analogy of cell walls as discussed in the previous sections, the primary cell wall is an exceptional mechanical structure. It provides mechanical support by being stiff and rigid, and at the same time, allows for the cell to grow and elongate. In order to achieve both conflicting properties, the cellulose–matrix interaction needs to be modulated on demand. During cell rest, the connection between cellulose fibrils and matrix polymers must remain tight to provide mechanical support; however, this connection must also be able to be loosened during cell growth.

It is well accepted that cell growth starts with a softening of the cell wall. This results in a loss of turgor pressure because the inner pressure of the cell and the stress in the cell wall counterbalance each other (Cosgrove 1993). In consequence, additional water will flow into the cell leading to a volume increase and plastic (irreversible) deformation of the cell wall (Peters et al. 2000). The cell volume increase by water uptake does not necessarily result in cell growth, but can also be utilized to pre-stress the cells (Fig. 6). In this case, the cell wall undergoes elastic rather than plastic deformation. Such systems are needed for reversible movements of living tissues in plants such as the opening and closure of stomata cells (Raschke 1975; Roelfsema and Hedrich 2002), the pulvinus cells actuating leaf movements (Toriyama and Jaffe 1972; Samejima and Sibaoka 1980; Iino et al. 2001; Moran 2007) or the leaf cells making the Venus flytrap snap (Williams and Bennett 1982; Hodick and Sievers 1989).

Hence, depending on maturity and functionality, primary walls can be either plastically or elastically deformed. This indicates that the cell to some extent must be able to control the mechanical response and deformation behaviour of its cell wall. From a mechanical perspective, it seems reasonable that this can only be achieved by adjusting and modulating the bonding pattern within and between macromolecules. A modification of the cell wall assembly takes place in order to allow the cell wall to plastically deform for cell growth (Cosgrove 2005). However, different theories exist about the biochemical agents responsible for this, e.g. expansins cleaving hydrogen bonds (McQueen-Mason et al. 1992; Cosgrove 2000; Cosgrove et al. 2002), enzymes opening and re-establishing

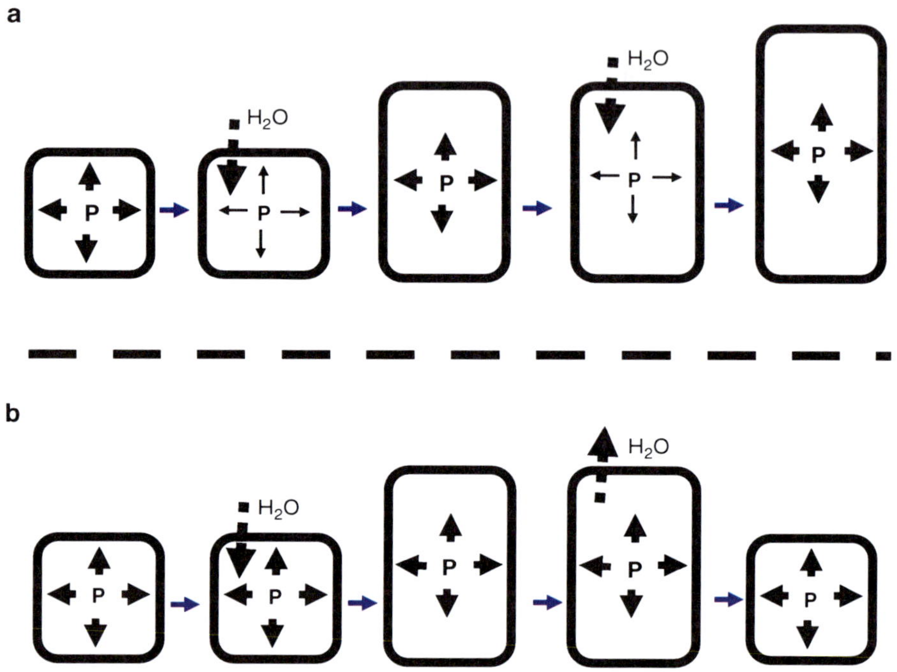

Fig. 6 (**a**) Plastic deformation of primary cell walls to facilitate growth (according to Peters et al. 2000) and (**b**) elastic deformation of primary cell walls to generate stresses. "P" in different font size illustrates the level of turgor

covalent bonds in xyloglucans (Fry et al. 1992) or hydroxyl radicals cleaving polymer linkages (Schopfer 2001; Liszkay et al. 2003). Since the mechanics of cell growth and cell wall elongation are not in the specific focus of this chapter, we refer readers to excellent review articles dealing with these questions (Cleland 1971; Taiz 1984; Cosgrove 1993, 2005; Schopfer 2006).

In terms of pure elastic deformation of cell walls for reversible organ movements, it is still unclear how relaxation and creep (see Fig. 3) of the cell wall can be avoided or limited under the pre-stressed conditions. Biological materials are normally viscoelastic, showing a time-dependent response to loading. This implies that pre-stresses are at least partly relaxed over time. This relaxation is probably due to reorientation of macromolecules and the opening and closure of labile cross-links. Therefore, in order to have a fully elastic response and retain the original self-stresses, the cell wall needs to react like a tightly bonded and fixed network.

Another important feature of either elastic or plastic deformation of cell walls is that the cell must be able to control its geometrical changes. This is achieved by controlling the cellulose fibril deposition via the microtubules which guide the movement of the cellulose synthase complexes (Paredez et al. 2006; Lindeboom et al. 2008; Wasteneys and Ambrose 2009). As a result a specific pattern of cellulose fibril orientation in the cell wall is formed which dictates the deformation behaviour of the wall (Baskin 2005). Due to the high aspect ratio of the parallel aligned stiff fibrils, the cell wall has an anisotropic response and can be more easily deformed in a direction perpendicular than parallel to the fibril orientation. This principle is utilized to allow for polarity of growth as well as pre-stressing of primary tissues. During cell growth, the cellulose fibrils are passively reoriented and shifted from a primarily perpendicular orientation to the cell axis toward a longitudinal orientation (Preston 1982; Anderson et al. 2010) (see also Fig. 5).

5.2 *Mechanical Behaviour Under Static Loading Conditions*

Research on the "static" mechanical behaviour of primary cell walls is mainly driven by understanding the mechanical role of individual cell wall components and their interaction. Since most of the recent work on primary cell wall mechanics has been performed on etiolated hypocotyls of *Arabidopsis* plants, their typical cell wall characteristics will be in the focus of this chapter.

Dark grown *Arabidopsis* hypocotyls display a slender cylindrical body with a rather constant diameter along their length making them suitable for microtensile testing. In standard microtensile tests, wild type hypocotyls show a stress–strain curve with an *s*-shape pattern which is illustrated in Fig. 7a.

In the initial phase of the test, one can observe a tensile stiffening with straining followed by a phase in which stress and strain increase linearly. Beyond this phase a yielding of the sample can be observed and the increase in strain exceeds the increase in stress. From a standard tensile test, such as that illustrated in Fig. 7a, the stiffness and the ultimate stress (~tensile strength) of a hypocotyl can be measured.

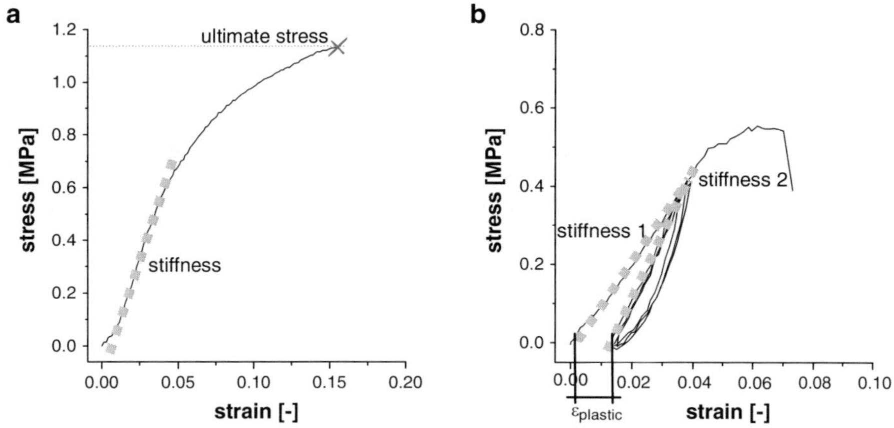

Fig. 7 Stress–strain curves of hypocotyls (**a**) of 6-day-old *Arabidopsis* wild type under static tensile loading; (**b**) of 6-day-old *Arabidopsis qua2* mutant under cyclic tensile loading

To gain further information on mechanical processes, cycling loading tests allow elastic and plastic deformations to be distinguished (Fig. 7b).

Usually the underlying objective of the hypocotyl tests is to learn about the mechanical properties of the cell walls from the mechanical performance of the hypocotyls. Here the interpretation of the initial phase of deformation in term of cell wall mechanics must be treated with care because other effects such as the hypocotyl re-alignment may play a role. But also for the other phases of the graphs one cannot conclude directly from the mechanical behaviour of the hypocotyls on cell wall mechanics. Further important parameters for the hypocotyl response which would affect the stress experienced by the cell wall are the density of the hypocotyls (e.g. how much cell wall material (mass) is agglomerated in a certain hypocotyl volume), parameters of cell geometry and the turgor pressure of the cells which stiffens the pliant primary cell wall structure. Even among *Arabidopsis* wild-type plants these parameters can vary considerably with hypocotyl age, as shown in a study on cell wall thickness (Derbyshire et al. 2007a). Therefore, structural and hydraulic parameters need to be considered in order to correlate hypocotyl properties with those of the cell wall.

5.3 Mechanical Tests of Genetically Modified Material

Various cell wall mutants have been produced to unravel crucial parameters in cell wall synthesis and for a better understanding of the mechanical role of the individual cell wall components.

Most of the mechanical tests on *Arabidopsis* hypocotyls have been focused on gaining further understanding of the interaction of the cellulose–xyloglucan network and in particular on the mechanical role of the xyloglucan side chain structure. The stiffness and strength properties of mutants with different severities of side chain modification (reduction) have been investigated. It was shown that impairing a fucosyltransferase as present in the *mur2* mutation only marginally alters the mechanical performance of the hypocotyls (Ryden et al. 2003). This observation is in good agreement with measurements by (Pena et al. 2004) and (Abasolo et al. 2009). However, a more drastic side chain modification by impairing a galactosyltransferase (*mur3*) resulted in severe and significant loss of stiffness and strength (Ryden et al. 2003). An even harsher alteration of the xyloglucan chain was achieved in the xylosyltransferase double mutant *xxt1 xxt2* (Cavalier et al. 2008). Mechanical tests revealed hypocotyl properties with reduced stiffness and strength for the *xxt1 xxt2* double mutant and even lower for the *mur3* mutant compared to the wild-type (col-0) (Fig. 8).

Another matrix component which may contribute to the primary cell wall properties is pectin. M*ur1*-mutants lack L-fucose in the shoot (Bonin et al. 1997) which leads to an alteration of xyloglucan and pectin, because the deficiency in

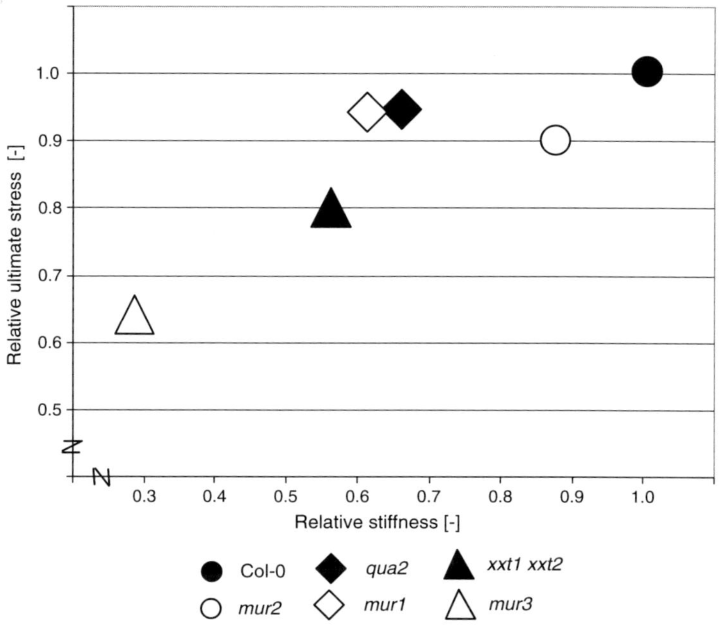

Fig. 8 Relative stiffness and ultimate stress of wild-type *Arabidopsis* hypocotyls, xyloglucan mutants and pectin mutants (4 days and 6 days old); arithmetic means are given in percentage of the wild type (col-0 = 1); detailed information on mechanical properties (arithmetic mean, SD) of col-0, *mur1*, *mur2*, *qua2* in Abasolo et al. (2009), of col-0, *mur3* in Burgert (2006) and of col-0, *xxt1 xxt2* in Cavalier et al. (2008)

fucose influences the side chain structure of the xyloglucan as well as the capability of the rhamnogalacturonan II (RGII) to form borate diesters. Interestingly, *mur1* shows different mechanical properties than the mutants which are altered exclusively on the xyloglucan side chain. The stiffness of *mur1*-hypocotyls is significantly reduced in comparison to the wild type but the hypocotyl strength is almost the same (Abasolo et al. 2009). A similar mechanical response was observed for another pectin mutant (*qua2*) which has 50% lower homogalacturonan content than the wild type (Mouille et al. 2007). Also here stiffness is reduced whereas strength is in the vicinity of the wild type (Fig. 8).

A reasonable explanation of the mechanical properties of the xyloglucan and pectin mutants may be that xyloglucan has a direct tethering function between cellulose fibrils. Pectin, however, stiffens the matrix and may prevent xyloglucan chains from unfolding in response to external tensile loading (Abasolo et al. 2009). In terms of pectins, the ion-mediated bonds may play a crucial role as they largely influence the cross-linking of pectin components and thereby the degree of matrix viscosity. This is in agreement with investigations which show that calcium ions and the degree of pectin methyl-esterification have a strong influence on pectin properties and in consequence on cell growth processes (Willats et al. 2001; Proseus and Boyer 2006, 2007, 2008; Derbyshire et al. 2007b). Also supporting this is a study by Fleischer and co-workers showing that cell wall pore size depends on the number of borate-ester cross-links in RGII (Fleischer et al. 1999).

5.4 Mechanical Tests of Cell Wall Analogues

As discussed in Sect. 4, an alternative way to identify the mechanical role of certain cell wall components is to resemble composite structures from cell wall analogues and determine their mechanical performance. Here the stiff cell wall component is mainly bacterial cellulose which can be embedded in different matrix polymers. The addition of xyloglucan to a cellulose fibril network results in reduced stiffness and strength (Whitney et al. 1999). A similar effect was observed when pectins were added to the cellulose network (Chanliaud and Gidley 1999). Although the pure cellulose network is a reasonable reference material to use in this bottom-up approach, it is not ideal since in plant cell walls a certain amount of matrix polymer is always associated with the cellulose. The same approach was used to characterize the influence of expansins on the mechanical performance of cell wall analogues. While the deformability of cellulose/xyloglucan networks increased due to the treatment with expansins, no effect was observed for pure cellulose composites or assemblies of cellulose with glucomannan or galactomannan (Whitney et al. 2000).

6 Structure and Mechanics of Secondary Cell Walls

6.1 Structure–Property Relationships

By far more research has been conducted on the mechanics of natural secondary cell walls, mainly wood cell walls, than on primary walls (Salmen and Burgert 2009). This is mainly due to the intensive use of wood for pulp and paper and as a construction material which demands detailed knowledge on the properties of the raw material.

A secondary cell wall deposition on the primary cell wall typically happens after the cell has terminated growth and a final size and shape is reached. Several cell types die at this stage which means that the turgor pressure which formerly stabilized the system is no longer retained. Therefore, secondary cell walls have to be much thicker than primary cell walls in order to provide mechanical stability to the dead cells (see also Fig. 5). However, there are also mechanical supporting cells which stay alive with secondary cell walls, such as some hardwood fibres in tropical species or the sclerenchyma fibres in monocots. An interesting aspect of the latter fibre type is that new secondary cell wall layers can be continuously added and therefore secondary cell wall thickness increases with ageing (Tomlinson 2006).

In the secondary xylem of dicotyledonous trees and gymnosperms, the secondary cell wall typically consists of three layers (S1–S3), with a dominance of the middle S2 layer, making up to ~80% of this sandwich structure (Fengel and Wegener 1984) (see Fig. 5b). The individual secondary cell wall layers can be distinguished because of different orientations of cellulose microfibrils. Typically, secondary cell wall layers show a strong parallel alignment of cellulose. The orientation of the fibrils toward the cell axis is called the cellulose microfibril angle. In wood, the surrounding cell wall layers (S1 and S3) show a large and rather steady angle between 50 and 80°, whereas the central S2 layer is varied between 0° and 50° depending on the mechanical function of the cell type. The ongoing consecutive deposition of secondary cell wall layers in monocots results in a multilamellar structure. Very characteristic is the alternation of thick and thin cell wall layers in bamboo, where thick cell wall layers possess a small cellulose fibril (<15°) and thin layers show a large microfibril angle (Parameswaran and Liese 1976; Liese 1987).

The cellulose microfibril angle has a strong influence on the mechanics of the cell wall; its adjustment enables the plant to control the mechanical performance of tissue according to need. The crucial role of the cellulose microfibril orientation in secondary cell walls becomes obvious when plotting the tensile stiffness of various plant tissues and fibres against the microfibril angle (Fig. 9).

For cell walls with cellulose microfibril orientation almost parallel to the cell axis in the S2 wall, high Young's moduli are measured, whereas tensile stiffness decreases with an increase of microfibril angle. This correlation also enables plants to adjust the mechanical properties of their tissues in the course of cellulose fibril deposition. This can be seen, for instance, for variations in microfibril angle between wood formed by young and mature trees (Lichtenegger et al. 1999). The juvenile wood has a high microfibril angle which makes the wood and the stem more flexible, whereas the mature wood has a low microfibril angle which results in high stiffness and a lower buckling risk.

The role of the matrix properties for the mechanical performance becomes more important with increasing microfibril angle. For instance, it has been shown for compression wood of gymnosperms with high microfibril angles (~45°) that the stiffness in the initial phase largely depends on the matrix stiffness. When the shear strength of the matrix is exceeded, then the stress–strain curve of the tissue shows a characteristic yield point and a long phase of plastic deformation typified in the typical biphasic stress–strain curves (Köhler and Spatz 2002; Keckes et al. 2003; Altaner and Jarvis 2008). A longitudinal stiffening effect by the matrix in case of high cellulose microfibril angles is also utilized in the dominating vascular bundle type of the palm tree, *Washingtonia robusta*. While the cellulose microfibril angle remains rather constant, the matrix properties are varied across the fibre cap due to changes in lignin content and lignin composition. Thereby, the plant creates a smooth transition in stiffness from the stiff inner fibre cap to its periphery, which is beneficial for the connection with the surrounding pliant parenchymatous tissue (Ruggeberg et al. 2008).

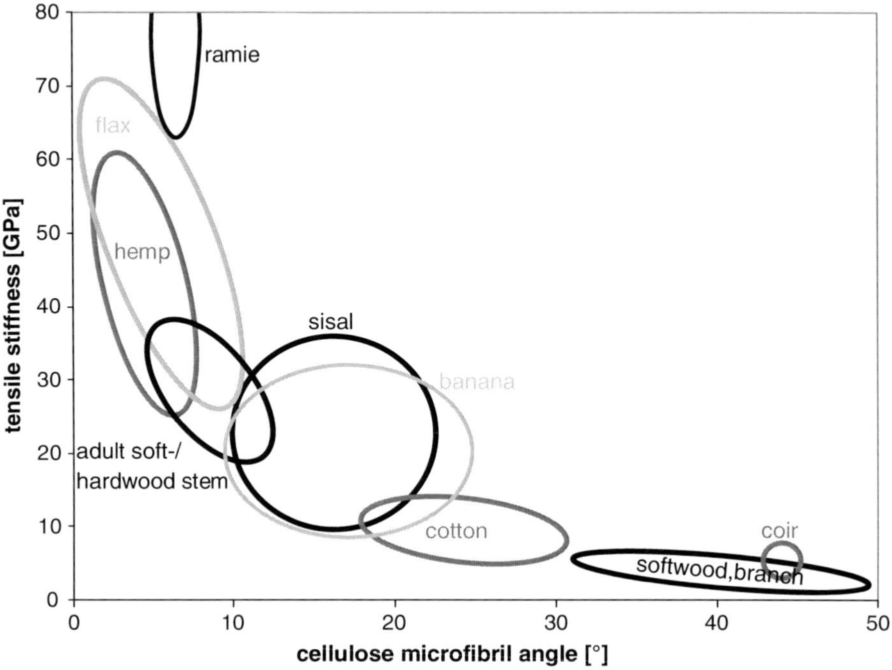

Fig. 9 Elastic modulus as a function of cellulose microfibril angle in the secondary cell wall for various species (for specific values and references see a table in Eder and Burgert (2010) giving a survey on plant fibre structure and properties)

6.2 Mechanical Tests of Genetically Modified Materials

Research on the mechanical properties of mutants with secondary cell wall alterations is still in its infancy. Most of the work has been done on tree species, mainly *Populus*. This is primarily motivated along the lines of wood modification to optimize cell wall composition for pulp and paper production. Trees with changes in lignin content and lignin composition can facilitate the chemical pulping processes (Boudet 2000; Pilate et al. 2002; Baucher et al. 2003; Boerjan et al. 2003). For instance, a downregulation of cinnamyl alcohol dehydrogenase (CAD) increased the lignin extractability (Baucher et al. 1996). In a study by Kasal and co-workers, young aspen trees with reduced lignin content and changes in lignin composition showed a lower compressive strength but no change in stiffness compared to the wild type. In contrast under tensile load, stiffness was reduced but strength was not. The bending properties were neither affected in stiffness or strength (Kasal et al. 2007). Macromechanical tests in tension, compression, bending and shear on wood from partially CAD-deficient pine trees did not reveal any significant changes compared to the wild type (Saralde et al. 2008).

7 Summary and Conclusions

The underlying structure–function relationships which result in the unique mechanical performance of plant cell walls have only been partly elucidated up to now. In this chapter, we showed current drawbacks and opportunities in terms of mechanical characterization. General mechanical terms and definitions as well composite theory have been briefly introduced to establish a solid basis for the mechanical characterization of biological materials, despite their inhomogeneous nature and complex mechanical response. For a better understanding of the mechanical role of individual components and possible adjustments of cell wall properties, the mechanical characterization of genetically modified plants appears to be a powerful approach. However, one has to bear in mind the complexity of cell walls due to the spatial organization of the polymers and the variety of weak and strong cross-links between them. Another problem is possible compensation strategies by the plant which may lead to alternative synthesis pathways as well as functional substitution of altered or diminished cell wall components.

In terms of application, the mechanical characterization of genetically modified plants will become an important step in the process of extracting optimized raw material properties for pulping as well as fuel production. A further application potential of cell walls can be foreseen in the field of biomimetics and biotemplating. The native fibre composite structure of the cell walls can inspire the improvement of technical composites by transferring design principles which are the basis for the exceptional mechanical properties (Fratzl 2006; Milwich et al. 2006). Biotemplating of plant cell walls has been utilized to create anisotropic ceramic structures (Greil 2001; Deshpande et al. 2006). Another interesting approach could be to retain the cellulose orientation but replace matrix components by nanoparticles or responsive polymers in order to build novel composites and hybrid materials as well as smart materials which have the sensor and the actuator on board.

Acknowledgements We would like to thank all those colleagues who have contributed with their research work to the results presented, in particular Michaela Eder and Peter Fratzl.

References

Abasolo W et al (2009) Pectin may hinder the unfolding of xyloglucan chains during cell deformation: implications of the mechanical performance of Arabidopsis hypocotyls with pectin alterations. Mol Plant 2:990–999

Akerholm M, Salmén L (2001) Interactions between wood polymers studied by dynamic FT-IR spectroscopy. Polymer 42:963–969

Altaner CM, Jarvis MC (2008) Modelling polymer interactions of the 'molecular Velcro' type in wood under mechanical stress. J Theor Biol 253:434–445

Anderson CT, Carroll A, Akhmetova L, Somerville C (2010) Real time imaging of cellulose reorientation during cell wall expansion in *Arabidopsis* roots. Plant Physiol 152:787–796

Ashby MF, Brechet YJM (2003) Designing hybrid materials. Acta Mater 51:5801–5821

Ashby MF, Jones D (2005) Engineering materials (vols 1 and 2), 3rd edn. Elsevier, Oxford

Baskin TI (2005) Anisotropic expansion of the plant cell wall. Annu Rev Cell Dev Biol 21: 203–222
Baucher M et al (1996) Red xylem and higher lignin extractability by down-regulating a cinnamyl alcohol dehydrogenase in poplar. Plant Physiol 112:1479–1490
Baucher M, Halpin C, Petit-Conil M, Boerjan W (2003) Lignin: genetic engineering and impact on pulping. Crit Rev Biochem Mol Biol 38:305–350
Boerjan W, Ralph J, Baucher M (2003) Lignin biosynthesis. Annu Rev Plant Biol 54:519–546
Bonin CP, Potter I, Vanzin GF, Reiter WD (1997) The MUR1 gene of *Arabidopsis thaliana* encodes an isoform of GDP-D-mannose-4,6-dehydratase, catalyzing the first step in the de novo synthesis of GDP-L-fucose. Proc Natl Acad Sci USA 94:2085–2090
Boudet AM (2000) Lignins and lignification: selected issues. Plant Physiol Biochem 38:81–96
Burgert I (2006) Exploring the micromechanical design of plant cell walls. Am J Bot 93: 1391–1401
Burgert I, Keckes J, Fruhmann K, Fratzl P, Tschegg SE (2002) A comparison of two techniques for wood fibre isolation evaluation by tensile tests on single fibres with different microfibril angle. Plant Biol 4:9–12
Burgert I, Fruhmann K, Keckes J, Fratzl P, Stanzl-Tschegg SE (2003) Microtensile testing of wood fibers combined with video extensometry for efficient strain detection. Holzforschung 57: 661–664
Caffall KH, Mohnen D (2009) The structure, function, and biosynthesis of plant cell wall pectic polysaccharides. Carbohydr Res 344:1879–1900
Callister WD (2003) Materials science and engineering, an introduction, 6th edn. Wiley, New York
Carpita NC (1996) Structure and biogenesis of the cell walls of grasses. Annu Rev Plant Physiol Plant Mol Biol 47:445–476
Carpita NC, Gibeaut DM (1993) Structural models of primary cell walls in flowering plants: consistency of molecular structure with the physical properties of the walls during growth. Plant J 3:1–30
Cavalier DM et al (2008) Disrupting two *Arabidopsis thaliana* xylosyltransferase genes results in plants deficient in xyloglucan, a major primary cell wall component. Plant Cell 20: 1519–1537
Chanliaud E, Gidley MJ (1999) In vitro synthesis and properties of pectin/*Acetobacter xylinus* cellulose composites. Plant J 20:25–35
Chanliaud E, Burrows KM, Jeronimidis G, Gidley MJ (2002) Mechanical properties of primary plant cell wall analogues. Planta 215:989–996
Chou T-W (1992) Microstructural design of fiber composites. Cambridge University Press, Cambridge
Clair B, Despaux G, Chanson B, Thibaut B (2000) Possible use of scanning acoustic microscopy to study local wood properties: preliminary study of experimental conditions. Ann For Sci 57: 335–343
Cleland R (1971) Cell wall extension. Annu Rev Plant Physiol 22:179–222
Cosgrove DJ (1993) Wall extensibility – its nature, measurement and relationship to plant cell growth. New Phytol 124:1–23
Cosgrove DJ (2000) Loosening of plant cell walls by expansins. Nature 407:321–326
Cosgrove DJ (2005) Growth of the plant cell wall. Nat Rev Mol Cell Biol 6:850–861
Cosgrove DJ, Chao Li L, Hyung-Taeg C, Hoffmann-Benning S, Moore RC, Blecker D (2002) The growing world of expansins. Plant Cell Physiol 43:1436–1444
Derbyshire P, Findlay K, McCann MC, Roberts K (2007a) Cell elongation in *Arabidopsis* hypocotyls involves dynamic changes in cell wall thickness. J Exp Bot 58:2079–2089
Derbyshire P, McCann MC, Roberts K (2007b) Restricted cell elongation in Arabidopsis hypocotyls is associated with a reduced average pectin esterification level. BMC Plant Biol 7:12
Deshpande A, Burgert I, Paris O (2006) Hierarchically structured ceramics by high precision nanoparticle casting of wood. Small 2:994–998

Dunlop JWC, Fratzl P (2010) Biological composites. Annu Rev Mater Res 40:1–24
Ebenstein DM, Pruitt LA (2006) Nanoindentation of biological materials. Nano Today 1:26–33
Eder M, Burgert I (2010) Natural fibres – function in Nature. In: Müssig J (ed) Industrial applications of natural fibres: structure, properties and technical applications. Wiley-VCH, New York
Eichhorn SJ, Hughes M, Snell R, Mott L (2000) Strain induced shifts in the Raman spectra of natural cellulose fibers. J Mater Sci Lett 19:721–723
Elbaum R, Zaltzman L, Burgert I, Fratzl P (2007) The role of wheat awns in the seed dispersal unit. Science 316:884–886
Fahlen J, Salmen L (2005) Pore and matrix distribution in the fiber wall revealed by atomic force microscopy and image analysis. Biomacromolecules 6:433–438
Fengel D, Wegener G (1984) Wood: chemistry, ultrastructure, reactions. de Gruyter, Berlin
Fleischer A, O'Neill MA, Ehwald R (1999) The pore size of non-graminaceous plant cell walls is rapidly decreased by borate ester cross-linking of the pectic polysaccharide rhamnogalacturonan II. Plant Physiol 121:829–838
Fratzl P (2006) Biomimetic materials research: what can we really learn from nature's structural materials? J R Soc Interf 4:637–642
Fratzl P, Weinkamer R (2007) Nature's hierarchical materials. Prog Mater Sci 52:1263–1334
Fratzl P, Burgert I, Gupta HS (2004) On the role of interface polymers for the mechanics of natural polymeric composites. Phys Chem Chem Phys 6:5575–5579
Fry SC, Smith RC, Renwick KF, Martin DJ, Hodge SK, Matthews KJ (1992) Xyloglucan endotransglycosylase, a new wall-loosening enzyme activity from plants. Biochem J 282:821–828
Geitmann A, McConnaughey W, Lang-Pauluzzi I, Franklin-Tong VE, Emons AMC (2004) Cytomechanical properties of *Papaver* pollen tubes are altered after self-incompatibility challenge. Biophys J 86:3314–3323
Gere JM (2002) Mechanics of materials, 5th edn. Nelson Thornes, Cheltenham
Gierlinger N, Schwanninger M, Reinecke A, Burgert I (2006) Molecular changes during tensile deformation of single wood fibers followed by Raman microscopy. Biomacromolecules 7: 2077–2081
Gindl W, Schoberl T (2004) The significance of the elastic modulus of wood cell walls obtained from nanoindentation measurements. Compos A Appl Sci Manuf 35:1345–1349
Gindl W, Gupta HS, Schoberl T, Lichtenegger HC, Fratzl P (2004) Mechanical properties of spruce wood cell walls by nanoindentation. Appl Phys A Mater Sci Process 79: 2069–2073
Goswami L et al (2008) Stress generation in the tension wood of poplar is based on the lateral swelling power of the G-layer. Plant J 56:531–538
Greil P (2001) Biomorphous ceramics from lignocellulosics. J Eur Ceram Soc 21:105–118
Groom L, Mott L, Shaler S (2002) Mechanical properties of individual southern pine fibers. Part I. Determination and variability of stress-strain curves with respect to tree height and juvenility. Wood Fiber Sci 34:14–27
Hashin Z, Shtrikman S (1963) A variational approach to the theory of the elastic behaviour of multiphase materials. J Mech Phys Solids 11:127–140
Hiller S, Bruce DM, Jeronimidis G (1996) Micro-penetration technique for mechanical testing of plant cell walls. J Texture Stud 27:559–587
Hodick D, Sievers A (1989) On the mechanism of trap closure of Venus flytrap (*Dionaea muscipula* Ellis). Planta 179:32–42
Hull D, Clyne T (1996) An introduction to composite materials, 2nd edn. Cambridge University Press, Cambridge
Iino M, Long C, Wang XJ (2001) Auxin- and abscisic acid-dependent osmoregulation in protoplasts of *Phaseolus vulgaris* pulvini. Plant Cell Physiol 42:1219–1227
Jakob HF, Fengel D, Tschegg SE, Fratzl P (1995) The elementary cellulose fibril in *Picea abies*: comparison of transmission electron microscopy, small-angle X-ray scattering, and wide-angle X-ray scattering results. Macromolecules 28:8782–8787

Kasal B, Peszlen I, Peralta P, Li L (2007) Preliminary tests to evaluate the mechanical properties of young trees with small diameter. Holzforschung 61:390–393
Keckes J et al (2003) Cell-wall recovery after irreversible deformation of wood. Nat Mater 2: 810–814
Kerstens S, Decraemer WF, Verbelen JP (2001) Cell walls at the plant surface behave mechanically like fiber-reinforced composite materials. Plant Physiol 127:381–385
Köhler L, Spatz HC (2002) Micromechanics of plant tissues beyond the linear-elastic range. Planta 215:33–40
Lawoko M, Henriksson G, Gellerstedt G (2005) Structural differences between the lignin-carbohydrate complexes present in wood and in chemical pulps. Biomacromolecules 6: 3467–3473
Lichtenegger H, Reiterer A, Stanzl-Tschegg SE, Fratzl P (1999) Variation of cellulose microfibril angles in softwoods and hardwoods: a possible strategy of mechanical optimization. J Struct Biol 128:257–269
Liese W (1987) Research on bamboo. Wood Sci Technol 21:189–209
Lindeboom J, Mulder BM, Vos JW, Ketelaar T, Emons AMC (2008) Cellulose microfibril deposition: coordinated activity at the plant plasma membrane. J Microsc 231:192–200
Liszkay A, Kenk B, Schopfer P (2003) Evidence for the involvement of cell wall peroxidase in the generation of hydroxyl radicals mediating extension growth. Planta 217:658–667
Martinschitz KJ, Boesecke P, Garvey CJ, Gindl W, Keckes J (2008) Changes in microfibril angle in cyclically deformed dry coir fibers studied by in-situ synchrotron X-ray diffraction. J Mater Sci 43:350–356
McQueen-Mason S, Durachko DM, Cosgrove DJ (1992) Two endogenous proteins that induce cell wall extension in plants. Plant Cell 4:1425–1433
Milwich M, Speck T, Speck O, Stegmaier T, Planck H (2006) Biomimetics and technical textiles: solving engineering problems with the help of nature's wisdom. Am J Bot 93:1455–1465
Moran N (2007) Osmoregulation of leaf motor cells. FEBS Lett 581:2337–2347
Mouille G et al (2007) Homogalacturonan synthesis in *Arabidopsis thaliana* requires a Golgi-localized protein with a putative methyltransferase domain. Plant J 50:605–614
Niklas K (1992) Plant biomechanics, an engineering approach to plant form and function. University of Chicago Press, London
Oliver WC, Pharr GM (1992) An improved technique for determining hardness and elastic-modulus using load and displacement sensing indentation experiments. J Mater Res 7: 1564–1583
O'Neill MA et al (1996) Rhamnogalacturonan-II, a pectic polysaccharide in the walls of growing plant cell, forms a dimer that is covalently cross-linked by a borate ester. J Biol Chem 271: 22923–22930
O'Neill MA, Ishii T, Albersheim P, Darvill AG (2004) Rhamnogalacturonan II: structure and function of a borate cross-linked cell wall pectic polysaccharide. Annu Rev Plant Biol 55: 109–139
Page DH, El-Hosseiny F (1983) The mechanical properties of single wood pulp fibres, Part IV. Fibril angle and the shape of the stress-strain curve. J Pulp Pap Sci 9:99–100
Parameswaran N, Liese W (1976) Fine-structure of bamboo fibers. Wood Sci Technol 10:231–246
Paredez AR, Somerville CR, Ehrhardt DW (2006) Visualization of cellulose synthase demonstrates functional association with microtubules. Science 312:1491–1495
Parre E, Geitmann A (2005a) More than a leak sealant. The mechanical properties of callose in pollen tubes. Plant Physiol 137:274–286
Parre E, Geitmann A (2005b) Pectin and the role of the physical properties of the cell wall in pollen tube growth of *Solanum chacoense*. Planta 220:582–592
Pauly M, Albersheim P, Darvill AG, York WS (1999) Molecular domains of the cellulose/xyloglucan network in the cell walls of higher plants. Plant J 20:629–639

Peetla P, Schenzel KC, Diepenbrock W (2006) Determination of mechanical strength properties of hemp fibers using near-infrared fourier transform Raman microspectroscopy. Appl Spectrosc 60:682–691

Pena MJ, Ryden P, Madson M, Smith AC, Carpita NC (2004) The galactose residues of xyloglucan are essential to maintain mechanical strength of the primary cell walls in *Arabidopsis* during growth. Plant Physiol 134:443–451

Peters WS, Hagemann W, Tomos D (2000) What makes plants different? Principles of extracellular matrix function in "soft" plant tissues. Comp Biochem Physiol A 125:151–167

Pilate G et al (2002) Field and pulping performances of transgenic trees with altered lignification. Nat Biotechnol 20:607–612

Preston RD (1982) The case for multinet growth in growing walls of plant cells. Planta 155: 356–363

Proseus TE, Boyer JS (2006) Calcium pectate chemistry controls growth rate of *Chara corallina*. J Exp Bot 57:3989–4002

Proseus TE, Boyer JS (2007) Tension required for pectate chemistry to control growth in *Chara corallina*. J Exp Bot 58:4283–4292

Proseus TE, Boyer JS (2008) Calcium pectate chemistry causes growth to be stored in *Chara corallina*: a test of the pectate cycle. Plant Cell Environ 31:1147–1155

Raschke K (1975) Stomatal action. Annu Rev Plant Physiol Plant Mol Biol 26:309–340

Raum K, Cleveland OR, Peyrin F, Laugier P (2006) Derivation of elastic stiffness from site-matched mineral density and acoustic impedance maps. Phys Med Biol 51:747–758

Roelfsema MRG, Hedrich R (2002) Studying guard cells in the intact plant: modulation of stomatal movement by apoplastic factors. New Phytol 153:425–431

Ruggeberg M et al (2008) Stiffness gradients in vascular bundles of the palm *Washingtonia robusta*. Proc R Soc B Biol Sci 275:2221–2229

Ryden P, Sugimoto-Shirasu K, Smith AC, Findlay K, Reiter W-D, McCann MC (2003) Tensile properties of *Arabidopsis* cell walls depend on both a xyloglucan cross-linked microfibrillar network and rhamnogalacturonan II-borate complexes. Plant Physiol 123:1033–1040

Salmén L (2001) Micromechanics of the wood cell wall: a tool for a better understanding of its structure. In: Navi P (ed) 1st International conference of the European Society for wood mechanics. EPFL, Lausanne, pp 385–398

Salmen L, Bergstrom E (2009) Cellulose structural arrangement in relation to spectral changes in tensile loading FTIR. Cellulose 16:975–982

Salmen L, Burgert I (2009) Cell wall features with regard to mechanical performance. A review COST Action E35 2004–2008: Wood machining – micromechanics and fracture. Holzforschung 63:121–129

Samejima M, Sibaoka T (1980) Changes in the extracellular ion concentration in the main pulvinus of *Mimosa pudica* during rapid movement and recovery. Plant Cell Physiol 21: 467–479

Saralde TCJ, Peralta PN, Peszlen I, Kasal B (2008) Mechanical properties of lumber from partially CAD-deficient loblolly pine (*Pinus taeda*). Wood Fiber Sci 40:657–662

Schopfer P (2001) Hydroxyl radical-induced cell wall loosening in vitro and in vivo: implications for the control of elongation growth. Plant J 28:679–688

Schopfer P (2006) Biomechanics of plant growth. Am J Bot 93:1415–1425

Somerville C et al (2004) Toward a systems approach to understanding plant cell walls. Science 306:2206–2211

Taiz L (1984) Plant cell expansion: regulation of cell wall mechanical properties. Annu Rev Plant Physiol 35:585–657

Thompson JE, Fry SC (2000) Evidence for covalent linkage between xyloglucan and acidic pectins in suspension-cultured rose cells. Planta 211:275–286

Tomlinson PB (2006) The uniqueness of palms. Bot J Linn Soc 151:5–14

Toriyama H, Jaffe MJ (1972) Migration of calcium and its role in regulation of seismonasty in motor cell of *Mimosa pudica* L. Plant Physiol 49:72

Vincent J (1990) Structural biomaterials. Princeton University Press, Princeton
Wainwright S, Biggs W, Currey J, Gosline J (1982) Mechanical design in organisms. Princeton University Press, Princeton
Wasteneys GO, Ambrose JC (2009) Spatial organization of plant cortical microtubules: close encounters of the 2D kind. Trends Cell Biol 19:62–71
Westbye P, Kohnke T, Glasser W, Gatenholm P (2007) The influence of lignin on the self-assembly behaviour of xylan rich fractions from birch (*Betula pendula*). Cellulose 14: 603–613
Whitney SEC, Brigham JE, Darke AH, Reid JSG, Gidley MJ (1995) In vitro assembly of cellulose/ xyloglucan networks: ultrastructural and molecular aspects. Plant J 8:491–504
Whitney SEC, Gothard MGE, Mitchell JT, Gidley MJ (1999) Roles of cellulose and xyloglucan in determining the mechanical properties of primary plant cell walls. Plant Physiol 121: 657–663
Whitney SEC, Gidley MJ, McQueen-Mason SJ (2000) Probing expansion action using cellulose/ hemicellulose composites. Plant J 22:327–334
Willats WGT et al (2001) Modulation of the degree and pattern of methyl-esterification of pectic homogalacturonan in plant cell walls – implications for pectin methyl esterase action, matrix properties, and cell adhesion. J Biol Chem 276:19404–19413
Williams SE, Bennett AB (1982) Leaf closure in the Venus flytrap – an acid growth-response. Science 218:1120–1122
Wimmer R, Lucas BN, Tsui TY, Oliver WC (1997) Longitudinal hardness and Young's modulus of spruce tracheid secondary walls using nanoindentation. Wood Sci Technol 31:131–141

Mechanics of the Cytoskeleton

Peter Nick

Abstract This chapter summarizes evidence for a cytoskeletal function in tensegral integration on both the organismal and the cellular levels. The plant cytoskeleton consists of two major elements, microtubules and actin filaments. The spatial organization of these elements is highly dynamic and changes fundamentally during the cell cycle, with conspicuous effects on the predicted stress–strain patterns. In interphase cells, microtubule bundles are thought to control the direction of cellulose deposition and thus to reinforce the axiality of cell growth. By microtubule–actin linkers such as the novel class of plant-specific kinesins with a calponin-homology domain, the rigid microtubules and the flexible actin bundles can be integrated into a system endowed with mechanical tensegrity. Because the plant cytoskeleton is relieved of the load-bearing task it fulfils in the non-walled animal cells, it has adopted sensory functions. Stretch-induced changes of protein conformation and stretch-activated ion channels seem to act in concert with the cytoskeleton, which acts either as a stress-focussing susceptor of mechanical force upon mechanosensitive ion channels or as a primary sensor that transduces mechanical force into differential growth of microtubule plus ends. This cytoskeletal tensegrity sensor is used both to integrate the growth of individual cells with mechanical load of tissues and organs and as an intracellular sensor used to control holistic properties of a cell such as organelle positioning. The distinct nonlinearity of microtubules in particular renders them an ideal tool for self-organization in response to mechanical input from the exterior.

P. Nick
Botanical Institute and Center of Functional Nanostructures, Karlsruhe Institute of Technology, Kaiserstr. 2, 76128 Karlsruhe, Germany
e-mail: peter.nick@kit.edu

P. Wojtaszek (ed.), *Mechanical Integration of Plant Cells and Plants*,
Signaling and Communication in Plants 9, DOI 10.1007/978-3-642-19091-9_3,

© Springer-Verlag Berlin Heidelberg 2011

1 Prologue: The Sensitive Cytoskeleton or the Hidden Face of Plant Tensegrity

Even before Ledbetter and Porter (1963) described "microtubules" in plant cells, which they had observed by transmission electron microscopy, the plant cytoskeleton was predicted to exist on the basis of biomechanical considerations. It was Paul Green who concluded from their geometry that growing plant cells repartition growth by an unknown "reinforcement mechanism" from the spontaneously preferred lateral expansion in favour of elongation, and he predicted that the cell can establish and maintain the mechanical anisotropy of cell walls through an as-yet-unknown lattice of tubular elements that are oriented in an ordered fashion (Green 1962). Thus, ever since its discovery, the plant cytoskeleton has been intimately linked with mechanical aspects of morphogenesis, and this chapter will therefore not follow the usual approach to describe the plant cytoskeleton from its molecular basis and then derive structures and functions. It will rather assume a view of the cytoskeleton that derives from its function in mechanical integration of cell and organ morphogenesis.

As a consequence of their photosynthetic lifestyle, plants increase their surface by folding outwards, producing a considerable degree of mechanical load. As long as they remained aquatic, this load was partially relieved by buoyancy, allowing considerable body sizes even for fairly simple architectures. However, when plants began to move into terrestrial habitats, they had to develop flexible yet robust mechanical supports. The invention of vasculature-based modules, the so-called *telomes* (Zimmermann 1965), was the decisive factor for the evolutionary success of the cormophytic land plants.

Mechanical load shaped plant architecture down to the cellular level. Plant cells are endowed with a rigid cell wall, and this affects cell division and cell expansion both specifically and fundamentally (see also chapter "Micromechanics of Cell Walls"). The deposition of a new cross wall will define the patterns of mechanical strain that, during subsequent cell expansion, will guide the complex interplay between the expanding protoplast and the yielding cell wall. It is even possible to describe the shape of individual cells in a plant tissue as a manifestation of minimal mechanical tension (Thompson 1959), emphasizing the strong influence of mechanical load on plant development.

When plants are challenged by mechanical stimuli, they respond by changing the architecture, which will allocate load-bearing elements (vessels and fibres on the organ level, cellulose microfibrils and lignin incrustations on the cellular level) guided by the imposed field of forces. An impressive example is the formation of tension and compression wood (for a recent review, see Funada 2008). This architectural response ensures that mechanical strains are balanced in an optimal fashion for minimal investment of energy and biomatter. Moreover, this mechanical balance is continuously adjusted to the current environment – a must in a system endowed with open morphogenesis, where the *Bauplan* is continually extended by addition of newly formed organs (see also chapter "Mechanical Force Responses of Plant Cells and Plants").

It was only in the 1920s when such self-supporting, flexible structures were deliberately constructed by mankind (Robby 1996). Starting from the Equilibrist Studies (consisting of three solid sticks interconnected by ropes) of the Soviet artist Karl Joganson, it was mainly the American architect and engineer Richard Buckminster Fuller (1895–1983) who systematically adopted biological principles and developed self-supportive structures that all consisted of a continuous network of tensile elements (which can transmit forces by pulling) linked to a discontinuous system of stiff elements (which can transmit forces by pushing). It was also Buckminster Fuller who later coined the term "tensegrity" as a combination of "tension" and "integrity".

A few years later it became clear that animal cells are shaped by tensegrity (for reviews see Ingber 2003a, b; chapter "Introduction: Tensegral World of Plants"). The part of the tensile elements is played by actin microfilaments, which are not only contractile, but are also mechanically comparable to silk fibres (Gittes et al. 1993). The part of the stiff elements is played by the microtubules, which are not only hollow cylinders, but are also mechanically much more rigid than actin filaments and can be approximated as very delicate glass fibres (Gittes et al. 1993). The actin filaments are connected through the membrane with a supporting scaffold – the extracellular matrix.

What is the secret that renders biological tensegrity so successful? Biological tensegrity is not constructed a priori, but emerges a posteriori from reorganization in response to ever-changing stress–strain patterns. To use a term of Jacob (1977), biological shape is not produced by intelligent design but rather from *bricolage*, and therefore it oscillates around the state minimal energy without every reaching it.

Plant cells are endowed with a cell wall that is built as a composite structure with elongate load-absorbing elements (cellulose microfibers) that are embedded in an amorphous matrix (hemicelluloses, pectins, proteins). It has been shown for technical applications that such composite materials optimally combine bending flexibility with mechanical stability (Niklas 1992). This means that the tensegrity function fulfilled by the interphase cytoskeleton in animal cells is replaced by tensegrity of the cell wall. The plant cytoskeleton is therefore not directly required to support cellular architecture and is therefore free to adopt other functions (Fig. 1).

Cellular architecture uses the tensegral principle to achieve maximal mechanical stability and, simultaneously, flexibility on the basis of parsimonious use of resources and load-bearing elements. In addition, it adapts continuously to the ever-changing conditions of growing and developing cells. This requires efficient sensing of forces and strains followed by appropriate reorganization of the tensegral elements. Thus, the tensegral cytoskeleton is not only a device to provide mechanical stability, it must also participate in the sensing of stress and strain patterns. This mechanical stimulation feeds back to the organization of the cytoskeleton in such a way that a stable minimum of mechanical energy is reached and continuously adjusted. It is this hidden face of tensegrity that becomes especially important in the walled plant cells that are under continuous turgor pressure and use this pressure for regulated expansion. The evolution of the interphasic plant cytoskeleton was therefore shaped by selective pressures towards optimized sensing and integration

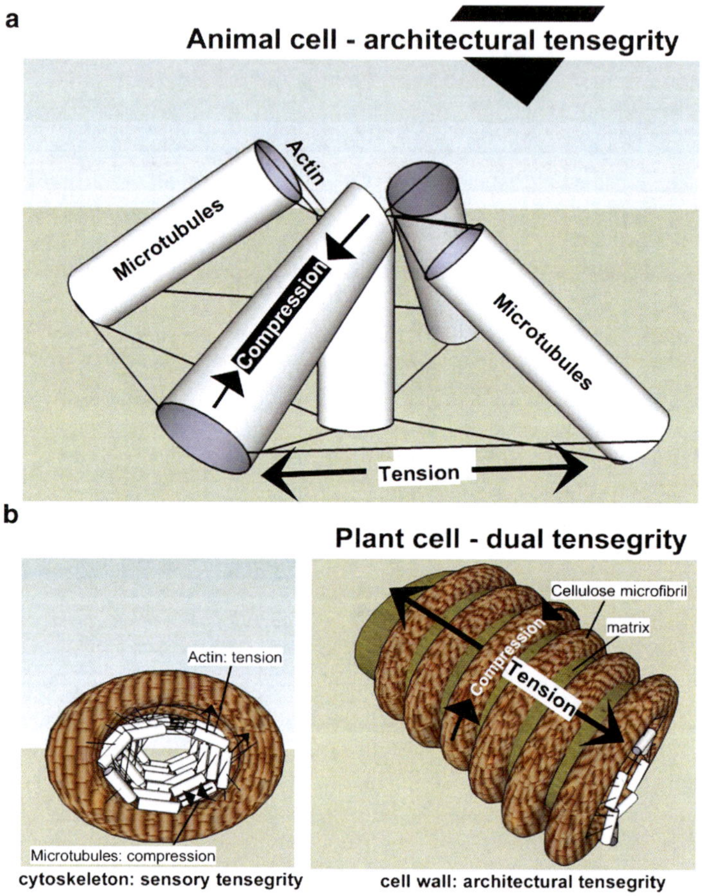

Fig. 1 Functional shift of the cytoskeleton from architectural tensegrity in animal cells towards sensory tensegrity in plant cells. In animal cells (**a**), architecture results from the interplay between stiff microtubules absorbing compressive forces and flexible actin filaments elastically tethering the cells through focal adhesions to the substrate. This set-up is used both to maintain cell shape and to sense and to adjust to mechanical force. In plant cells (**b**), cellulosic microfibers in combination with a flexible cell wall matrix maintain cell shape, such that the tensegral cytoskeleton is released from its architectural function and can be optimized for sensory integration of mechanical stimuli

of mechanical stimuli. The aim of this review is to give a survey of the role of the cytoskeleton in mechanical integration of plant cells, tissues and organs.

2 Cellular Players: The Microtubule–Actin Tensegrity System

The plant cytoskeleton consists of two major elements, microtubules and actin filaments – intermediate filaments have remained elusive in plants. The spatial organization of these elements is highly dynamic and changes fundamentally during the cell cycle, with conspicuous effects on the predicted stress–strain patterns. Since microtubules and actin filaments are functionally, but probably also structurally, interconnected (Wasteneys and Galway 2003; Collings 2008), it seems more appropriate to describe their dynamics as a functional entity. The molecular players have been extensively reviewed elsewhere (for recent reviews see Hamada 2007; Sedbrook and Kaloriti 2008), and thus the focus of this chapter is on cellular function, rather than on molecular composition.

2.1 Cell Expansion

2.1.1 Microtubules

In interphase cells, microtubules are organized in arrays of parallel bundles perpendicular to the axis of preferential cell expansion (Fig. 2a). These bundles are thought to control the direction of cellulose deposition and thus to reinforce the axiality of cell growth. Cortical microtubules can change their orientation in response to various external and internal stimuli, and this reorientation will shift the preferential direction of cellulose deposition and thus the mechanical anisotropy of the yielding cell wall such that the proportionality of cell expansion can be altered in response to the stimulus (for a recent review see Nick 2008a). The cell wall in cells that are not endowed with tip growth is formed by apposition of cellulose on the inner surface of the cell wall. Cellulose is synthesized by specialized enzyme complexes that, in freeze-fracture preparations, appear as rosettes of six subunits of about 25–30-nm diameter around a central pore (Kimura et al. 1999) and are therefore designated as terminal rosettes. The terminal rosettes are integrated into the membrane by fusion of exocytotic vesicles. UDP-glucose is transported towards the central pore and polymerized in a β-1,4 configuration. Each subunit of the cellulose synthase will produce six cellulose chains that will be integrated by hydrogen bonds into a long and fairly stiff cellulose microfibril. These enzyme complexes are thought to move within the fluid membrane and leave a "trace" of crystallizing cellulose behind them. This movement will therefore determine the orientation of cellulose microfibrils and thus the anisotropy of the cell wall. It is at this point that the microtubules come into play. In fact, it was cell-wall anisotropy that led Green (1962) to predict that microtubules must exist even before they were actually discovered microscopically by Ledbetter and Porter (1963).

The intimate link between cortical microtubules and the preferential axis of growth is supported by the following main arguments (for a review see Nick 2008a) (1) upon plasmolysis, direct contact between cortical microtubules and newly formed cellulose microfibrils can be demonstrated by electron microscopy; (2) when the axis of cell expansion changes in response to a stimulus or during development, this is accompanied by a corresponding switch in the preferential axis of cellulose deposition, preceded by a corresponding reorientation of cortical microtubules; (3) elimination of cortical microtubules by inhibitors produces a

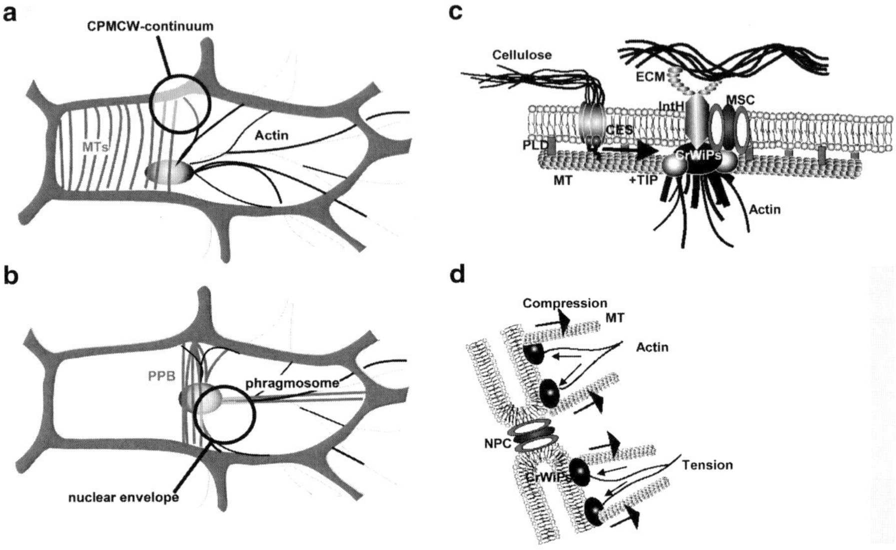

Fig. 2 Molecular players of mechanointegration in expanding (**a**, **c**) and dividing (**b**, **d**) plant cells. Microtubules and actin filaments cooperate to form tensegral structures together with other proteins. (**a**) Organization of microtubules (*MTs*) and actin filaments in an expanding cell; molecular details of the cell wall–plasma membrane–cytoskeleton (*WMC*) continuum are shown in (**c**). (**b**) Organization of the microtubular preprophase band *(PPB)*, and the actin-based phragmosome in a cell preparing for mitosis; molecular details of the nuclear envelope are shown in (**d**). (**c**) Molecular details of the cell wall–plasma membrane–cytoskeleton continuum. *PLD* phospholipase D anchor for microtubules, *CES* cellulose synthase complexes that are pulled by kinesins of the KIF4 family along microtubules towards the plus-end complexes (*+TIP*) that are linked via cross-wiring proteins (*CrWiPs*) such as kinesins containing a calponin-homology domain with actin, putative plant homologues of integrins (*IntH*) and mechanosensitive ion channels (*MSC*). The functional integrin homologues are linked with the extracellular matrix (*ECM*), e.g. arabinogalactan proteins, and cellulose microfibrils. (**d**) Molecular details of the link between the nuclear envelope and the cytoskeleton. Through cross-wiring proteins, a tensegral link between actin filaments and microtubules is generated, whereby microtubules might confer compression forces between the periphery and the nucleus, whereas the flexible actin filaments transfer tension forces. *NPC* nuclear pore complex

progressive loss of ordered cellulose texture and the axiality of cell expansion, leading, in extreme cases, to lateral swelling and bulbous growth. The mode of action of several herbicide classes, such as the phenyl carbamates or the dinitroanilines, is based on the elimination of cortical microtubules and the subsequent inhibition of elongation growth.

The striking parallelism between cortical microtubules and newly deposited cellulose microfibrils stimulated two alternative models: The original "monorail" model postulated that cortical microtubules adjacent to the plasma membrane guide the movement of the cellulose-synthesizing enzyme complexes and thus generate a pattern of microfibrils that parallels the orientation of microtubules (Heath 1974). The driving force for the movement of cellulose synthases in the monorail model would be an active transport through microtubule motors. Alternatively, the interaction between microtubules and cellulose synthases could be more indirect, whereby the microtubules act as “guardrails” inducing small folds of the plasma membrane that confine the movement of the enzyme complexes (Giddings and Staehelin 1991). The driving force for the movement would be the crystallization of cellulose. The solidifying microfibril would thus push the enzyme complex through the fluid plasma membrane, and the role of microtubules would be limited to delineating the direction of this movement.

The practical discrimination between these two models is far from straightforward because experimental evidence was mostly based on electron microscopy observation and thus prone to fixation artefacts. Moreover, great luck was required to locate the right section. For instance, the newly synthesized cellulose microfibrils formed after treatment with taxol were sometimes found to be directly adjacent to individual microtubules, for instance in tobacco BY-2 cells (Hasezawa and Nozaki 1999), supporting the monorail model. On the other hand, the cellulose synthase complexes were observed “in gap” between adjacent microtubules in the alga *Closterium* (Giddings and Staehelin 1988), favouring the guardrail model. The situation was further complicated by situations where the orientations of microtubules and cellulose microfibrils differ (for a review see Baskin 2001; Wasteneys 2004), leading to a debate on the role of microtubules in guiding cellulose synthesis. This debate stimulated a key experiment exploiting the potential of live-cell imaging in *Arabidopsis thaliana* (Paredez et al. 2006). A component of the terminal rosette, cellulose synthase subunit A6 (CESA6), was expressed as fusion with yellow fluorescent protein under the native promoter in the background of a *cesa6*-null mutant, such that overexpression artefacts could be excluded. The resulting punctuate signal was observed to be localized adjacent to the plasma membrane and to move along parallel pathways that resembled cortical microtubules. By crossing this line into a background, where one of the α-tubulins was expressed as fusion with a blue fluorescent protein, it became possible to follow this movement under simultaneous visualization of CESA6 and microtubules. This dual visualization demonstrated very clearly that CESA6 was moving along individual microtubule bundles. Moreover, in a recent publication, a central problem of the monorail model, i.e. the existence of polylamellate walls with layers of differing microfibril orientation, could be plausibly explained by a rotary movement of groups of microtubules (Chan et al. 2007; for a recent review see Lucas and Shaw 2008).

The original monorail model postulated a microtubule motor that pulls the cellulose synthase complex along the microtubules. If this motor were defective, cellulose microfibrils would deviate from the orientation of microtubules. A screen for reduced mechanical resistance in *A. thaliana* yielded a series of so-called *fragile fiber* mutants (Burk et al. 2001; Burk and Ye 2002) that were shown to be completely normal in terms of cell wall thickness or cell wall composition, but were affected in terms of wall texture. One of these mutants, *fragile fiber 2*, allelic to the mutant *botero* (Bichet et al. 2001), was affected in the microtubule-severing protein katanin, leading to swollen cells and increased lateral expansion. A second mutant, *fragile fiber 1*, was mutated into a kinesin-related protein belonging to the KIF4 family of microtubule motors (Fig. 2c). As expected, the array of cortical microtubules was completely normal; however, the helicoidal arrangement of cellulose microfibrils was messed up in these mutants. This suggests that this KIF4 motor is involved in guiding cellulose synthesis and might be a component of the monorail complex. Thus, the original monorail model for microtubule guidance of the terminal rosettes (Heath 1974) was rehabilitated after more than three decades of dispute.

However, the microtubule–microfibril model is still far from complete. In addition to occasionally discordant orientations of microtubules, there are cell wall textures that are difficult to reconcile with a simple monorail model. For instance, cellulose microfibrils are often observed to be intertwined (Preston 1988). This has stimulated views that claim that microtubules are dispensable for the correct texture of microfibrils. The self-organization of cellulose synthesis would be sufficient to perpetuate the pattern because the geometrical constraints from microfibrils that are already laid down would act as templates for the synthesis of new microfibrils (for a review see Mulder et al. 2004). There are two problems with this model. First, it ignores that microtubules and microfibrils are actually parallel in most cases, at least if cells in a tissue context are analysed. Second, it ignores that disruption of microtubules either by inhibitors (for a review see Nick 2008a, b) or by mutations that impair the formation of ordered microtubule arrays (Burk et al. 2001; Bichet et al. 2001 for katanin; Whittington et al. 2001 for *mor1*) is accompanied by a progressive loss of ordered cell wall texture and a loss of growth axiality. Nevertheless, the issue of cellulose self-organization highlights that the original microtubule–microfibril model has to be extended by a feedback control of microfibrils upon cortical microtubules.

2.1.2 Actin Filaments

Similar to microtubules, actin is organized into several distinct arrays that presumably exert different functions. For cells with pronounced tip growth, such as pollen tubes or root hairs, actin functions as a track for the transport of vesicles with cell-wall material that are inserted into the tip by intussusception (reviewed in Hepler et al. 2001; chapter "Generating a Cellular Protuberance. Mechanics of Tip Growth") and are probably conveyed by actin polymerization (Gossot and Geitmann 2007; Cárdenas et al. 2008). However, cells growing in a tissue context grow by apposition to the stretched cell wall rather than by intussusception. The role of actin therefore must be different and is not as obvious as for tip growth. During the diffuse elongation of tissue cells, longitudinal actin bundles prevail, especially in vacuolated cells (Parthasarathy et al. 1985; Sonobe and Shibaoka 1989). The rigidity of these transvacuolar strands and the degree of their bundling is regulated by signals such as plant hormones (Grabski and Schindler 1996), kinase cascades (Grabski et al. 1998) or light (Waller and Nick 1997). In addition to the transvacuolar bundles, a fine network of highly dynamic microfilaments can be detected in the cortical cytoplasm of elongating cells often accompanying cortical microtubules (for a review see Collings 2008). This cortical network can be rendered visible after pretreatment with protein cross-linkers (Sonobe and Shibaoka 1989), upon very mild fixation (Waller and Nick 1997) or by fusing binding domains of actin-binding proteins to fluorescent proteins (Voigt et al. 2005; Nick et al. 2009).

The transvacuolar actin cables were suggested to limit cell expansion by their rigidity and auxin was thought to stimulate growth by releasing this rigidity (Grabski and Schindler 1996). This mechanical model for the growth control through actin was supported by experiments where growth was modulated by light (Waller and Nick 1997). In the dark, when cells underwent rapid elongation, actin was organized into fine strands that became bundled in response to light-induced inhibition of growth. This actin reorganization was rapid and preceded the changes in growth rate (Waller and Nick 1997). Moreover, this response was confined to the epidermis, the target tissue for the signal control of growth. However, the mechanical model of actin function was shattered by experiments with actin inhibitors. The mechanical model predicted that elimination of actin cables should release the rigidity that limits growth. However, experiments with cytochalasin D (Thimann et al. 1992; Wang and Nick 1998) and latrunculin B (Baluška et al. 2001) revealed that even mild elimination of actin inhibited, rather than promoted, cell elongation. Thus, actin, possibly in combination with directional vesicle transport (Baskin and Bivens 1995), is a positive regulator of growth.

Subsequently, two actin populations could be separated owing to differences in sedimentability (Waller et al. 2002). Whereas the fine actin filaments correlated with a cytosolic fraction of actin, actin became progressively trapped on the endomembrane system and partitioned into the microsomal fraction when bundling was induced by light (perceived by phytochrome), by fluctuations of auxin content or by brefeldin A. This bundling of actin was accompanied by a reduced auxin sensitivity of cell elongation. This led to a model where auxin-signalling triggered the reorganization of actin bundles into finer filaments that more efficiently transported auxin-signalling/transport components towards the cell pole. The debundling in response to auxin predicted by this model was later demonstrated in intact rice coleoptiles in vivo using the actin-binding domain of mouse talin in fusion with yellow fluorescent protein first upon transient expression (Holweg et al. 2004) and later after stable expression (Nick et al. 2009).

To understand the link between actin and the auxin response of growth, excessive actin bundling was induced by overexpression of the actin-binding domain of talin in tobacco BY-2 cells (Maisch and Nick 2007) and in rice plants (Nick et al. 2009). In both systems, the reversion of a normal actin configuration can be restored by addition of exogenous auxin and this fully restores the respective auxin-dependent functions. These findings led to a model of a self-referring regulatory circuit between polar auxin transport and actin organization. Thus, although actin can stimulate growth by virtue of its mechanical properties in tip-growing cells (Gossot and Geitmann 2007), within a tissue context it does not act through mechanics, but acts by controlling the proper localization and thus activity of the signalling machinery that regulates cell expansion.

2.1.3 Actin–Microtubule Interaction

For cell growth, coordination and cross talk between microtubules and actin filaments have been inferred from their close coalignment and structural interaction (for reviews see Wasteneys and Galway 2003; Collings 2008) that seem to be ubiquitous and have been observed in different species and cell types. This conclusion has been supported by experiments involving pharmacological manipulation, where both microfilament- and microtubule-eliminating agents reduced cell elongation, often accompanied by a loss of growth anisotropy (*Arabidopsis*: Baskin and Bivens 1995; Collings et al. 2006; Graminean seedlings: Gianí et al. 1998; Wang et al. 2004; Blancaflor 2000; cotton fibres: Seagull 1990). Recently, it has even been proposed that microtubule reorientation is caused by detachment of microtubules from membrane-associated contact points (controlled by specific isoforms of phospholipase D; Fig. 2c) followed by their realignment with the direction of actin-based streaming (Sainsbury et al. 2008). Despite extensive studies on microtubular association of actin filaments, the proteins that mediate these interactions have remained elusive. Interactions between microtubules and actin filaments could be mediated either by bifunctional proteins that can bind to both cytoskeletal elements or, alternatively, by a connecting complex of two or more monofunctional proteins harbouring a microtubule-binding or an actin-binding domain, respectively. In animal and fungal cells, a number of proteins have been identified that mediate such interactions, either in a bifunctional way or as complexes consisting of monofunctional proteins (Goode et al. 2000; Rodriguez et al. 2003). In plants, however, only a few candidates such as MAP190 have been proposed so far (Igarashi et al. 2000).

In animal cells, the microtubular minus-end motor dynein is connected with and activated through the dynactin complex (for a review see Karki and Holzbaur 1999). The dynactin complex is further linked to the microtubule tip component EB1 and thus regulates the stability of microtubules (for a review see Tirnauer and Bierer 2000). Dyneins as central elements of the dynactin complex are not found in plants. This is probably the consequence of the loss of flagellate cells and centrioles (for a review see Schmit and Nick 2008). Plants must have evolved a functional compensation for the loss of dyneins in the form of other minus-end-directed motors that are able to interact, either directly or indirectly, with actin filaments. Recently, kinesins with a calponin-homology domain (KCH kinesins) were identified as plant-specific subset of the kinesin-14 family (Tamura et al. 1999; Preuss et al. 2004; Frey et al. 2009; Xu et al. 2009). In addition to the characteristic microtubule-binding kinesin motor domain, these proteins possess a conserved calponin-homology domain, well known as an actin-binding domain from a variety of actin-associated proteins such as α-actinin, spectrin and fimbrin. Thus, KCHs are strong candidates for a bifunctional mediation between both cytoskeletal elements, and several studies confirmed that they can bind, in fact, both elements of the cytoskeleton (cotton: Preuss et al. 2004; Xu et al. 2009; rice: Frey et al. 2009). These cross-linking microtubule motors are present in higher plants, but also in *Physcomitrella patens* (Richardson et al. 2006; Frey et al. 2009). Their cellular function is not really understood, but for one of the cotton KCHs, a role in cell elongation through cross-linking of microtubules and microfilaments has been proposed, however without experimental evidence (Xu et al. 2009). For the rice homologue OsKCH1, the phenotype of insertion mutants and a KCH overexpression line generated in tobacco BY-2 cells (Frey et al. 2010) suggest that this kinesin stimulates cell elongation, although this stimulation might be a secondary phenomenon caused by changes in the timing of cell division.

2.1.4 The Cell-Wall Cytoskeletal Continuum

A continuum between the cytoskeleton and the extracellular matrix is central for mechanosensing in animal cells (for a review see Geiger and Bershadsky 2001) and involves interaction between integrins and extracellular matrix proteins that contain Arg–Gly–Asp (RGD) motifs (for a review see Giancotti and Ruoslahti 1999). Plants seem to lack integrin homologues, but there is evidence for cytoskeletal reorganization in response to treatment with RGD peptides (Canut et al. 1998; Wang et al. 2007). As a molecular basis for the cytoskeleton–plasma membrane–cell wall continuum in plants (Fig. 2c), cell-wall-associated kinases, arabinogalactan proteins, pectins and cellulose synthases themselves have been discussed (for a review see Baluška et al. 2003; chapter "Introduction: Tensegral World of Plants"). Recently, the rich but circumstantial evidence for such transmembrane interactions was assembled into a model for a so-called plasmalemmal reticulum as a third element of the plant cytoskeleton (Pickard 2008). This plasmalemmal reticulum is considered to be a tensile structure and to participate in the control of cellulose deposition. It can be visualized as a reticulate structure by antibodies raised against components of the mammalian extracellular matrix and also contains arabinogalactan proteins. The reticulum is reorganized in the context of cell elongation in a manner similar to cellulose deposition and is suggested to represent a morphological manifestation of lipid rafts.

2.2 *Cell Division*

2.2.1 Microtubules

In dividing cells, the ensuing mitosis is heralded by a displacement of the nucleus to the cell centre, where the prospective cell-plate will be formed (for a review see Nick 2008a). Simultaneously, radial microtubules emanate from the nuclear surface and merge with the cortical cytoskeleton, tethering the nucleus to its new position (Fig. 2b). In the next step, the preprophase band is organized by the nucleus as a broad band of microtubules around the cell equator, marking the site where after completed mitosis the new cell plate will be formed. In fern protonemata, where the formation of the preprophase band can be manipulated by centrifugation of the nucleus to a new location (Murata and Wada 1991), a causal relationship between the preprophase band and cell-plate orientation was elegantly demonstrated. Moreover, in cells where the axis or symmetry of cell division changes, this change is always predicted by a corresponding localization of the preprophase band. The division spindle is always laid down perpendicular to the preprophase band, with the spindle equator located in the plane of the preprophase band. As soon as the chromosomes have separated, a new array of microtubules, the phragmoplast, appears at the site that had already been marked by the preprophase band. This microtubular structure consists of a double ring of interdigitating microtubules that increases in diameter with increasing size of the cell plate. New microtubules are organized along the edge of the growing phragmoplast (Vantard et al. 1990). The phragmoplast targets vesicle transport to the periphery of the expanding cell plate. Microtubules seem to pull at tubular–vesicular protrusions emanating from the endoplasmic reticulum (Samuels et al. 1995). The guiding function of the preprophase band is supported by evidence from *Arabidopsis* mutants (*tonneau/fass*), where the preprophase band is absent owing to a mutation in a phosphatase PP2A regulatory subunit (Camilleri et al. 2002). In these mutants, the ordered pattern of cell divisions that characterizes the development of the wild type is replaced by a completely randomized pattern of cross walls (Traas et al. 1995; McClinton and Sung 1997). It should be mentioned, however, that during meiosis the division plane can be controlled in the absence of a preprophase band (for a recent review see Brown and Lemmon 2007), suggesting that there exist additional mechanisms of spatial control.

2.2.2 Actin Filaments

In contrast to the obvious and dramatic reorganization of the microtubule during mitosis, actin filaments seem to be more persistent. Two decades ago, actin filaments were shown to accompany mitotic microtubule arrays such as preprophase band, spindle and phragmoplast (Kakimoto and Shibaoka 1987; Lloyd and Traas 1988). However, there is still some controversy as to the exact behaviour, persistence and orientation of actin filaments during M phase (for a recent review see Panteris 2008). The microtubular preprophase band that disappears with the breakdown of the nuclear envelope leaves a so-called actin-depleted zone as a negative imprint that later, upon reestablishment of the daughter nuclei, will be the site where the new cell plate is formed. Despite some debate regarding to what extent this zone is completely void of actin or whether it just contains fewer actin filaments, there is a clear correlation between the actin-depleted zone and the site of the prospective cell plate. When the actin filaments lining this depletion zone are eliminated by inhibitor treatment, this will affect the subsequent cell division when the treatment occurs during the presence of the microtubular preprophase band. However, actin inhibitors will have only a marginal effect once the preprophase band has disappeared (Sano et al. 2005). What is the actin-dependent function that defines the cell plate? It might be linked to a belt composed of endosomes that is laid down adjacent to the preprophase band by joint action of microtubule-driven and actin-driven transport (Dhonukshe et al. 2005). This belt persists during mitosis, and is, upon completed separation of the chromosome, "read out" by a new set of microtubules emerging from the spindle poles that "explore" the cell periphery in different directions. The lifetime of these exploratory microtubules is increased when they hit the endosomal belt, whereas microtubules that fail to interact with the endosomes are prone to undergo catastrophic decay (Dhonukshe et al. 2005). Thus, the actin-depleted zone is rather a manifestation of and not the cause for the correct positioning of the cell plate.

2.2.3 Nuclear Migration

In cells that prepare for mitosis, the nucleus is tethered by the so-called phragmosome forming the characteristic "Maltese cross" to its new position in the cell centre, and persists during the subsequent mitosis (Fig. 2b). The cytoskeletal reorganizations accompanying mitosis assign a central role to nuclear migration in the control of division symmetry, and to the preprophase band for the control of division axis. Nuclear migration can be blocked by actin inhibitors such as cytochalasin B (Katsuta and Shibaoka 1988). However, microtubules also seem to be involved in nuclear positioning, since antimicrotubular compounds such as colchicine (Thomas et al. 1977) and pronamide (Katsuta and Shibaoka 1988) have been found to loosen the nucleus such that it can be displaced by mild centrifugation. This indicates again that proteins mediating actin–microtubular interaction are relevant for nuclear positioning. In fact, the plant-specific KCHs (Frey et al. 2009) were found to be dynamically repartitioned during the cell cycle: in premitotic cells, KCH1 was clearly aligned in a punctuate pattern along filamentous, mesh-like structures on both sides of the nucleus and on perinuclear filaments spanning over and surrounding the nucleus. At the onset of mitosis, KCH1 retracted to both sides of the nucleus, but not in preprophase bands nor in the spindle apparatus nor at the division plate. During late telophase and the beginning of cytokinesis, KCH1 was shifted towards the newly forming nuclei and lined the filaments that tethered these nuclei to the periphery and the new cell wall (Frey et al. 2010). Tos17 *kch1* knockout in rice showed increased cell numbers in the seedling shoot, whereas overexpression of KCH1 in tobacco BY-2 cells reduced the cell number. This effect was assigned to a delay in the premitotic nuclear migration towards the cell centre, suggesting that KCH regulates nuclear positioning and thus the progression of mitosis (Fig. 2d).

3 Molecular Players: Protein Conformation Versus Stretch-Activated Channels

Mechanical integration requires the perception of stress–strain patterns. In turgescent plant cells, the expanding protoplast exerts considerable turgor pressure upon the yielding cell wall. This cell-autonomous component of mechanical load is complemented by mechanical tension across the tissue that is caused by the limiting extensibility of the epidermis (for a recent review see Kutschera 2008). Thus, any mechanism for plant mechanosensing has to cope with this strong, but tonic background stimulus.

Essentially, there are two basic models for the molecular basis of mechanoperception:

1. Stretching of proteins will change their conformation and create new binding sites for the recruitment of associated proteins (for reviews see Janmey and Weitz 2004; Orr et al. 2006).
2. Mechanosensitive channels are able to directly detect and respond to forces from the lipid bilayer. Such channels will open when the plasma membrane is deformed or when the channel is pulled by a tether (for a review see Kung 2005).

Mechanosensing by stretch-induced conformational changes is well supported for the adhesion of mammalian cells (for a review see Geiger and Bershadsky 2001). Here, the growth of focal contacts, where the actin cytoskeleton is tethered to the extracellular matrix through a complex of associated proteins and integrins, is promoted by local mechanical force (Riveline et al. 2001). A similar mechanosensing network was proposed for plant cells, where analogues of integrins link the cytoskeleton at the inner face with the cell wall at the outer face of the plasma membrane (Jaffe et al. 2002). However, the transfer of this model from mammalian cells to plants is not straightforward because the molecular components differ considerably (for a review see Baluška et al. 2003). It is the cell wall with a completely different set of molecules that replaces the extracellular matrix of mammalian cells. Furthermore, canonical integrins are obviously absent from plants, suggesting that the link between actin filaments and the extracellular binding sites must use different groups of molecules. Further, although plants and animals share several actin-binding proteins, important components of focal contacts, such as talin, do not exist in plants. On the other hand, there seem to exist integrin analogues that can bind to RGD tripeptides in a way similar to the way integrins do in animal cells. It seems that class VIII myosins and formins might act as linkers between the actin cytoskeleton and the plant analogues of the extracellular matrix, such as cell-wall-associated kinases and arabinogalactan proteins (Baluška and Hlavačka 2005). Additionally, cellulose synthases that move along microtubules (Paredez et al. 2006) tether the cytoskeleton to the cell wall, and several microtubule-associated proteins, for instance phospholipase D (Gardiner et al. 2001) and MAP18 (Wang et al. 2007), act as linkers between cortical microtubules and the plasma membrane Thus, although the molecular components differ considerably from their animal counterpart, a contiguous link between the cytoskeleton and the cell wall does exist and is commonly referred to as the cell wall–plasma membrane-cytoskeleton interface (Telewski 2006).

The existence of mechanosensitive channels in plants was originally discovered in specialized cells, where a touch stimulus induced an action potential (Shibaoka 1966) such as in the seismonastic leaves of *Mimosa pudica* (Toriyama and Jaffe 1972), or internodal cells of *Chara* (Kishimoto 1968). From electrophysiological and pharmacological evidence, a model emerged where these touch-sensitive channels mediate an influx of calcium (for a review see Jaffe et al. 2002). In fact, with use of aequorin-transformed plants, mechanical stimulation was demonstrated to trigger calcium influx with stimulus-specific signatures (Knight et al. 1991). A causative role of calcium fluxes was supported by the isolation of touch-insensitive *Arabidopsis* mutants affected in calmodulin genes (Braam and Davis 1990), and inhibition of touch responses by inhibitors of calmodulin (Jones and Mitchell 1989). A mechanosensitive calcium channel was demonstrated for epidermal cells of onion (Ding and Pickard 1993), but the molecular identity of mechanosensitive calcium channels has remained elusive. Molecular identification is also hampered by the highly artificial conditions required to identify stretch-activated ion fluxes by patch-clamp techniques. Removal of the cell wall, isotonic conditions, and suction by the holding electrode create conditions where most ion channels would be defined as mechanosensitive (Gustin et al. 1991).

Although both mechanisms, stretch-induced conformational changes and stretch-activated ion channels, are often discussed separately, they might, in fact, act in concert, as components of a so-called plasmalemmal reticulum (for a review see Pickard 2008). This integrative structure comprises adhesive components (among others arabinogalactan proteins and wall-associated kinases) that link the plasma membrane with the cell wall, and are also connected with mechanosensory calcium channels. This structure has been demonstrated and described for tobacco BY-2 as a cell biological model system (Gens et al. 2000; Pickard and Fujiki 2005). The plant integrin analogues have been reported to connect microtubules, plasma membrane, actin filaments and stretch-activated membrane channels (Telewski 2006). It is highly conceivable that such a network could act, on the one hand, as a tensegral entity that can convey and focus mechanical force upon stretch-activating membrane channels and, simultaneously, transduce forces into conformational changes that can result in differential decoration with associated proteins that can act as a trigger for signalling. The necessity for stress-focussing is supported by estimations of the activation energies for mechanosensitive channels (around 1 mN m^{-1}, Sachs and Morris 1998) in a range not far below the lytic tension of plant membranes (around 4 mN m^{-1}, Kell and Glaser 1993).

If the tensegral cytoskeleton is linked to the cell wall through such integrative linkers, this should become manifest as organizing influences of the cell wall upon the cytoskeleton. This has been observed. For instance, removal of the cell wall renders microtubules cold-sensitive in tobacco cells (Akashi et al. 1990). When in the same cells the incorporation of UDP-glucose into the cell wall was blocked by the herbicide isoxaben (Fisher and Cyr 1993), this impaired the axiality of cell expansion, resulting in isodiametric cells and disordered cortical arrays of microtubules. Thus, the mechanical strains produced by cellulose microfibrils align cortical microtubules, closing a regulatory circuit between the cell wall and the cytoskeleton. Since expansion is reinforced in a direction perpendicular to the orientation of microtubules and microfibrils, forces will be generated parallel to the major strain axis. These forces are then relayed back through the plasma membrane upon cortical microtubules that are aligned in relation to these strains. Since individual microtubules mutually compete for tubulin heterodimers, and since the number of microfibrils is limited by the quantity of cellulose synthase rosettes, this regulatory circuit should follow the rules of a reaction–diffusion system (Turing 1952) and should therefore be capable of self-organization and patterning.

How could mechanical strains from the cell wall cause a corresponding alignment of cortical microtubules? The so-called microtubule plus-end tracking proteins (+TIP proteins) seem to play an important role in this context. These proteins associate with growing plus ends of microtubules and form complexes that control microtubule dynamics, organization and the interaction of microtubules with membranes, organelles and proteins (for a review see Akhmanova and Steinmetz 2008). EB1, a central component of this complex, is important in searching for so-called exploratory microtubules for intracellular capture sites that are often marked by specific actin structures. For instance, such capture sites are laid down by the preprophase band and the phragmosome prior to mitosis and are recognized by exploratory microtubules emanating from the cell poles during telophase (Dhonukshe et al. 2005). Because of this function, several members of +TIP proteins interact with the actin cytoskeleton. Some +TIP proteins, such as adenomatous polyposis coli (Moseley et al. 2007) and CLIP-associated protein (Tsvetkov et al. 2007), interact with actin filaments directly, others interact through the actin-binding formins. A third group, including EB1, interact with the dynactin complex linking the minus-end microtubule motor dynein with actin filaments (for a review see Tirnauer and Bierer 2000). EB1 binds to microtubule plus ends at the seam that joins the tubulin protofilaments (Sandblad et al. 2006) and is therefore a good candidate for a conformational mechanosensor. During microtubule catastrophe, the protofilaments bend outwards, which means that they have to be actively tied together to sustain microtubule growth. The +TIP complex, in general, and EB1, in particular, are therefore subject to mechanical tension and must be considered as primary targets for mechanical strains on microtubules. In fact, *Arabidopsis* mutants in members of the EB1 family have been found to be touch-insensitive (Bisgrove et al. 2008).

Summarizing, in plant cells both stretch-induced changes of protein conformation and stretch-activated ion channels seem to act in concert during the perception of mechanical stimuli. The cytoskeleton can participate in both pathways, either as a stress-focussing susceptor of mechanical force upon mechanosensitive ion channels or as a primary sensor that transduces mechanical force into differential growth of microtubule plus ends. Microtubules are endowed with nonlinear dynamics, leading to phase transitions between growth and catastrophic shrinkage. In addition, they have to compete for a limited pool of free heterodimers. Microtubules are therefore ideal devices to amplify the minute inputs from mechanical stimulation (small deformations of the perceptive membranes, changes in the dynamic equilibrium between assembly and disassembly of microtubules at the microtubule plus end) into clear and nearly qualitative outputs that can then be processed by downstream signalling cascades (for a review see Nick 2008b).

4 The Cytoskeleton as a Sensor: Intercellular Sensing

It was the loss of buoyancy as a supporting force that drove the evolution of mechanical integration in terrestrial plants. Signalling through a mechanical signal is much faster than any diffusion-based process and its velocity equals or even exceeds that of electric signalling. Moreover, mechanical integration is holistic in nature, since it allows the sensing almost simultaneously of the presence or absence of building elements even if they are remote from the sensing cell (as long as they are mechanically coupled). The stiffer this mechanical coupling, the less energy is dissipated during signalling. Microtubules are endowed with a high degree of rigidity (Gittes et al. 1993) and therefore represent ideal transducers for mechanical integration even across the borders of individual cells.

Such microtubule orientations that transcend the borders of individual cells have been reported during phyllotaxis (Hardham et al. 1980; Hamant et al. 2008). In wounded pea roots, a supracellular alignment of microtubules heralds corresponding changes of cell axis and cell divisions such that the wound is efficiently closed (Hush et al. 1990). A curious case of microtubule patterning was discovered in the *Arabidopsis* mutants *spiral*, *lefty* and *tortifolia* (Furutani et al. 2000; Thitamadee et al. 2002; Buschmann et al. 2004). In these mutants, microtubules are aligned over many cells in the distal elongation zone of the root (*spiral* and *lefty*) or the petiole (*tortifolia*), accompanied by twisted growth.

The twisted growth phenotypes of these mutants are conventionally explained on the basis of uniformly oblique arrays of microtubules (and consequently microfibrils). In the *spiral*, *lefty* and *tortifolia* mutants, it is either tubulin itself or microtubule-associated proteins that are affected by these mutations. Moreover, spiral growth can be phenocopied in the wild type by inhibitors of microtubule assembly (Furutani et al. 2000). As pointed out earlier, the microtubule–microfibril circuit is endowed with self-amplification linked to mutual inhibition. A typical systemic property of such a self-organizing morphogenetic system is an oscillating output (Gierer 1981). Any factor that alters the lifetime of microtubules will alter the relay times within this feedback circuit. Since neighbouring cells are mechanically coupled by tissue tension, even a weak coupling will result in a partial synchronization of the individual circuits (Campanoni et al. 2003), and the degree of synchrony will depend on the velocity of the feedback circuit. Thus, mutations in an associated protein such as the *tortifolia* gene product (Buschmann et al. 2004), mutations in tubulin itself, as in case of *lefty* (Thitamadee et al. 2002), or treatment with microtubule inhibitors (for a review see Hashimoto and Kato 2006) is expected to enhance synchrony, leading to the observed oscillations of growth. In contrast, the synchrony should be reduced when microtubule lifetimes are increased, which seems to be the case for the mutant *radially swollen 6* (Bannigan et al. 2006), where microtubule arrays are ordered within individual cells, but deviate strongly between neighbouring cells, suggesting that supracellular alignment is affected.

The impact of microtubules for mechanointegration can be exemplarily studied in the context of gravity responses (see chapter "Mechanical Aspects of

Gravity-Controlled Growth, Development and Morphogenesis"). To compensate for mechanical load by gravity, plants have to optimize the arrangement of force-bearing elements in space in a manner such that they provide optimal mechanical support, but simultaneously consume minimal biomass and are as light as possible. This optimization task can only be achieved when the arrangement of supportive structures is guided by the pattern of mechanical strain. Thus, gravity has to be perceived very efficiently and it has, in addition, to be linked to morphogenesis. When the orientation of a plant is changed with respect to gravity, the plant will respond by bending in such a way as to restore the original orientation and thus to minimize mechanical stress (gravitropism). When new organs develop, they are often adjusted with respect to gravity (gravimorphosis). When the mechanointegrative role of the cytoskeleton is discussed with respect to gravity, it is important to separate so-called susception from perception in the strict sense. "Susception" means transformation of physical energy into a different type of energy that can be perceived by the perceptive system (Björkman 1988). For instance, the difference in gravitational field strength between the two flanks of a misoriented plant would be certainly far too small to be sensed by any biochemical process. It is generally accepted that gravity is first transformed into mechanical force by acting on heavy particles, the so-called statoliths. These statoliths (as well as their accessory structures) themselves are not gravisensitive, but they assist sensing by acting as susceptors.

4.1 Microtubules and Gravitropism

For the rhizoid of *Chara*, experiments by Johannes Buder (1961) demonstrated that vesicles filled with barium sulfate, the *Glanzkörperchen*, are necessary and sufficient for gravisusception. For higher plants, the classical starch-statolith theory (Nemec 1900; Haberland 1900) postulated that amyloplasts in the perceptive tissues (e.g. root cap or bundle sheath cells) are responsible for the susception of the gravitropic stimulus. A long tradition of experimentation demonstrated that amyloplasts are necessary for efficient gravitropism. For instance, gravitropic sensitivity was reduced in starch-deficient mutants. However, it took almost a century until it was shown that susception of energy by amyloplasts is sufficient to trigger a curvature response. By using high-gradient magnetic fields, Kuznetsov and Hasenstein (1996) succeeded in inducing bending in vertically oriented roots and thus were able to prove very elegantly that the generation of mechanical force by statoliths is sufficient for gravisusception. It is an irony of science history that this breakthrough was not achieved by the elaborate and expensive microgravity experiments in the context of space research, but instead through a very cheap, but well-designed ground experiment. Thus, in higher plants as well, the primary stimulus is produced by statolithic particles (the amyloplasts), although the actual perception event has remained elusive so far.

For the rhizoid of *Chara* the sedimentation of the *Glanzkörperchen* to the lower flank of the rhizoid was found to divert vesicle flow towards the upper side such that more material is intussuscepted into the upper flank, resulting in a growth differential driving downward bending (Sievers and Schröter 1971). This hypothesis was later extended to negative gravitropism by combining sedimentation with a different mode of growth (Hodick 1994). According to this model, the actual perception of gravity would rely upon a proximity mechanism. It is doubtful that proximity is used for graviperception in higher plants because classical studies (Rawitscher 1932) using intermittent stimulation showed that perception can occur in the absence of amyloplast sedimentation. Moreover, dose–response studies employing centrifugation have shown that the output (gravitropic curvature) is dose-dependent even for stimuli that completely saturate amyloplast sedimentation. Even for the rhizoid of *Chara*, for which the proximity mechanism was originally postulated, it was demonstrated that strong stimuli that saturate the sedimentation of the *Glanzkörperchen* can still be discriminated (Hertel and Friedrich 1973). This suggests that the actual perception of gravity is not based on proximity, but is based on the force exerted by the statoliths on a mechanosensor such as stretch-activated ion channels and/or the cell wall–plasma membrane–cytoskeleton interface.

If gravity is perceived not by proximity but by pressure, this poses a big challenge for the sensing mechanism. Since gravity is sensed by individual cells (in contrast to the direction of light in phototropism; Buder 1920; Nick and Furuya 1996), the maximal energy available for stimulation is the potential energy of the sensing cell. This energy barely exceeds thermal noise if it is not focussed upon small areas. These considerations stimulated research on a potential role of microtubules as amplifiers of gravitropic perception. In fact, gravitropism can be blocked by antimicrotubular drugs in the rhizoid of *Chara* (Hertel and Friedrich 1973) as well as in moss protonemata (Schwuchow et al. 1990; Walker and Sack 1990) or in coleoptiles of maize (Nick et al. 1991) and rice (Godbolé et al. 2000; Gutjahr and Nick 2006) at concentrations that leave the machinery for growth and bending essentially untouched. Conversely, when the dynamics of microtubules is reduced as a consequence of either a mutation (Nick et al. 1994) or treatment with taxol, this results in a strong inhibition of gravitropic responses (Nick et al. 1997; Godbolé et al. 2000; Gutjahr and Nick 2006). The gravitropically induced reorientation of cortical microtubules has been observed for both shoot gravitropism (Nick et al. 1991) and root gravitropism (Blancaflor and Hasenstein 1993). In maize coleoptiles, the microtubules in the epidermal cells of the upper flank of the stimulated organ assumed a longitudinal orientation, whereas the microtubules in the lower flank remained transverse. By microinjection of fluorescent tubulin into epidermal cells of intact maize coleoptiles, it was later even possible to demonstrate the gravitropic microtubule reorientation in vivo (Himmelspach et al. 1999). The time course of this response was consistent with a model where gravitropic stimulation induced a lateral shift of auxin transport towards the lower organ flank and, as a consequence, a depletion of auxin in the upper flank. The microtubular response was thought to be primarily by this decrease in auxin concentration rather than by gravity itself. In maize roots, however, where a similar reorientation was observed in the cortex (Blancaflor and Hasenstein 1993), the time course of reorientation was found to be slower than the changes in growth rate induced by gravity.

This leads to the question of whether the gravitropic response of microtubules is direct or whether microtubules merely respond to changes in growth rate. In fact, it is possible to induce microtubule reorientation by bending coleoptiles with manual force (Zandomeni and Schopfer 1994) – the microtubules will then become longitudinal in the concave flank, but remain transverse in the convex flank. To dissect the gravitropic response and a potential response to changed growth rate, microtubule behaviour was followed in coleoptiles that were prevented by a surgical adhesive from elongating and were kept either in a horizontal orientation (such that a gravitropic stimulation occurred) or in a vertical orientation (such that growth was inhibited in the absence of a gravitropic stimulus). In this set-up, microtubule reorientation from transverse to longitudinal was observed only in the horizontal orientation (Himmelspach and Nick 2001), demonstrating unequivocally that microtubules, at least in this system, responded to gravity rather than to the inhibition of growth.

4.2 *Microtubules and Gravimorphosis*

It seems trivial that roots form at the lower pole of a plant organ, but this is a manifestation of gravimorphosis. Although a considerable amount of phenomenological work was dedicated to this problem at the turn of the nineteenth century (Vöchting 1878; Sachs 1880; Goebel 1908), the underlying mechanisms are still far from being understood. This is partially due to the use of adult organs, where polarity has already been fixed and is hard to invert. However, in germinating fern spores, the first asymmetric division that separates a larger, vacuolated rhizoid precursor from a smaller and denser thallus precursor can be oriented by gravity (Edwards and Roux 1994). This first cell division is clearly of formative character; when it is rendered symmetric by treatment with antimicrotubular herbicides (Vogelmann et al. 1981), the two daughter cells both give rise to thalloid tissue. When this spore is tilted after the axis of the first division has been determined, the rhizoid will grow in the wrong direction and cannot adjust for this error (Edwards and Roux 1994). Prior to division, at the time when the spore is competent to the aligning influence of gravity, a vivid migration of the nucleus towards the lower half of the spore is observed. This movement is not a simple sedimentation process because it is oscillatory and interrupted by short periods of active sign reversal, indicating that the nucleus is tethered to a motive force (Edwards and Roux 1997). The action of antimicrotubular compounds strongly suggests that this guiding mechanism is based on microtubules that probably align with the gravity vector, resembling the determination of the grey crescent in amphibian eggs (Gerhart et al. 1981), where the dorsoventral axis is determined by an interplay among gravity-dependent sedimentation of yolk particles, sperm-induced nucleation of microtubules and self-amplifying alignment of newly formed microtubules that drive cortical rotation (Elinson and Rowning 1988).

4.3 Microtubules and the Sensing of Gravity

Since microtubules guide the anisotropic deposition of cellulose in the cell wall, it is not trivial to discriminate their function in gravity sensing from their role in the response to gravity during the development of tropistic curvature. When gravitropic bending is inhibited by antimicrotubular agents, this might be caused by a block of either the sensory function or the effector function of microtubules. To discern these microtubular functions, the lateral transport of auxin can be used as a response upstream of differential growth. With use of radioactively labelled auxin in rice coleoptiles (Godbolé et al. 2000), lateral auxin transport was found to be blocked by ethyl-*N*-phenylcarbamate, a herbicide that binds to the carboxy terminus of α-tubulin and inhibits assembly of tubulin heterodimers to the growing ends of microtubules (Wiesler et al. 2002). Interestingly, Taxol inhibited lateral transport partially without any inhibition of longitudinal transport of auxin. This indicates that sensory microtubules have to be not only present, but, in addition, also dynamic to fulfil their function. The high dynamics of this sensory microtubule population might also explain the extreme sensitivity of gravisensing to low temperature that would be difficult to explain otherwise (Taylor and Leopold 1992). The necessity of high microtubular turnover favours a model where microtubules sense gravity-induced forces actively rather than merely acting as gravisusceptors. The gravisensory function of microtubules can be specifically blocked by acrylamide (Gutjahr and Nick 2006), a widely used inhibitor of intermediate-filament function in mammalian cells (Eckert and Yeagle 1988). Similar to ethyl-*N*-phenylcarbamate, acrylamide interrupts a very early step in the gravitropic response chain, clearly upstream of auxin redistribution and differential growth. No clear homologues of intermediate-filament proteins are known in the plant kingdom, but acrylamide treatment specifically disrupts microtubules, leaving actin filaments, for instance, untouched (Gutjahr and Nick 2006). The immediate target of acrylamide in mammalian cells seems to be a kinase that phosphorylates keratin (Eckert and Yeagle 1988). Since kinases and phosphatases have been shown to regulate the organization of plant microtubules (Baskin and Wilson 1997), the inhibition of gravitropism by acrylamide might be caused by interference with the regulatory circuits active in the highly dynamic microtubule population responsible for gravisensing.

4.4 Microtubules and Mechanosensing

Gravity is not the only source of mechanical stimulation used to integrate plant architecture. In contrast to terrestrial animals, the cells of land plants are not surrounded by an isotonic medium, but are surrounded by a hypotonic medium, with the consequence that their plasma membrane is under continuous tension from the expanding cytoplasm that is counterbalanced by the cell wall. On the level of organs, considerable tissue tensions develop that can be used for mechanointegration when, for instance, new organs emerge and will thus generate local tension. This phenomenon has been intensively studied and modelled for phyllotaxis by Paul Green and co-workers, who showed that models of stress-strain patterns could perfectly predict the positions of incipient primordia (for a review see Green 1980). In the growing meristem, the formation of new primordia is suppressed by the older primordia. The tissue tension present in an expanding meristem produces mechanical stresses resulting from buckling of the preceding primordia. One of the earliest events of primordial initiation is a reorientation of cortical microtubules that are perpendicular with respect to the microtubules of their non-committed neighbours. This difference is sharp, but later it is smoothed by a transitional zone of cells with oblique microtubules, such that eventually a gradual, progressive change in microtubular reorientation emerges over several rows of cells (Hardham et al. 1980). This phenomenon has been revisited recently making use of microtubule marker lines labelled with green fluorescent protein in the developing shoot apex of *A. thaliana* (Hamant et al. 2008), where a feedback between microtubule orientation and organ growth was demonstrated. Mechanical modelling of the expanding shoot meristem predicted the transcellular pattern of microtubule orientation that was predicted and observed to be aligned with the directions of maximal stress. By ablation of specific cells in the outer meristem layer of the meristem, a redistribution of stress was induced and modelled. As expected, this redistribution caused a corresponding redistribution of microtubular orientation.

4.5 Candidates for the Underlying Mechanism

The sensing of gravity relies on the mechanical forces suscepted by the amyloplasts. This would, at first sight, suggest that the sensing of gravity and the sensing of mechanical stimuli should run in parallel. This can be tested by antagonistic application of artificial bending stress in antagonism to a gravitropic stimulus; it is possible to separate the response of gravity from the secondary mechanical stimulus that is induced by the differential growth during gravitropic bending (Ikushima and Shimmen 2005). When, under these conditions, the activity of mechanosensitive channels was suppressed by gadolinium in hypocotyls of adzuki beans, this suppressed the (mechanically induced) reorientation of microtubules in the effector tissue, whereas gravitropic curvature proceeded unaltered (indicating that the microtubule population resident in the inner tissues of the apical hook that is responsible for gravisensing remained functional). Thus, at least in this system, mechanosensing is sensitive to gadolinium, whereas gravisensing is not. Similarly, when the gravitropic bending of maize coleoptiles was inhibited by a surgical adhesive, the gravity-induced reorientation of microtubules was nevertheless developed (Himmelspach and Nick 2001), suggesting different signal chains for gravisensing and mechanosensing. Although the responses of plant cells to gravity and mechanical stimuli are generally discussed in the context of stretch-activated ion channels (Ding and Pickard 1993), the protein conformation paradigm of sensing should not be neglected. This is emphasized by the phenotype of *Arabidopsis* mutants, where the microtubule plus-end protein EB1 is affected (Bisgrove et al. 2008). In these mutants, both the initiation of gravitropic curvature and a specific touch-induced waving of roots on inclined agar plates were affected. Although our knowledge of the primary events of mechanosensing and gravisensing in plants is extremely limited, it is clear even at this stage that the roles of microtubules might differ qualitatively. In mechanosensing, microtubules seem to act as susceptor structures that focus deformation stress towards ion channels. In contrast, in gravisensing, the necessity for high dynamics and dimer turnover favours a direct sensory role of microtubules. Thus, nature might utilize both mechanisms simultaneously to sense (and possibly to discriminate) different stimuli. The challenge for future research in this field will be to design experimental approaches with clear outputs based on clear concepts of the sensing mechanism. Only in a second step will it become possible to define and test molecular and cellular candidates.

5 The Cytoskeleton as a Sensor: Intracellular Sensing

The previous section dealt with the mechanical integration of individual cells into a coordinated response of the entire organ. However, mechanosensing is also used to integrate the different components of a cell into an individual. This becomes evident as redistribution of organelles in response to mechanical stimulation (thigmomorphogenesis; chapter "Mechanical Force Responses of Plant Cells and Plants"), but also involves responses that are less evident, such as the adaptation to cold, the induction of plant defence, osmoregulation (chapter "Osmosensing") and the regulation of division symmetry. The fundamental role of intracellular mechanosensing has emerged in recent years, but its full impact is still far from being recognized.

5.1 Thigmomorphogenesis

Morphological responses to mechanical stimulation have been demonstrated not only on the supracellular level, but also on the sub-cellular level. When fern protonemata are squeezed by a needle, chloroplasts avoid the site of contact (Sato et al. 1999). In epidermal cells (Kennard and Cleary 1997) or in suspension cells of parsley (Gus-Mayer et al. 1998) such a local pressure can induce nuclear movement towards the contact point. When regenerating protoplasts or cells are challenged by either mild centrifugation or touch, the axis of cell division is aligned with the force vector (Lintilhac and Vesecky 1984; Wymer et al. 1996; Zhou et al. 2007). Using tension-free protoplasts, Wymer et al. (1996) aligned microtubules by a short centrifugation and thus oriented the axis of cell expansion in a direction perpendicular to the force vector. They used this system to demonstrate a role of microtubules in mechanosensing. To separate this sensory role from the microtubular function in cell-wall synthesis, microtubules were eliminated transiently during the application of force by amiprophos-methyl and allowed to recover by washing out the herbicide, such that cellulose synthesis could occur and thus the cell axis could develop. When this transient microtubule elimination was performed either immediately before or immediately after the centrifugation, the alignment of cell division was not observed. In a control, microtubules were eliminated subsequent to the centrifugation, which did not impair the alignment of the cell axis by the mechanical stimulus. This demonstrated clearly that microtubules are necessary for the sensing of this mechanical stimulus. A recent study in agarose-embedded walled cells of *Chrysanthemum* (Zhou et al. 2007) extends these findings by the interaction with the cell-wall cytoplasmic continuum. Here, cell divisions were aligned by compression force (Lintilhac and Vesecky 1984). When microtubules were removed by oryzalin prior to the treatment or when the cell-wall cytoplasmic continuum was impaired by treatment with RGD peptides, this alignment response was interrupted. In contrast, elimination of actin filaments by cytochalasin B was not effective.

5.2 *Cold Sensing*

The sensing of temperature must occur cell-autonomously. This is generally ascribed to a reduced fluidity of membranes that will alter the activity of ion channels or the balance of metabolites (Lyons 1973). For instance, overexpression of desaturases has been shown repeatedly to modify chilling sensitivity in plants (Murata et al. 1992). Since microtubules disassemble in the cold, they have long been discussed as alternative sensors for low temperatures. In fact, when microtubules were manipulated pharmacologically, this was accompanied by changes in cold hardiness (Kerr and Carter 1990). Microtubule disassembly of plants and of animals in the cold differ depending on the type of organism. Whereas mammalian microtubules disassemble at temperatures below 20°C, the microtubules from poikilothermic animals remain intact far below that temperature (Modig et al. 1994). In plants, the cold stability of microtubules is generally more pronounced than in animals, reflecting the higher developmental plasticity. However, the critical temperature where microtubule disassembly occurs varies between different plant species, which is correlated with differences in chilling sensitivity (Jian et al. 1989). The close correlation between microtubular cold sensitivity and general chilling sensitivity is supported by the observation that abscisic acid, a hormonal inducer of cold hardiness (Holubowicz and Boe 1969; Irving 1969; Rikin et al. 1975; Rikin and Richmond 1976), can stabilize cortical microtubules against low temperature (Sakiyama and Shibaoka 1990; Wang and Nick 2001). Tobacco mutants, where microtubules are more cold-stable owing to expression of an activation tag, are endowed with cold-resistant leaf expansion (Ahad et al. 2003). Conversely, destabilization of microtubules by assembly blockers such as colchicine and podophyllotoxin increased the chilling sensitivity of cotton seedlings, and this effect could be rescued by addition of abscisic acid (Rikin et al. 1980). Conversely, gibberellin, a hormone that has been shown in several species to reduce cold hardiness (Rikin et al. 1975; Irving and Lanphear 1968), renders cortical microtubules more cold-susceptible (Akashi and Shibaoka 1987).

It is possible to increase the cold resistance of an otherwise chilling-sensitive species by pre-cultivation at moderately cool temperature. Cold hardening can be detected on the level of microtubules as well. Microtubules of cold-acclimated spinach mesophyll cells coped better with the consequences of a freeze–thaw cycle (Bartolo and Carter 1991a). Microtubules are not only the target of cold stress, they seem, in addition, to participate in cold sensing itself, triggering a chain of events that culminates in increased cold hardiness. When microtubule disassembly is suppressed by Taxol, this can suppress cold hardening (Kerr and Carter 1990; Bartolo and Carter 1991b). This indicates that microtubules have to disassemble to a certain degree to trigger cold hardening. To test this hypothesis, cold hardening was followed in three cultivars of winter wheat that differed in freezing tolerance (Abdrakhamanova et al. 2003). During cultivation at 4°C, the growth rate of roots recovered progressively as a manifestation of cold hardening. In parallel, the roots acquired progressive resistance to a challenging freezing shock (-7°C) which would impair growth irreversibly in non-acclimated roots. When microtubules were monitored during cold hardening, a rapid, but transient partial disassembly was observed in cultivars that were freezing-tolerant, but not in a cultivar that was freezing-sensitive. However, when a transient disassembly was artificially generated by a pulse treatment with the antimicrotubular herbicide pronamide in the sensitive cultivar, this induced freezing tolerance. This demonstrates that a transient, partial disassembly of microtubules is necessary and sufficient to trigger cold hardening, suggesting that microtubules act as "thermometers".

Similar to mechanosensing and gravisensing, this leads to the question of whether microtubules act as susceptors or as true receptors for low temperature. The primary signal for cold perception is thought be increased membrane rigidity (Los and Murata 2004; Sangwan et al. 2001). For instance, the input of low temperature can be mimicked by chemical compounds that increase rigidity, such as dimethyl sulfoxide, whereas benzyl alcohol, a compound that increases membrane fluidity, can block cold signalling (Sangwan et al. 2001). With use of aequorin as a reporter in transgenic plants, rapid and transient increases of intracellular calcium levels in response to a cold shock were demonstrated by monitoring changes of bioluminescence (Knight et al. 1991). Pharmacological data (Monroy et al. 1993) confirmed that this calcium peak is not only a by-product of the cold response, but is also necessary to trigger cold acclimation. This peak is generated through calcium channels in conjunction with calmodulin. Calcium and calmodulin in turn are intimately linked to microtubule dynamics. Immunocytochemical data show that microtubules are decorated with calmodulin depending on the concentration of calcium (Fisher and Cyr 1993). It was further suggested that the dynamics of microtubules is regulated via a calmodulin-sensitive interaction between microtubules and microtubule-associated proteins such as the bundling protein EF-1α (Durso and Cyr 1994). However, the interaction could be even more direct, because cleavage of the carboxy terminus of maize tubulin was shown to render microtubules resistant to both low temperature and calcium (Bokros et al. 1996). When the release of calcium from intracellular pools was blocked by treatment with lithium, an inhibitor of polyphosphoinositide turnover, this resulted in increased cold stability of microtubules in spinach mesophyll (Bartolo and Carter 1992). With use of a cold-responsive reporter system it was demonstrated that disassembly of microtubules by oryzalin or treatment with the calcium ionophore A23187 could mimic the effect of low temperature, whereas the calcium channel inhibitor gadolinium or suppression of microtubule disassembly by taxol prevented the activation of this promoter by low temperature (Sangwan et al. 2001). These data favour a model where microtubules act as receptors that limit the permeability of calcium channels that are triggered by membrane rigidification. When microtubules function as modulators of calcium channel activity and when microtubule integrity is regulated through calcium/calmodulin, this would set-up a regulatory circuit capable of self-amplification: Stable microtubules that limit the activity of cold-induced voltage-dependent calcium channels would, upon disassembly, release this constraint and this would elevate the activity of the channels, resulting in an increased influx of calcium. This calcium influx, in turn, would result in further disintegration of the microtubular cytoskeleton and thus trigger by this positive-feedback the influx of additional calcium. A very small initial calcium influx might thus be amplified into a strong signal that can be easily processed by the activation of calcium-dependent signalling cascades. The resulting signalling cascade will activate cold hardening as an adaptive response to cold stress. Interestingly, microtubules will be rendered cold-stable as a consequence of this cold hardening (Pihakaski-Maunsbach and Puhakainen 1995; Abdrakhamanova et al. 2003), which, in turn, should result in reduced activity of the calcium channels that respond to membrane rigidification. Thus, microtubules would not only endow cold sensing with high sensitivity, but, in addition, would also endow it with the ability to downregulate sensitivity upon prolonged stimulation, a key requirement for any biological sensory process.

5.3 Plant Defence

The interaction of pathogens and their hosts is subject to an evolutionary arms race, where the pathogens develop various strategies to circumvent or suppress defence responses of the host, whereas the host develops various strategies to sense and attack the invading pathogen or its effector molecules. Cellular responses to elicitors include formation of cell-wall papillae around sites of pathogen penetration. The formation of these papillae is preceded by a reorganization of the cytoskeleton causing a redistribution of vesicle traffic and a cytoplasmic aggregation towards the penetration site (for reviews see Takemoto and Hardham 2004; Kobayashi and Kobayashi 2008), and a somewhat slower migration of the nucleus (for a review see Schmelzer 2002). By localized mechanical stimulation of parsley cells, it was possible to partially mimic an attack by *Phytophthora sojae* and to induce several aspects of a non-host resistance, including nuclear migration, cytoplasmic reorganization, formation of reactive oxygen species and the induction of several defence-related genes (Gus-Mayer et al. 1998). In contrast, localized application of the corresponding elicitor (pep-13) failed to induce the morphological changes, although it induced the full set of defence-related genes and the formation of reactive oxygen species. Interestingly, the elicitor completely inhibited cytoplasmic aggregation and nuclear migration in response to the mechanical stimulus. Since pep-13 induces in this system the activity of a mechanosensitive calcium channel (Zimmermann et al. 1997), it seems that chemical and mechanical signalling converge during the cytoskeletal response to pathogen attack. Neither the mechanical stimulus nor the elicitor nor their combination was able to induce hypersensitive cell death in these experiments, leading the investigators to conclude that additional chemical signals are required to obtain the complete pathogen response. This suggests an interaction between microtubules and mechanosensitive ion channels that are important for the induction of defence.

5.4 *Osmoregulation*

The ability to maintain ionic balance represents a basic capacity of all living beings. Prokaryotes have already developed osmoregulation. In plants, the mechanical tensions produced in the context of an expanding tissue have to be balanced by osmotic pressure (chapter "Osmosensing"). Microtubules seem to be directly involved in osmoadaptation. By application of osmotic stress to root tips of *Triticum turgidum*, microtubules were induced to disassemble and to reorganize into massive bundles (Komis et al. 2002). When the formation of these so-called macrotubules was suppressed by treatment with oryzalin, the protoplasts were no longer able to adapt to osmotic stress by controlled swelling and perished. A pharmacological study (Komis et al. 2006) revealed that inhibitors of phospholipase D, such as butan-1-ol and *N*-acetylethanolamine, suppressed osmotic adaptation as well as the formation of the macrotubules. In contrast, phosphatidic acid, a product of the action of phospholipase D, enhanced osmoadaptation and macrotubule formation and was able to overcome the inhibitory effect of butan-1-ol. These observations demonstrate that the microtubule response (formation of macrotubules) is essential for osmoadaptation, and that signalling through phospholipase D acts upstream of microtubules in this response.

5.5 *Division Symmetry*

A homologue of the bacterial mechanosensitive channel MscS, MSL3, localized in discrete patches in the plastid envelope and co-localized with the plastid division factor MinE, indicating an interaction with plastid division (Vitha et al. 2003). When this bona fide channel was mutated, this resulted in chloroplasts that were irregular in size, shape and partially number. Thus, these channels regulate morphogenesis and development of plastids, suggesting a functional shift from osmoregulation towards regulation of plastid morphogenesis. However, it is not only in the symmetry of organelle division where mechanosensing seems to play a role. Mechanosensing seems to be the integrator that allows the nucleus to determine the correct position prior to mitosis – this premitotic nuclear movement is an active process and is driven by the cytoskeleton. It will determine the symmetry of the ensuing cell division and thus the basic morphology of the prospective daughter cells. The discovery that overexpression or knockout of the plant-specific kinesin motor KCH1 (Frey et al. 2010), which binds to actin filaments, retards or accelerates premitotic nuclear positioning, indicates that cytoskeletal tensegrity is used to determine the correct position of the nucleus. Two principal modes are conceivable that are not necessarily mutually exclusive. Microtubules and actin filaments might transmit forces that are generated by the KCH1 motor at the perinuclear contact sites to the cortex such that the nucleus is either pulled or pushed, or both (Fig. 2c, d). Alternatively, KCH1 might simply anchor the perinuclear network at the cell cortex and move the nucleus by mutual sliding of actin filaments and microtubules in the cortical cytoplasm. From studies in yeast, filamentous fungi and a variety of animal cells, the molecular mechanisms that orient and move nuclei were found to be moderately conserved and involve as key players dynein, dynactin and other proteins at the plus ends of astral microtubules, mediating interaction with the cell cortex and actin filaments (Morris 2003; Yamamoto and Hiraoka 2003). Both repulsive and attractive forces are generated by a combination of microtubule polymerization and de-polymerization events, complemented by dynein-mediated sliding of microtubules along the cell cortex (Adames and Cooper 2000). In plants, which lack dynein and its associated proteins (Lawrence et al. 2001), the mechanisms for nuclear movement must involve fundamentally different players that are able to interact with both premitotic microtubules and actin filaments. Could KCH proteins be these missing links as functional homologues of dyneins by anchoring minus-end-directed motor activity to the cortex?

6 Evolutionary Perspective: The Cytoskeleton as a Central Integrator

This chapter summarized evidence for a cytoskeletal function in tensegral integration on both the organismal and the cellular level. We are still far from understanding the molecular set-up of tensegral sensing. But even at this stage, the first differences between the different stimulus qualities have emerged.

For mechanoperception, microtubules seem to interact with stretch-activated ion channels, probably focussing minute deformations of the membrane or changes in membrane fluidity towards specific membrane areas. Because the demonstration of mechanosensitivity by patch-clamp experiments is experimentally very problematic and prone to artefacts (Gustin et al. 1991), the molecular identity of stretch-activated channels has remained elusive in plants. When mechanosensitivity is not an intrinsic property of such channels, but is conferred by accessory structures (such as cytoskeletal tensegrity), the identification of these channels will go beyond simple expression in heterologous systems (such as frog oocytes), and will require the reconstitution of the entire structure, i.e. it will rely on synthetic biology.

The situation might be different for the sensing of gravity. Here microtubules themselves could act as primary sensors. The findings of the few experiments where the involvement of ion channels has been addressed experimentally (Ikushima and Shimmen 2005) suggest that these channels might be dispensable for gravity sensing. The necessity of microtubule turnover in the sensing of gravity indicates a true perceptive function rather than stimulus susception.

The sensing of cold seems to represent a third mechanism, where the gating of cold-sensitive channels (which probably respond to membrane rigidity as an input) is limited by microtubules. When these microtubules disassemble in response to cold, the constraints upon the activity of the ion channels will be released such that calcium can enter, which will facilitate, through interaction with calmodulin, further disassembly of microtubules and thus trigger a positive-feedback loop. In this system microtubules would play a dual function – first as a perceptive device and later as accessory structures for the perceptive channels.

Why is the cytoskeleton central for mechanical integration? The reason is probably linked to the innate properties of cytoskeletal dynamics that render microtubules and actin filaments ideal for the sensing of minute and noisy inputs. These dynamics are nonlinear and endowed with autocatalytic properties. In the cell, the abundance of monomers is limited, which means that different polymers compete for the incorporation of free monomers. For instance, in all organisms investigated so far, tubulin synthesis is tightly regulated by an elaborate system of transcriptional and post-transcriptional controls, probably to avoid the accumulation of (highly toxic) supernumerous free heterodimers (for a review see Breviario 2008). Although the term "cytoskeleton" was coined in the model of a rigid framework that stabilizes the structure of a cell, such associations are far from reality. The half-time of a plant microtubule, for example, has been estimated to be in the range of 30–60 s (Hush et al. 1990). Therefore, it is more appropriate to conceive of microtubules and actin filaments as states of dynamic equilibrium between assembly and disassembly of monomers. It is this dynamic equilibrium that provides the major source for the characteristic nonlinearity of cytoskeletal dynamics.

Interestingly, the relation between assembly and disassembly is practically never balanced –one always dominates over its antagonist. This statement is valid in both space and time: in space, because dimer addition and dispersal define a distinct polarity of each individual cytoskeletal polymer, with dimer addition dominating at the plus end and dimer dissociation dominating at the minus end; in time, because each polymer can switch between a growing state, when dimer addition at the plus end predominates over dimer dissociation at the minus end, and a shrinking state, when dimer dissociation at the minus end exceeds dimer addition at the plus end. The switch between both states is so swift and dramatic that it has been termed "microtubule catastrophe". These conversions depend on associated proteins that can increase or decrease the frequency of transition between growth and shrinkage.

Because of its nonlinear growth, the cytoskeleton is an ideal tool for developmental patterning. This has been exemplarily shown for the induction of the grey crescent in developing frog eggs. This manifestation of dorsoventrality is produced in an epigenetic process, where a gravity-dependent gradient of developmental determinants in the central yolk interacts with a second, displaced, gradient in the egg cortex (Gerhart et al. 1981). The displacement of the egg cortex is driven by microtubules and is triggered by the penetration of the sperm. The sperm induces the nucleation of microtubules that act as tracks for a kinesin-driven movement (Elinson and Rowning 1988). The movement, in turn, triggers shear forces that align the nucleation of additional microtubules in a direction parallel to the movement, whereas deviant microtubules more frequently undergo catastrophic transitions. The resulting net alignment of tracks increases the efficiency of movement and thus the aligning force. This culminates in a rapid rotation of the cortical plasma in a direction from the sperm towards the more remote equator of the egg. This movement will then cause an overlap of upper cortex with a small region of the lower core and eventually trigger inductive events that lay down the Spemann organizer.

The combination of nonlinear, autocatalytic dynamic states of individual cytoskeletal polymers with the tight control of free monomers accentuating mutual competition generates system properties that are highly relevant for sensory processes. Microtubules and actin filaments fulfil all formal criteria of a reaction–diffusion system (Turing 1952). This means that they can be understood as ideal pattern generators that are able to produce qualitatively clear, neat outputs from minute and highly noise contaminated inputs. One can model how owing to their innate dynamic properties microtubules will spontaneously self-organize in response to even a weak external factor such as gravity or mechanical fields (Tabony et al. 2004). It thus seems that nature has made ample use of these unique molecular properties to build sensory systems that are both sensitive and robust against stochastic noise and can be used to integrate even minute mechanical strains on the level of individual cells as well as on the level of entire organs.

References

Abdrakhamanova A, Wang QY, Khokhlova L, Nick P (2003) Is microtubule assembly a trigger for cold acclimation? Plant Cell Physiol 44:676–686

Adames NR, Cooper JA (2000) Microtubule interactions with the cell cortex causing nuclear movements in Saccharomyces cerevisiae. J Cell Biol 149:863–874

Ahad A, Wolf J, Nick P (2003) Activation-tagged tobacco mutants that are tolerant to antimicrotubular herbicides are cross-resistant to chilling stress. Transgenic Res 12:615–629

Akashi T, Shibaoka H (1987) Effects of gibberellin on the arrangement and the cold stability of cortical microtubules in epidermal cells of pea internodes. Plant Cell Physiol 28:339–348

Akashi T, Kawasaki S, Shibaoka H (1990) Stabilization of cortical microtubules by the cell wall in cultured tobacco cells. Effect of extensin on the cold stability of cortical microtubules. Planta 182:363–369

Akhmanova A, Steinmetz MO (2008) Tracking the ends: a dynamic protein network controls the fate of microtubule tips. Nat Rev Mol Cell Biol 9:309–322

Baluška F, Hlavačka A (2005) Plant formins come of age: something special about cross-walls. New Phytol 168:499–503

Baluška F, Jasik J, Edelmann HG, Salajová T, Volkmann D (2001) Latrunculin B-induced plant dwarfism: plant cell elongation is F-actin-dependent. Dev Biol 231:113–124

Baluška F, Šamaj J, Wojtaszek P, Volkmann D, Menzel D (2003) Cytoskeleton-plasma membrane-cell wall continuum in plants. Emerging links revisited. Plant Physiol 133:482–491

Bannigan A, Wiedemeier AMD, Williamson RE, Overall RL, Baskin TI (2006) Cortical microtubule arrays lose uniform alignment between cells and are oryzalin resistant in the *Arabidopsis* mutant, *radially swollen 6*. Plant Cell Physiol 47:949–958

Bartolo ME, Carter JV (1991a) Microtubules in the mesophyll cells of nonacclimated and cold-acclimated spinach. Plant Physiol 97:175–181

Bartolo ME, Carter JV (1991b) Effect of microtubule stabilization on the freezing tolerance of mesophyll cells of spinach. Plant Physiol 97:182–187

Bartolo ME, Carter JV (1992) Lithium decreases cold-induced microtubule depolymerization in mesophyll cells of spinach. Plant Physiol 99:1716–1718

Baskin TI (2001) On the alignment of cellulose microfibrils by cortical microtubules: a review and a model. Protoplasma 215:150–171

Baskin TI, Bivens NJ (1995) Stimulation of radial expansion in *Arabidopsis* roots by inhibitors of actomyosin and vesicle secretion but not by various inhibitors of metabolism. Planta 197:514–521

Baskin TI, Wilson JE (1997) Inhibitors of protein kinases and phosphatases alter root morphology and disorganize cortical microtubules. Plant Physiol 113:493–502

Bichet A, Desnos T, Turner S, Grandjean O, Höfte H (2001) BOTERO1 is required for normal orientation of cortical microtubules and anisotropic cell expansion in *Arabidopsis*. Plant J 25:137–148

Bisgrove SR, Lee YRJ, Liu B, Peters NT, Kropf DL (2008) The microtubule plus-end binding protein EB1 functions in root responses to touch and gravity signals in Arabidopsis. Plant Cell 20:396–410

Björkman T (1988) Perception of gravity by plants. Adv Bot Res 15:1–4

Blancaflor EB (2000) Cortical actin filaments potentially interact with cortical microtubules in regulating polarity of cell expansion in primary roots of maize (*Zea mays* L.). J Plant Growth Regul 19:406–414

Blancaflor EB, Hasenstein KH (1993) Organization of cortical microtubules in graviresponding maize roots. Planta 191:230–237

Bokros CL, Hugdahl JD, Blumenthal SSD, Morejohn LC (1996) Proteolytic analysis of polymerized maize tubulin: regulation of microtubule stability to low temperature and Ca^{2+} by the carboxyl terminus of β-tubulin. Plant Cell Environ 19:539–548

Braam J, Davis RW (1990) Rain-, wind-, and touch-induced expression of calmodulin and calmodulin-related genes in Arabidopsis. Cell 60:357–364

Breviario D (2008) Plant tubulin genes: regulatory and evolutionary aspects. Plant Cell Monogr 11:207–232

Brown RC, Lemmon BE (2007) The pleiomorphic plant MTOC: An evolutionary perspective. J Int Plant Biol 49:1142–1153

Buder J (1920) Neue phototropische Fundamentalversuche. Ber Dtsch Bot Ges 38:10–19

Buder J (1961) Der Geotropismus der Characeenrhizoide. Ber Dtsch Bot Ges 74:14–23

Burk DH, Ye ZH (2002) Alteration of oriented deposition of cellulose microfibrils by mutation of a katanin-like microtubule severing protein. Plant Cell 14:2145–2160

Burk DH, Liu B, Zhong R, Morrison WH, Ye ZH (2001) A katanin-like protein regulates normal cell wall biosynthesis and cell elongation. Plant Cell 13:807–827

Buschmann H, Fabri CO, Hauptmann M, Hutzler P, Laux T, Lloyd CW, Schäffner AR (2004) Helical growth of the Arabidopsis mutant *tortifolia1* reveals a plant-specific microtubule-associated protein. Curr Biol 14:1515–1521

Camilleri C, Azimzadeh J, Pastuglia M, Bellini C, Grandjean O, Bouchez D (2002) The Arabidopsis *TONNEAU2* gene encodes a putative novel PP2A regulatory subunit essential for the control of cortical cytoskeleton. Plant Cell 14:833–845

Campanoni P, Blasius B, Nick P (2003) Auxin transport synchronizes the pattern of cell division in a tobacco cell line. Plant Physiol 133:1251–1260

Canut H, Carrasco A, Galaud J-P, Cassan C, Bouyssou H, Vita N, Ferrara P, Pont-Lezica R (1998) High affinity RGD-binding sites at the plasma membrane of *Arabidopsis thaliana* links the cell wall. Plant J 16:63–71

Cárdenas L, Lovy-Wheeler A, Kunkel JG, Hepler PK (2008) Pollen tube growth oscillations and intracellular calcium levels are reversibly modulated by actin polymerization. Plant Physiol 146:1611–1621

Chan J, Calder G, Fox S, Lloyd C (2007) Cortical microtubule arrays undergo rotary movements in *Arabidopsis* hypocotyl epidermal cells. Nat Cell Biol 9:171–175

Collings DA (2008) Crossed-wires: interactions and cross-talk between the microtubule and micro. lament networks in plants. Plant Cell Monogr 11:47–79

Collings DA, Lill AW, Himmelspach R, Wasteneys GO (2006) Hypersensitivity to cytoskeletal antagonists demonstrates microtubule-microfilament cross-talk in the control of root elongation in *Arabidopsis thaliana*. New Phytol 170:275–290

Dhonukshe P, Mathur J, Hülskamp M, Gadella TWJ (2005) Microtubule plus-ends reveal essential links between intracellular polarization and localized modulation of endocytosis during division-plane establishment in plant cells. BMC Biol 3:11

Ding JP, Pickard BG (1993) Mechanosensory calcium-selective cation channels in epidermal cells. Plant J 3:83–110

Durso NA, Cyr RJ (1994) A calmodulin-sensitive interaction between microtubules and a higher plant homolog of elongation factor 1α. Plant Cell 6:893–905

Eckert BS, Yeagle PL (1988) Acrylamide treatment of PtK1 cells causes dephosphorylation of keratin polypeptides. Cell Motil Cytoskelet 11:24–30

Edwards ES, Roux SJ (1994) Limited period of graviresponsiveness in germinating spores of *Ceratopteris richardii*. Planta 195:150–152

Edwards ES, Roux SJ (1997) The influence of gravity and light on developmental polarity of single cells of *Ceratopteris richardii* gametophytes. Biol Bull 192:139–140

Elinson RP, Rowning B (1988) Transient array of parallel microtubules in frog eggs: potential tracks for a cytoplasmic rotation that specifies the dorso-ventral axis. Dev Biol 128:185–197

Fisher DD, Cyr RJ (1993) Calcium levels affect the ability to immunolocalize calmodulin to cortical microtubules. Plant Physiol 10:543–551

Frey N, Klotz J, Nick P (2009) Dynamic bridges – a calponin-domain kinesin from rice links actin filaments and microtubules in both cycling and non-cycling cells. Plant Cell Physiol 50: 1493–1506

Frey N, Klotz J, Nick P (2010) A kinesin with calponin-homology domain is involved in premitotic nuclear migration. J Exp Bot 61:3423–3437

Funada R (2008) Microtubules and the control of wood formation. Plant Cell Monogr 11:83–119

Furutani I, Watanabe Y, Prieto R, Masukawa M, Suzuki K, Naoi K, Thitamadee S, Shikanai T, Hashimoto T (2000) The *SPIRAL* genes are required for directional control of cell plates elongation in *Arabidopsis thaliana*. Development 127:4443–4453

Gardiner JC, Harper JDI, Weerakoon ND, Collings DA, Ritchie S, Gilroy S, Cyr RJ, Marc J (2001) A 90-kD phospholipase D from tobacco binds to microtubules and the plasma membrane. Plant Cell 13:2143–2158

Geiger B, Bershadsky A (2001) Assembly and mechanosensory function of focal contacts. Curr Opin Cell Biol 13:584–592
Gens JS, Fujiki M, Pickard BG (2000) Arabinogalactan protein and wall-associated kinase in a plasmalemmal reticulum with specialized vertices. Protoplasma 212:115–134
Gerhart J, Ubbeles G, Black S, Hara K, Kirschner M (1981) A reinvestigation of the role of the grey crescent in axis formation in *Xenopus laevis*. Nature 292:511–516
Giancotti FG, Ruoslahti E (1999) Integrin signaling. Science 285:1028–1032
Giani S, Qin X, Faoro F, Breviario D (1998) In rice, oryzalin and abscisic acid differentially affect tubulin mRNA and protein levels. Planta 205:334–341
Giddings TH, Staehelin A (1988) Spatial relationship between microtubules and plasmamembrane rosettes during the deposition of primary wall microfibrils in *Closterium spec*. Planta 173: 22–30
Giddings TH, Staehelin A (1991) Microtubule-mediated control of microfibril deposition. A re-examination of the hypothesis. In: Lloyd CW (ed) The cytoskeletal basis of plant growth and form. Academic, London, pp 85–99
Gierer A (1981) Generation of biological patterns and form: some physical, mathematical, and logical aspects. Progr Biophys Mol Biol 37:1–47
Gittes F, Mickey B, Nettleton J, Howard J (1993) Flexual rigidity of microtubules and actin filaments measured from thermal fluctuations in shape. J Cell Biol 120:923–934
Godbolé R, Michalke W, Nick P, Hertel R (2000) Cytoskeletal drugs and gravity-induced lateral auxin transport in rice coleoptiles. Plant Biol 2:176–181
Goebel K (1908) Einleitung in die experimentelle Morphologie der Pflanzen. Teubner, Leipzig, pp 218–251
Goode BL, Drubin DG, Barnes G (2000) Functional cooperation between the microtubule and actin cytoskeletons. Curr Opin Cell Biol 12:63–71
Gossot O, Geitmann A (2007) Pollen tube growth: coping with mechanical obstacles involves the cytoskeleton. Planta 226:405–416
Grabski S, Schindler M (1996) Auxins and cytokinins as antipodal modulators of elasticity within the actin network of plant cells. Plant Physiol 110:965–970
Grabski S, Arnoys E, Busch B, Schindler M (1998) Regulation of actin tension in plant cells by kinases and phosphatases. Plant Physiol 116:279–290
Green PB (1962) Mechanism for plant cellular morphogenesis. Science 138:1404–1405
Green PB (1980) Organogenesis – a biophysical view. Annu Rev Plant Physiol 31:51–82
Gus-Mayer S, Naton B, Hahlbrock K, Schmelzer E (1998) Local mechanical stimulation induces components of the pathogen defense response in parsley. Proc Natl Acad Sci USA 95:8398–8403
Gustin MC, Sachs F, Sigurdson WJ, Ruknudin A, Bowman C (1991) Technical comments. Single channel mechanosensitive currents. Science 253:1195–1197
Gutjahr C, Nick P (2006) Acrylamide inhibits gravitropism and destroys microtubules in rice coleoptiles. Protoplasma 227:211–222
Haberland G (1900) Über die Perzeption des geotropischen Reizes. Ber Dtsch Bot Ges 18:261–272
Hamada T (2007) Microtubule-associated proteins in higher plants. J Plant Res 120:79–98
Hamant O, Heisler MG, Jönsson H, Krupinski P, Uyttewaal M, Bokov P, Corson F, Sahlin P, Boudaoud A, Meyerowitz EM, Couder Y, Traas J (2008) Developmental patterning by mechanical signals in Arabidopsis. Science 322:1650–1655
Hardham AR, Green PB, Lang JM (1980) Reorganization of cortical microtubules and cellulose deposition during leaf formation of *Graptopetalum paraguayense*. Planta 149:181–195
Hasezawa S, Nozaki H (1999) Role of cortical microtubules in the orientation of cellulose microfibril deposition in higher-plant cells. Protoplasma 209:98–104
Hashimoto T, Kato T (2006) Cortical control of plant microtubules. Curr Opin Plant Biol 9:5–11
Heath IB (1974) A unified hypothesis for the role of membrane bound enzyme complexes and microtubules in plant cell wall synthesis. J Theor Biol 48:445–449

Hepler PK, Vidali L, Cheung AY (2001) Polarized cell growth in higher plants. Annu Rev Cell Dev Biol 17:159–187
Hertel R, Friedrich U (1973) Abhängigkeit der geotropischen Krümmung der *Chara*-Rhizoide von der Zentrifugalbeschleunigung. Z Pflanzenphysiol 70:173–184
Himmelspach R, Wymer CL, Lloyd CW, Nick P (1999) Gravity-induced reorientation of cortical microtubules observed in vivo. Plant J 18:449–453
Himmelspach R, Nick P (2001) Gravitropic microtubule reorientation can be uncoupled from growth. Planta 212:184–189
Hodick D (1994) Negative gravitropism in *Chara* protonemata: a model integrating the opposite gravitropic responses of protonemata and rhizoids. Planta 195:43–49
Holubowicz T, Boe AA (1969) Development of cold hardiness in apple seedlings treated with gibberellic acid and abscisic acid. J Am Soc Hortic Sci 94:661–664
Holweg C, Süßlin C, Nick P (2004) Capturing in-vivo dynamics of the actin cytoskeleton. Plant Cell Physiol 45:855–863
Hush JM, Hawes CR, Overall RL (1990) Interphase microtubule re-orientation predicts a new cell polarity in wounded pea roots. J Cell Sci 96:47–61
Igarashi H, Orii H, Mori H, Shimmen T, Sonobe S (2000) Isolation of a novel 190 kDa protein from tobacco BY-2 cells: possible involvement in the interaction between actin filaments and microtubules. Plant Cell Physiol 41:920–931
Ikushima T, Shimmen T (2005) Mechano-sensitive orientation of cortical microtubules during gravitropism in azuki bean epicotyls. J Plant Res 118:19–26
Ingber DE (2003a) Tensegrity I: cell structure and hierarchical systems biology. J Cell Sci 116:1157–1173
Ingber DE (2003b) Tensegrity II: how structural networks influence cellular information processing networks. J Cell Sci 116:1397–1408
Irving RM (1969) Characterization and role of an endogenous inhibitor in the induction of cold hardiness in *Acer negundo*. Plant Physiol 44:801–805
Irving RM, Lanphear FO (1968) Regulation of cold hardiness in *Acer negundo*. Plant Physiol 43:9–13
Jacob F (1977) Evolution and tinkering. Science 196:1161–1166
Jaffe MJ, Leopold AC, Staples RA (2002) Thigmo responses in plants and fungi. Am J Bot 89:375–382
Janmey PA, Weitz DA (2004) Dealing with mechanics: mechanisms of force transduction in cells. Trends Biochem Sci 29:364–370
Jian LC, Sun LH, Lin ZP (1989) Studies on microtubule cold stability in relation to plant cold hardiness. Acta Bot Sin 31:737–741
Jones RS, Mitchell CA (1989) Calcium ion involvement in growth inhibition of mechanically stressed soybean *Glycine max* seedlings. Physiol Plant 76:598–602
Kakimoto T, Shibaoka H (1987) Actin filaments in the preprophase band and phragmoplast of tobacco cells. Protoplasma 140:151–156
Karki S, Holzbaur EL (1999) Cytoplasmic dynein and dynactin in cell division and intracellular transport. Curr Opin Cell Biol 1:45–53
Katsuta J, Shibaoka H (1988) The roles of the cytoskeleton and the cell wall in nuclear positioning in tobacco BY-2 cells. Plant Cell Physiol 29:403–413
Kell A, Glaser RW (1993) On the mechanical and dynamic properties of plant-cell membranes: their role in growth, direct gene transfer and protoplast fusion. J Theor Biol 160:41–62
Kennard JL, Cleary AL (1997) Pre-mitotic nuclear migration in subsidiary mother cells of *Tradescantia* occurs in G1 of the cell cycle and requires F-actin. Cell Motil Cytoskelet 36:55–67
Kerr GP, Carter JV (1990) Relationship between freezing tolerance of root-tip cells and cold stability ofmicrotubules in rye (*Secale cereale* L. Cv. Puma). Plant Physiol 93:77–82
Kimura S, Laosinchai W, Itoh T, Cui X, Linder CR, Brown RM (1999) Immunogold labeling of rosette terminal cellulose-synthesizing complexes in the vascular plant *Vigna angularis*. Plant Cell 11:2075–2086

Kishimoto U (1968) Response of *Chara* internodes to mechanical stimulation. Ann Rep Biol Works Fac Sci Osaka Univ 16:61–66

Knight MR, Campbell AK, Smith SM, Trewavas AJ (1991) Transgenic plant aequorin reports the effects of touch and cold-shock and elicitors on cytoplasmic calcium. Nature 352:524–526

Kobayashi I, Kobayashi Y (2008) Microtubules and pathogen defence. Plant Cell Monogr 11:121–140

Komis G, Apostolakos P, Galatis B (2002) Hyperosmotic stress induces formation of tubulin macrotubules in root-tip cells of *Triticum turgidum*: their probable involvement in protoplast volume control. Plant Cell Physiol 43:911–922

Komis G, Quader H, Galatis B, Apostolakos P (2006) Macrotubule-dependent protoplast volume regulation in plasmolysed root-tip cells of *Triticum turgidum*: involvement of phospholipase D. New Phytol 171:737–750

Kung C (2005) A possible unifying principle for mechanosensation. Nature 436:647–654

Kutschera U (2008) The outer epidermal wall: design and physiological role of a composite structure. Ann Bot 101:615–621

Kuznetsov OA, Hasenstein KH (1996) Magnetophoretic induction of root curvature. Planta 198:87–94

Lawrence CJ, Morris NR, Meagher RB, Dawe RK (2001) Dyneins have run their course in plant lineage. Traffic 2:362–363

Ledbetter MC, Porter KR (1963) A microtubule in plant cell fine structure. J Cell Biol 12:239–250

Lintilhac PM, Vesecky TB (1984) Stress-induced alignment of division plane in plant tissues grown in vitro. Nature 307:363–364

Lloyd CW, Traas JA (1988) The role of F-actin in determining the division plane of carrot suspension cells. Drug Stud Dev 102:211–221

Los DA, Murata N (2004) Membrane fluidity and its roles in the perception of environmental signals. Biochim Biophys Acta 1666:142–157

Lucas J, Shaw SL (2008) Cortical microtubule arrays in the *Arabidopsis* seedling. Curr Opin Plant Biol 11:94–98

Lyons JM (1973) Chilling injury in plants. Annu Rev Plant Physiol 24:445–466

Maisch J, Nick P (2007) Actin is involved in auxin-dependent patterning. Plant Physiol., Plant Physiol 143:1695–1704.

McClinton RS, Sung ZR (1997) Organization of cortical microtubules at the plasma membrane in *Arabidopsis*. Planta 201:252–260

Modig C, Strömberg E, Wallin M (1994) Different stability of posttranslationally modified brain microtubules isolated from cold-temperate fish. Mol Cell Biochem 130:137–147

Monroy AF, Sarhan F, Dhindsa RS (1993) Cold-induced changes in freezing tolerance, protein phosphorylation, and gene expression. Plant Physiol 102:1227–1235

Morris NR (2003) Nuclear positioning: the means is at the ends. Curr Opin Cell Biol 15:54–59

Moseley JB, Bartolini F, Okada K, Wen Y, Gundersen GG, Goode BL (2007) Regulated binding of adenomatous polyposis coli protein to actin. J Biol Chem 282:12661–12668

Mulder B, Schell J, Emons AM (2004) How the geometrical model for plant cell wall formation enables the production of a random texture. Cellulose 11:395–401

Murata T, Wada M (1991) Effects of centrifugation on preprophase-band formation in *Adiantum* protonemata. Planta 183:391–398

Murata N, Ishizaki-Nishizawa O, Higashi H, Tasaka Y, Nishida I (1992) Genetically engineered alteration in chilling sensitivity of plants. Nature 356:710–713

Nemec B (1900) Über die Art der Wahrnehmung des Schwerkraftreizes bei den Pflanzen. Ber Dtsch Bot Ges 18:241–245

Nick P (2008a) Control of cell axis. Plant Cell Monogr 11:3–46

Nick P (2008b) Microtubules as sensors for abiotic stimuli. Plant Cell Monogr 11:175–203

Nick P, Furuya M (1996) Buder revisited – cell and organ polarity during phototropism. Plant Cell Environ 19:1179–1187

Nick P, Schäfer E, Hertel R, Furuya M (1991) On the putative role of microtubules in gravitropism of maize coleoptiles. Plant Cell Physiol 32:873–880

Nick P, Yatou O, Furuya M, Lambert AM (1994) Auxin-dependent microtubule responses and seedling development are affected in a rice mutant resistant to EPC. Plant J 6:651–663

Nick P, Godbolé R, Wang QY (1997) Probing rice gravitropism with cytoskeletal drugs and cytoskeletal mutants. Biol Bull 192:141–143

Nick P, Han M, An G (2009) Auxin stimulates its own transport by actin reorganization. Plant Physiol 151:155–167

Niklas KJ (1992) Plant biomechanics. An engineering approach to plant form and function. University of Chicago Press, Chicago

Orr AW, Helmke BP, Blackman BR, Schwartz MA (2006) Mechanisms of mechanotransduction. Dev Cell 10:11–20

Panteris E (2008) Cortical actin filaments at the division site of mitotic plant cells: a reconsideration of the 'actin-depleted zone'. New Phytol 179:334–341

Paredez AR, Somerville CR, Ehrhardt DW (2006) Visualization of cellulose synthase demonstrates functional association with microtubules. Science 312:1491–1495

Parthasarathy MV, Perdue TD, Witztum A, Alvernaz J (1985) Actin network as a normal component of the cytoskeleton in many vascular plant cells. Am J Bot 72:1318–1323

Pickard BG (2008) "Second extrinsic organizational mechanism" for orienting cellulose: modeling a role for the plasmalemmal reticulum. Protoplasma 233:7–29

Pickard BG, Fujiki M (2005) Ca^{2+} pulsation in BY-2 cells and evidence for control of mechanosensory Ca^{2+}-selective channels by the plasmalemmal reticulum. Funct Plant Biol 32:863–879

Pihakaski-Maunsbach K, Puhakainen T (1995) Effect of cold exposure on cortical microtubules of rye (*Secale cereale*) as observed by immunocytochemistry. Physiol Plant 93:563–571

Preston RD (1988) Cellulose-microfibril-orienting mechanisms in plant cell walls. Planta 174:67–74

Preuss ML, Kovar DR, Lee YR, Staiger CJ, Delmer DP, Liu B (2004) A plant-specific kinesin binds to actin microfilaments and interacts with cortical microtubules in cotton fibers. Plant Physiol 136:3945–3955

Rawitscher F (1932) Der Geotropismus der Pflanzen. Fischer, Jena

Richardson D, Simmons M, Reddy A (2006) Comprehensive comparative analysis of kinesins in photosynthetic eukaryotes. BMC Genomics 7:18

Rikin A, Richmond AE (1976) Amelioration of chilling injuries in cucumber seedlings by abscisic acid. Physiol Plant 38:95–97

Rikin A, Waldman M, Richmond AE, Dovrat A (1975) Hormonal regulation of morphogenesis and cold resistance. I. Modifications by abscisic acid and gibberellic acid in alfalfa (*Medicago sativa* L.) seedlings. J Exp Bot 26:175–183

Rikin A, Atsmon D, Gitler C (1980) Chilling injury in cotton (*Gossypium hirsutum* L.): effects of antimicrotubular drugs. Plant Cell Physiol 21:829–837

Riveline D, Zamir E, Balaban NQ, Schwarz US, Ishizaki T, Narumiya S, Kam Z, Geiger B, Bershadsky AD (2001) Focal contacts as mechanosensors: externally applied local mechanical force induces growth of focal contacts by an mDia1-dependent and ROCK-independent mechanism. J Cell Biol 153:1175–1185

Robby T (1996) A new architecture. Yale Academic Press, New Haven

Rodriguez OC, Schaefer AW, Mandato CA, Forscher P, Bement WM, Waterman-Storer CM (2003) Conserved microtubule-actin interactions in cell movement and morphogenesis. Nat Cell Biol 5:599–609

Sachs J (1880) Stoff und Form der Pflanzenorgane. Arb Bot Inst Würzburg 2:469–479

Sachs F, Morris CE (1998) Mechanosensitive ion channels in nonspecialized cells. Rev Physiol Biochem Pharmacol 132:1–77

Sainsbury F, Collings DA, Mackun K, Gardiner J, Harper JDI, Marc J (2008) Developmental reorientation of transverse cortical microtubules to longitudinal directions: a role for actomyosin-based streaming and partial microtubule-membrane detachment. Plant J 56:116–131

Sakiyama M, Shibaoka H (1990) Effects of abscisic acid on the orientation and cold stability of cortical microtubules in epicotyl cells of the dwarf pea. Protoplasma 157:165–171
Samuels AL, Giddings TH, Staehelin LA (1995) Cytokinesis in tobacco BY-2 and root tip cells – a new model of cell plate formation in higher plants. J Cell Biol 130:1345–1357
Sandblad L, Busch KE, Tittmann P, Gross H, Brunner D, Hoenger A (2006) The *Schizosaccharomyces pombe* EB1 homolog Mal3p binds and stabilizes the microtubule lattice seam. Cell 127:1415–1424
Sangwan V, Foulds I, Singh J, Dhindsa RS (2001) Cold-activation of *Brassica napus BN115* promoter is mediated by structural changes in membranes and cytoskeleton, and requires Ca^{2+} influx. Plant J 27:1–12
Sano T, Higaki T, Oda Y, Hayashi T, Hasezawa S (2005) Appearance of actin microfilament 'twin peaks' in mitosis and their function in cell plate formation, as imaged in tobacco BY-2 cells expressing GFP-fimbrin. Plant J 44:595–605
Sato Y, Kadota A, Wada M (1999) Mechanically Induced Avoidance Response of Chloroplasts in Fern Protonemal Cells. Plant Physiol 121:37–44
Schmelzer E (2002) Cell polarization, a crucial process in fungal defence. Trends Plant Sci 7:411–415
Schmit AC, Nick P (2008) Microtubules and the evolution of mitosis. Plant Cell Monogr 11: 1500 233–266
Schwuchow J, Sack FD, Hartmann E (1990) Microtubule disruption in gravitropic protonemata of the moss *Ceratodon*. Protoplasma 159:60–69
Seagull R (1990) The effects of microtubule and microfilament disrupting agents on cytoskeletal arrays and wall deposition in developing cotton fibers. Protoplasma 159:44–59
Sedbrook JC, Kaloriti D (2008) Microtubules, MAPs and plant directional cell expansion. Trends Plant Sci 13:303–310
Shibaoka T (1966) Action potentials in plant organs. Symp Soc Exp Biol 20:165–184
Sievers A, Schröter K (1971) Versuch einer Kausalanalyse der geotropischen Reaktionskette im *Chara*-Rhizoid. Planta 96:339–353
Sonobe S, Shibaoka H (1989) Cortical fine actin filaments in higher plant cells visualized by rhodamine-phalloidin after pretreatment with m-maleimidobenzoyl-N-hydrosuccinimide ester. Protoplasma 148:80–86
Tabony J, Glade N, Papaseit C, Demongeot J (2004) Microtubule self-organization as an example of the development of order in living systems. J Biol Phys Chem 4:50–63
Takemoto D, Hardham AR (2004) The cytoskeleton as a regulator and target of biotic interactions in plants. Plant Physiol 136:3864–3876
Tamura K, Nakatani K, Mitsui H, Ohashi Y, Takahashi H (1999) Characterization of katD, a kinesin-like protein gene specifically expressed in floral tissues of *Arabidopsis thaliana*. Gene 230:23–32
Taylor DP, Leopold AC (1992) Offset of gravitropism in maize roots by low temperature. ASGSB Bull 6:75
Telewski FW (2006) A unified hypothesis of mechanoperception in plants. Am J Bot 93:1466–1476
Thimann KV, Reese K, Nachmikas VT (1992) Actin and the elongation of plant cells. Protoplasma 171:151–166
Thitamadee S, Tuchihara K, Hashimoto T (2002) Microtubule basis for left-handed helical growth in *Arabidopsis*. Nature 417:193–196
Thomas DDS, Dunn DM, Seagull RW (1977) Rapid cytoplasmic responses of oat coleoptiles to cytochalasin B, auxin, and colchicine. Can J Bot 55:1797–1800
Thompson DW (1959) On growth and form. Cambridge University Press, Cambridge, pp 465–644
Tirnauer JS, Bierer BE (2000) EB1 proteins regulate microtubule dynamics, cell polarity, and chromosome stability. J Cell Biol 149:761–766
Toriyama H, Jaffe MJ (1972) Migration of calcium and its role in the regulation of seismonasty in the motor cell of *Mimosa pudica* D. Plant Physiol 49:72–81
Traas J, Bellini C, Nacry P, Kronenberger J, Bouchez D, Caboche M (1995) Normal differentiation patterns in plants lacking microtubular preprophase bands. Nature 375:676–677

Tsvetkov AS, Samsonov A, Akhmanova A, Galjart N, Popov SV (2007) Microtubule-binding proteins CLASP1 and CLASP2 interact with actin filaments. Cell Motil Cytoskelet 64:519–530

Turing AM (1952) The chemical basis of morphogenesis. Philos Trans R Soc Lond Ser B 237:37–72

Vantard M, Leviliiers N, Hill AM, Adoutte A, Lambert AM (1990) Incorporation of *Paramecium* axonemal tubulin into higher plant cells reveals functional sites of microtubule assembly. Proc Natl Acad Sci USA 87:8825–8829

Vitha S, Froehlich JE, Koksharova O, Pyke KA, van Erp H, Osteryoung KW (2003) ARC6 is a J-domain plastid division protein and an evolutionary descendant of the cyanobacterial cell division protein Ftn2. Plant Cell 15:1918–1933

Vöchting H (1878) Über Organbildung im Pflanzenreich. Cohen, Bonn

Vogelmann TC, Bassel AR, Miller JH (1981) Effects of microtubule-inhibitors on nuclear migration and rhizoid formation in germinating fern spores (*Onoclea sensibilis*). Protoplasma 109:295–316

Voigt B, Timmers ACJ, Šamaj J, Müller J, Baluška F, Menzel D (2005) GFP-FABD2 fusion construct allows *in vivo* visualization of the dynamic actin cytoskeleton in all cells of *Arabidopsis* seedlings. Eur J Cell Biol 84:595–608

Walker LM, Sack FD (1990) Amyloplasts as possible statoliths in gravitropic protonemata of the moss *Ceratodon purpureus*. Planta 181:71–77

Waller F, Nick P (1997) Response of actin microfilaments during phytochrome-controlled growth of maize seedlings. Protoplasma 200:154–162

Waller F, Riemann M, Nick P (2002) A role for actin-driven secretion in auxin-induced growth. Protoplasma 219:72–81

Wang QY, Nick P (1998) The auxin response of actin is altered in the rice mutant *Yin-Yang*. Protoplasma 204:22–33

Wang QY, Nick P (2001) Cold acclimation can induce microtubular cold stability in a manner distinct from abscisic acid. Plant Cell Physiol 42:999–1005

Wang YS, Motes CM, Mohamalawari DR, Blancaflor EB (2004) Green fluorescent protein fusions to *Arabidopsis* fimbrin 1 for spatio-temporal imaging of F-actin dynamics in roots. Cell Motil Cytoskelet 59:79–93

Wang X, Zhua L, Liu B, Wang C, Jin L, Zhao Q, Yuan M (2007) Arabidopsis microtubule-associated protein18 functions in directional cell growth by destabilizing cortical microtubules. Plant Cell 19:877–889

Wasteneys GO (2004) Progress in understanding the role of microtubules in plant cells. Curr Opin Plant Biol 7:651–660

Wasteneys GO, Galway ME (2003) Remodeling the cytoskeleton for growth and form: an overview with some new views. Annu Rev Plant Biol 54:691–722

Whittington AT, Vugrek O, Wei KJ, Hasenbein NG, Sugimoto K, Rashbrooke MC, Wasteneys GO (2001) MOR1 is essential for organizing cortical microtubules in plants. Nature 411:610–613

Wiesler B, Wang QY, Nick P (2002) The stability of cortical microtubules depends on their orientation. Plant J 32:1023–1032

Wymer C, Wymer SA, Cosgrove DJ, Cyr RJ (1996) Plant cell growth responds to external forces and the response requires intact microtubules. Plant Physiol 110:425–430

Xu T, Qu Z, Yang X, Qin X, Xiong J, Wang Y, Ren D, Liu G (2009) A cotton kinesin GhKCH2 interacts with both microtubules and microfilaments. Biochem J 421:171–180

Yamamoto A, Hiraoka Y (2003) Cytoplasmic dynein in fungi: insights from nuclear migration. J Cell Sci 116:4501–4512

Zandomeni K, Schopfer P (1994) Mechanosensory microtubule reorientation in the epidermis of maize coleoptiles subjected to bending stress. Protoplasma 182:96–101

Zhou J, Wang B, Li Y, Wang Y, Zhu L (2007) Responses of *Chrysanthemum* cells to mechanical stimulation require intact microtubules and plasma membrane-cell wall adhesion. J Plant Growth Regul 26:55–68

Zimmermann W (1965) Die Telomtheorie. Fischer, Stuttgart

Zimmermann S, Nürnberger T, Frachisse JM, Wirtz W, Guern J, Hedrich R, Scheel D (1997) Receptor-mediated activation of a plant Ca^{2+}-permeable ion channel involved in pathogen defense. Proc Natl Acad Sci USA 94:2751–2755

Intracellular Movements: Integration at the Cellular Level as Reflected in the Organization of Organelle Movements

Weronika Krzeszowiec and Halina Gabryś

Abstract This chapter presents updated information on organelle movements in plant cells, with attention focused on photosynthesizing cells. The first part summarizes current knowledge on cytoskeletal and motor proteins which integrate the intracellular communication and transport phenomena. Next, the relationship between cytoplasmic streaming and organelle motility is discussed, including recent models of hydrodynamic flow in plant cells. Subsequently, organelle movements are characterized, starting from nonphotosynthesizing organelles: Golgi network, peroxisomes, mitochondria and nuclei, followed by the light-controlled redistribution of chloroplasts. Recent evidence pertaining to the mechanisms is presented, with a special focus on the involvement of the cytoskeletal elements/motor proteins.

1 Introduction

Plant cells have to be divided by a complicated system of membranes into discrete compartments to allow for the simultaneous occurrence of different, often antagonistic, metabolic processes. Having said this, tight integration and control of these processes are essential to the survival of the cell, and consequently, to the growth and development of the organism. Such integration requires highly efficient transport of metabolites. It also requires signaling, not only between intracellular compartments but also between cells. The movement of organelles largely facilitates both transport and signaling, and thereby represents a crucial part of this integration.

In this overview of organelle movements, attention will be focused on interphase, photosynthesizing plant cells. Special, fast growing cells, e.g., root hairs, are dealt with in chapter "Generating a Cellular Protuberance. Mechanics of Tip

W. Krzeszowiec and H. Gabryś (✉)
Department of Plant Biotechnology, Faculty of Biochemistry, Biophysics and Biotechnology, Jagiellonian University, Gronostajowa 7, 30-387 Krakow, Poland
e-mail: weronika.krzeszowiec@uj.edu.pl; halina.gabrys@uj.edu.pl

P. Wojtaszek (ed.), *Mechanical Integration of Plant Cells and Plants*,
Signaling and Communication in Plants 9, DOI 10.1007/978-3-642-19091-9_4,

© Springer-Verlag Berlin Heidelberg 2011

Growth." First, the cytoskeleton involved in the intracellular transport is described. Second, the motor proteins are presented. And third, organelle movement is characterized, including the current, more sophisticated discussion of the mechanisms in the light of the most up-to-date evidence.

A much more advanced knowledge of animal cells is often used to explain phenomena occurring in plant cells. The application of this method to describe organelle movements requires caution because of major dissimilarities between these two types of cells. In spite of the obvious difference caused by the existence of the cell wall, problems appear at different levels (as discussed by Wasteneys and Galway 2003): First of all, protein domains that are known from animal, fungal, or yeast cells, very often exist in plants in particular combinations creating unique proteins. This ambiguity may lead to errors in ascribing roles to putative proteins deduced from sequences present in plant genomes. Secondly, diverse proteins are employed in similar phases of organismal growth. Therefore, an expression pattern established for an organism belonging to one kingdom is inappropriate for a representative of the other. Thirdly, some gene families have not been found in sequenced plant genomes, whereas other gene families have many more members in plants than in animals. Further differences can also be found at higher levels of organization. For example, in animal cells microtubules are anchored at discrete points, i.e., sites of microtubule organization, which results in radial growth. In plant cells, microtubules are dispersed in the cytoplasm, due to the absence of centrosomes. Moreover, the main tracks for organelle movement are provided by microfilaments in plant cells, whereas the dominant mechanism for organelle trafficking involves microtubules in animal cells.

2 Tracks and Motors

2.1 A Detailed Look at Scaffolding

If transport inside a cell were limited only to diffusion it would be much too slow to sustain the processes essential to a healthy living cell. Easy, active, and highly orchestrated transport is indispensable. Almost all intracellular movements depend on cytoskeleton – protein polymers which compose a highly dynamic scaffolding.

2.1.1 The Actin Cytoskeleton

The actin cytoskeleton is a complex, three-dimensional assembly of monomeric actin and of a large group of actin-binding proteins (ABP). Actins make up between 1 and 10% of all the proteins in eukaryotic cells [1–2% in tobacco suspension cells as reported by Wang et al. (2005), 5–10% in maize and field poppy reported by Gibbon et al. (1999); Sowman et al. (2002)]. Actins occur in the cell as monomers called G-actins (globular actins) or as polymers – F-actins (filamentous actins).

G-actin consists of a single polypeptide chain of about 374 amino acids; it has an isoelectric point of approximately 5.4. The molecular mass of a monomer extracted from tomato is 42 kDa (Vahey et al. 1982) and that of pea tendrils is 43 kDa (Ma and Yen 1989). The actin monomer shows similarities with hexokinase. For details of X-ray structure, see Kabsch et al. (1990). Each monomer is composed of two domains, and each domain contains two subdomains. The monomer binds one molecule of ATP or ADP and a calcium or magnesium ion. The nucleotide and the ion enter into a deep cleft between these domains. Monomers dominate at low salt concentrations. An increase in salt concentration is followed by the hydrolysis of ATP and the rapid formation of filaments. Filamentous actin (F-actin) is a helical structure with 13 actin monomers per 6 (left-handed) turns and a repeat of about 360 Å. The distance added per G-actin is about 2.7 nm with a rotation of 166°. The maximum diameter of the filament is about 90–95 Å. An atomic model of F-actin has been published by Holmes et al. (1990). For an overview of actin localization and function, see the old but excellent reviews by Staiger and Schliwa (1987). All monomers are oriented in the same direction, thus the whole filament shows polarity. The filament ends are referred to as fast and slow growing ends or plus (barbed) and minus (pointed) ends, respectively. A critical concentration of actin is required to start filament assembly. An equal number of monomers is added to or dissociated from the same end at such a concentration. A net polymerization takes place when the concentration of G-actin is higher than the critical concentration and a net depolymerization occurs when it is lower. The critical concentration characteristic of the minus (slow growth) end is higher than that of the plus end. From the classical point of view, actin dynamics results from a treadmilling process in which the number of subunits added at the plus end is equal to the number disassembled at the minus end. A filament turnover starts with the addition of ATP-G-actin at the plus end. Next, hydrolysis of ATP occurs within the filament, followed by a dissociation of the ADP-G-actin at the minus end. Finally, the ADP associated with the G-actin is exchanged with ATP. The in vivo turnover of a filament is 100–200 times faster than one in vitro, which shows that the cellular process is facilitated by ABPs (Chen et al. 2000). Staiger et al. (2009) proposed a more stochastic mechanism of actin dynamics in plant cells. They examined the dynamics of the cortical actin array in hypocotyl epidermal cells from *Arabidopsis* seedlings. Interestingly, the maximum elongation rate was 1.7 $\mu m\ s^{-1}$. In consequence, filaments were able to span over the cell diameter within 10–20 s. The frequency of the breakage of actin filaments has also been measured. The frequency of severing, defined as breaks $\mu m^{-1} s^{-1}$ was 0.01. The authors suggested that a stochastic process explains better than treadmilling what takes place during elongation. Severing, rather than depolymerization also better explains what happens during shrinkage.

As in animal cells, bundles of actin exist in plant cells (Thomas et al. 2009). Four types of actin-bundling proteins have been found in plants: fimbrins, formins, villins, and LIM domain-containing proteins. Fimbrins belong to the fimbrin/plastin class (Staiger and Hussey 2004). One protein possesses two actin-binding domains (ABP) that allow the crosslink of two actin filaments by a single fimbrin. The actin binding domain contains two calponin homology (CH) domains. AtFim1, isolated by Kovar et al. (2000), binds F-actin in a calcium-independent manner. AtFim1 prevents actin depolymerization mediated by profilin. Moreover, it has been used for a visualization of the actin cytoskeleton as a fusion protein with GFP.

Formins are another class of actin-bundling proteins. They contain formin homology domains (FH), especially FH1 and FH2 domains. Two demonstrated functions of formins in plants are the nucleating of actin (Ingouff et al. 2005; Michelot et al. 2005) and bundling (Michelot et al. 2005). Villins belong to the villin/gelsolin/fragmin superfamily (Su et al. 2007). Villins possess six gelsolin homology domains. They are multifunctional proteins able to cap the plus ends of the actin filament, to sever, bundle, and nucleate actin. The domain that allows a villin to crosslink filaments into bundles is a headpiece domain. Each villin may attach to two actin filaments. In the presence of a micromolar concentration of calcium, most villins sever F-actin and cap the barbed ends. The proteins containing LIM domains are another group of actin bundling proteins. The LIM domain is a tandem zinc finger motif that functions in protein–protein interactions. In plants, the group of LIM proteins has a limited number of members (Arnaud et al. 2007). The LIM proteins WLIM1 in *Nicotiana* (Thomas et al. 2008) and LlLIM1 in *Lilium* (Wang et al. 2008) have been shown to promote actin bundle formation.

Although there is a lack of ultimate experimental proof that actin bundles are important for cellular processes in plants, it seems very probable that they are. As discussed by Thomas et al. (2009), the involvement of bundles in intracellular processes is especially clear in nuclear movement and in the transport of vesicles in tip growing cells. Other indirect evidence that bundles of microfilaments play a role in intracellular transport comes from studies on chloroplast movements by Anielska-Mazur et al. (2009). Actin bundles widened under strong light and this widening was reversible by weak light. This suggests that the rearrangement of bundles is precisely controlled. The control mechanism is as yet unknown.

It has repeatedly been illustrated that the actin cytoskeleton in leaves consists of a spatial network of actin bundles and of fine actin meshwork named "baskets" around chloroplasts. For example, such a structure of the cytoskeleton has been demonstrated with Alexa fluor 488 phalloidin in the fixed mesophyll of *Arabidopsis thaliana* (Krzeszowiec et al. 2007). The same arrangement of microfilaments has been demonstrated in vivo using a fusion reporter protein plastin-GFP in *Nicotiana tabacum* leaves (Anielska-Mazur et al. 2009).

2.1.2 Microtubules

The second component of the scaffolding involved in transport processes in the plant cell is the microtubule cytoskeleton (for reviews see book by Nick 2008). Tubulin, the principal microtubule protein, has a molecular mass of about 50 kDa. A high-resolution model shows that tubulin is a protein of 46 Å width × 40 Å height × 65 Å depth. It contains a nucleotide-binding domain at the N-terminus, a small domain in the middle, and a helical domain at the C-terminus (Goddard et al.

1994; Nogales et al. 1999). The basic element of microtubules is a heterodimer of α/β subunits. The heterodimer is 8 nm in length. During the self-association process, heterodimers stick together in head-to-tail fashion forming a protofilament. Microtubules are cylindrical structures usually composed of 13 protofilaments. Lateral junctions between neighboring protofilaments stabilize the microtubule. The internal diameter of a microtubule is 14 nm and the external diameter is 25 nm. A length of 2–4 nm is reported by Hardham and Gunning (1978). The polarity of the heterodimer is translated into a polarity of the whole microtubule. The end with a vigorous addition of dimers is called the plus end. Three dynamic events are possible: polymerization, depolymerization, and rapid disassembly. The latter is called dynamic instability and is a stochastic process. The half-life of a single microtubule is of the order of several minutes (Hardham and Gunning 1978).

2.2 *A Detailed Look at Motor Proteins*

2.2.1 Myosins

Myosins are defined as Mg^{2+}-dependent ATPases. They transport various cargoes, e.g., vesicles or organelles, along actin filaments. In addition, they are able to transport microfilaments along a plasma membrane. Myosins belong to two groups: conventional and unconventional. Conventional myosins are involved in producing muscle contraction. Unconventional myosins, present in every cell, are involved in transport processes along microfilaments. They possess a characteristic, highly conservative motor domain. This domain, situated at the N-terminal end of a protein, has a molecular weight of about 80 kDa. Apart from the motor region called the head domain, two other domains can be distinguished: a neck domain which contains IQ motives and a tail domain at the C-terminal end of the protein. The part of the head domain that binds ATP is the most conservative part of the protein. Another subdomain is involved in the binding of actin in the N-terminal part. As shown in animals, the step of myosin on the microfilament is dependent on the hydrolysis of the nucleotide.

A processive motion is a characteristic feature of myosins, fundamental to intracellular movement in plants. The current model of the movement of unconventional myosins is based on the observation that the proteins form dimers. Movement of myosin starts when ATP binds to myosin which is attached to actin. Next, hydrolysis of ATP takes place, followed by a release of inorganic phosphate and ADP. Thereafter, ATP rebinding to the acto-myosin complex starts a new event (De La Cruz et al. 1999).

Myosin VIII was the first myosin cloned by Knight and Kendrick-Jones (1993). This 131.2-kDa protein is composed of 1,166 amino acids. It includes a head domain and a neck domain with four IQ motives followed by a coiled coil region. The N-terminal part encloses a unique 150 amino acid domain of unknown function. Detailed studies have shown a localization of class VIII myosins in plasmodesmata, on the surface of statoliths, chloroplasts, and a colocalization with callose. They were also found in the regions of the dividing cell which are involved in the formation and maturation of cell walls.

Myosin XI from BY-2 cells has 175 kDa and its total length is about 33 nm. It is a two-headed myosin. The neck domain contains six IQ motives. Myosin XI possesses a DIL domain (dilute domain) and a coiled coil region (Tominaga et al. 2003). Its structure is similar to that of myosin V. Myosin XI is involved in cytoplasmic streaming and organelle movements, including chloroplast movements (Hashimoto et al. 2005; Li and Nebenführ 2007; Wang and Pesacreta 2004). Myosin XI is the fastest processive motor protein known (~7 $\mu m\ s^{-1}$) with a 35-nm step size. On the basis of a model postulated for myosin V, Tominaga et al. (2003) proposed a mechanism for myosin XI movement. The leading head attaches to an actin filament and binds ADP-P_i. Following the release of P_i, the head domain is tightly attached to actin. At the same time, a trailing head binds actin and may bind ADP. After the liberation of ADP, ATP attaches to the trailing head and concomitantly this head is detached from the microfilament. The tension accumulated in the neck domain of the leading head moves forward the detached trailing head. The head hydrolyzes ATP to ADP-P_i and rebinds the microfilament and becomes the leading head. Following the release of P_i, the domain is tightly fastened to actin. The mechanism of detaching from and attaching to actin and of swinging the trailing head causes a one-way movement toward the plus end of the filament. The release of ADP from the trailing head is the rate-limiting step.

2.2.2 Kinesins

Motor proteins that use microtubules as tracks are called kinesins. They carry cargo directionally along microtubules using the energy of ATP hydrolysis (Vale 2003). There are 14 classes of kinesins plus orphan kinesins according to the standardized nomenclature (Lawrence et al. 2004). Kinesins contain three regions: a motor domain, a coiled coil stalk, and a globular tail (Kao et al. 2000). The motor domain is composed of a catalytic domain of about 350 amino acids and a neck of about 40 amino acids. It contains an ATP-binding domain and a microtubule binding site. Interaction between kinesin and microtubule is mediated via the binding of the beta subunits of microtubule heterodimers. The motor domain has a conserved structure. A common classification of kinesins is based on the localization of their motor domains (Hirokawa et al. 2009). Accordingly, there are N-kinesins, M-kinesins, or C-kinesins. Their motor domains are located at the amino terminus, in the middle or at the carboxyl terminus, respectively. Generally, the direction of the movement of kinesins depends on the location of the motor domain. While N-kinesins move toward the plus and C-kinesins move toward the minus end, M-kinesins depolymerize microtubules. Stalk and tail domains share little sequence homology and are family specific. The coiled coil stalks are disrupted. These brakes probably increase the flexibility of the region (Vale and Fletterick 1997).

Crystallographic studies show a structural homology between the motor domain of kinesin and of myosin. As discussed by Woehlke and Schliwa (2000) the similarity in the catalytic core (the place of ATP hydrolysis) indicates that the events immediately following ATP hydrolysis are probably very similar in both motor proteins. The movement of kinesin along a microtubule is called "walking." It should be mentioned that most kinesins form dimers. One of the heads binds tightly to the filament (the heads either contain ATP or no nucleotide) and at the same time the other head carries ADP. For a short time, both heads are attached to the filament. Then the cycle repeats itself (Woehlke and Schliwa 2000).

Unlike plant myosins, plant kinesins do not belong to specific classes. Sixty-one kinesin sequences have been found in the *Arabidopsis* genome (Reddy and Day 2001). Among them 21 are predicted to be a minus-end-directed. Interestingly, seven of these kinesins contain a CH domain characteristic of actin binding proteins (Lee and Liu 2004).

Dyneins are a second type of motor protein that moves cargo along microtubules. They are minus end-directed motors. Surprisingly, they are absent from higher plants (Lawrence et al. 2001). It has been suggested that minus end-directed kinesins replace them.

3 Intracellular Movements as Examples of Cell Integration

3.1 Cytoplasmic Streaming

A common strategy used to characterize the mechanism of a physiological phenomenon is to dissect it from coincident effects. This approach has hardly ever been implemented in the case of organelle movements and cytoplasmic streaming. In fact, cytoplasmic streaming has been from the beginning identified with organelle movements. Or, to be more precise, this streaming has been seen as the basis of such movements, or as a corollary of them (e.g. Shimmen 2007; Esseling-Ozdoba et al. 2008; Verchot-Lubicz and Goldstein 2009).

Cytoplasmic streaming, also termed cyclosis, is particularly conspicuous in the internodal cells of characean algae, due to their macroscopic size and characteristic organization (for a review see Williamson 1992). The length of these giant cylindrical cells is in the order of cm, whereas the diameter is in the order of millimeters. Over 90% of the cell volume is occupied by the central vacuole, a trait which is also typical of mature plant cells. The cortical, stationary part of the cytoplasm contains motionless chloroplasts arranged in helical rows. The internal endoplasm containing the nuclei and other organelles rotates continuously along the chloroplast files. Their large size and simple geometry make characean cells amenable to experimental techniques that are unusable for typical plant cells: cutting, perfusion, permeabilization, and a reconstitution of the cytoplasmic flow by using cytoplasmic fractions or isolated proteins. These techniques greatly facilitated work on the mechanism and regulation of streaming in Characeae. With the demonstration that *Chara* and *Nitella* cells contain actin filaments, followed by their localization at the interface between stationary ectoplasm and moving endoplasm (Williamson 1974; Palevitz and Hepler 1975), both genera became models for actin-based cytoplasmic streaming and organelle motility. Parallel bands of stable actin bundles running next to the chloroplast files stretch throughout the entire length of the internodal cell. All microfilaments have the same polarity in the bundle, with barbed ends pointing in the direction of streaming (Palevitz and Hepler 1975). The flow of cytoplasm is generated by an ATP-dependent sliding between the bundles and myosin motors attached to the cytoplasmic organelles (Sugi and Chaen 2003). Characean motors are among the fastest myosins identified; the velocities of up to 100 $\mu m\ s^{-1}$ which are generated in these algal cells exceed by far the maximum velocity of actin–myosin sliding in muscle (Sugi and Chaen 2003). The results of five-decade-long research on the role of myosins in cytoplasmic streaming and organelle movements in Characeae have been summarized in an extensive review by Shimmen (2007).

Along with problems that have attracted much attention, such as the identification of motor proteins, others still await experimental investigation. One of them is the old question about the extent to which individual organelles reacting with actin situated at the surface of cortical chloroplasts contribute to moving the bulk of the cytoplasm (Williamson 1992). Taking into account the large size of Characeae, part of the cytoplasm can be several micrometers away from the force-generating area. Moreover, the shear generated in the cytoplasm is transmitted to the vacuole and sets in motion the entire vacuolar fluid, as shown in the pioneering experiments of Kamiya and Kuroda over half a century ago (1956).

The fluid dynamics of streaming was recently analyzed theoretically in a model characean internodal cell taking into account the helical character of flow tracks (Goldstein et al. 2008; van de Meent et al. 2008). The contribution of diffusion and advection (fluid flow) to the transport of molecular species were assessed. Transverse and helical flow components were shown to be generated between two indifferent zones devoid of chloroplasts. These flows are due to the particular geometric features of rotational streaming driven by forces acting helically in the cytoplasm. Flows across the cell diameter result in a mixing of the vacuolar contents, which is much faster than by diffusion.

Pollen tubes and root hairs represent two other classic objects for the study of cytoplasmic streaming, being easily usable in microscopic analyses. The pattern of streaming in these elongated, rapidly growing cells is different from characean rotation and may be described as a reverse fountain, with cytoplasm flowing towards the apex alongside the cell walls and returning to the base in the cell center (Hepler et al. 2001). The organization of the actin cytoskeleton differs substantially depending on the region of the tip-growing cell. Whereas longitudinal arrays of relatively thick actin bundles run along the pollen tube and the length of root hair, the subapical region contains very dynamic, short and thin filaments (Ketelaar et al. 2003; Lovy-Wheeler et al. 2005). In this region, the cytoskeleton evidently undergoes intense remodeling in both pollen tubes and root hairs (Voigt et al. 2005; Cheung et al. 2008). Such organization of actin serves to deliver Golgi-derived vesicles to the growing apex: the long actin bundles function as tracks for the myosin-mediated transport of the vesicles along the cell, and the dynamic subapical actin zone traps them and targets the tip (Hepler et al. 2001). The two pools of actin show different sensitivity to low concentrations of depolymerizing agents, with subapical actin filaments much more susceptible than long bundles (Ketelaar et al. 2003; Cárdenas et al. 2008). Consequently, cytoplasmic streaming remains unaffected by concentrations that hinder tip growth in pollen tube (Cárdenas et al. 2008).

A still different pattern of streaming is characteristic of the water angiosperms *Elodea* and *Vallisneria*. It is seen in most cells as a cyclic movement around the cell walls with velocities varying from one cell to another, peaking at 12 $\mu m\ s^{-1}$ in *Elodea* (Forde and Steer 1976). Unlike in Characeae, chloroplasts move around the cell with the cytoplasm. Another particular feature of cytoplasmic streaming in these organisms is its strong dependence on environmental signals, in particular on light. Extensive studies have been carried out on light dependence of cytoplasmic motion in *Vallisneria*, mainly in the context of chloroplast movement regulation. The motility of cytoplasm is suppressed in the dark and transiently accelerated upon illumination. Initially, light accelerates random movements of cytoplasmic particles and only after a few minutes is a continuous, directional flow of cytoplasm (photodinesis) observed. Based on red-far red light reversibility and on the inhibitory activity of DCMU, an inhibitor of photosynthetic electron transport, a combined regulation of the cytoplasmic flow has been proposed, by phytochrome on the one hand, and by photosynthesis on the other (Takagi and Nagai 1985; Takagi 1997; Dong et al. 1998; Takagi et al. 2003).

Tobacco BY-2 suspension cells are another object of choice for the study of cytoplasmic streaming and organelle movement (Ruthardt et al. 2005). Although not as big as the internodal cells of Characeae, they are up to 1 mm long and contain numerous transvacuolar strands which display vivid movement of cytoplasm/organelles. The role of organelles in creating the driving force for cytoplasmic flow has recently been investigated using this model (Esseling-Ozdoba et al. 2008). The rate of fluorescence recovery after photobleaching (FRAP) was determined in BY-2 cells expressing cytosolic GFP. Free GFP molecules were shown to move faster in cells with active organelle transport than in those where this transport was inhibited with BDM (2,3-butanedione monoxime), an inhibitor of the myosin ATP-ase function. The results have been interpreted in terms of the active, motor-driven transport of organelles giving rise to the hydrodynamic flow of fluid cytosol, in line with other recent analyses (Verchot-Lubicz and Goldstein 2009).

The process of cytoplasmic streaming relies on the acto-myosin system. Prominent bundles of actin filaments have been found in most experimental models. Only *Elodea* leaf and *Tradescantia* stamen cells were reported to contain mostly fibrillar–granular material scattered throughout the cytoplasm, and the rare occurrence of microfilaments, irrespective of the fixation method used for electron microscopy

(Forde and Steer 1976). An alignment of actin bundles with the cytoplasmic flow and a uniform polarity of microfilaments within the bundles are characteristic of all types of cells exhibiting cytoplasmic streaming, and apparently define the direction of flow (Palevitz and Hepler 1975; Tominaga et al. 2003; Lenartowska and Michalska 2008).

An undisturbed contact between plasma membrane and cell wall at the end of *Vallisneria* cell is required to maintain the organization of actin bundles necessary for a proper pattern of cytoplasmic streaming (Ryu et al. 1995). Microfilaments proved to be much more resistant to cytochalasin D, an actin depolymerizing factor, at the transverse than at the longitudinal walls. The application of trypsin broke the contact and disrupted the filaments. After restoration of the cytoskeleton, the direction of streaming was reversed in 50% of cells. These experiments were complemented by the prolonged application of a synthetic hexapeptide with an RGD motif known to be a recognition site in molecules required for adhesion to the substratum at focal contact sites (Ryu et al. 1997). This peptide, localized to the cell wall, caused serious disturbances in both streaming and the cytoskeleton organization. Thus, the organization of the actin bundles appears to depend on the protein(s) present in the extracellular matrix.

Cytoplasmic streaming is strictly regulated by Ca^{2+} ions. High extracellular Ca^{2+} concentrations act as inhibitors in various cases (Forde and Steer 1976; Dorée and Pickard 1980; Williamson and Ashley 1982; Woods et al. 1984; Kohno and Shimmen 1988; Takagi 1997; Sugi and Chaen 2003). This regulation by Ca^{2+} was meticulously analyzed for *Vallisneria*, in a series of experiments with the combined application of red and far-red light, EGTA, calcium channel blockers nifedipine and La^{3+}, calcium-sensitive dye murexide and pyroantimonate to evaluate the calcium level in the cells by precipitation (Takagi and Nagai 1985; Takagi 1997). The phytochrome-mediated stimulation of streaming by red light, mimicked also by the extracellular addition of EGTA in the dark, has been attributed to an outflow of Ca^{2+}, and the inhibition by far-red light to an influx through the plasma membrane via Ca^{2+} channels.

So far, the massive flow of cytoplasm in higher land plants has been substantiated only for specialized cells (pollen tubes, root and stamen hairs, trichomes, epidermal cells, and seedlings, i.e., mainly their organs – hypocotyls and coleoptiles). This is consistent with the generally accepted role of cytoplasm, which is to accelerate the distribution of metabolites in the cell and to enhance exchange across organelle membranes, particularly important in elongated and fast growing cells. However, cytoplasmic streaming is commonly claimed to occur universally in plant cells. Such generalizations are unfounded, not least because of our ignorance about cytoplasm motility in the photosynthetic mesophyll, the key tissue in vascular land plants. A recent analysis of thousands of individual organelles in *Nicotiana benthamiana* leaf cells has revealed independent movement patterns for Golgi stacks, mitochondria, and peroxisomes (Avisar et al. 2008). The authors conclude that the notion of coordinated cytoplasmic streaming is not generally applicable to higher plants.

3.2 *Movements of Nonphotosynthesizing Organelles*

3.2.1 Golgi Network in Plants

In animal cells, Golgi complexes with an ordered structure are localized in close proximity to the nucleus. This localization is controlled by the microtubular cytoskeleton (Burkhardt 1998). Golgi membranes are assembled around the centrosome with *cis* cisternae anchored to the minus ends of microtubules. Vesicles which target this central Golgi complex from the peripheral endoplasmic reticulum are transported along microtubules. Post-Golgi transport towards the periphery of the cell is also based on microtubule tracks. The Golgi apparatus of plant cells, however, consists of numerous small, independent units scattered throughout the cytoplasm (Boevink et al. 1998; Nebenführ et al. 1999). These units are approximately 1 μm in diameter when stained fluorescently. As a central organelle in the secretory pathway, the plant Golgi not only distributes proteins processed in ER but also specializes in the synthesis and export of complex glycans targeted to the cell wall. The molecular regulation of plant Golgi functions was reviewed by Hawes and Satiat-Jeunemaitre (2005). The random distribution of units reduces the distance both from ER export sites to the Golgi and from the Golgi to the plasma membrane and/or vacuole. The plant Golgi apparatus has also been shown to participate in the transport of proteins to chloroplasts (Chen et al. 2004; Villarejo et al. 2005).

Each Golgi unit is a stack of cisternae organized in a polar way, changing from the *cis*, ER adjacent side, through medial- and *trans* cisternae, to TGN (*trans*-Golgi network) cisternae on the *trans* side of the stack. Impressive three-dimensional models have recently been presented, based on electron tomography analysis of plant Golgi stacks (Staehelin and Kang 2008). The structural polarity reflects also their polar function. The Golgi stacks move rapidly along the cortical ER network (Boevink et al. 1998; Nebenführ et al. 1999). Each stack moves independently with periods of fast, up to 2–4 μm s^{-1}, translational movements alternating with phases of slower oscillatory (wiggling) activity. The movements are based on actin. Firstly, cytochalasin D stopped the stacks which clumped on small areas of cortical ER lamellae. Secondly, F-actin was shown to closely match the ER array. It has been suggested that myosin provides the motive force, as concluded from movement inhibition by BDM (Nebenführ et al. 1999).

Both the high mobility of Golgi stacks on one hand and its close association with the ER network have raised controversies as to the mode of vesicular trafficking between them. Two different models have been proposed to explain this transport. An earlier "stop and go" model developed by Nebenführ et al. (1999) provides an explanation of both the alternating phases of movement and the precise targeting of ER-derived transport vesicles to the *cis*-Golgi stack. Upon a hypothetical stop signal sent out by an active ER export site, the closest Golgi stack uncouples from the actin track and stops at the site to pick up vesicles. This activity is accompanied by an oscillatory motion of the unit around a stationary position. After vesicle transfer the unit resumes its fast movement. Similar stop signs are postulated at sites of Golgi product delivery. This model gained credibility from the observation of very short-lived (with a half-life of several seconds) punctuate fluorescent structures formed randomly over the ER in BY-2 cells express Sec13:GFP (a COPII coat protein) and an ER export site marker (Yang et al. 2005). Golgi stacks were shown to associate sporadically with these structures. An alternative model has been proposed that puts forward a stable coupling between ER export sites and Golgi units, in other words, the existence of a stable ER–Golgi secretory unit (daSilva et al. 2004). The latter model is based on immunocytochemical and live cell images which show the close association of ER with Golgi stacks, and which suggest membrane continuity between both organelles (daSilva et al. 2004; Hawes and Satiat-Jeunemaitre 2005). Later, using photoactivable GFP, Golgi units were demonstrated to move in the same direction and at the same speed as the surface ER proteins (Runions et al. 2006). This provided additional support for a postulated continuity between the ER membrane and Golgi units.

New light on this controversial issue has recently been shed by high-resolution structural information obtained by electron tomography of plant material cryo-fixed under high pressure and freeze substituted (summarized in: Staehelin and Kang 2008). An ultra-fast fixation technique made possible visualization of several short-lived COPII vesicles budding from ER export sites. Thanks to their high-resolution three-dimensional reconstructions of these vesicles have located them in different ER tubular domains. Imaging such as this is almost impossible with conventional thin section electron microscopy. The electron tomograms also reveal scaffold-type molecules that accumulate on the budding vesicles and relocate together with them to the *cis* side of Golgi. The COPII scaffold has been suggested to contain proteins which mediate the stop signal and provide physical coupling between ER export sites and Golgi units. These findings are in line with the stop-and-go model of Golgi motility and extend it by a new "dock-pluck-and-go" model of ER-to-Golgi vesicle trafficking (Staehelin and Kang 2008).

3.2.2 Peroxisomes and Mitochondria

In 2002, two independent groups described peroxisome motility in various *Arabidopsis* cell types. These cells express GFP/YFP which are fused to peroxisome targeting signals (Jedd and Chua 2002; Mathur et al. 2002). Unlike animal peroxisomes, which move using microtubules as tracks and dynein/dynactin motors, plant organelles move on actin bundles. This was shown in double-labeled cells where peroxisomes colocalized with actin. By contrast, no association with microtubules was demonstrated in onion epidermal cells (Mathur et al. 2002). The involvement of actin was additionally confirmed by actin drug treatments; cytochalasin D and latrunculin B arrested movement, while neither the microtubule depolymerizing drugs, oryzalin and propyzamide, nor paclitaxel affected peroxisomal motility. Three types of motile behavior were singled out. About 70% of peroxisomes were inactive, showing only weak oscillations around a stationary position. Others performed slow, short-range movements, and 10% exhibited fast, long-distance saltations, with velocities of up to 4 μm s^{-1} (or 6 μm s^{-1}, as reported by Jedd and Chua 2002). Individual peroxisomes moved independently, with varied speeds and directions, and they could suddenly change, even reverse, direction or speed. The organelles frequently interacted with each other and transient fusions were observed. The changes of direction were caused by crossing over to neighboring actin filaments. Although both groups worked with similar plant material, Mathur et al. (2002) emphasize that peroxisomes did not follow the general flow of cytoplasm and could move in opposite directions within one transvacuolar cytoplasmic strand. In particular, peroxisome trajectories did not follow the typical reverse fountain cytoplasmic streaming pattern in *Arabidopsis* root hairs. By contrast, Jedd and Chua (2002) conclude that "like giant algae, most large cells in higher plants probably use cytoplasmic streaming as a means of transporting organelles, vesicles and solutes over long distances."

A similar experimental approach was used in concomitant studies of mitochondrial movements (Van Gestel et al. 2002). Mitochondria were visualized with GFP, and both types of cytoskeletal structures, F-actin and microtubules, were disturbed by depolymerizing/stabilizing drugs. Cultured cells of *N. tabacum* regenerating from protoplasts and elongating were used as an experimental model. Mitochondria exhibited two types of behavior in these cells: they either moved actively in the cytoplasmic strands or remained immobile in the cortical layer of cytoplasm. The fast, active movement was F-actin based. The microtubule disruptor oryzalin did not disturb this movement but slightly stimulated the recruitment of cytoplasmic strands, pointing to negative control of actin-based mobility by microtubules. Similar negative control had earlier been reported by Nebenführ et al. (1999) for Golgi stacks. Immobile mitochondria were situated along parallel lines, oblique or transverse to the long cell axis. The positioning pattern suggested the contribution of microtubules, which was verified by an inhibitory action of oryzalin.

Last year, the intracellular positioning of mitochondria in *A. thaliana* mesophyll palisade cells was shown to depend on light (Islam et al. 2009). Mitochondria were visualized using GFP fused to a mitochondrion-targeting signal. In darkness, they were distributed randomly in the cells. While mitochondria accumulated along the periclinal walls under weak blue light, they gathered along the anticlinal walls under strong blue light. The light-induced responses were similar to the accumulation and the avoidance response of chloroplasts, respectively, except for strong red light which induced the accumulation of mitochondria along the inner periclinal walls. Red light does not induce the avoidance response of chloroplasts in *Arabidopsis* (Trojan and Gabryś 1996; Sztatelman et al. 2010). The mode of movement of individual mitochondria along the outer periclinal walls was analyzed by time-lapse fluorescence microscopy under different light conditions. The number of static mitochondria located in the vicinity of chloroplasts increased gradually with irradiation time. This light-induced colocalization of mitochondria with chloroplasts strongly suggests mutual metabolic interactions between both organelles.

3.2.3 Nucleus

Mechanisms of nuclear positioning have previously been studied in plant cells in the context of developmental processes (Britz 1979; Katsuta and Shibaoka 1988; Katsuta et al. 1990). In 1993, the positioning of nuclei in the fern *Adiantum capillus veneris* was shown to be controlled by light, an environmental signal of key importance for green plants (Kagawa and Wada 1993). Recently, strong light was also reported to control the positioning of nuclei in mature cells of higher plant leaves (Iwabuchi et al. 2007, 2010). The nucleus occupies different positions in *A. thaliana* leaf cells in a light dependent way. In the dark-adapted mesophyll, tissue nuclei are located at the center of the cell bottom. Under blue light of more than 50 μmol m^{-2} s^{-1}, the nuclei relocate to the anticlinal walls. Relocation from the dark to the light position is relatively slow. The response requires several hours of blue light illumination and is reversible. An analogous response also takes place in epidermal cells which do not contain chloroplasts. Thus, the mechanism of nuclear positioning appears to operate independently of chloroplast positioning.

Mutants deficient in phototropins, which are blue light receptors (for more detailed characteristics see Sect. 3.3), were used to identify the photoreceptor responsible for light absorption (Iwabuchi et al. 2007, 2010). Responses to blue light were induced in *phot1* but not in *phot2* and *phot1phot2* mutants. The response in the *phot2* mutant was restored after effect of the mutation had been neutralized by introducing the *PHOT2* gene into the plant (Iwabuchi et al. 2007). These results indicate that phototropin 2 mediates blue light-dependent nuclear positioning in *Arabidopsis*.

The involvement of the actin, not microtubular, cytoskeleton in nuclear relocation was confirmed in experiments employing the inhibitors, latrunculin B and propyzamide (Iwabuchi et al. 2010). The actin organization, as imaged by immunofluorescence microscopy, differed in dark and light-adapted epidermal cells of *Arabidopsis*. Thick actin bundles, periclinally arranged parallel to the longest axis of the cell, were shown to be associated with the nucleus in darkness. The actin bundles adjacent to the nucleus reorganized closer to the anticlinal walls under strong blue light. These light-mediated changes in actin organization were evident in *phot1* but not in *phot2* and *phot1phot2* mutants. These findings led the authors to hypothesize that blue-light-dependent nuclear positioning in *Arabidopsis* is regulated by a phototropin2-dependent reorganization of the actin cytoskeleton (Iwabuchi et al. 2010).

3.3 Chloroplast Redistribution in Mesophyll Tissue

Uncovering the mechanisms controlling chloroplast motility appears to be a still more complicated task because the behavior of chloroplasts is more diversified than that of other organelles. It ranges from nonmotility, the trait of most chloroplasts in Characeae, through light-inducible movement in higher terrestrial plants, light-inducible movement along with swimming cytoplasm in water angiosperms, to swimming with the bulk cytoplasm in *Elodea* (see reviews: Haupt and Scheuerlein 1990; Haupt 1999; Takagi et al. 2003; Wada et al. 2003; Wada and Suetsugu 2004). Chloroplasts have a special place among plant organelles as the site of photosynthesis. Understandably, light is the main environmental cue which controls their movement. A general feature of light-controlled chloroplast positioning is accumulation in weakly illuminated regions of the cell and the avoidance of regions exposed to strong light. This distribution pattern helps to maximize light harvesting under energy-limiting conditions (Zurzycki 1955; Takemiya et al. 2005), while it acts as a protection from excess energy in strong light (Park et al. 1996; Kasahara et al. 2002; Sztatelman et al. 2010). The distribution is brought about in various ways. At least two distinct basic mechanisms seem to operate in chloroplast positioning, one in mosses, ferns, and water plants and another one in higher terrestrial plants.

One representative type of behavior has been studied in detail in the water angiosperms, *Vallisneria* sp. (Izutani et al. 1990; Dong et al. 1996, 1998; Sakurai et al. 2005). The chloroplasts are motionless in the dark. Under continuous weak light they are set in motion together with the cytoplasmic matrix and migrate to the periclinal side of the cell. The migration is slow: the full accumulation response requires several hours. Irradiation with strong light rapidly drives the chloroplasts to the anticlinal sides. The avoidance response is much faster, lasting tens of minutes. While red and blue spectral regions are active in stimulating the accumulation response, with an approximately triple quantum efficiency of red as compared to blue light, the avoidance response is activated almost exclusively by blue light (Izutani et al. 1990). The effects of red light can be negated either by far-red light or by the presence of DCMU, which has been interpreted in terms of cooperative regulation by the P_{FR} form of phytochrome and photosynthesis (Dong et al. 1996, 1998). From the point of view of the mechanism, the most striking feature is a very special rearrangement of the actin cytoskeleton accompanying the accumulation response (Fig. 1, left panel). In dark-adapted cells, actin bundles are arranged in a network array. A characteristic honeycomb formation appears upon irradiation with red light, with a concomitant decline in the motility of chloroplasts, which come into sight inside the cavities of the honeycomb structure. In consequence, the accumulation of chloroplasts at the periclinal wall of *Vallisneria* has been interpreted as the trapping of the flowing chloroplasts into the unique structure built of light-reorganized actin (Dong et al. 1996, 1998). Strong blue light, evidently absorbed by a blue-specific photoreceptor, causes a fast reconstruction of F-actin into an ordinary network of linearly arranged bundles. Chloroplasts regain mobility, and, after a short phase of random small-range activity, they start to move unidirectionally along the actin tracks (Sakurai et al. 2005).

Various elements of the above mechanism can be found in light-mediated chloroplast positioning of several species of cryptogam plants (for review see Wada et al. 2003). For example, the orientation movements of chloroplasts in protonemal cells of the fern *A. capillus veneris* are under the control of phytochrome and blue light receptor(s). The accumulation response has been shown to entail characteristic rearrangements of actin filaments associated with single

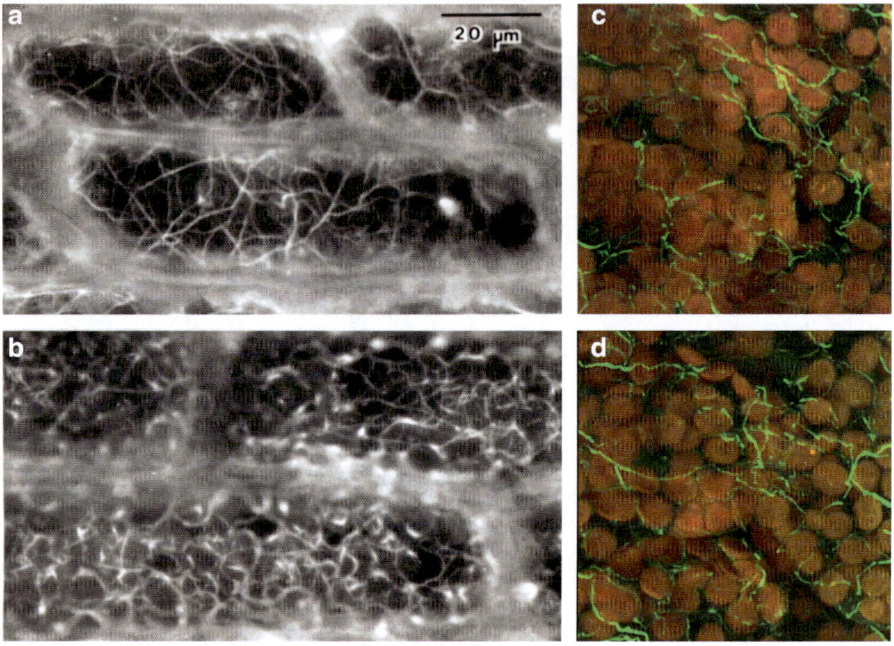

Fig. 1 (**a**) and (**b**): Changes in the organization of actin cytoskeleton caused by red light in *Vallisneria gigantea* epidermal cells, visualized with FITC-conjugated phalloidin. (**a**) Dark-adapted tissue; (**b**) Tissue irradiated with weak red light for 5 min and kept in darkness for 40 min. Note the honeycomb structure of the actin network (Dong et al. 1998). (**c**) and (**d**) Actin cytoskeleton visualized with Alexa Fluor 488-phalloidin in mesophyll cells of *Arabidopsis thaliana* (**c**) irradiated with strong blue light; (**d**) irradiated with weak blue light. Both *bars* are 20 μm

chloroplasts in this species. Chloroplast responses are also mediated by phytochrome and a blue light receptor also in the moss *Physcomitrella patens* (Sato et al. 2001). The cytoskeletal basis of chloroplast movements differs, however, in this moss in that it combines both microfilament and microtubular systems.

Chloroplast movements in the mesophyll of terrestrial angiosperms appear to be considerably more autonomous than those in water angiosperms. Whereas in the model organisms belonging to the latter group, the directional redistribution of chloroplasts depends mutually on cytoplasmic streaming, and the role of light can be defined as regulatory, in higher land plants light itself does induce the directional movement of chloroplasts. The light signal is perceived by phototropins (phot1 and phot2), blue-light receptors associated with the plasma membrane (Sakai et al. 2001; Briggs and Christie 2002). Both phototropins redundantly control chloroplast accumulation, whereas phot2 alone controls the avoidance response (Jarillo et al. 2001; Sakai et al. 2001). Like other plant organelles, chloroplasts move along actin tracks in terrestrial angiosperms. The nature of the involvement of the actin cytoskeleton has not been resolved and the available data is rather controversial.

Seeking light-induced changes in the organization of the actin cytoskeleton analogous to those observed in cryptogams and water angiosperms, Krzeszowiec et al. (2007) found blue light-induced changes neither in *Arabidopsis* (Fig. 1, right panel) nor in tobacco. Although these experiments were carried out with fixed tissue, irradiation with strong and weak red light induced tangible differences between the mesophyll cells of *phot2* mutants devoid of phototropin 2, i.e., differences in the shape and distribution of F actin formations. This line of investigation was continued using live mesophyll cells of transgenic tobacco expressing a fragment of human plastin, an actin bundling protein fused to GFP (Anielska-Mazur et al. 2009). While strong light caused a reversible widening of actin bundles in this model, the changes were neither directional nor blue light specific. Similar effects were observed in blue and red light, although the latter region does not activate chloroplast positioning in tobacco, a terrestrial angiosperm. Myosins, previously shown to localize at the surface of the *Arabidopsis* chloroplasts (Wojtaszek et al. 2005), were tested as another potential target for light signals. Investigations using polyclonal antibodies against plant myosin VIII and animal myosins provide evidence that myosins may indeed serve as a target for a blue-light chloroplast-orientation signal (Krzeszowiec and Gabryś 2007). Arrangements of myosins observed in the mesophyll of wild-type *Arabidopsis* differed depending on light intensity (Fig. 2). Myosins were associated with chloroplast envelopes in tissue irradiated with weak blue light (Fig. 2c). By contrast, in tissue irradiated with strong blue light chloroplasts were almost myosin free (Fig. 2f). No effect occurred either in red light or in the *phot2* mutant which lacked phototropin 2 and, in consequence, any avoidance response of chloroplasts (Fig. 2e, h). In contrast, the redistribution of myosins was similar to wild type at the surface of chloroplasts in a *phot1* mutant (Fig. 2d, g). Both accumulation and avoidance responses are normal in this mutant, and only slightly shifted towards higher light intensities as a result of impaired expression of phototropin 1. Thus, myosins have been proposed as the final step of the blue light signal transduction pathway starting with phototropin 2 and leading to chloroplast movements.

Another mechanism has recently been proposed by Kadota et al. (2009). Using transgenic *Arabidopsis* plants expressing the GFP-mouse fusion protein they identified short actin filaments situated at the chloroplast periphery on the plasma membrane site. The amount of these filaments at the front and the rear of chloroplasts depended on light intensity. In response to strong light, the filaments on irradiated chloroplasts transiently disappeared, and chloroplasts demonstrated increased motility in random directions. The presence of specific actin filaments was shown to depend on an actin binding protein, unusual chloroplast positioning (CHUP1), localized on the chloroplast envelope (Kasahara et al. 2002; Oikawa et al. 2003). On the basis of molecular genetic analysis of the *Arabidopsis* mutant *chup1* which showed a total immobility of chloroplasts, this protein was proposed as a possible factor linking chloroplasts to actin filaments. Short actin filaments were not detected in the *chup1* mutant. In phototropin mutants, their dynamics differed according to the mutant type and light conditions (Kadota et al. 2009). Meticulous analysis of all these findings led the authors to propose a model of blue

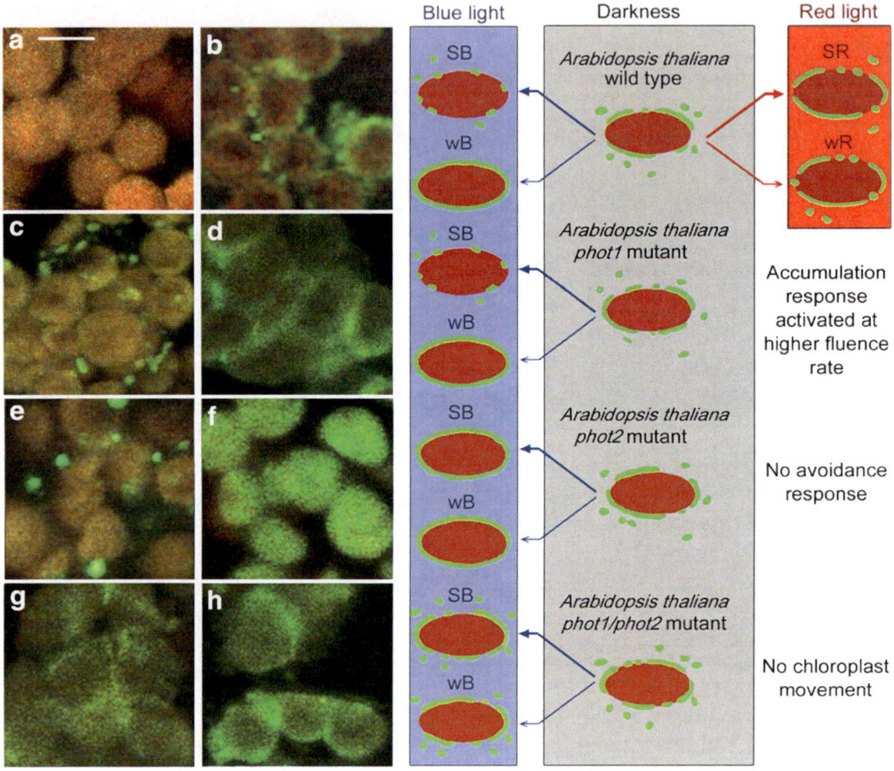

Fig. 2 *Left*: Confocal immunofluorescence images of *Arabidopsis thaliana* mesophyll, wild type (WT), *phot1* and *phot2* mutant plants. Myosins are labeled with antirabbit (Smooth and Skeletal) antibodies and with secondary FITC-conjugated antibodies (*yellow–green color*). *Orange color* comes from chloroplast autofluorescence. (**a**)–(**d**) Localization of myosins in wild type tissue: (**a**) labeled with secondary antibodies only – control; (**b**) dark-adapted; (**c**) strong blue light-irradiated. (**d**) weak blue light-irradiated. (**e**) Strong and (**f**) weak blue light-irradiated *phot1* mutant. (**g**) Strong and (**h**) weak blue light-irradiated *phot2* mutant. Note the difference in the localization of myosins between strong and weak blue-irradiated WT and *phot1* tissues, and the similarity of *phot2* images irrespective of light intensity. The *bar* is 10 μm. Right: Schematic representation of myosin localization at the chloroplasts in irradiated leaves of *Arabidopsis thaliana* wild type and in phot 1, phot2, and phot1/phot2 mutants. SB(SR) – strong blue(red) light, wB(wR) – weak blue(red) light. (Krzeszowiec and Gabryś 2007, modified)

light-controlled chloroplast positioning which is based entirely on rearrangements of short actin filaments on the surface of chloroplasts. In their opinion, plants have evolved a unique, actin-based mechanism for chloroplast movements, different from those identified to date for other organelles.

Although the unique character of the mechanism which lies behind chloroplast movements gains support from other results (see discussion of myosin involvement in the next section), this is so far restricted to terrestrial angiosperms. Obvious differences in the regulation of the movements in various plant groups have to be explained before a general mechanism can be acceptable. For example, the regulation of chloroplast movements by Ca^{2+} differs fundamentally in water and higher land plants. Whereas extracellular Ca^{2+} at millimolar concentrations acts as an inhibitor in *Vallisneria gigantea* (Takagi and Nagai 1985) and *Elodea densa* (Forde and Steer 1976), it does not affect chloroplast redistribution in *N. tabacum* (Anielska-Mazur et al. 2009) nor in *A. thaliana* (unpublished results). As mentioned earlier, inhibition of cytoplasmic streaming by Ca^{2+} has been reported in pollen tubes and the hair cells of higher plants. This supports the suggestion and the opinion that chloroplast positioning in water angiosperms is much more strongly connected with cytoplasmic streaming than in terrestrial angiosperms.

4 Involvement of Cytoskeletal Elements and Motor Proteins in Plant Intracellular Movements

Organelle movements in plants largely rely on the actin cytoskeleton. This has been documented in countless experiments showing the inhibition of movement upon the application of actin-stabilizing and disrupting drugs. In other experiments, the contribution of particular actin structures was deduced from their localization with regard to moving objects.

Experimental evidence for the involvement of myosin, the natural partner of actin, is much weaker. In most cases, such evidence as there is has been obtained by drug inhibition. There are drawbacks to both drugs commonly used to verify myosin participation, NEM (N-ethylmaleimid) and BDM. The former, acting as an SH-blocker, is unspecific and the latter gives inconsistent results. For example, the involvement of myosin in peroxisome movement was questioned on the basis of their incomplete inhibition with BDM and the high concentration of the inhibitor (30 mM) that had to be used (Mathur et al. 2002). Instead, the authors discussed the possibility that movement may be based on an actin polymerization process. Unlike the inconsistent behavior of the peroxisomes, mitochondria were arrested by BDM (20 mM) and this was interpreted as evidence of myosin involvement (Van Gestel et al. 2002).

Only recently has myosin been convincingly shown to participate in movements of peroxisome, mitochondria, and Golgi stacks (Avisar et al. 2008; Peremyslov et al. 2008). A dominant negative inhibition strategy was employed to determine the role of six *N. benthamiana* class XI and class VIII myosins in organelle movements. Suppression of the myosin XI-K function and/or RNA interference dramatically reduced the movement of peroxisomes, mitochondria, and Golgi stacks in the tobacco leaf cells. Lesser or negligible effects were observed when similar approaches were used to inhibit the functions of myosin XI-2 and XI-F. Organelle movements were unaffected by the inhibition of each of the three class VIII myosins. Remarkably, none of the six myosins tested appeared to be involved in light-induced movements of chloroplasts. These data point to a principal role for myosin XI-K in the movement/trafficking of peroxisomes, mitochondria, and Golgi stacks.

A question in need of even more work is the role of the microtubular cytoskeleton. Unlike the situation in animal cells, this cytoskeleton seems to play a secondary role in the mechanism of organelle movements in plants. Microtubules are mostly regarded as stabilizing structures, backing up the actin cytoskeleton in its task. In other words, while organelle motility is considered to be based on microfilaments, the positioning of stationary organelles is considered to be based on microtubules. As mentioned in Sect. 3.2.2, GFP-tagged mitochondria, which normally move along microfilaments, align along microtubules when the actin-based movement is inhibited (Van Gestel et al. 2002). This is consistent with observations in pollen tubes which suggest that microtubules and kinesins decrease the speed of mitochondria, thus contributing to their positioning (Romagnoli et al. 2007). An association of Golgi units with microtubules and a Golgi-linked kinesin has been documented in *Arabidopsis*. Kinesin, however, does not appear to contribute to the movement of Golgi (Liu et al. 2005).

A more direct involvement of microtubules has also been documented. In a recent study using a double-labeling technique together with variable-angle total internal reflection fluorescence microscopy and spinning disk confocal microscopy, a mechanism was proposed for the regulation of mitochondrial movements in the root hairs of *Arabidopsis* seedlings (Zheng et al. 2009). This mechanism combines the coordinated activity of myosin with the rate of actin turnover on the one hand, and with microtubule dynamics on the other. De novo actin polymerization and depolymerization are proposed to take part in the transition of mitochondria from a state of rest to motion. The microtubule dynamic directs the arrangement of actin filaments, thereby affecting the velocity, the trajectory, and the positioning of mitochondria.

The cooperation of microfilament- and microtubule-based systems has been demonstrated in the chloroplast positioning of the moss *P. patens* by using a combination of microbeam irradiations with blue and/or red light, and two inhibitors, latrunculin B and cremart which disrupts microtubules (Sato et al. 2001). Chloroplasts can utilize both microtubule and microfilament tracks for traveling. These two systems are differently regulated by two types of photoreceptors. Apart from a blue light regulatory pathway involving actin filaments, similar to the system operating in higher plants, two novel pathways have been proposed, based on microtubules, one pathway controlled by phytochrome and the other by a blue light receptor. The authors suggest that chloroplast movement in the moss may represent an evolutionary intermediate between the microtubule-dominated movement system operating in algal cells and the microfilament-based system operating in higher plant cells.

5 Concluding Remarks

Significant progress has been made in elucidating the mechanisms of organelle movements, but numerous questions still remain unanswered. Some of them, such as the relationship between cytoplasmic streaming and organelle motility, date back to the initial studies of these processes. Others, like the problem of myosin involvement in chloroplast positioning, are more recent. From a broader perspective, although the mechanochemical processes underlying the activity of motor proteins have been extensively studied at biochemical and biophysical levels, their application to the mechanical properties of the cytoplasmic matrix is far from understood. Another problem, hardly addressed up till now, is the interaction between the moving organelles and the coordination of their movements. The recent development of a transgenic *Arabidopsis* plant with nuclei, plastids, mitochondria, and plasma membranes tagged genetically with four different varieties of (G)FP, along with a system for quantifying the interaction between these organelles at a submicron resolution, is a step towards resolving this question (Kato et al. 2008).

References

Anielska-Mazur A, Bernaś T, Gabryś H (2009) *In vivo* reorganization of the actin cytoskeleton in leaves of *Nicotiana tabacum* L. transformed with plastin-GFP: correlation with light-activated chloroplast responses. BMC Plant Biol 9:64

Arnaud D, Déjardin A, Leplé JC, Lesage-Descauses MC, Pilate G (2007) Genome-wide analysis of LIM gene family in *Populus trichocarpa*, *Arabidopsis thaliana*, and *Oryza sativa*. DNA Res 14:103–116

Avisar D, Prokhnevsky AI, Makarova KS, Koonin EV, Dolja VV (2008) Myosin XI-K is required for rapid trafficking of Golgi stacks, peroxisomes, and mitochondria in leaf cells of *Nicotiana benthamiana*. Plant Physiol 146:1098–1108

Boevink P, Oparka K, Santa Cruz S, Martin B, Bettridge A, Hawes C (1998) Stacks on tracks: the plant Golgi apparatus traffics on actin/ER network. Plant J 15:441–447

Briggs WR, Christie JM (2002) Phototropins 1 and 2: versatile plant blue-light receptors. Trends Plant Sci 7:204–210

Britz S (1979) Chloroplast and nuclear migration. Encycl Plant Physiol 7:170–205

Burkhardt JK (1998) The role of microtubule-based motor proteins in maintaining the structure and function of the Golgi complex. Biochim Biophys Acta 1404:113–126

Cárdenas L, Lovy-Wheeler A, Kunkel JG, Hepler PK (2008) Pollen tube growth oscillations and intracellular calcium levels are reversibly modulated by actin polymerization. Plant Physiol 146:1611–1621

Chen H, Bernstein BW, Bamburg JR (2000) Regulating actin-filament dynamics in vivo. Trends Biochem Sci 25:19–23

Chen MH, Huang LF, Li HM, Chen YR, Yu SM (2004) Signal peptide-dependent targeting of rice α-amylase and cargo proteins to plastids and extracellular compartments of plant cells. Plant Physiol 135:1367–1377

Cheung AY, Duan QH, Costa SS, de Graaf BHJ, Di Stilio VS, Feijo J, Wu HM (2008) The dynamic pollen tube cytoskeleton: live cell studies using actin-binding and microtubule-binding reporter proteins. Mol Plant 1:686–702

daSilva LL, Snapp EL, Denecke J, Lippincott-Schwartz J, Hawes C, Brandizzi F (2004) Endoplasmic reticulum export sites and Golgi bodies behave as single mobile secretory units in plant cells. Plant Cell 16:1753–1771

De La Cruz EM, Wells AL, Rosenfeld SS, Ostap EM, Sweeney HL (1999) The kinetic mechanism of myosin V. Proc Natl Acad Sci USA 96:13726–13731

Dong XJ, Ryu JH, Takagi S, Nagai R (1996) Dynamic changes in the organization of microfilaments associated with the photocontrolled motility of chloroplasts in epidermal cells of *Vallisneria*. Protoplasma 195:18–24

Dong XJ, Nagai R, Takagi S (1998) Microfilaments anchor chloroplasts along the outer periclinal wall in *Vallisneria* epidermal cells through cooperation of P_{FR} and photosynthesis. Plant Cell Physiol 39:1299–1306

Dorée M, Pickard A (1980) Release of Ca^{2+} from intracellular pools stops cytoplasmic streaming in *Tradescantia* staminal hairs. Experientia 36:1291–1292

Esseling-Ozdoba A, Houtman D, Van Lammeren AAM, Eiser E, Emons AMC (2008) Hydrodynamic flow in the cytoplasm of plant cells. J Microsc 231:274–283

Forde J, Steer MW (1976) Cytoplasmic streaming in Elodea. Can J Bot 54:2688–2694

Gibbon BC, Kovar DR, Staiger CJ (1999) Latrunculin B has different effects on pollen germination and tube growth. Plant Cell 12:2349–2363

Goddard RH, Wick SM, Silflow CD, Snustad DP (1994) Microtubule components of the plant cell cytoskeleton. Plant Physiol 104:1–6

Goldstein RE, Tuval I, van de Meent JW (2008) Microfluidics of cytoplasmic streaming and its implications for intracellular transport. Proc Natl Acad Sci USA 105:3663–3667

Hardham AR, Gunning BE (1978) Structure of cortical microtubule arrays in plant cells. J Cell Biol 77:14–34

Hashimoto K, Igarashi H, Mano S, Nishimura M, Shimmen T, Yokota E (2005) Peroxisomal localization of a myosin XI isoform in *Arabidopsis thaliana*. Plant Cell Physiol 46:782–789

Haupt W (1999) Chloroplast movement: from phenomenology to molecular biology. Prog Bot 60:3–36

Haupt W, Scheuerlein R (1990) Chloroplast movement. Plant Cell Environ 13:595–614

Hawes C, Satiat-Jeunemaitre B (2005) The plant Golgi apparatus – going with the flow. Biochim Biophys Acta 1744:93–107

Hepler PK, Vidali L, Cheung AY (2001) Polarized cell growth in higher plants. Annu Rev Cell Dev Biol 17:159–187

Hirokawa N, Noda Y, Tanaka Y, Niwa S (2009) Kinesin superfamily motor proteins and intracellular transport. Nat Rev Mol Cell Biol 10:682–696

Holmes KC, Popp D, Gebhard W, Kabsch W (1990) Atomic model of the actin filament. Nature 347:44–49

Ingouff M, Fitz Gerald JN, Guérin C, Robert H, Sørensen MB, Van Damme D, Geelen D, Blanchoin L, Berger F (2005) Plant formin AtFH5 is an evolutionarily conserved actin nucleator involved in cytokinesis. Nat Cell Biol 4:374–380

Islam MdS, Niwa Y, Takagi S (2009) Light dependent intracellular positioning of mitochondria in *Arabidopsis thaliana* mesophyll cells. Plant Cell Physiol 50:1032–1040

Iwabuchi K, Sakai T, Takagi S (2007) Blue light-dependent nuclear positioning in *Arabidopsis thaliana* leaf cells. Plant Cell Physiol 48:1291–1298

Iwabuchi K, Minamino R, Takagi S (2010) Actin reorganization underlies phototropin-dependent positioning of nuclei in *Arabidopsis thaliana* leaf cells. Plant Physiol 152:1309–1319

Izutani Y, Takagi S, Nagai R (1990) Orientation movements of chloroplasts in *Vallisneria* epidermal cells: different effects of light at low- and high-fluence rate. Photochem Photobiol 51:105–111

Jarillo JA, Gabryś H, Capel J, Alonso JM, Ecker JR, Cashmore AR (2001) Phototropin-related NPL-1 controls chloroplast relocation induced by blue light. Nature 410:952–954

Jedd G, Chua N-H (2002) Visualization of peroxisomes in living plant cells reveals acto-myosin-dependent cytoplasmic streaming and peroxisome budding. Plant Cell Physiol 43: 384–392

Kabsch W, Mannherz HG, Suck D, Pai EF, Holmes KC (1990) Atomic structure of the actin: DNase I complex. Nature 347:37–44

Kadota A, Yamada N, Suetsugu N, Hirose M, Saito C, Shoda K, Ichikawa S, Kagawa T, Nakano A, Wada M (2009) Short actin-based mechanism for light-directed chloroplast movement in *Arabidopsis*. Proc Natl Acad Sci USA 106:13106–13111

Kagawa T, Wada M (1993) Light-dependent nuclear positioning in prothallial cells of Adiantum capillus-veneris. Protoplasma 177:82–85

Kamiya N, Kuroda K (1956) Velocity distribution of the protoplasmic streaming in *Nitella* cells. Bot Mag Tokyo 69:544–554

Kao YL, Deavours BE, Phelps KK, Walker RA, Reddy AS (2000) Bundling of microtubules by motor and tail domains of a kinesin-like calmodulin-binding protein from *Arabidopsis*: regulation by Ca^{2+}/Calmodulin. Biochem Biophys Res Commun 267:201–207

Kasahara M, Kagawa T, Oikawa K, Suetsugu N, Miyao M, Wada M (2002) Chloroplast avoidance movement reduces photodamage in plants. Nature 420:829–832

Kato N, Reynolds D, Brown ML, Boisdore M, Fujikawa Y, Morales A, Meisel Lee A (2008) Multidimensional fluorescence microscopy of multiple organelles in *Arabidopsis* seedlings. Plant Methods 4:1–14

Katsuta J, Shibaoka H (1988) The roles of the cytoskeleton and the cell wall in nuclear positioning in tobacco BY-2 cells. Plant Cell Physiol 29:403–413

Katsuta J, Hashiguchi Y, Shibaoka H (1990) The role of the cytoskeleton in positioning of the nucleus in premitotic tobacco BY-2 cells. J Cell Sci 95:413–422

Ketelaar T, de Ruijter NC, Emons AM (2003) Unstable F-actin specifies the area and microtubule direction of cell expansion in *Arabidopsis* root hairs. Plant Cell 15:285–292

Knight AE, Kendrick-Jones J (1993) A myosin-like protein from a higher plant. J Mol Biol 231:148–154

Kohno T, Shimmen T (1988) Mechanism of Ca^{2+} inhibition of cytoplasmic streaming in lily pollen tubes. J Cell Sci 91:501–509

Kovar DR, Staiger CJ, Weaver EA, McCurdy DW (2000) AtFim1 is an actin filament crosslinking protein from *Arabidopsis thaliana*. Plant J 24:625–636

Krzeszowiec W, Gabryś H (2007) Blue light-induced reorganization of myosins in *Arabidopsis thaliana*. Plant Signal Behav 2:333–336

Krzeszowiec W, Rajwa B, Dobrucki J, Gabryś H (2007) Actin cytoskeleton in *Arabidopsis thaliana* under blue and red light. Biol Cell 99:251–260

Lawrence CJ, Dawe RK, Christie KR, Cleveland DW, Dawson SC, Endow SA, Goldstein LS, Goodson HV, Hirokawa N, Howard J, Malmberg RL, McIntosh JR, Miki H, Mitchison TJ, Okada Y, Reddy AS, Saxton WM, Schliwa M, Scholey JM, Vale RD, Walczak CE, Wordeman L (2004) A standardized kinesin nomenclature. J Cell Biol 167:19–22

Lawrence CJ, Morris NR, Meagher RB, Dawe RK (2001) Dyneins have run their course in plant lineage. Traffic 2:362–363

Lee YR, Liu B (2004) Cytoskeletal motors in *Arabidopsis*. Sixty-one kinesins and seventeen myosins. Plant Physiol 136:3877–3883

Lenartowska M, Michalska A (2008) Actin filament organization and polarity in pollen tubes revealed by myosin II subfragment I decoration. Planta 228:891–896

Li JF, Nebenführ A (2007) Organelle targeting of myosin XI is mediated by two globular tail subdomains with separate cargo binding sites. J Biol Chem 282:20593–20602

Liu L, Lee YR, Pan R, Maloof JN, Liu B (2005) An internal motor kinesin is associated with the Golgi apparatus and plays a role in trichome morphogenesis in *Arbidopsis*. Mol Biol Cell 16:811–823

Lovy-Wheeler A, Wilsen KL, Baskin TI, Hepler PK (2005) Enhanced fixation reveals the apical cortical fringe of actin filaments as a consistent feature of the pollen tube. Planta 221:95–104

Ma YZ, Yen LF (1989) Actin and myosin in pea tendrils. Plant Physiol 89:586–589

Mathur J, Mathur J, Hülskamp M (2002) Simultaneous visualization of peroxisomes and cytoskeletal elements reveals actin and not microtubule-based peroxisome motility in plants. Plant Physiol 128:1031–1045

Michelot A, Guérin C, Huang S, Ingouff M, Richard S, Rodiuc N, Staiger CJ, Blanchoin L (2005) The formin homology 1 domain modulates the actin nucleation and bundling activity of *Arabidopsis* FORMIN1. Plant Cell 17:2296–2313

Nebenführ A, Gallagher LA, Dunahay TG, Frohlick JA, Mazurkiewicz AM, Meehl JB, Staehelin LA (1999) Stop-and-go movements of plant Golgi stacks are mediated by the acto-myosin system. Plant Physiol 121:1127–1141

Nick P (2008) Plant microtubules: development and flexibility. Springer, Berlin
Nogales E, Whittaker M, Milligan RA, Downing KH (1999) High-resolution model of the microtubule. Cell 96:79–88
Oikawa K, Kasahara M, Kiyosue T, Kagawa T, Suetsugu N, Takahashi F, Kanegae T, Niwa Y, Kadota A, Wada M (2003) Chloroplast unusual positioning 1 is essential for proper chloroplast positioning. Plant Cell 15:2805–2815
Palevitz BA, Hepler PK (1975) Identification of actin *in situ* at the ectoplasm-endoplasm interface of *Nitella,* Microfilament-chloroplast association. J Cell Biol 65:29–38
Park YI, Chow WS, Anderson JM (1996) Chloroplast movement in the shade plant *Tradescantia albiflora* helps protect photosystem II against light stress. Plant Physiol 111:867–875
Peremyslov VV, Prokhnevsky AI, Avisar D, Dolja VV (2008) Two class XI myosins function in organelle trafficking and root hair development in *Arabidopsis.* Plant Physiol 146:1109–1116
Reddy AS, Day IS (2001) Analysis of the myosins encoded in the recently completed *Arabidopsis thaliana* genome sequence. Genome Biol 7:0024
Romagnoli S, Cai G, Faleri C, Yokota E, Shimmen T, Cresti M (2007) Microtubule- and actin filament-dependent motors are distributed on pollen tube mitochondria and contribute differently to their movement. Plant Cell Physiol 48:345–361
Runions J, Brach T, Kühner S, Hawes C (2006) Photoactivation of GFP reveals protein dynamics within the endoplasmic reticulum membrane. J Exp Bot 57:43–50
Ruthardt N, Gulde N, Spiegel H, Fischer R, Emans N (2005) Four-dimensional imaging of cytoplasmic strand dynamics in tobacco BY-2 cells. Protoplasma 225:205–215
Ryu JH, Takagi S, Nagai R (1995) Stationary organization of the actin cytoskeleton in *Vallisneria*: the role of stable microfilaments at the end walls. J Cell Sci 108:1531–1539
Ryu JH, Mizuno K, Takagi S, Nagai R (1997) Extracellular components implicated in the stationary organization of the actin cytoskeleton in mesophyll cells of *Vallisneria.* Plant Cell Physiol 38:420–432
Sakai T, Kagawa T, Kasahara M, Swartz TE, Christie JM, Briggs WR, Okada K (2001) Arabidopsis nph1 and npl1: blue light receptors that mediate both phototropism and chloroplast relocation. Proc Natl Acad Sci USA 98:6969–6974
Sakai Y, Takagi S (2005) Reorganized actin filaments anchor chloroplasts along the anticlinal walls of Vallisneria epidermal cells under high-intensity blue light. Planta 221:823–830
Sakurai N, Domoto K, Takagi S (2005) Blue-light-induced reorganization of the actin cytoskeleton and the avoidance response of chloroplasts in epidermal cells of *Vallisneria gigantea.* Planta 221:66–74
Sato Y, Wada M, Kadota A (2001) Choice of tracks, microtubules and/or actin filaments for chloroplast photo-movement is differentially controlled by phytochrome and a blue light receptor. J Cell Sci 114:269–279
Shimmen T (2007) The sliding theory of cytoplasmic streaming: fifty years of progress. J Plant Res 120:31–43
Sowman BN, Kovar DR, Shevchenko G, Franklin-Tong VE, Staiger CJ (2002) Signal-mediated depolymerization of actin in pollen during the self-incompatibility response. Plant Cell 14:2613–2626
Staehelin AL, Kang BH (2008) Nanoscale architecture of endoplasmic reticulum export sites and of Golgi membranes as determined by electron tomography. Plant Physiol 147:1454–1468
Staiger CJ, Hussey PJ (2004) Actin and actin-modulating proteins. In: Hussey PJ (ed) The plant cytoskeleton in cell differentiation and development. Blackwell, London, pp 32–80
Staiger CJ, Schliwa M (1987) Actin localization and function in higher plants. Protoplasma 41:1–12
Staiger CJ, Sheahan MB, Khurana P, Wang X, McCurdy DW, Blanchoin L (2009) Actin filament dynamics are dominated by rapid growth and severing activity in the *Arabidopsis* cortical array. J Cell Biol 184:269–280
Su H, Wang T, Dong H, Ren H (2007) The Villin/Gelsolin/Fragmin superfamily proteins in plants. J Integr Plant Biol 49:1183–1191

Sugi H, Chaen S (2003) Force-velocity relationships in actin-myosin interactions causing cytoplasmic streaming in algal cells. J Exp Biol 206:1971–1976

Sztatelman O, Waloszek A, Banaś AKB, Gabryś H (2010) Photoprotective function of chloroplast avoidance movement: In vivo chlorophyll fluorescence study. J Plant Physiol 167:709–716

Takagi S (1997) Photoregulation of cytoplasmic streaming: cell biological dissection of signal transduction pathway. J Plant Res 110:299–303

Takagi S, Nagai R (1985) Light-controlled cytoplasmic streaming in *Vallisneria* mesophyll cells. Plant Cell Physiol 26:941–951

Takagi S, Kong SG, Mineyuki Y, Furuya M (2003) Regulation of actin-dependent cytoplasmic motility by type II phytochrome occurs within seconds in *Vallisneria gigantea* epidermal cells. Plant Cell 15:331–345

Takemiya A, Inoue S, Doi M, Kinoshita T, Shimazaki K (2005) Phototropins promote plant growth in response to blue light in low light environments. Plant Cell 17:1120–1127

Thomas C, Dieterle M, Gatti S, Hoffmann C, Moreau F, Papuga J, Steinmetz A (2008) Actin bundling via LIM domains. Plant Signal Behav 3:320–321

Thomas C, Tholl S, Moes D, Dieterle M, Papuga J, Moreau F, Steinmetz A (2009) Actin bundling in plants. Cell Motil Cytoskelet 66:940–957

Tominaga M, Kojima H, Yokota E, Orii H, Nakamori R, Katayama E, Anson M, Shimmen T, Oiwa K (2003) Higher plant myosin XI moves processively on actin with 35 nm steps at high velocity. EMBO J 22:1263–1272

Trojan A, Gabryś H (1996) Chloroplast distribution in *Arabidopsis thaliana* depends on light conditions during growth. Plant Physiol 111:419–425

Vahey M, Titus M, Trautwein R, Scordilis S (1982) Tomato actin and myosin: contractile proteins from a higher plant. Cell Motil 2:131–147

Vale RD (2003) The molecular motor toolbox for intracellular transport. Cell 112:467–480

Vale RD, Fletterick RJ (1997) The design plan of kinesin motors. Annu Rev Cell Dev Biol 13:745–777

van de Meent JW, Tuval I, Goldstein RE (2008) Nature's microfluidic transporter: rotational cytoplasmic streaming at high Peclet numbers. Phys Rev Lett 101:178102

Van Gestel K, Köhler RH, Verbelen J-P (2002) Plant mitochondria move on F-actin, but their positioning in the cortical cytoplasm depends on both F-actin and microtubules. J Exp Bot 53:659–667

Verchot-Lubicz J, Goldstein RE (2009) Cytoplasmic streaming enables the distribution of molecules and vesicles in large plant cells. Protoplasma 240:99–107

Villarejo A, Buren S, Larsson S, Dejardin A, Monne M, Rudhe C, Karlsson J, Jansson S, Lerouge P, Rolland N et al (2005) Evidence for a protein transported through the secretory pathway en route to the higher plant chloroplast. Nat Cell Biol 7:1124–1131

Voigt B, Timmers ACJ, Samaj J, Muller J, Baluska F, Menzel D (2005) GFP-FABD2 fusion construct allows in vivo visualization of the dynamic actin cytoskeleton in all cells of *Arabidopsis* seedlings. Eur J Cell Biol 84:595–608

Wada M, Suetsugu N (2004) Plant organelle positioning. Curr Opin Plant Biol 7:626–631

Wada M, Kagawa T, Sato Y (2003) Chloroplast movement. Annu Rev Plant Biol 54:455–468

Wang Z, Pesacreta TC (2004) A subclass of myosin XI is associated with mitochondria, plastids, and the molecular chaperone subunit TCP-1alpha in maize. Cell Motil Cytoskelet 57: 218–232

Wang HY, Yu Y, Chen ZL, Xia GX (2005) Functional characterization of *Gossypium hirsutum* profilin 1 gene (GhPFN1) in tobacco suspension cells. Characterization of in vivo functions of a cotton profilin gene. Planta 222:594–603

Wang HJ, Wan AR, Jauh GY (2008) An actin-binding protein, LlLIM1, mediates calcium and hydrogen regulation of actin dynamics in pollen tubes. Plant Physiol 147:1619–1636

Wasteneys GO, Galway ME (2003) Remodeling the cytoskeleton for growth and form: an overview with some new views. Annu Rev Plant Biol 54:691–722

Williamson RE (1974) Actin in the alga *Chara corallina*. Nature 248:801–802

Williamson RE (1992) Cytoplasmic streaming in characean algae: mechanism, regulation by Ca^{2+}, and organization. In: Melkonian M (ed) Algal cell motility. Chapman and Hall, New York, pp 73–97

Williamson RE, Ashley CC (1982) Free Ca^{2+} and cytoplasmic streaming in the alga *Chara*. Nature 296:647–651

Woehlke G, Schliwa M (2000) Directional motility of kinesin motor proteins. Biochim Biophys Acta 1496:117–1127

Wojtaszek P, Anielska-Mazur A, Gabryś H, Baluška F, Volkmann D (2005) Recruitment of myosin VIII towards plastid surfaces is root cap-specific and provides the evidence for actomyosin involvement in root osmosensing. Funct Plant Biol 32:721–736

Woods CM, Polito VS, Reid MS (1984) Response to chilling stress in plant cells II. Redistribution of intracellular calcium. Protoplasma 121:17–24

Yang Yd, Elamawi R, Bubeck J, Peperkok R, Ritzenthaler C, Robinson DG (2005) Dynamics of COPII vesicles and the Golgi apparatus in cultured *Nicotiana* tabacum BY-2 cells provides evidence for transient association of Golgi stacks with endoplasmic reticulum exit sites. Plant Cell 17:1513–1531

Zheng M, Beck M, Müller J, Chen T, Wang X, Wang F, Wang Q, Wang Y, Baluška F, Logan DC, Šamaj J, Lin J (2009) Actin turnover is required for myosin-dependent mitochondrial movements in *Arabidopsis* root hairs. PloS ONE 4:e5961

Zurzycki J (1955) Chloroplast arrangements as a factor in photosynthesis. Acta Soc Bot Pol 24:27–63

Generating a Cellular Protuberance: Mechanics of Tip Growth

Anja Geitmann

Abstract A plant cell grows by expansive deformation of its surface, the cell wall. Global cellular elongation growth and the mechanical principles governing this process are intensively studied, but the generation of cellular protuberances, a fundamental process required for the formation of complex plant cell geometries, remains poorly understood. Pollen, the male gametophyte stage of the flowering plants, is an excellent model system for the investigation of the mechanics of protuberance formation. The initiation of pollen tube growth requires the spatially confined formation of a bulge, followed by the elongation of the forming tube through tip growth. Since turgor is a nonvectorial force, this process must be controlled by the mechanical properties of the cell wall. In the elongating tube, cell wall expansion is confined to the apex of the cell, requiring the tubular region to be stabilized against turgor-induced tensile stress. How this is achieved, and why the pollen tube is so successful in invading other tissues, is elucidated from the point of view of cell mechanics.

1 Introduction

Growth processes in plant cells are fundamentally different from those in animal cells due to the different mechanical features determining cell shape in these two cell types. While the forces employed by animal cells to grow and crawl are generated by polymerizing and sliding cytoskeleton arrays, the force required to expand plant cells is exerted by the turgor pressure. This hydrostatic pressure is significantly higher than the cytoskeletal polymerizing force that could theoretically be applied per surface area. This is necessary, since in addition to the pliable plasma membrane surrounding animal cells, plants also possess a polymer wall that needs to be deformed to generate growth. On the other hand, contrary to the cytoskeletal forces which are directed and

A. Geitmann
Département de sciences biologiques, Institut de recherche en biologie végétale, Université de Montréal, 4101 rue Sherbrooke est, Montréal, Québec, Canada H1X 2B2
e-mail: anja.geitmann@umontreal.ca

P. Wojtaszek (ed.), *Mechanical Integration of Plant Cells and Plants*, Signaling and Communication in Plants 9, DOI 10.1007/978-3-642-19091-9_5,
© Springer-Verlag Berlin Heidelberg 2011

occur parallel to the orientation of the cytoskeletal array, the force exerted by the turgor pressure is non-vectorial and it is applied to the entire cellular surface. Cellular geometry determines the resulting stress pattern in the cellular surface, but it is the distribution of mechanical properties in the wall in combination with the geometry that determines the strain pattern that ultimately leads to a particular cellular geometry.

In order to morph into anything other than the approximately spherical or polyhedric shape of the original meristematic cell, the surface of the growing cell needs to expand unevenly, either with a preferential direction of expansion (caused by an anisotropy in

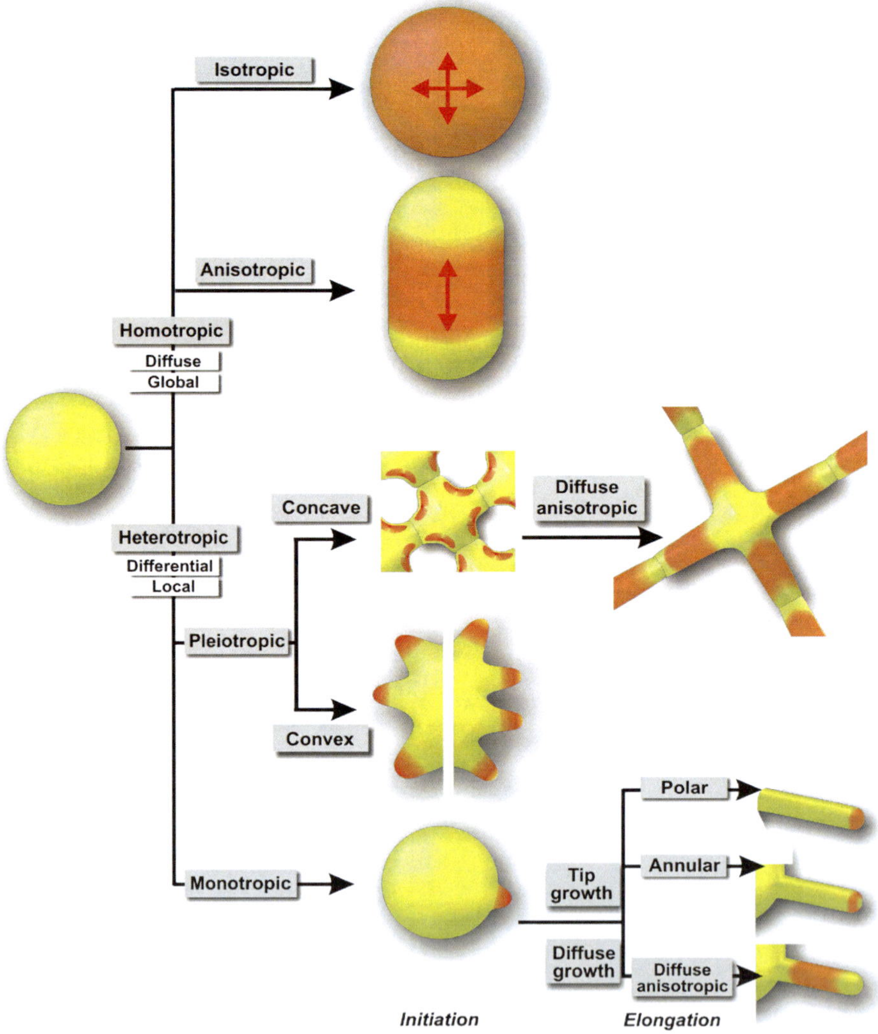

Fig. 1 Different shapes of plant cells generated by heterotropic growth. *Red* indicates expanding areas on the cellular surface. Modified after Geitmann and Ortega (2009) with permission

extensibility) or with different degrees of expansion at different sites of the surface (caused by spatial variations of the degree of extensibility). It is the former (anisotropic expansion) that leads to cell shapes such as the cylindrical cells of the palisade parenchyma, whereas the latter (local growth) causes the formation of protuberances required to shape cells such as star-shaped trichomes or jigsaw puzzle shaped leaf epidermis cells (Fig. 1). Any combination of these two processes, occurring simultaneously or during subsequent phases of development, are possible within an individual cell.

The formation of a finger-shaped cellular protuberance, or monotropic growth (Geitmann and Ortega 2009), is a growth pattern that can be produced through varying combinations of anisotropic and locally confined growth (Fig. 1). At least two phases can be distinguished during monotropic growth (1) the initiation of the growth event and (2) the elongation phase. In some cell types, such as root hairs, one can also distinguish a well-defined termination phase. The elongation phase of a monotropically growing cell protuberance can proceed in various ways, ranging from diffuse anisotropic (e.g., *Arabidopsis* trichome branches, cotton fibers) to truly polar (e.g., fungal hyphae). In the former, cell wall expansion occurs over the entire length of the cellular protuberance, whereas in the latter, the highest rate of expansion is localized at the apical pole of the cell. Root hairs and pollen tubes seem to represent an intermediary type since while the pole does expand, the highest rates of surface expansion are observed in an annular region around the pole of the cell (Geitmann and Dumais 2009). In both truly polar and annular growth, the cylindrical shank of the cell does not expand in any direction and hence both types are defined as tip growth. Because of the speed of the monotropic growth process in pollen tubes, this cell type has become a model system for the investigation of plant cell growth.

2 Initiation of a Bulge

2.1 Spatially Confined Surface Expansion Requires Local Cell Wall Softening

The hydrostatic pressure generated by the turgor is nonvectorial and uniform throughout the cytoplasm of each individual cell. Therefore, in a spherical cell the resulting stress pattern in the wall is isotropic and homogeneous. Consequently, the generation of a local protuberance must rely on local softening of the cell wall that allows the strain to vary over short distances despite the presence of a uniform stress pattern in the surface. Pollen germination requires such a generation of a spatially confined surface expansion and most pollen grains have precisely such local regions of mechanically different cell wall – the apertures. It is generally through these openings in the exine layer that pollen tubes emerge. The intine layer forming the aperture region usually appears ultrastructurally different from the intine of the other regions of the grain (Heslop-Harrison 1987). During normal pollen grain hydration (either in vitro or on the surface of a receptive stigma) the intine layer undergoes structural changes, frequently including local thickening (Heslop-Harrison and Heslop-Harrison 1992; Márquez et al. 1997). This would seem to be counterintuitive given that cell wall stiffness is proportional to cell wall thickness and would thus be expected to increase at the aperture (Geitmann and Ortega 2009). However, from transmission electron micrographs it is evident that the material composing the apertural intine seems to loosen during the increase in thickness (Márquez et al. 1997). As a result, the tensile modulus and with it the stretching stiffness of the aperture intine are likely to decrease overall. It is largely unknown which biochemical changes in the cell wall are responsible for this loosening of the apertural intine. Pollen germination is associated with production of reactive oxygen species (ROS) at the germination site (Smirnova et al. 2009; Speranza et al. 2011) and the expression of expansins and enzymes such as xyloglucan endotransglycosylase/hydrolase, polygalacturonases, pectate lyases, and pectin esterases in hydrating or germinating pollen (Holmes-Davis et al. 2005; Noir et al. 2005; Pina et al. 2005; Dai et al. 2007; Russell et al. 2008; Sheoran et al. 2009) suggests possible involvement of these proteins in cell wall softening. However, the exact timing and subcellular location of these protein activities in relation to pollen germination are poorly understood and only little information about their targets – the pollen grain's own cell wall or the apoplast of the stigmatic and transmitting tissues – is available (Suen and Huang 2007; Valdivia et al. 2009).

The fact that a local outgrowth must be preceded by a spatially confined softening has been demonstrated on spherical swellings of the pollen tube apex. Various pharmacological or enzymatic treatments are known to transform the unidirectional growth pattern of the pollen tube apex into an isotropic pattern leading to a spherical swelling. This swelling of the pollen tube apex is easier to handle experimentally and analytically since it eliminates the complicating architecture of the pollen grain. Microindentation on different locations of this drug-induced swelling was able to predict the location of the emergence of a new tubular outgrowth (Zerzour et al. 2009). Analysis of time course experiments revealed that this softening clearly occurs prior to any visible outgrowth and is thus likely to be a prerequisite for its onset (Fig. 2).

2.2 Pectin Chemistry

Among the enzymes suspected to be crucial during pollen germination are pectin modifying enzymes, since the chemistry of pectin controls the behavior of this polymer under a mechanical load. Highly methyl-esterified pectin (as is typical for newly deposited pectin) is known to be softer or even liquefied compared with de-esterified pectins. De-esterification results from the activity of the enzyme pectin methylesterase. The ensuing increase in the abundance of negatively charged groups causes calcium ions to gel the polymer (Carpita and Gibeaut 1993). This process has been shown experimentally to rigidify the pollen tube cell wall (Parre and Geitmann 2005b). In *Zygophyllum* pollen, the apertural intine is richer in methyl-esterified pectin than the other regions of the intine (Castells et al. 2003).

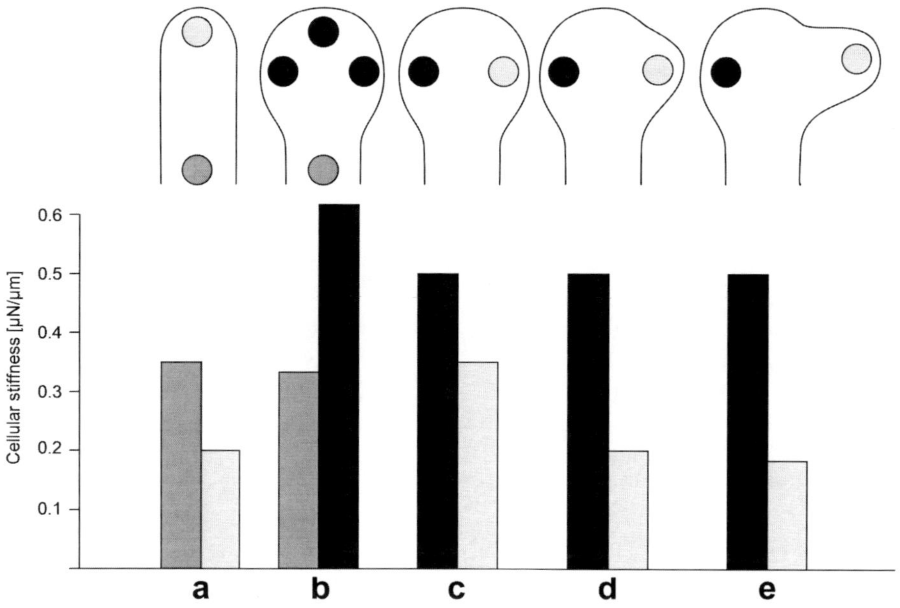

Fig. 2 Temporal changes in cellular stiffness at different locations on *Papaver* pollen tube during administration and after removal of cytochalasin D. The *graph* represents the recovery of a typical tube after cytochalasin D treatment. *Circles* indicate the position of the microindenter on the tube. *Circle colors* correspond to *bar colors* in the graph. (**a**) Normally growing pollen tube. The apex is softer than the distal region. (**b**) The swelling induced by a 10-min treatment with 10 µM cytochalasin D is significantly stiffer than the apex of the normally growing tube or the distal region. (**c**) Before the swelling becomes visibly asymmetric the stiffness on one side is significantly lowered. (**d**) The future outgrowth is recognizable. The stiffness in the outgrowth is reduced to the same level as that of a normally growing tube apex. (**e**) The outgrowth forms into a normally growing tube. Reprinted from Zerzour et al. (2009) with permission

Within the relatively large colpus area of the ungerminated *Lilium* pollen grain, a small spot displaying strong accumulation of esterified pectin develops at the future site germination (Parre and Geitmann 2005b). These highly localized accumulations of esterified pectins are likely generated through spatially controlled exocytosis, as is supported by the accumulation of cytoplasmic vesicles subjacent to the aperture (Cresti et al. 1977, 1985; Heslop-Harrison and Heslop-Harrison 1992). A recently developed conceptual model explains how the addition of pectate molecules to a preexisting cell wall may cause wall loosening and growth. Newly inserted pectate might remove calcium ions from the existing wall and thus break load-bearing bonds (Proseus and Boyer 2006, 2007). So if pectate deposition can be appropriately confined in space, the resulting growth event should also be restricted to a limited surface area.

It is interesting to note in this context that inviable stored pollen of rye is able to produce a short protuberance upon contact with growth medium, but that this protuberance is isodiametric and not able to elongate in polar manner (Heslop-Harrison 1979).

This suggests that in rye at least the initial bulging of the intine is a passive, turgor-driven deformation process and that the subsequent, polar elongation requires addition of new cell wall material. It remains unknown, however, which residual metabolic processes are ongoing in these "inviable" grains that may be involved in the "passive" initiation of the isodiametric outgrowth in the first place.

2.3 Cellulose in the Aperture

A somewhat surprising observation is that calcofluor white labeling indicates an accumulation of cellulose at the future site of pollen tube emergence (Heslop-Harrison and Heslop-Harrison 1992; Parre and Geitmann 2005a). Because of their crystalline nature, the formation of cellulose microfibrils is generally associated with rigidification and the loss of extensibility of the cell wall. Why would a potentially stiffening cell wall component be present in the site of the future pollen tube emergence? Several considerations may explain this apparent contradiction. Firstly, in order to confer stability to the cell wall, microfibrils need to be attached to something or linked to each other, either by a gel-like matrix consisting of other cell wall polymers or directly by hemicellulose bridges. Their presence per se does not therefore necessarily rigidify the wall, and in *Narcissus* apertures, the cellulose fibrils have indeed been found to be short and randomly disposed (Heslop-Harrison and Heslop-Harrison 1992). Secondly, the dye used for these experiments, calcofluor white, does not distinguish between amorphous and crystalline cellulose, and it might be even less specific and also bind to other cell wall components (Hughes and McCully 1975). Therefore, in order to interpret the mechanical role of the cellulose at the site of pollen tube emergence it will be necessary to confirm its presence with a specific antibody, and then determine the degree of crystallinity, cross-linking, and orientation of the microfibrils.

2.4 Mechanics of the Collar of the Protuberance

The emerging pollen tube often has a diameter that is significantly smaller than the size of the aperture (Fig. 3a). From a mechanical point of view this is intriguing, since geometry causes the area immediately surrounding the base of the emerging tube to experience locally elevated tensile stress in circumferential direction. The exine is generally several microns removed from this critical area and therefore does not seem to contribute to its reinforcement (Fig. 3b). Structural support from this area seems to be provided by the intine itself, since in many species the base of the pollen tube is bordered by a thickening pointing into the inside of the cell (Heslop-Harrison and Heslop-Harrison 1992; Suarez-Cervera et al. 2002; Castells et al. 2003). These collar-shaped thickenings can be of pectic or callosic nature. Despite its amorphous character in cytomechanical tests, callose has been shown to

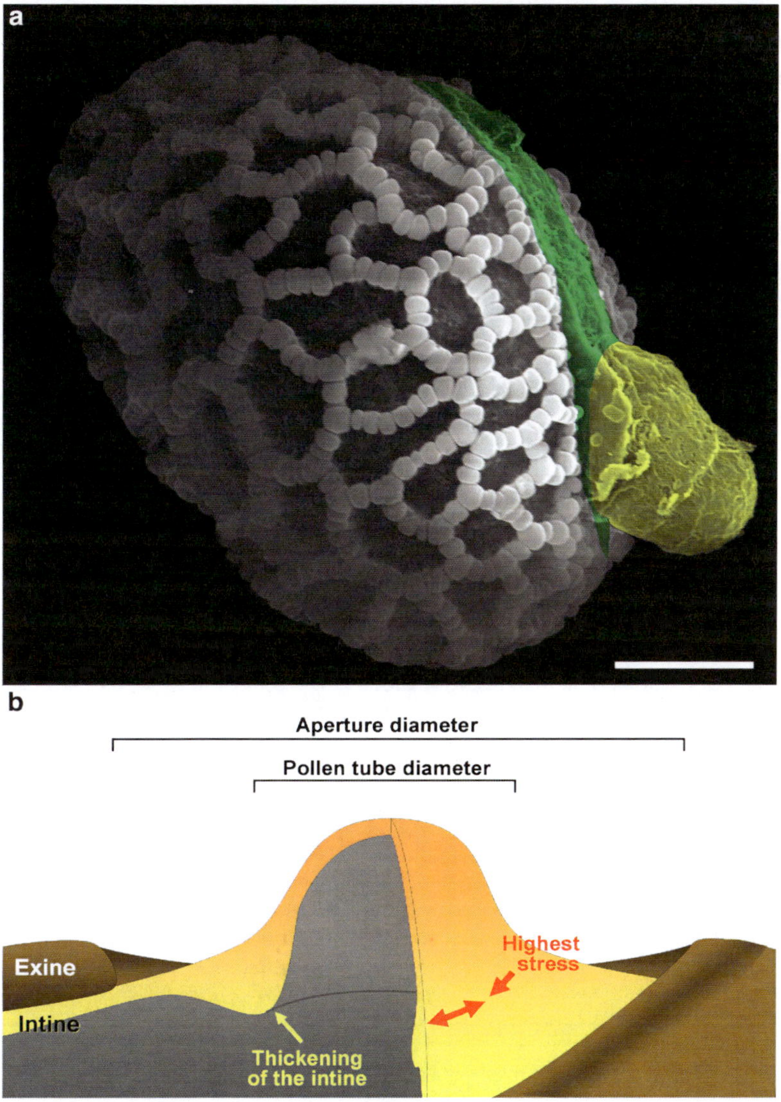

Fig. 3 (**a**) Scanning electron micrograph of a lily pollen grain germinated in vitro. The colpus area (*green*) is significantly larger than the cross-section of the emerging pollen tube (*yellow*). Bar = 20 μm. (**b**) Schematic drawing of the tensile stress pattern in the apertural region of a germinating pollen grain. The borders of the exine are located at a significant distance from the intine region experiencing the highest tensile stress (*red arrows*). Because of the distance, the exine is unlikely to stabilize the base of the emerging pollen tube. Instead, the intine shows a collar-like thickening, likely to be responsible for mechanical support. Reprinted from Geitmann (2010) with permission

possess the ability to resist tensile stress (Parre and Geitmann 2005a), and the spatiotemporal dynamics of its appearance prior to and during pollen germination is consistent with its having an important mechanical role in the formation of the protuberance.

3 Elongation of the Protuberance

3.1 Spatial Distribution of Cell Wall Extensibility

Once the pollen tube has germinated and assumes a cylindrical shape, the turgor generated stress pattern in the cell wall is not uniform anymore. Similar to cylindrical pressure vessels with hemisphere-shaped ends, the tensile stress in circumferential direction is twice as high as the stress in longitudinal direction or the stress in hemispherical ends (Geitmann and Steer 2006). For this reason, the cell wall of the cylindrical shank must be mechanically reinforced in order not to balloon under pressure.

The nonuniform distribution of callose, which is present only in the distal region of the tube, is at least partly responsible for this mechanical reinforcement. Resistance against tensile stress of the amorphous cell wall component callose has been demonstrated by enzymatic digestion of the callose layer resulting in an increased cellular diameter. Furthermore, using microindentation the cellular stiffness of pollen tubes is reduced and the viscous component in the viscoelastic behavior measured using microindentation is more pronounced (Parre and Geitmann 2005a).

However, since visible amounts of callose are only present starting at a significant distance from the apex (Ferguson et al. 1998; Parre and Geitmann 2005a), the spatial distribution of this polymer does not correspond to the critical increase of tensile stress at the transition region between hemisphere shaped apex and shank (Fig. 4a, b). This transition region is rather characterized by a steep gradient in the degree of esterification of pectin polymers (Fayant et al. 2010). Pectins are deposited by exocytosis into the apical cell wall in a highly methylesterified form (Bosch et al. 2005; Parre and Geitmann 2005b). Their de-esterification during cell wall maturation is the result of the activity of pectin methylesterases that are also secreted at the apex (Bosch et al. 2005). Apically localized pectin methylesterase inhibitors do prevent the precocious enzymatic activity of these enzymes until the newly assembled wall section has undergone expansion and forms the cylindrical tube which requires such rigidification (Röckel et al. 2008). The disturbance of pectin methylesterase activity through mutation (Tian et al. 2006) and the application of ectopic pectin-methyl esterase (Parre and Geitmann 2005b) have dramatic consequences for pollen tube growth and morphology illustrating the importance of this enzyme in this process.

The morphology of the apical region of the pollen tube can be affected by administration of inhibitors of cellulose synthesis which cause apical swelling (Anderson et al. 2002; Lazzaro et al. 2003; Aouar et al. 2010). This suggests that

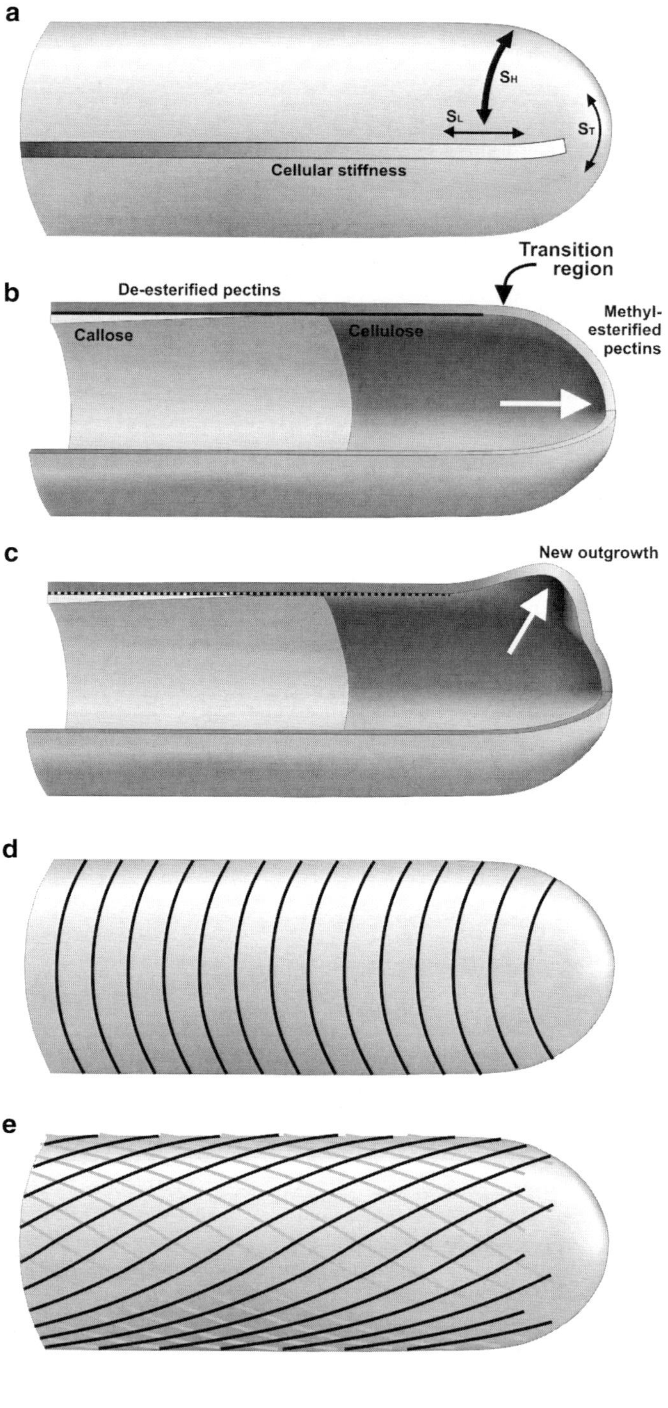
a
SH
SL
ST
Cellular stiffness
Transition region
b
De-esterified pectins
Methyl-esterified pectins
Callose
Cellulose
c
New outgrowth
d
e

cellulose is involved in controlling cell wall stability in the subapical transition region. Remarkably, recovery of growth after temporary arrest through inhibitor treatment often occurs in a direction different from the original growth direction (Fig. 4c). Only a mechanical model will be able to provide a conclusive answer, but it is possible that due to the geometry-based stress pattern in combination with altered distribution of mechanical properties are responsible for a deviation in the new outgrowth direction.

3.2 *Helical Arrangement of Cellulose*

In most plant cells exhibiting primary growth, patterns of cellulose orientation play a pivotal role in producing an anisotropically reinforced wall. It is this anisotropy that causes polyhedric meristematic cells to elongate along a particular axis, such as is typically the case in stem and root tissues, in which cellulose microfibrils are oriented perpendicular to the growth axis (Baskin 2005). Even though compared with other primary plant cells the pollen tube cell wall has very little cellulose (Rae et al. 1985) it is worth investigating its configuration to understand the mechanics of the cellular architecture. The principal difference between a cylindrical root cell and a pollen tube is of course the fact that the former expands over its entire length, whereas the latter expands only at the apical end. Nevertheless, to prevent passive radial expansion of the cylindrical region of the tube, the wall needs to be reinforced and cellulose could play a role in this resistance against circumferential stress. If this was the case, microfibrils should be arranged perpendicularly to the length axis of the cell (Fig. 4d). This is not the case, however. In pollen tubes from *Petunia* and *Pinus*, the principal direction of cellulose orientation was determined to be 45° to the long axis (Sassen 1964; Derksen et al. 1999), and in *Lilium* and *Solanum* pollen tubes, the pitch of the helical arrangement is even lower with 20° and 15°, respectively (Aouar et al. 2010) (Fig. 4e). It is therefore not the turgor-induced stress in circumferential direction against which microfibrils provide resistance in the distal tubular region of the pollen tube. Why then is there deposition of cellulose, albeit in low amounts? Comparison with fiber-reinforced pressure vessels

Fig. 4 Schematic drawing of the cell wall architecture in the apical region of a pollen tube. (**a**) According to the laws of thin walled pressure vessels the tensile stress in longitudinal direction (S_L) is twice as high as the tensile stress in circumferential or hoop direction. (S_H) which in turn is equal to the tensile stress at the tip (S_T). (**b**) The nonuniform distribution of cell wall components and consequently of the mechanical properties in the normally growing pollen tube drives growth to occur in the direction parallel to the length axis of the cell (*white arrow*). (**c**) Through temporary growth arrest the apical pectin hardens and the relationship between cell wall extensibility and tensile stress favors the formation of an outgrowth in the subapical transition region between hemisphere-shaped apex and cylindrical shank. (**d**) Hypothetical ideal orientation of a fiber with reinforcing function in circumferential direction on the surface of a cylinder. (**e**) Microfibril orientation with a pitch of approximately 25° observed in *Lilium* pollen tubes. Modified after Aouar et al. (2010) with permission

provides an interesting explanation. Obliquely oriented fiber reinforcement stabilizes cylindrical hydroskeletal systems against flexural bending stress (Koehl et al. 2000). Above a pitch angle of 54°, mechanical resistance against buckling is limited, but lower pitch angles confer significant resistance to longitudinal compression stress and prevent buckling. Manipulating in vitro growing pollen tubes using micromechanical techniques reveals indeed that these cells display relatively high, elastic resistance against bending stress, thus supporting this hypothesis.

It is interesting to try to interpret this mechanical property in the context of the pollen tube's biological purpose. Cylindrical cells in stem tissues are certainly exposed to longitudinal compressive stress, but their buckling is prevented by the hydroskeleton of the surrounding tissue cells. The pollen tube on the other hand grows individually and through air or mucilage filled spaces. On its path toward the ovule, the pollen tube has to invade the apoplastic space of the stigmatic and micropylar tissues which are likely to present mechanical obstacles despite the enzyme-mediated digestive attacks of the pollen tube on the mechanics of these tissues. Therefore, the pollen tube needs to exert a mechanical force that permits this invasive activity. A relatively stiff rod is more likely to succeed in this endeavor than a cell that easily bends or buckles upon application of a longitudinal compressive load. Therefore, the cellulose-mediated stiffening against bending stress may facilitate the invasion of the pollen tube into the pistillar tissues. A similar strategy may be employed by root hairs. In these cells, microfibrils are oriented in helicoidal manner (Emons and van Maren 1987). Although this makes for an overall isotropic orientation of tensile stress resistance, the helical component in the microfibril population may nevertheless have the role of stabilizing the cell against longitudinal compression stress.

4 Temporal Dynamics of the Elongation Process

Having established that the mechanical properties of the cell wall determine the spatial parameters of pollen tube growth, similar considerations impose themselves regarding the temporal dynamics of the process. It has been described for numerous plant species that pollen tube growth does rarely occur in steady manner, but that the growth rate undergoes oscillatory changes. The periods of these oscillations are typically between 30 and 120 s and the growth rate changes typically by a factor 6 (Cárdenas et al. 2008; Michard et al. 2008), but this value can reach up to 50 (Geitmann 1997). The oscillations can be sinusoidal (typical for lily pollen tubes) or occur in pulse-likes peaks (tobacco pollen tubes). Since their first description it has been proposed repeatedly that these changes in the growth rate might be caused by changes in the intracellular turgor pressure (Plyushch et al. 1995; Derksen 1996; Messerli et al. 2000; Messerli and Robinson 2003; Zonia and Munnik 2007, 2009).

Although intriguing, this hypothesis is not consistent with the widely recognized biophysical principles governing growth in other types of plant cells. In most plant cells the onset of growth is not associated with increases in turgor pressure, but rather with a relaxation of the cell wall, for example, through enzymatic action

(Schopfer 2006). This relaxation leads to a small and temporary reduction in the turgor pressure, which will cause influx of water to reestablish turgor. It is this influx resulting in volume increase that drives the irreversible deformation of the cell wall and thus growth. The geometry of tip growth is of course different from that of diffusely growing cells. However, tip growing cells nevertheless have to obey biophysical laws, and therefore there is no real reason why their growth should be governed by principles that are different from those of other plant cells. A first indication that general principles of plant cell growth apply to pollen tubes was the observation that pollen tube turgor does not change significantly over time when measured with a turgor pressure probe (Benkert et al. 1997). Unfortunately, pollen tube growth rates in that particular study were not monitored, and although the age of the tubes would lead to expect them to oscillate (since in lily, oscillations appear about 2 h after germination), there is no proof that the measured cells actually did.

If turgor pressure does not change during oscillations, the mechanical properties of the cell wall have to. Therefore, looking at the pollen tube cell wall could provide conclusive answers instead. A recent study demonstrated that the thickness of the apical cell wall region in pollen tubes undergoes changes in time (McKenna et al. 2009). The cell wall is thickest preceding peaks in the growth rate by 100° on a scale on which 360° corresponds to a full period between two growth peaks, typically 40 s for lily. This thickening may appear puzzling at first, because cell wall extensibility is directly dependent on cell thickness and a thicker wall should thus be stiffer. However, similarly to pollen germination, if the newly added cell wall material is significantly softer, or if it causes a softening of the existing wall, this thickening preceding a growth peak actually represents a reduction in overall cell wall extensibility which is thought to be the reason for the growth event (Kroeger et al. 2011). This is corroborated by the finding that the degree of cell wall thickness prior to individual growth peaks seems to predict the size of these peaks (McKenna et al. 2009).

Clear mechanical proof for a softening event was then provided by microindentation (Zerzour et al. 2009). These measurements revealed that the apical cell wall is softest at a moment preceding the growth peak by 10–90°. It hardens during the rapid growth phase, which is consistent with, albeit no proof for a strain hardening process occurring during rapid expansion. Even more conclusive, when carried out in the distal region of the cell, microindentation demonstrated clearly that overall turgor pressure does not change measurably. Since growth rate was monitored to be truly oscillating in the measured cells, these data provide proof for the concept that turgor does not build up prior to a growth event, but that it is the softening of the apical cell wall, which causes rapid growth (Winship et al. 2010).

5 Overcoming Mechanical Obstacles

Confining cellular expansion to a small area at the very end of the cell requires very precise spatial control and highly active intracellular transport processes. Therefore, there must be a significant advantage to tip growth that is associated with the biological function of the cells exhibiting this mode of elongation. The common feature of all tip-growing cells is their ability to invade a matrix, be it biological or inorganic. Pollen tubes invade the pistillar tissues, root hairs penetrate the soil, and fungal hyphae force their way into anything providing nutrients. Invasion requires overcoming the mechanical impedance of the invaded matrix by inducing a fracture in the matrix or by liquefying the material. To be a successful invader, any waste of energy needs to be avoided. The larger the contact area between the invading cell and the invaded matrix, the greater the potential for friction due to slippage of the cell wall of the invader against the invaded matrix. The reduction of the expanding cellular surface region to a minimum (by restricting all expansion to the tip of the cell) reduces the size of the area experiencing friction, and thus minimizes potential energy loss.

Focusing expansion to a single site has an additional advantage. It allows the advancing cell to react rapidly and precisely to directional signals. A new growth direction can be adopted within minutes or even seconds (Higashiyama and Hamamura 2008). The change of direction can be at a significant angle; turns of 90° are easily possible if required. In the case of the pollen tube, this capacity is crucial, since several changes in growth direction are necessary to transverse the pistillar maze and find the target of the growth process, the female gametes. Therefore, in addition to being energetically advantageous in vivo, tip growth also allows for optimal flexibility thus increasing the effectiveness of the response to environmental signals.

6 Conclusions

The complex cellular geometries that can be encountered in plants are often the result of the formation of one or several protuberances from the main body of the cell. The initiation of a protuberance requires spatially confined softening of the cell wall and its elongation can proceed in various ways, diffusely or tip-focused. In both cases, the cylindrical shank of the protuberance needs to be reinforced against circumferential tension stress caused by the turgor pressure, but the cell wall structure employed for this purpose differs between the cell types. Tip-focused growth activity represents an adaptation to the invasive lifestyle of the cells displaying this growth pattern, such as pollen tubes. Confinement of the growth activity to the apex reduces friction, and the invasive force is ensured through a helically reinforced hydroskeleton. Analyzing the cellular mechanics of different plant cell types is therefore highly informative and reveals important aspects of cellular functioning.

Acknowledgments Research in the Geitmann lab is funded by the Natural Sciences and Engineering Research Council of Canada (NSERC), the *Fonds Québécois de la Recherche sur la Nature et les Technologies* (*FQRNT*), and the Human Frontier Science Program (HFSP). Thanks to Youssef Chebli and Louise Pelletier for preparing the scanning electron micrographs. Selected text passages are taken from or modified after (Geitmann 2010) with permission.

References

Anderson JR, Barnes WS, Bedinger P (2002) 2,6-Dichlorbenzonitrile, a cellulose biosynthesis inhibitor, affects morphology and structural integrity of petunia and lily pollen tubes. J Plant Physiol 159:61–67

Aouar L, Chebli Y, Geitmann A (2010) Morphogenesis of complex plant cell shapes – the mechanical role of crystalline cellulose in growing pollen tubes. Sex Plant Reprod 23:15–27

Baskin T (2005) Anisotropic expansion of the plant cell wall. Annu Rev Cell Dev Biol 21:203–222

Benkert R, Obermeyer G, Bentrup FW (1997) The turgor pressure of growing lily pollen tubes. Protoplasma 198:1–8

Bosch M, Cheung AY, Hepler PK (2005) Pectin methylesterase, a regulator of pollen tube growth. Plant Physiol 138:1334–1346

Cárdenas L, Lovy-Wheeler A, Kunkel JG, Hepler PK (2008) Pollen tube growth oscillations and intracellular calcium levels are reversibly modulated by actin polymerization. Plant Physiol 146:1–11

Carpita NC, Gibeaut DM (1993) Structural models of primary cell walls in flowering plants: consistency of molecular structure with the physical properties of the walls during growth. Plant J 3:1–30

Castells T, Seoane-Camba JA, Suàrez-Cervera M (2003) Intine wall modifications during germination of *Zygophyllum fabago* (Zygophyllaceae) pollen grains. Can J Bot 81:1267–1277

Cresti M, Pacini E, Ciampolini F, Sarfatti G (1977) Germination and early tube development in vitro of *Lycopersicum peruvianum* pollen: ultrastructural features. Planta 136:239–247

Cresti M, Ciampolini F, Mulcahy DLM, Mulcahy G (1985) Ultrastructure of *Nicotiana alata* pollen, its germination and early tube formation. Am J Bot 72:719–727

Dai S, Chen T, Chong K, Xue Y, Liu S, Wang T (2007) Proteomics identification of differentially expressed proteins associated with pollen germination and tube growth reveals characteristics of germinated *Oryza sativa* pollen. Mol Cell Prot 6:207–230

Derksen J (1996) Pollen tubes: a model system for plant cell growth. Bot Acta 109:341–345

Derksen J, Li Y-Q, Knuiman B, Geurts H (1999) The wall of *Pinus sylvestris* L. pollen tubes. Protoplasma 208:26–36

Emons AMC, van Maren N (1987) Helicoidal cell-wall texture in root hairs. Planta 170:145–151

Fayant P, Girlanda O, Aubin C-E, Villemure I, Geitmann A (2010) Finite element model of polar growth in pollen tubes. Plant Cell 22:2579–2593

Ferguson C, Teeri TT, Siika-aho M, Read SM, Bacic A (1998) Location of cellulose and callose in pollen tubes and grains of *Nicotiana tabacum*. Planta 206:452–460

Geitmann A (1997) Growth and formation of the cell wall in pollen tubes of *Nicotiana tabacum* and *Petunia hybrida*. PhD thesis, Vol PhD thesis. Hänsel-Hohenhausen, Egelsbach Frankfurt Washington

Geitmann A (2010) How to shape a cylinder: pollen tube as a model system for the generation of complex cellular geometry. Sex Plant Reprod 23:63–71

Geitmann A, Dumais J (2009) Not-so-tip-growth. Plant Signal Behav 4:136–138

Geitmann A, Ortega JKE (2009) Mechanics and modeling of plant cell growth. Trends Plant Sci 14:467–478

Geitmann A, Steer MW (2006) The architecture and properties of the pollen tube cell wall. In: Malhó R (ed) The pollen tube: a cellular and molecular perspective, Plant Cell Monographs, vol 3. Springer, Berlin, pp 177–200

Heslop-Harrison J (1979) Aspects of the structure, cytochemistry and germination of the pollen of rye (*Secale cereale* L.). Ann Bot 44(Suppl):1–47

Heslop-Harrison J (1987) Pollen germination and pollen-tube growth. Int Rev Cytol 107:1–78

Heslop-Harrison Y, Heslop-Harrison J (1992) Germination of monocolpate angiosperm pollen: evolution of actin cytoskeleton and wall during hydration, activation and tube emergence. Ann Bot 69:385–394

Higashiyama T, Hamamura Y (2008) Gametophytic pollen tube guidance. Sex Plant Reprod 21:17–26

Holmes-Davis R, Tanaka CK, Vensel WH, Hurkman WJ, McCormick S (2005) Proteome mapping of mature pollen of *Arabidopsis thaliana*. Proteomics 5:4864–4884

Hughes J, McCully ME (1975) The use of an optical brightener in the study of plant structure. Stain Technol 50:319–329

Koehl MAR, Quillin KJ, Pell A (2000) Mechanical design of fiber-wound hydraulic skeletons: The stiffening and straightening of embryonic notochords. Am Zool 40:28–41

Kroeger JH, Zerzour R, Geitmann A (2011) Regulator or driving force? The role of turgor pressure in oscillatory plant cell growth. PLoS One 6:e18549

Lazzaro MD, Donohue JM, Soodavar FM (2003) Disruption of cellulose synthesis by isoxaben causes tip swelling and disorganizes cortical microtubules in elongating conifer pollen tubes. Protoplasma 220:201–207

Márquez J, Seoane-Camba JA, Suárez-Cervera M (1997) Allergenic and antigenic proteins released in the apertural sporoderm during the activation process in grass pollen grains. Sex Plant Reprod 10:269–278

McKenna ST, Kunkel JG, Bosch M, Rounds CM, Vidali L, Winship LJ, Hepler PK (2009) Exocytosis precedes and predicts the increase in growth in oscillating pollen tubes. Plant Cell 21:3026–3040

Messerli MA, Robinson KR (2003) Ionic and osmotic disruption of the lily pollen tube oscillator: testing proposed models. Planta 217:147–157

Messerli MA, Creton R, Jaffe LF, Robinson KR (2000) Periodic increases in elongation rate precede increases in cytosolic Ca^{2+} during pollen tube growth. Dev Biol 222:84–98

Michard E, Dias P, Feijó JA (2008) Tobacco pollen tubes as cellular models for ion dynamics: improved spatial and temporal resolution of extracellular flux and free cytosolic concentration of calcium and protons using PHluorin and YC3.1CaMelon. Sex Plant Reprod 21:169–181

Noir S, Bräutigam A, Colby T, Schmidt J, Panstruga R (2005) A reference map of the *Arabidopsis thaliana* mature pollen proteome. Biochem Biophys Res Commun 337:1257–1266

Parre E, Geitmann A (2005a) More than a leak sealant – the physical properties of callose in pollen tubes. Plant Physiol 137:274–286

Parre E, Geitmann A (2005b) Pectin and the role of the physical properties of the cell wall in pollen tube growth of *Solanum chacoense*. Planta 220:582–592

Pina C, Pinto F, Feijó JA, Becker JD (2005) Gene family analysis of the *Arabidopsis* pollen transcriptome reveals biological implications for cell growth, division control, an gene expression regulation. Plant Physiol 138:744–756

Plyushch TA, Willemse MTM, Franssen-Verheijen MAW, Reinders MC (1995) Structural aspects of in vitro pollen tube growth and micropylar penetration in *Gasteria verrucosa* (Mill.) H. Duval and *Lilium longiflorum* Thunb. Protoplasma 187:13–21

Proseus T, Boyer J (2006) Calcium pectate chemistry controls growth rate of *Chara corallina*. J Exp Bot 57:3989–4002

Proseus T, Boyer J (2007) Tension required for pectate chemistry to control growth in *Chara corallina*. J Exp Bot 58:4283–4292

Rae AL, Harris PJ, Bacic A, Clarke AE (1985) Composition of the cell walls of *Nicotiana alata* Link et Otto pollen tubes. Planta 166:128–133

Röckel N, Wolf S, Kost B, Rausch T, Greiner S (2008) Elaborate spatial patterning of cell-wall PME and PMEI at the pollen tube tip involves PMEI endocytosis, and reflects the distribution of esterified and de-esterified pectins. Plant J 53:133–143

Russell SD, Bhalla PL, Singh MB (2008) Transcriptome-based examination of putative pollen allergens of rice (*Oryza sativa* ssp. japonica). Mol Plant 1:751–759

Sassen MMA (1964) Fine structure of *Petunia* pollen grain and pollen tube. Acta Bot Neerl 13:175–181

Schopfer P (2006) Biomechanics of plant growth. Am J Bot 93:1415–1425

Sheoran IS, Pedersen EJ, Ross ARS, Sawhney VK (2009) Dynamics of protein expression during pollen germination in canola (*Brassica napus*). Planta 230:779–793
Smirnova AV, Matveyeva NP, Polesskaya OG, Yermakov IP (2009) Generation of reactive oxygen species during pollen grain germination. Russ J Develop Biol 40:345–353
Speranza A, Crinelli R, Scoccianti V, Geitmann A (2011) Reactive oxygen species are involved in pollen tube initiation in kiwifruit. Plant Biol (in press)
Suarez-Cervera M, Arcalis E, Le Thomas A, Seoane-Camba J (2002) Pectin distribution pattern in the apertural intine of *Euphorbia peplus* L. (Euphorbiaceae) pollen. Sex Plant Reprod 14:291–298
Suen DF, Huang AH (2007) Maize pollen coat xylanase facilitates pollen tube penetration into silk during sexual reproduction. J Biol Chem 282:625–636
Tian G-H, Chen M-H, Zaltsman A, Citovsky V (2006) Pollen-specific pectin methylesterase involved in pollen tube growth. Dev Biol 294:83–91
Valdivia ER, Stephenson AG, Durachko DM, Cosgrove D (2009) Class B β-expansins are needed for pollen separation and stigma penetration. Sex Plant Reprod 22:141–152
Winship LJ, Obermeyer G, Geitmann A, Hepler PK (2010) Under pressure, cell walls set the pace. TiPS 15:363–369
Zerzour R, Kroeger JH, Geitmann A (2009) Polar growth in pollen tubes is associated with spatially confined dynamic changes in cell mechanical properties. Dev Biol 334:437–446
Zonia L, Munnik T (2007) Life under pressure: hydrostatic pressure in cell growth and function. Trends Plant Sci 12:90–97
Zonia L, Munnik T (2009) Uncovering hidden treasures in pollen tube growth mechanics. Trends Plant Sci 14:318–327

Mechanics of the Meristems

Dorota Kwiatkowska and Jerzy Nakielski

Abstract In this chapter, the structure, function, and growth of apical meristems and cambium are discussed from a perspective of mechanics. We first characterize the meristems and point to implications of the symplasm, apoplasm, and organismal concepts for our understanding of plant morphogenesis. Then we discuss the symplastic (coordinated) growth and a putative role of principal directions of growth and mechanical stress tensor in the meristem function, also explaining how the principal directions are manifested in cellular pattern and cell behavior. The present knowledge on the mechanics of meristems, in particular on the distribution of mechanical stress and on the mechanical properties of the meristems, is to a large extent speculative. Our objectives are to present and discuss the available empirical data and hypotheses on the meristem mechanics, and the evidence on the role of mechanical factors in plant morphogenesis.

1 Introduction

Plant meristems are embryonic tissues dedicated to growth and morphogenesis, usually active throughout the individual plant lifetime. Typically for plants, and unlike animals, they enable continuous growth and production of new organs, which is an important adaptation for settled habitus. Understanding mechanisms of meristem functioning is therefore crucial for understanding plant development and morphogenesis. A lot is known on the role of biochemical, molecular, or genetic factors in the regulation of meristem function. However, physical factors, especially mechanical, may also play a role but this role is not yet quite elucidated and often remains underestimated (Hamant and Traas 2009).

D. Kwiatkowska (✉) and J. Nakielski
Department of Biophysics and Morphogenesis of Plants, University of Silesia, Jagiellońska 28, 40-032 Katowice, Poland
e-mail: dorota.kwiatkowska@us.edu.pl; jerzy.nakielski@us.edu.pl

P. Wojtaszek (ed.), *Mechanical Integration of Plant Cells and Plants*, 133
Signaling and Communication in Plants 9, DOI 10.1007/978-3-642-19091-9_6,
© Springer-Verlag Berlin Heidelberg 2011

The meristems, like other plant organs grow symplastically (Priestley 1930; Erickson 1986), i.e., in a continuous and coordinated way. Such growth is of a tensor nature (Hejnowicz and Romberger 1984, see Sects. 3 and 4). From the mechanics point of view, the plant growth can be treated as an irreversible deformation of the cell wall system, which is its plastic strain (Green 1962). Accordingly, the growth rate, which is equivalent to the strain rate (Nakielski and Hejnowicz 2003), must be a result of tensile stresses acting in cell walls. Both growth and stress are usually anisotropic, which means that the value of growth rate or stress at a point is different in different directions.

It is well known that the primary cause for the stress in the cell wall is the turgor pressure (Green 1962; Romberger et al. 1993). At the cellular level, the stress component resulting from turgor depends on the cell geometry (Castle 1937, cited after Schopfer 2006). At the organ level, the stress distribution is influenced by the organ geometry (e.g., Dumais and Steele 2000) and can be modified also by tissue stresses (see Sect. 4.2). It can thus be assumed that the stress distribution in the cell wall system of a growing plant organ is a function of position and depends on the above-mentioned factors that may change in space and influence the stress in different proportions. In meristems, which are self-perpetuating organs, the stress distribution is most likely rather steady in time as long as the organ growth and geometry do not change. Mathematically, the mechanical stress as the second rank tensor (Fung 1984, see Sect. 4.1) defines three mutually orthogonal principal directions of stress (PDSs) in which the stress considered in 3D attains its extreme values. These directions form a spatial pattern of PDS trajectories. Knowing this pattern one may estimate a variation of stress within the organ.

A tensorial analysis of growth led to a hypothesis that the meristem cells are able to perceive directional signals. These signals participate in growth regulation at the tensorial level (anisotropy) and are used by cells to define the orientation of cell division planes (Hejnowicz 1984). One may expect that such signals originate from PDSs. Lynch and Lintilhac (1997) elegant experiments support this view. They cultured isolated tobacco (*Nicotiana tobacum*) protoplasts in agarose medium shaped in the form of a block to which a mechanical load was applied, generating a particular (nearly planar) stress distribution, and assessed principal stress trajectories with the aid of photoelastic method. Such treated cells orient their division walls perpendicularly to one of the two PDSs recognizable in the block. In the case of lower and axisymmetric (i.e., of a rotational symmetry) stress, the preferential division plane is perpendicular to the direction of maximal compression (referred to as compression tensor by Lynch and Lintilhac 1997). In the case of higher but most likely not axisymmetric stress, the division plane is parallel to the direction of maximal compression. At least in the case of uniaxial stress resulting from the load applied to the agarose block, this division plane is a shear-free plane.

The question arises how the agarose experiments on protoplasts or isolated cells are related to the stress distribution in a meristem. Apparently, mutually orthogonal trajectories of PDSs recognized with photoelastic method by Lynch and Lintilhac (1997) are curvilinear and confocal. Similar pattern of trajectories may be expected in a median section of shoot or root apical meristem (SAM or RAM, respectively) (the two patterns are compared in Fig. 9 of Nakielski 2008). The difference is in that in the agarose medium the stresses are compressive and result from an externally applied force, while in the apical meristem the stresses are tensile and originate from turgor as the primary factor. They are therefore a result of internal forces, which are also related to cell wall growth. Despite these differences, in both cases the stress pattern is general in a sense that the stresses operate on the entire agarose block or at the organ level. The distribution of stresses in cell walls within an organ is hardly known. A possibility to assess them empirically is strongly limited because any physical interference most likely changes the stress distribution. However, according to the so-called Laplace rule (Hejnowicz 1997; Dumais et al. 2006), a dome-shaped surface is a so-called principal surface in terms of PDSs. It means that at each point on the apical meristem surface, two of the PDSs are tangent to this surface while the third one is perpendicular. That meristem cells seem to perceive local stress fields and use this mechanical signal while orienting the microtubules has been shown recently for the SAM of *Arabidopsis thaliana* by Hamant et al. (2008; see also Dumais 2009).

In this chapter, the structure, function, and growth of plant meristems are discussed from a perspective of mechanics. We first characterize different types of meristems and point to implications of the symplasm, apoplasm, and organismal concepts for our understanding of plant morphogenesis. Then we discuss the symplastic, i.e., coordinated, growth and a putative role of principal directions of growth and mechanical stress tensor in the meristem function, also explaining how the principal directions are manifested in cellular pattern and cell behavior. The role of mechanics in plant development and morphogenesis has been postulated a long time ago (e.g., Schwendener 1874 and Rasdorsky 1925, cited after Romberger et al. 1993; Thompson D'Arcy 1942). Recently, a number of very interesting and comprehensive reviews and opinion papers have been published in which the problem of significance of mechanics, especially the mechanical stress, in plant development is revisited from the perspective of new empirical evidence (e.g., Dumais 2009; Hamant and Traas 2009; Szymanski and Cosgrove 2009; Uyttewaal et al. 2009). However, our knowledge on the mechanics of meristems, in particular on the distribution of mechanical stress and on the mechanical properties of the meristems, is to a large extent speculative. Therefore, our objectives are first to present and discuss the available empirical data and hypotheses on the meristem mechanics, and then the evidence on the role of mechanical factors in plant morphogenesis.

2 Meristem Types in Plants

In general, three types of meristems, apical, lateral, and intercalary, are distinguished on the basis of their localization in a plant (Romberger et al. 1993). Apical meristems are located at tips of shoots and roots or their appendages, such as for example SAM and RAM or an apical meristem of a leaf primordium (which usually plays an important role only in the early stages of leaf development). Lateral meristems are in turn located at organ peripheries, beneath the organ surface. These are cambium (also called the vascular cambium) and phellogen, which are formed in mature root and stem portions. Intercalary meristems are located between non-meristematic organ segments, e.g., at the bases of young leaves or stem internodes. These differences in meristem localization are related to major differences in meristem structures, shapes, and a putative stress-mechanical environment (Lintilhac and Vesecky 1984).

The duration of the meristem activity (the meristem lifetime) depends on the type of organ growth that may be either determinate or indeterminate. For example, a SAM in a conifer tree may be active throughout the very long plant life. The activity of flower or leaf meristems in turn is restricted to early developmental stages of these organs. The common feature of all the meristems is the ability of their cells to expand and divide, while cells leaving the meristems eventually cease to divide and start to differentiate. In case of apical meristems of organs exhibiting indeterminate growth and in the case of cambium, initial cells (or initials) can be distinguished. Being the source of all the other cells, initial cells remain within the meristem and occupy a special position. These unique cells are often called stem cells, analogous to stem cells crucial in animal development.

We will focus on the apical meristems of root and shoot and on the cambium. These meristems contribute to indeterminate organ growth, primary or secondary, and are fundamental in plant morphogenesis. We will also limit our interest to seed plants, so that the topic may be discussed within the available space.

2.1 Function, Geometry, and Organization of Apical Meristems

Apical meristems of roots and shoots are the source of all the cells building these organs. Because of major differences between root and shoot architecture, the fundamental functions of the two types of apical meristems are also not the same. The major function of the RAM is its self-perpetuation, meaning that the meristem maintains its shape and size while continuously providing new cells to the root. The SAM functions are not only the self-perpetuation but also the formation of primordia of lateral organs like leaves, flower organs, or flowers. This difference between the two meristems has a profound influence on their geometry. The RAM is nearly axisymmetric, i.e., exhibits a rotational symmetry. In a short time frame, both shape and size of the RAM are steady. The SAM geometry is more complex. Firstly, although the most common SAM shape is a dome, there are also flat meristems, and the SAM shape is not always axisymmetric. Secondly, even in a short time frame the SAM size and often also shape are changing in plastochron cycles, i.e., in the time intervals between initiation of consecutive primordia (Erickson and Mitchelini 1957). The size-only change in the plastochron cycle may be visualized as a gradual increase of a dome height while its shape (surface curvature) remains the same, as if we were looking at a flooded hill which gradually comes out when the water retreats. The increase of dome height is due to the meristem rebuilding after its portion has been used up for primordium formation. If the meristem produces relatively large primordia, often also the meristem shape is changing during the plastochron, i.e., changes in the dome curvature take place often together with changes in size (Kwiatkowska 2008). This is clearly the case in *Arabidopsis* and also *Anagallis arvensis* (Fig. 1a) and can be visualized by local geometry quantification. The local geometry is represented by principal directions of curvature, which are the directions in which the normal curves described on a surface exhibit maximal or minimal curvatures (Struik 1988). The local geometry

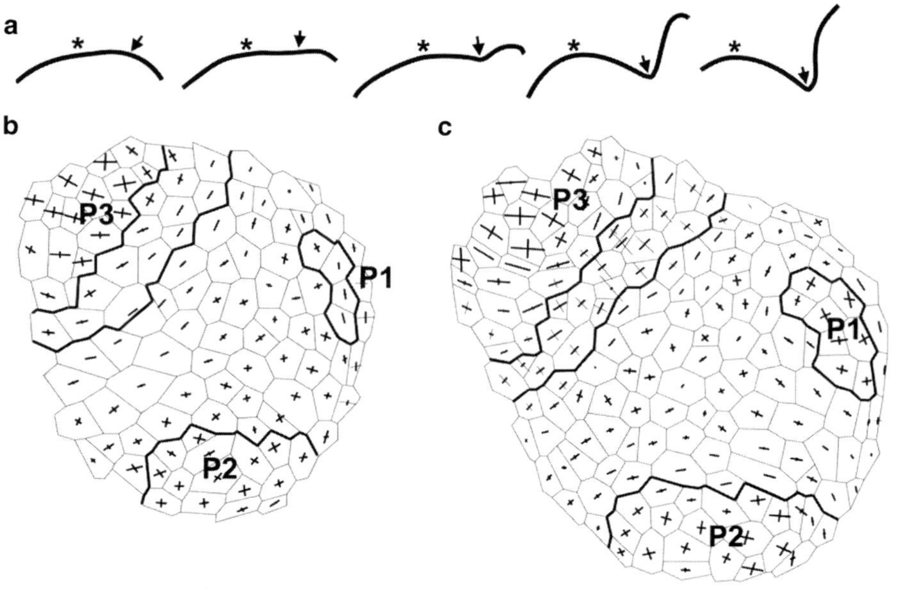

Fig. 1 Ontogenetic changes of geometry at *Anagallis arvensis* vegetative shoot apex visualized in the apex profiles and curvature plots prepared on the basis of sequential replicas and stereoscopic reconstruction of the apex surface. (**a**) The series of profiles showing the leaf primordium development from the earliest stage at which it can be recognized in replicas to the stage when it is separated from the SAM by a deep and distinct axil, i.e., adaxial primordium boundary (prepared on the basis of reconstructions shown in Kwiatkowska and Routier-Kierzkowska 2009). *Asterisk* points to the SAM region of the slowest growth (the central zone surface), *arrow* points to the region where the adaxial primordium boundary is formed. (**b, c**) Two curvature plots for replicas taken from the same apex at 36-h interval (prepared from sequential replicas in a way described by Kwiatkowska and Dumais 2003). The plots are overlaid on cell wall patterns of the apex shown in top view. Cross arms point to the principal directions of curvature, the arm length is proportional to the curvature in its direction. The arm is a thin line if in this direction the surface is concave. *Thick lines* outline leaf primordium (P1–3) boundaries that are first recognized in the plot (**c**), then the same cells are recognized in (**b**). Primordia P1 and P2 are distinguished as regions of increased and nearly isotropic curvature, while P3 is separated from the SAM by a distinct saddle-shaped axil (between the two thick lines on the adaxial side of P3). The stage of primordium development shown in the first profile corresponds to P1 in (**c**); second profile – P3 in (**b**); third and fourth – P3 in (**c**). Note the changes in the shape of the SAM periphery contacting primordia of different developmental stages

quantification enables recognition of SAM regions unique in terms of curvature that are boundaries between the SAM and primordia. These regions are saddle shaped, i.e., concave in one and convex in the other principal curvature direction (Fig. 1b; Kwiatkowska and Dumais 2003).

In both RAM and SAM, a unique distal position is occupied by initial cells. This position allows them to remain within the meristem while all their derivatives are eventually displaced away from the meristem and undergo differentiation. In a median longitudinal section of an apical meristem, this displacement of derivatives from the initial cells is manifested in cell files convergence at a region where initial cells are located. In the case of root apex, the cell file convergence at this region is a manifestation of the fact that the RAM generates the organ composed of two segments, i.e., the root proper and the root cap. Both the cell files visible in the root proper and those visible in the root cap converge at the initial cells. Initial cell derivatives contributing to the root proper are displaced proximally from the initial cells in the course of RAM growth (similar to initial cell derivatives in the SAM). However, the derivatives that contribute to the root cap are displaced distally from their initial cells. In some meristems, like in *Arabidopsis* and other Brassicaceae members, the rhizodermis has common initial cells with the lateral root cap.

In the SAM of seed plants, initial cells are arranged in one up to several layers in the most distal portion of the meristem. Proximal to these cells an organizing center (called the zone of central mother cells in older literature) is located. The initial cells and the organizing center form a central zone that is characterized by relatively low mitotic activity. The mitotic activity of other SAM zones (the rib meristem and the peripheral zone where primordia are formed) is higher. Such cytohistological zonation of the SAM has been proposed already by Foster (1939, 1943) who constructed the zonation around the central mother cells zone. It is now known that this zonation is manifested also in expression domains of SAM regulatory genes (reviewed by Traas and Doonan 2001).

A specific feature of the RAM in a majority of seed plants is the presence of a quiescent center (QC), first recognized by Clowes (1959, 1961). This is a hemisphere or disk shaped region located around the convergence site of cell files, just beneath the root cap. The QC plays a unique role in the root maintenance. For example, a small fragment of the maize RAM composed mostly of the QC cells, isolated and cultured in vitro, directly regenerates the whole root (Feldman and Torrey 1976). Mitotic activity of QC cells is the lowest of the whole RAM, though some cell divisions do occur in this region. It is thus convenient to distinguish two types of initial cells in the RAM (reviewed in Jiang and Feldman 2005). Structural initial cells are arranged in one to three layers that are placed at the site of cell file convergence. This region is located within the QC. Functional initial cells in turn surround the QC and are the actual source of root cells. Some authors regard structural initial cells as the stem cells (Jiang and Feldman 2005) while the others refer to functional initials as stem cells (Scheres 2007). The reason for it may be in the fact that the most intensively studied RAM of *Arabidopsis* is built of a relatively very small number of cells. Since the lifetime of the root is short in this species cell divisions in the QC virtually never take place. This is unlike many other roots, for example, maize (*Zea mays*) root in which the individual RAMs function for a longer time and are composed of numerous cells. Nevertheless, it has been shown also for the *Arabidopsis* RAM that the QC-derived cells sometimes replace functional initial cells (Kidner et al. 2000). Structural organization of RAM is different in different taxonomic groups. These differences can be reduced to the number of layers of initial cells and the origin of various root parts (stele, cortex, rhizodermis, and root cap). The origin of the different root parts is, however, not fully separate since sometimes cell divisions of one initial layer give rise to a derivative contributing to the other layer (Kidner et al. 2000).

2.2 Function, Geometry, and Organization of Cambium

Cambium provides cells to the secondary xylem and phloem. It forms a hollow cylinder of meristematic tissue surrounding the solid cylinder of secondary xylem. Geometry of cambium changes during its ontogeny since the cylinder perimeter is gradually increasing, i.e., the curvature of the cambium cylinder in the circumferential direction is decreasing. This is due to increasing perimeter of secondary xylem cylinder. The process is called dilatation of cambium. Simultaneously, changes in the thickness of the meristem may take place in seasonal cycles in the temperate climate zone (Romberger et al. 1993).

Cambium comprises two types of cells: fusiform cells, profoundly elongated along the stem or root axis; ray cells that are elongated not along the organ axis but along its radius, and to a much smaller extent than fusiform cells. Both periclinal (division plane is parallel to the organ surface) and anticlinal (perpendicular to the organ surface) cell divisions occur in cambium. Periclinal divisions increase the number of cells in radial files and provide cells differentiating into secondary vascular tissues. Differences in wall thickness of cells in each radial file enable identification of a single putative initial cell (fusiform or ray), and mother cells of xylem or phloem. Anticlinal divisions contribute to the cambium dilatation increasing the number of the above mentioned radial files thus also a number of initial cells (Romberger et al. 1993).

3 Growth of Plant Cells and Organs

3.1 Apoplasm, Symplasm, and the Organismal Concept of Multicellularity

Meristems, like all plant organs, are made of two complementary interweaving systems: apoplasm (apoplast), which is the complex cell wall system; symplasm (symplast), i.e., the network of protoplasts connected by plasmodesmata (Erickson 1986; Romberger et al. 1993). The two systems are not isolated one from the other but connected in the so-called cytoskeleton–plasma membrane–cell wall continuum (Baluška et al. 2003; see also chapter "Introduction: Tensegral World of Plants"). The existence of apoplasm is due to the adherence of adjacent cell walls. Noteworthy, the majority of plant cells are formed already in an adherent state, i.e., the intercellular adhesion arises in plants in a way different from adhesion in animals (Jarvis et al. 2003). Cytokinesis accompanying the mitotic cell division in plants can be thought of as the process of internal cytoplasmic differentiation (Peters et al. 2000).

Thinking of plant organs as composed of symplasm and apoplasm implies that a supracellular level is necessary for understanding the plant organ function and growth. This is in agreement with the organismal concept of multicellularity stating that multicellularity arises by the secondary chambering of the organ or plant body into cells and not by cell aggregation (reviewed by Kaplan 1992). If this concept is adopted in a study of plant organ development, like, for example, morphogenesis at the SAM, properties of an organ as a whole become more significant than those of individual cells. Also in line with the organismal concept it is postulated that the cell proliferation on its own is not sufficient to drive morphogenesis at the SAM (reviewed recently by Fleming 2006). This postulate is supported by the observations that local manipulation of cell division frequency or orientation in the SAM does not affect shoot morphogenesis (Wyrzykowska et al. 2002; Wyrzykowska and Fleming 2003) while local manipulation of the cell wall properties directly affecting cell wall growth has a significant influence (Pien et al. 2001).

Adopting the organismal concept enables also a better understanding of how the organ size is controlled. Numerous experiments show that manipulation of cell proliferation or cell size within quite wide ranges does not affect an organ size, both in plants and in animals. This is true also for organs being mosaics of cells differing in size (reviewed in Day and Lawrence 2000). The altered cell numbers seem to be compensated by their altered size, and vice versa, so that the organ size is not affected. It means that the extent of an organ growth is more likely regulated by variables describing its absolute dimensions than those measuring the cell number. Examples of the former variables are the surface area to volume ratio or perimeter to surface area ratio as has been proposed for leaf blade size control (Day and Lawrence 2000; Anastasiou et al. 2007).

3.2 Symplastic Growth as a Deformation of Cell Wall System Observing Continuum Condition

Plant growth, i.e., an increase in body dimensions, can be interpreted as an irreversible deformation (plastic strain) of cell walls (Green 1962). Fundamental attributes of the plant organ growth are continuity and coordination. Though the structure of a plant organ is complex and not continuous, growth continuity is related to the apoplasm. It is known that plant cells, in contrast to animal cells, are surrounded by relatively rigid walls that determine shapes of individual cells. The walls of neighboring cells are joined together by middle lamellas composed usually of a solid pectic material, while the neighboring protoplasts are joined by plasmodesmata lined by plasma membrane. Therefore, a displacement of organ building cells, resulting from its growth, has to be coordinated. Such a coordinated growth of cells, during which contiguous walls do not slide or slip with respect to each other, has been called symplastic growth (Priestley 1930; Erickson 1986). The growth of adjacent cells may be coordinated by a biophysical feedback loop in which adhering walls of two neighboring cells coordinate their growth by physical interactions so that there is no shear strain at the middle lamella. It may thus be appealing to look at the growth as quantized not into individual cells but rather into pairs of adherent cell walls, growth pattern of the two walls in a pair being the same (Jarvis et al. 2003).

Plant growth is driven by turgor in a majority of cases. To induce the cell enlargement, it is necessary to relax its wall tension by chemical modification of cell wall that leads to its loosening. This results in a decrease of the turgor pressure and thus also the water potential of the protoplast, providing a driving force for the passive influx of water and a corresponding increase of the cell volume. At the same time the original turgor pressure is restored. A steady-state growth occurs when all these processes take place simultaneously. Biochemical factors mainly are at the basis of such growth interpretation (discussed by Schopfer 2006). Wei and collaborators (2003, 2006) point out that biophysical factors may possibly also be involved in the process of the wall stress relaxation. They suggest that the stress in the wall of growing cell that is under turgor pressure increases smoothly only to some critical point determined by material properties and cell geometry. Once the critical level of stress corresponding to this point is reached, the gradually increasing stress results with abruptly facilitated deformation and a loss of stability that manifest themselves as the wall extension. Not questioning the role of turgor pressure and biochemically mediated cell wall loosening, such postulate provides a complementary biophysical interpretation of the process of stress relaxation in the cell wall (see also chapter "Plants as Mechano-Osmotic Transducers").

Although it is still under some dispute whether it is the cell wall loosening or increased osmotic pressure that is the initial event of the cell wall growth (Zonia and Munnik 2007), it is accepted that growth depends on mechanical properties of the cell wall. The cell wall is often both structurally and mechanically anisotropic. Structural anisotropy of the cell wall is mainly in its cellulose microfibril arrangement (Baskin 2005). Three different types of wall architecture can be distinguished on this basis: full anisotropy in which microfibrils are aligned preferentially in one direction; transverse isotropy in which cell wall is organized in layers each with randomly arranged microfibrils; and full isotropy meaning that the wall is not organized in layers and microfibrils are oriented randomly in all directions in space (Dumais et al. 2006). The cell wall is mechanically anisotropic both in its plastic and in elastic properties as it has been shown in theoretical consideration and experiments (e.g., Preston 1964; Probine and Barber 1966; Kerstens et al. 2001; see also chapter "Micromechanics of Cell Walls"). The experiments confirm that the mechanical anisotropy of the cell wall is related to the structural anisotropy although the role of additional factors has also been proven (reviewed in Baskin 2005).

According to the microtubule/microfibril paradigm, the orientation of cellulose microfibrils is directed by cortical microtubules (Green 1962; Green and Selker 1991; Paredez et al. 2006). It means that cytoskeleton affects the structural cell wall anisotropy and, as a consequence, also its mechanical anisotropy. This relationship, however, turns out to be very complex. Firstly, the paradigm has been extended by the postulate that forces generated by anisotropic growth, resulting indirectly from the structural cell wall anisotropy, affect the alignment of cortical microtubules (Fisher and Cyr 1998). Other observations point that microtubule organization is affected both by biomechanical factors and by proteins involved in cell wall synthesis (Paredez et al. 2008). Secondly, microtubule activity is postulated to affect not only the cellulose microfibril orientation but also the length of cellulose microfibrils that has an impact on growth anisotropy (Wasteneys 2004), and the delivery of cellulose synthase complexes to plasma membrane (Crowell et al. 2009). Taking together, the growth rates and growth anisotropy are regulated by complex interactions of a feedback nature in which cytoplasmic (cytoskeleton) and cell wall components are involved and in which at least a part of interactions is via the biomechanical signals (see also chapter "Mechanics of the Cytoskeleton").

3.3 The Growth Tensor and Principal Growth Directions

Let us consider a growing organ with arbitrary chosen elements of its cell walls system. Due to the organ growth the elements are displaced. This displacement per time unit defines the displacement velocity $\boldsymbol{V}$. Both theoretical considerations (Gandar 1980, 1983a, b) and empirical data coming from earlier investigations with streak photography (Erickson and Sax 1956) and at present mainly from video and image analysis (Van der Weele et al. 2003) indicate that this velocity changes in a continuous way if one moves from one point to another. It implies that the shape, growth, and development of individual cells are subordinated, at least partially, to growth of the organ as the whole.

Let us assume that we know the vector field $\boldsymbol{V}$ for a growing organ. Let a linear element of the cell wall move along the path m with the displacement velocity $\boldsymbol{V}_m$ tangent to this path. Then along this direction, the relative elemental rate of growth in length is $\mathrm{RERG}_{l(m)} = \mathrm{d}\boldsymbol{V}_m/\mathrm{d}m$ (Richards and Kavanagh 1943; Silk 1984). The same quantity can be expressed by a vector formula as: $\mathrm{RERG}_{l(m)} = \mathrm{d}(\boldsymbol{V}\cdot\boldsymbol{e}_m)/\mathrm{d}m$, where $\boldsymbol{e}_m$ is a unit vector tangent to m, and dot refers to the scalar product. For any direction $\boldsymbol{e}_s$, we then have (Hejnowicz and Romberger 1984): $\mathrm{RERG}_{l(s)} = \nabla(\boldsymbol{V}\cdot\boldsymbol{e}_s)\cdot\boldsymbol{e}_s$, where ∇ is a gradient.

In orthogonal curvilinear coordinate system ($\boldsymbol{e}_1$, $\boldsymbol{e}_2$, $\boldsymbol{e}_3$), the relative elemental rate of growth in length along $\boldsymbol{e}_s$ can be given as $\mathrm{RERG}_{l(s)} = T_p^q \boldsymbol{e}^p \boldsymbol{e}_q$, for $p, q = 1,$

2, 3 (summation convention), where T_p^q represents a second rank tensor called the tensor of growth rate or growth tensor (GT) (Hejnowicz and Romberger 1984). Accordingly, the field of growth rate of an organ (GT field) is of the tensor type.

If RERG_l is not the same in all directions (anisotropic growth), then there must exist three mutually orthogonal principal directions of growth rate (PDGs), distinguished in this sense that RERG_l along them attains extreme values (Hejnowicz and Romberger 1984). In two of them, RERG_l attains its maximum and minimum. These are the $\mathrm{PDG}_{\mathrm{max}}$ and $\mathrm{PDG}_{\mathrm{min}}$, respectively. If growth is not transversely isotropic, the third principal direction representing the saddle type extreme appears. It is the direction of the highest RERG_l in a plane normal to the $\mathrm{PDG}_{\mathrm{max}}$, and simultaneously the direction of the lowest RERG_l in a plane normal to the $\mathrm{PDG}_{\mathrm{min}}$. The directional variation of growth rates at a point can be visualized by means of the GT indicatrix that is a surface plotted around the point. In any direction, a distance from this point to the surface is proportional to the growth rate in this particular direction (Kwiatkowska 2004a). Between successive points of the organ, the PDGs change in a continuous way giving rise to PDG trajectories. Three such mutually orthogonal trajectories pass through every point in the organ. The surfaces tangent to pairs of PDGs (and normal to the third PDG) are the principal surfaces. There are three families of such surfaces differing in their orientation in 3D. In the case of an organ exhibiting an elongated shape, these are periclinal, anticlinal longitudinal, and anticlinal transverse principal surfaces. The periclinal surface is normal to anticlinal direction, anticlinal longitudinal to periclinal transverse direction, and anticlinal transverse to periclinal longitudinal (Fig. 2). The surface of the organ coincides with one of periclinal principal surfaces whereas surfaces of the two remaining families are normal to the organ surface (Kwiatkowska 2004a).

Knowing the PDGs one can find a curvilinear coordinate system which lines coincide with trajectories of these directions. In such a system, referred to as a natural coordinate system for a given organ (Hejnowicz 1984), specification of the fields $\boldsymbol{V}$ and GT becomes simplified. Examples of the natural coordinate systems dedicated for SAM and RAM are shown in Fig. 2. A distribution of $\mathrm{RERG}_{l(s)}$ in the interior and at the surface of apical meristems has been modeled (Hejnowicz et al. 1984a, b; Nakielski 1987, 1991; Hejnowicz and Karczewski 1993).

3.4 *Classical Rules of Cell Divisions and Principal Growth Directions*

Meristem growth is accompanied by cell divisions in such a way that generally uniform cell dimensions and shapes are maintained. Cells that have left the meristem start to differentiate and may increase their dimension tremendously as, for example, in the elongation zone of a root. In some plant structures, the cellular pattern generated or maintained by cell divisions seems to be repetitive, as it is the case in the *Arabidopsis* root apex, i.e., the division planes in particular cells are to a

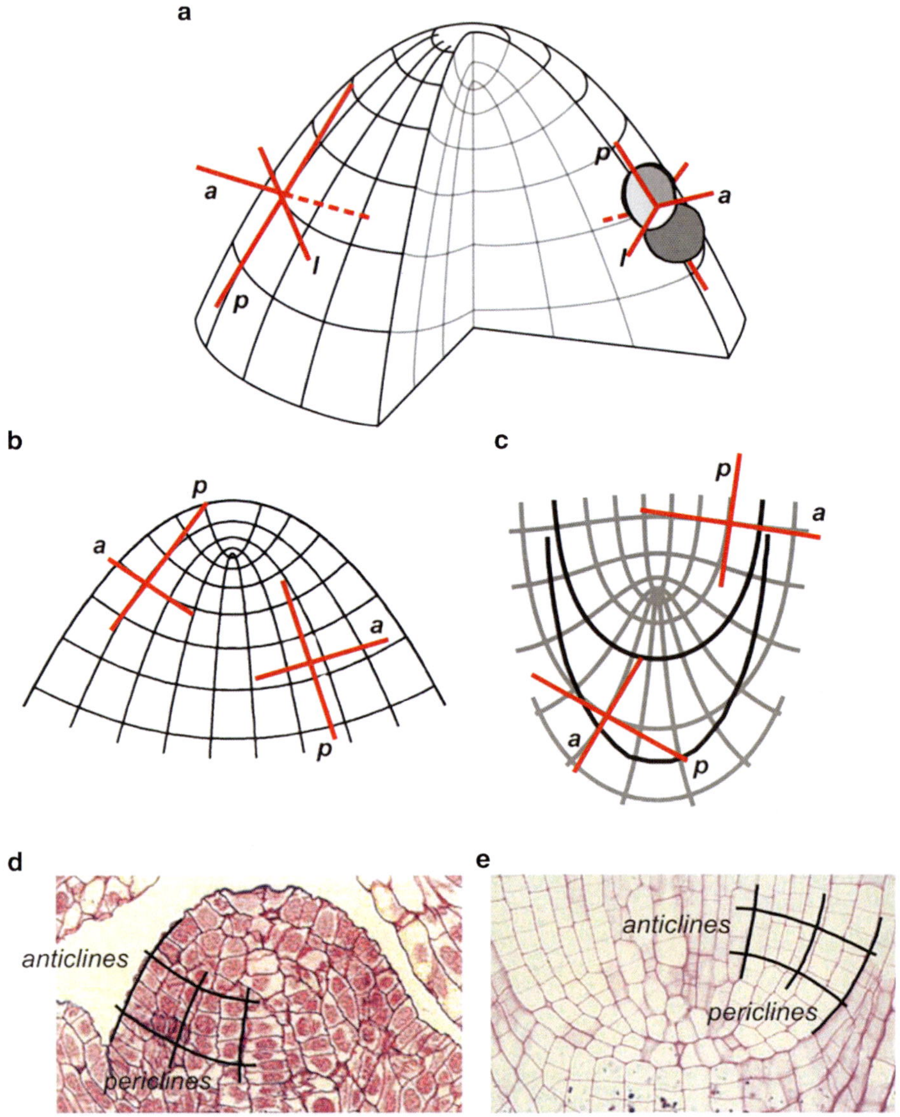

Fig. 2 Principal growth directions and the system of periclines and anticlines describing the cell wall system in exemplary apical meristems. The natural curvilinear orthogonal coordinate systems are defined by trajectories of PDGs (*p* – periclinal, *a* – anticlinal, *l* – latitudinal). (**a**) The exemplary system dedicated to the SAM shown in 3D. At one point the $RERG_l$ indicatrix is plotted (1/8 of the plot has been cut off by the planes defined by pairs of PDGs). (**b**) The PDG trajectories in the axial plane corresponding to the periclines and anticlines overlaid on the median longitudinal section of spruce (*Picea abies*) seedling SAM shown in (**d**). (**c**) The trajectories corresponding to periclines and anticlines in the RAM of radish shown in (**e**)

large extent predictable. The exact cell division planes and division sequence in the majority of meristems is, however, not so predictable. This is true especially for proliferative cell divisions. Proliferative divisions are divisions in which a progeny resembles the parental cells, unlike the formative divisions that are the divisions in which daughter cells differ biochemically or morphologically from their parental cells as in the case of an apical initial cell in the RAM or SAM of ferns (Gunning et al. 1978; Fosket 1994).

Dividing cells "observe" a number of classical rules on the orientation of division planes, called Hofmeister's, Errera's, and Sachs' rules (Sinnott 1960). The Hofmeister's rule states that the cell division plane is perpendicular to the direction of the fastest growth of an organ; Errera's rule – the new wall follows the shortest path that will halve the parental cell; and Sachs' rule – new cell wall meets the parental cell wall at a right angle (Kwiatkowska 2004a). It is noteworthy that geometry of the dividing cell is an important factor in these three rules. The last rule is that division walls tend not to align with walls between their neighbors thus avoiding four-way junction formation (Sinnott and Bloch 1941; reviewed in Lloyd 1991a, b). The explanation for the last rule proposed for vacuolated cells is that the tensile microtubule bundles of transvacuolar strands seek the minimal path thus avoiding the already existing three-way junctions (Flanders et al. 1990; Lloyd 1991a, b). Similar mechanism may be operating in the meristematic cells.

According to the postulate put forward by Hejnowicz (1984), cell division planes in meristems are in principal planes of the GT, i.e., normal to one of the PDGs. This is especially pronounced in the case of a strongly anisotropic growth. The postulate is based on the observation that cell walls in apical meristems follow lines of periclines and anticlines (see below). Further support comes from the similarity between real and computer-simulated cell wall patterns (Hejnowicz 1989; Nakielski and Barlow 1995). In the case of isotropic or nearly isotropic growth, the principal directions are not obvious and the rule is not observed. Then cell division planes appear to be oriented more randomly.

The PDG pattern operates at the organ level. It is thus not surprising that classical rules of cell divisions can be generalized by the Hejnowicz postulate (Hejnowicz 2002; Kwiatkowska 2004a). The Hofmeister's rule is a special case of the relationship between division planes and PDGs. Hejnowicz postulate explains also exceptions to the Hofmeister's rule, i.e., when the division plane is perpendicular to the PDG_{min} as it is the case of cortex initial cells in the *Arabidopsis* RAM. The Errera's rule is usually observed when the division plane is normal to the PDG_{max}. Exceptions to this rule like divisions of fusiform cambium initials (see Sect. 6) can be explained by the PDGs/division plane relationship, since in this case the cell is elongated along the PDG_{min}. Finally, since PDGs are mutually orthogonal the consecutively formed walls will meet at right angle as stated in the Sachs' rule.

3.5 Periclines, Anticlines, and a Natural Coordinate System of Apical Meristems

The cell arrangement in meristems is regular. If one focuses on a relatively small fragment of a longitudinal section of the RAM or SAM, cell wall arrangement in zigzags is apparent, which results from shapes of individual cells. At the whole meristem level, however, it becomes visible that these zigzags can be satisfactorily approximated by two types of smooth, mutually orthogonal lines (Fig. 2d, e). These lines, known as periclines and anticlines (Sachs 1887), preserve their orthogonal intersections during growth. This unique property implies that the wall tangent to a given pericline or anticline, though displaced as a result of a continuous flow of cells from the distal portion of the meristem, retains its periclinal or anticlinal orientation. Furthermore, a majority of new walls formed as a result of cell divisions are either periclinal or anticlinal (Hejnowicz 1984), i.e., they "adjust themselves" to the pattern of periclines and anticlines in the organ. The explanation of this phenomenon is related to PDGs. Namely, in the GT field of a steady type, two mutually orthogonal line elements preserve their orthogonal intersection during growth only if they are oriented along PDGs. Otherwise, the angle between these elements changes into acute or obtuse. On the basis of an analysis of cellular patterns in sections of SAM and RAM of various plant species, a conclusion has been drawn (Hejnowicz 1984, 1989) that periclines and anticlines represent PDG trajectories. This further leads to the already mentioned hypothesis that cells typically divide in the planes defined by two PDGs, i.e., the division wall is oriented in one of principal planes (Hejnowicz 1984, 1989). Accordingly, the apparent cellular regularity may be understood if we assume that individual cells are somehow able to perceive directional cues coming from PDGs at their present position and make use of them when setting the division planes.

The hypothesis that PDGs trajectories are manifested in the cellular pattern as periclines and anticlines has a practical meaning. While adapting a natural coordinate system for SAM or RAM observed in a median longitudinal section, we overlay the coordinate system lines on the visible pattern of periclines and anticlines. It is relatively easy in the case of steady growth when the shape of the meristem is not changing. Then the outermost pericline that describes the meristem surface in the section is also one of the coordinate lines of the system. It, however, becomes difficult to recognize the periclines and anticlines if the shape of the meristem and its GT field are changing in time. Moreover, if the changes of the GT field are accompanied by a rearrangement of cellular pattern, as for example during the regeneration of a damaged meristem portion, the cells may be strongly deformed (sheared) so that the new walls formed during these changes as perpendicular to one of PDGs are oblique with respect to parental cell walls. Such "anomalies" and some cellular pattern modifications observed in mutants can be explained in terms of PDGs (Nakielski 1992; Nakielski and Barlow 1995) if we realize that the pattern of the organ PDGs is changing. The occurrence of anomalies may thus point to an important role played by PDGs in generation of a cellular pattern of a growing organ in the case of both steady and unsteady growth.

Observations questioning the role of principal directions in setting the cell division plane can also be interpreted in terms of temporal changes of PDGs. Green and Poethig (1982) assessed principal directions of strain (strain crosses) of protodermal cells of the residual meristem of *Graptopetalum paraguayense* (Crassulaceae). This meristem is located at the base of a detached mature leaf and participates in the de novo shoot formation. The protodermal cells at the shoot formation site are strongly sheared. The orientation of computed strain cross arms that correspond to PDGs is in some cases at 45° with respect to cell division planes. On the basis of this observation, Green and Poethig (1982) ruled out the putative role of principal directions of strain (growth) in orienting cell division planes. The strain crosses were, however, computed for the 24-h time intervals. Since the growth pattern changes dramatically in the investigated region it is not obvious what the actual PDGs were at the time of cell division (Kwiatkowska 2004a).

4 Mechanical Stress in Plant Organs

4.1 Principal Directions of Stress in Cell Walls and Organs

It has been explained above that plant cell growth results from tensile stresses in its walls. Let us consider a plane that passes through a cell wall and in general does not have to be transverse to the wall (Fig. 3a). The orientation of the plane is given by the unit vector $\boldsymbol{n}$, perpendicular (normal) to the plane. The plane divides the cell wall into two parts. If "stretching" forces are applied to the wall these two parts interact with each other. We assume that for an element ΔA of the plane that includes a point B a resultant of the forces is $\Delta \boldsymbol{P}$. We will now consider a stress at the point B. The stress vector is defined as $\boldsymbol{t} = \mathrm{d}\boldsymbol{P}/\mathrm{d}A$, i.e., a surface density of forces acting locally at B, for $\Delta A \rightarrow 0$. Although this may be against our intuition the direction of the vector $\boldsymbol{t}$ is usually not the same as the direction of the force $\Delta \boldsymbol{P}$. It is because $\boldsymbol{t}$ depends not only on the B position but also on the orientation of the plane under consideration, i.e., it is assigned both to the point B and to the normal vector $\boldsymbol{n}$. Moreover, vectors $\boldsymbol{t}$ and $\boldsymbol{n}$ are generally not collinear. It means that if we were computing stress vectors for the same point B but for different planes, described by different $\boldsymbol{n}$, we would obtain different $\boldsymbol{t}$ for each plane. Among these stress vectors only three are significantly different, all the other vectors may be considered as their linear combination (the number of linearly independent directions equals the dimension of a space). These three vectors $\boldsymbol{t}$ define the state of stress at a point.

Let us now come back to the stress vector $\boldsymbol{t}$ obtained for the $\boldsymbol{n}$, normal to the plane under consideration. The vector $\boldsymbol{t}$ may be divided into two components (Fig. 3b): one of them (σ) defines the normal stress, i.e., the stress in the direction

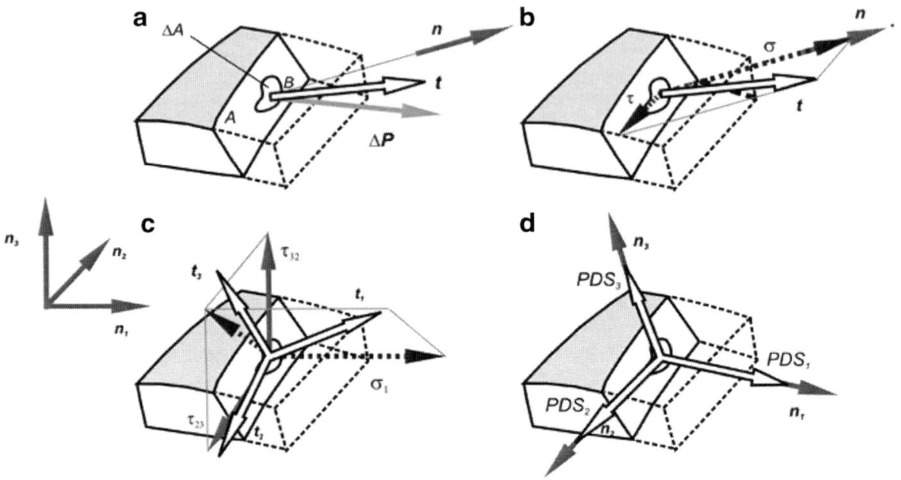

Fig. 3 Schematic representation of mechanical stresses that act in a cell wall: (**a**) The stress vector $\boldsymbol{t}$ obtained for forces acting on the plane given by the vector $\boldsymbol{n}$ normal to this plane. (**b**) The vector $\boldsymbol{t}$ as that shown in **a** divided into normal (σ) and tangential (τ) stress components. (**c**) Three stress vectors $\boldsymbol{t}_1$, $\boldsymbol{t}_2$, $\boldsymbol{t}_3$, assigned to the vectors $\boldsymbol{n}_1$, $\boldsymbol{n}_2$, $\boldsymbol{n}_3$ corresponding to axes of the Carthesian coordinate system. The $\boldsymbol{t}_1$, is divided into one normal (σ_1) and two tangential (τ_2, τ_3) components. (**d**) Three mutually orthogonal directions for which the stress vectors $\boldsymbol{t}_1$, $\boldsymbol{t}_2$, $\boldsymbol{t}_3$ are collinear. They are principal stress directions (PDS_1, PDS_2, PDS_3). Notice that if the state of stress at the point is characterized using PDSs all tangential components disappear

of $\boldsymbol{n}$; the other (τ) is the tangent stress, in the direction perpendicular to $\boldsymbol{n}$. These components depend on $\boldsymbol{t}$ and on the orientation of the plane. We may describe these components in any Carthesian coordinate system, like that in the Fig. 3c. Assuming that the directions of the system axes are $\boldsymbol{n}_1$, $\boldsymbol{n}_2$, $\boldsymbol{n}_3$, the stress vectors assigned to them are $\boldsymbol{t}_1$, $\boldsymbol{t}_2$, $\boldsymbol{t}_3$ (Fig. 3c). Each of the three stress vectors, like any vector, may be divided into three components. For example, $\boldsymbol{t}_1$ (obtained for the direction $\boldsymbol{n}_1$) has the following components: σ_1, representing the normal stress in $\boldsymbol{n}_1$; two others (τ_{23}, τ_{32}) representing tangent stresses in the plane defined by $\boldsymbol{n}_2$ and $\boldsymbol{n}_3$. The same operation can be done with the other stress vectors, $\boldsymbol{t}_2$ and $\boldsymbol{t}_3$. Their components are σ_2, τ_{13}, τ_{31} for $\boldsymbol{t}_2$ and σ_3, τ_{12}, τ_{21} for $\boldsymbol{t}_3$. Accordingly, the state of stress at the point B is given by a 3×3 matrix composed of nine elements, where each stress vector has its "own" three components (one normal and two tangent). The normal components are diagonal elements of this matrix. The tangent components represent shear stresses. The operator represented by the matrix is called the stress tensor.

The above description of stress can be greatly simplified. This possibility is related to one of the most interesting features of the stress tensor. Namely, at every point of a deformed body there are three mutually orthogonal directions, like $\boldsymbol{n}_1$, $\boldsymbol{n}_2$, $\boldsymbol{n}_3$ for which the stress vectors $\boldsymbol{t}_1$, $\boldsymbol{t}_2$, $\boldsymbol{t}_3$ are collinear (Fig. 3d). These are PDSs. The normal stresses σ_1, σ_2, σ_3 in these directions are called the principal stresses. Since the stress tensor field is usually not homogenous PDSs at different points are not the

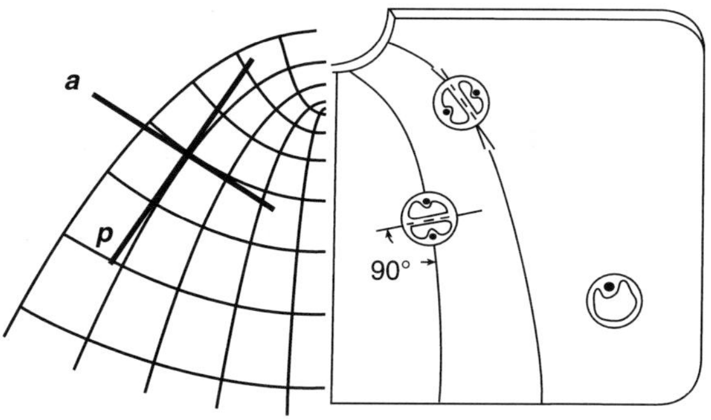

Fig. 4 The scheme showing stress lines recognized by Lynch and Lintilhac (1997) experiment (*on the right*) compared with the pattern of PDG trajectories of the SAM. Notice that the planes of divisions in compressed cells correspond to those represented by the pattern of PDG trajectories. Directions *p* and *a* are PDGs

same. They arrange themselves into PDS trajectories. Stress trajectories are paths comprised of points at which the PDSs are tangent to the path. An orthogonal network is formed by three families of such trajectories, one everywhere tangent to σ_1, and the other two tangent either to σ_2, or to σ_3. Three of such mutually orthogonal trajectories meet at every point creating a spatial pattern of three families of PDS trajectories. Note that this pattern concerns only the normal stresses since in the reference (coordinate) system where axes are in PDSs there are no tangent stress components. This fact, obviously, greatly simplifies an analysis. Moreover, also the form of the stress tensor matrix is then the simplest since the only nonzero entries are diagonal entries σ_1, σ_2, σ_3.

The pattern of PDS trajectories may be a valuable source of directional information. According to Lintilhac (1974b) in a cell under the axial stress there is one plane that can be completely free from shear stress. Plant cells seem to avoid shear, i.e., tangent, stresses. Thus, a reference (coordinate) system based on PDSs may be a good candidate for a natural coordinate system of organs (Fig. 4) in a sense proposed by Hejnowicz (1984). It is worth to realize that even in an intuitively simple single cell case, the PDSs may deviate from relevant cell axes as it has been shown for the cell wall of *Chara* (Wei et al. 2006).

4.2 Tissue Stresses

The turgor pressure is direct and the only cause of stresses in an isolated cell wall. At the organ level, however, essential role may be played by tissue stresses (TSs) that only indirectly result from turgor (see also chapter "Plants as Mechano-Osmotic Transducers"). These two stress components are superimposed (Hejnowicz and Sievers 1995, 1996).

The turgor pressure is a hydrostatic pressure that arises as an osmotic phenomenon in a cell. In a steady state it equals the difference between an osmotic potential inside the cell and in the cell wall (i.e., an element of the apoplasm). The living cell is pressurized by its turgor pressure that causes an equivalent tension in the cell wall. The turgid cell is therefore prestressed like a tensegral structure (Ingber 2003; see chapter "Introduction: Tensegral World of Plants"), meaning that the protoplast is compressed whereas the cell wall is stretched.

Turgor pressure acts uniformly in all directions. However, tensile stresses in the cell wall system are usually anisotropic. The magnitude of the tensile stress in most cases varies with direction in the wall and this variation depends on the cell shape (Castle 1937, cited after Schopfer 2006). In the spherical cell, tensile stress in the cell wall does not depend on a direction. In the cylindrical cell such dependence occurs: the transversal stress is twice the longitudinal one. This, however, is the case only for isolated cells, while in situ (cells within an organ) also TSs, which are dependent on direction, are superimposed.

TSs in a cell wall are the stresses, tensile and compressive, which occur in tissues of an intact turgid organ, like a stem, and which are not due to external forces. Thus, they represent prestresses in the organ (Hejnowicz et al. 2000). The plural form (TSs) is used with reference to the organ, while the singular form (TS) refers to a particular organ tissue. The TSs disappear when the tissue is isolated. If, for example, a prestressed stem was cut along an interface between the two tissue types (like the interface between more rigid peripheral tissues and less rigid inner tissues), one tissue would shrink while the other would expand (e.g., Niklas and Paolillo 1998). Such behavior is explained by the fact that in situ the two tissues were prestressed: the shrinking one was under tension, while the expanding one – under compression.

In static equilibrium, the sum of forces involved in the generation of TSs is zero. This means that the tensile TS in a given direction is accompanied by the compressive TS in this direction (but in other tissues). However, because the stress is force per unit cross-sectional area (we consider here normal stresses only), the magnitudes of the tensile and compressive TSs may differ depending on the cross-sectional areas of the tissue layers or strands upon which the forces act (Hejnowicz 1997). If the organ is composed of turgid tissues that differ in mechanical parameters (e.g., cell wall thickness, cell diameter, elastic moduli of cell walls), then an unavoidable physical consequence of turgor pressure is the occurrence of TSs. It means that the TSs are a consequence of turgor pressure and differences in structural properties of tissues comprising an organ. We have thus two stress effects of the turgor pressure. One is a tensile stress in the cell wall as the primary effect that always occurs in the wall, irrespective of whether the cell or tissue is isolated or remains in an organ. The TSs are the secondary effect of turgor pressure occurring only when cells or tissues are within an organ.

The effect of TSs is well recognized for shoot-type organs like a stem or hypocotyl (Hejnowicz and Sievers 1995, 1996; Hejnowicz et al. 2000). In sunflower hypocotyl, there occur two groups of tissues that can be distinguished from the point of view of TSs: the peripheral tissue (mainly epidermis) and the inner tissues (mostly the ground tissue). The former, being under tensile longitudinal TS, is composed of cells characterized by thicker walls, while the latter, being under the compressive longitudinal TS, consists of relatively thin-walled cells. The forces involved in the generation of TSs are equal in magnitudes, however, the cross-sectional areas of tissues differ. Therefore, the tensile TS in the peripheral tissue is approximately five times higher than the turgor pressure, whereas at the same time the compressive TS in the ground tissue is much less than the turgor pressure. Due to the tensile TS, the stress anisotropy in cylindrical epidermal cells in situ is reversed with respect to that in isolated cells. As mentioned earlier, in isolated cells of such a shape the transverse stress in cell walls is twice the longitudinal stress. In situ the TS is superimposed and as a consequence the longitudinal stresses are much higher than transversal ones. Such example shows clearly that TSs should not be underestimated.

Measurements of stresses and comparison of the results with theoretical values calculated on the basis of structural and mechanical properties of the tissues allow one to conclude (Hejnowicz and Sievers 1996) that TSs result from and can be fully explained by the structural variation of tissues. However, another origin of TSs may be a differential growth of tissues (Peters and Tomos 1996). If there is so, it would be necessary to distinguish between structure-based TSs and differential growth-based TSs. The dispute, however, is beyond the scope of this chapter.

The occurrence of TSs may be expected also in meristems although the structural differences between meristem building cells, obviously, are much less pronounced than in differentiating or mature organs. Nevertheless, for example, in the SAM the outermost periclinal walls of the tunica cells are thicker than inner walls (e.g., Dumais and Steele 2000). Some authors postulate also that forces are generated in the SAM due to differential growth (Schüepp 1926). However, the empirical data on the TSs in meristems are missing.

4.3 Principal Stress Directions and Orientation of Cell Division Planes

The apoplasm architecture depends on the orientation of cell division planes. Since the stress tensor field operating in the apoplasm is complex and depends on organ geometry and structure, it seems natural that in setting the division planes the cells would use the directional cues coming from stress. Indeed experimental and theoretical investigations show that the orientation of cell walls in tissues or organs is related to the PDSs. Lintilhac (1974a) and Lintilhac and Vesecky (1980) have worked on planar (2D) models made of a plastic appropriate for the photoelastic stress analysis that enables recognition of principal stress trajectories (in some cases the load was applied to the models in order to mimic loads existing in the organ). The visualized principal stress trajectories were strikingly similar to the real cellular patterns. Theoretical consideration of putative relationships between stress trajectories and cell wall pattern led to the postulate that shear-free planes are preferential planes of cell division (Lintilhac 1974b). The advantage for a dividing cell to choose such a plane for the new wall is in that shear stresses accumulating in the tissue could tear the daughter cells apart. This preference for new cell wall orientation has been confirmed in experiments on in vitro growing plant tissues that have been under compressive stress (Lintilhac and Vesecky 1984). Lintilhac and Vesecky (1984) postulated that one of the reasons for a disorganized growth (the callus formation) of in vitro cultured tissue is in the loss of their in vivo mechanical environment that cannot be restored exclusively by nutritional or hormonal manipulations in vitro.

In search for a common mechanism underlying the classical cell division rules, Lynch and Lintilhac (1997) have performed the already mentioned experiments on isolated protoplasts or walled cells from tobacco NT1 suspension culture. The protoplasts or cells were cultured in agarose medium to which a load was applied and principal stress trajectories (compressive stress) were assessed with the aid of photoelastic method. The shape of cells subjected to the compressive stress of the medium is ellipsoid, i.e., they undergo strain. Such treated isolated cells (both walled and unwalled) orient their division walls normal to one of PDSs. In case of lower and axisymmetric stress/strain, the preferential division plane is perpendicular to the direction of maximal compression and parallel to the long cell axis. In case of higher but most likely not axisymmetric stress/strain, it is parallel to the direction of maximal compression and perpendicular to the long cell axis.

Similar experimental system has been used also by Zhou et al. (2006, 2007) on mesophyll cells from suspension culture of *Dendranthema morifolium*. There the stress distribution in the agarose gel was theoretically estimated on the basis of monitored load application. Nondividing protoplast put under compressive stress exhibited the anisotropic growth (elongation) in a direction perpendicular to the principal compressive stress direction, although immediately after the load application the shape of protoplasts has not been changed. Supplementing the growth medium with compounds inhibiting cell contacts with the agarose medium, Zhou et al. (2007) showed that cell contact with the medium is necessary for the observed effect on elongation. Further experiments with dividing walled cells showed that the cell division planes were oriented in the direction perpendicular to the elongation axis and that intact microtubule cytoskeleton is necessary for this mechanical stress response (Zhou et al. 2006, 2007). They also confirmed that the chemical disturbance of plasma membrane–cell wall adhesion domains affects the response. Interestingly, the influence of extracellular matrix on the cell division axis has been experimentally confirmed also for in vitro cultured animal cells (Théry et al. 2005).

4.4 Stress, Growth Rate, and Positional Information

An attractive explanation on the way in which cells recognize their position in an organ comes from the concept of positional information formulated by Wolpert (reviewed by Wolpert 1994). According to this concept, every cell acquires a positional identity in a coordinate system, Carthesian or polar depending on organ geometry. This information is interpreted by the cell according to its genetic constitution and developmental history. Wolpert assumed that the coordinate system is specified by morphogen concentrations. Specific interpretation of positional information explains the pattern formation in such different cases as a *Drosophila* embryo and a vertebrate limb. Recently, the concept of positional information has been modified by introducing the feedback between the cell activity and gradient-encoded information (Jaeger et al. 2008). The concept of positional information has also been adopted, though with some generalization, to various aspects of plant development (Barlow and Carr 1984).

Tensorial nature of plant organ growth and the relationships between growth and stress imply another possibility of positional information. Gradient of a morphogen concentration is a vector and as such is not able to provide either spatial variation or directional preferences (anisotropy), like those provided by signals coming from PDGs or PDSs. Directional signals coming from principal directions are perceived by cells. The periclinal and anticlinal alignment of cell walls in meristems and experiments with isolated cells (Lynch and Lintilhac 1997; Zhou et al. 2006, 2007) support this statement. Trajectories of principal directions create a natural system of orthogonal coordinate lines. Since growth as an irreversible deformation results from stresses, the coordinate system created by PDS trajectories seems to be the most suitable natural coordinate system (Hejnowicz 1984). Moreover, the directional information included in PDSs can possibly be used in creating a symmetry of organs or organisms (Hejnowicz 1984). The symmetry may then be a natural consequence of the tensorial level of growth and morphogenesis regulation.

In belief that individual cells perceive directional signals of a tensorial nature, specific for their position, the computer model was worked out, in which growth is generated by the GT and cells divide in PDGs (Nakielski and Hejnowicz 2003; Nakielski 2008). The model is dedicated to simulate the development of cellular pattern in apical meristems, assuming the control at the organ level. Its application is presented on the example of an idealized root apex (Fig. 5). Growth of the root apex is determined by the GT field in which the QC is accounted for (Fig. 5a). Taking the axial section as the simulation plane the cellular pattern is visualized by a 2D network in which cells are represented by polygons. The cells grow and divide under the control of GT field in which PDGs are recognizable. New walls that appear as a result of cell divisions are formed perpendicularly either to *a* or *p* (meaning anticlinal or periclinal longitudinal PDGs, Fig. 5b). The simulated cellular pattern develops quite realistically (Fig. 5c, d). Since dividing cells take PDGs into account, new walls add to the general pericline/anticline organization of the cell wall system in the root apex as the whole. The formation of new walls perpendicularly to either *a* or *p* guarantees the pattern stability. Notice that the group of cells colored in the first time step (light and dark blue cells in Fig. 5c) gives rise to the whole apex shown in the second time step (Fig. 5d). It means that all the noncolored cells not belonging to the QC were "pushed away" from the simulation region. The cell group, similar to the one colored in the first time step, can be recognized also in the second time step (dark blue cells in Fig. 5d). The progeny of these cells replaces all the light blue cells during the following simulation (not shown in the figure). This way the cellular pattern is preserved and the meristem can

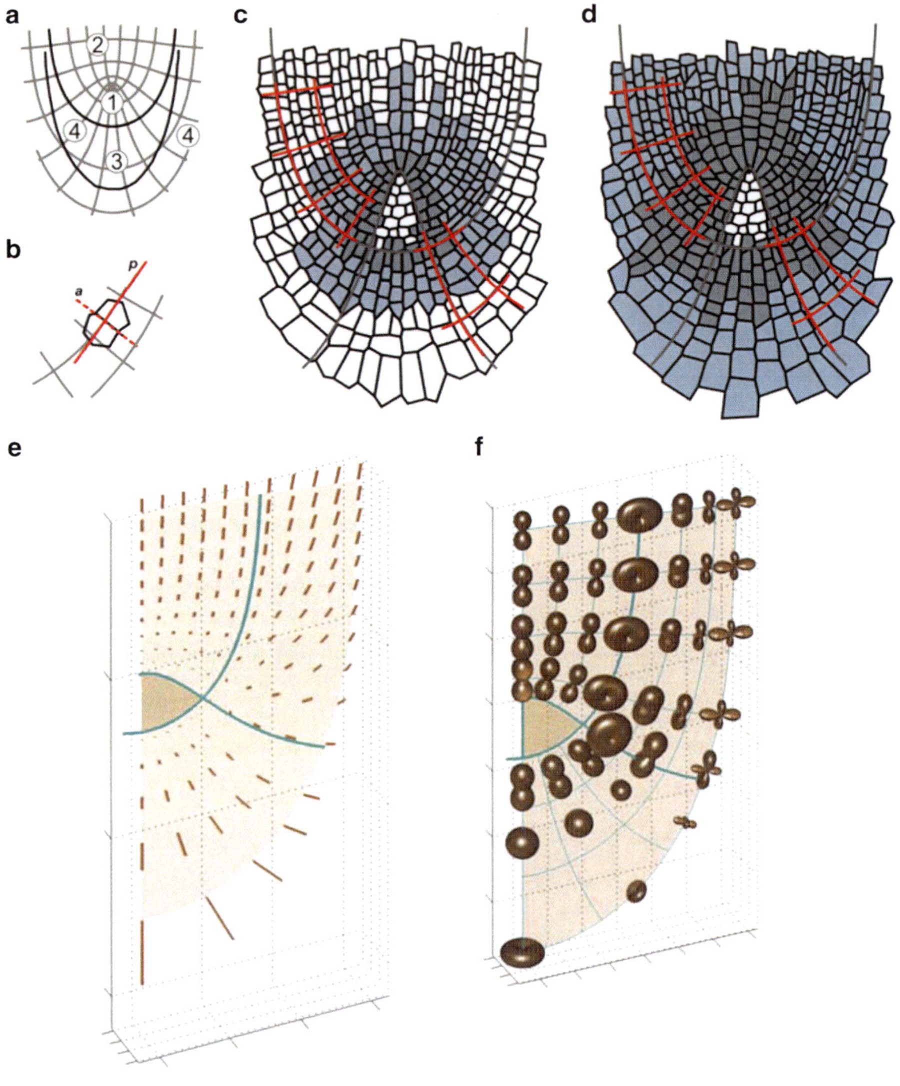

Fig. 5 Growth of the idealized root apex generated by the GT-based simulation model in which cells divide with respect to PDGs (only the axial plane is shown). (**a**) GT field dedicated for the root apex. In this root natural coordinate system, there are four zones that define growth: the root proper (1, 2) and the root cap (3, 4). The zone 1 represents QC; *gray lines* are PDG trajectories. (**b**) The cell division in which PDGs are taken into account. The division wall may be normal to *p* or *a* but *p*, as the direction for which its length can be shorter, is chosen. (**c**, **d**) Two time-steps of the simulated growth. *Red lines* are periclines and anticlines that represent PDG trajectories. (**e**, **f**) The displacement velocity field (**e**) and the variation of $RERG_l$ generating the development of cellular patterns shown in (**c**, **d**). *Dark brown* plots represent $RERG_l$ indicatrices

be considered as self-perpetuating. The same model applied for description of the real root apex of radish (*Raphanus sativus*) provided satisfactory results (Nakielski 2008). The computer-generated cellular pattern was realistic and rather steady. Computer-generated division walls, perpendicular to PDGs, formed periclinal and anticlinal zigzags as regular as those observed in microscopic sections. It was concluded that cells may perceive PDGs and obey them in the course of cell divisions. The control of growth at the tensor level, via the GT field with PDGs, may thus be necessary for the RAM function.

In the presented approach, GT defines a field of growth rates at the organ level. The QC, fundamental for this field, determines the group of quiescent initial cells surrounded by active functional initials, from which all tissues of the root apex originate. The simulations have shown (Nakielski 2008) that the cellular pattern stability depends on whether the group of the functional initials is permanent. If this is not the case, the cellular pattern changes in accordance with PDGs.

The suggestion that coordinate system of a positional information could be also defined by stress or growth rate, which are known to have a specific distribution in an organ, is in agreement with Lintilhac and Vesecky (1984) postulate that plant cells require special "stress-mechanical environment" for proper growth and divisions. The stress distribution depends on the organ geometry (Dumais and Steele 2000). Thus, the coordinate system described by PDSs trajectories would be especially attractive as a positional information system since the system is self-adapting to changes taking place during development. A similar role could be played by the natural coordinate system of periclines and anticlines, and the related PDGs. Moreover, organ boundaries in plants, like a boundary between the primordium and SAM, are analogous to compartment boundaries pointed out as a signal source by Wolpert (1994). In plants, these boundaries are the regions of unique gene expression and also geometry, and growth (reviewed in Aida and Tasaka 2006), so they may provide a specific mechanical stress signal. This special geometry is disturbed in the boundary gene mutants. Such disturbance most likely also affects the mechanical stress at malformed boundaries.

5 Apical Meristems

5.1 Growth of Apical Meristems and Methods of Its Measurement

Theoretically, the number of possible growth patterns that would allow the dome-shaped apical meristem to self-perpetuate is unlimited (Green et al. 1970). It is thus not trivial to perform detailed measurements of the meristem growth. Although as a rule any growth measurements are discrete, any finite growth may be regarded as an integration of the differentiable growth (Skalak et al. 1982).

If cell sizes are nearly uniform, which is often generally the case in apical meristems, the regions within the meristem that differ in volumetric growth rates (or area growth rates in the case of meristem surface) can be estimated on the basis of Mitotic Index or other parameters characterizing cell division distribution (e.g., Lyndon 1968; Breuil-Broyer et al. 2004, methods reviewed in Fiorani and Beemster 2006). Such assessment, however, provides no information on the growth anisotropy. The complete characteristics of growth need to account not only for area or volumetric growth rates but also for growth rate anisotropy and PDGs. The RAM and SAM differ in technical aspects of such growth assessment. This is mainly because of the differences in the availability of the two meristems. Also the RAM growth may be assumed to be steady in time at least for a short time frame, while in the SAM growth is often unsteady in time, especially if relatively big primordia are formed. It is nevertheless helpful to resolve SAM growth into growth accompanying the self-perpetuation (nearly steady in time) and primordia formation (unsteady).

A number of indirect methods of growth assessment are available if the condition that growth is steady in time is fulfilled. Silk et al. (1989) presented mathematical formulas that enable to accurately estimate growth trajectories, displacement velocity field, and general strain rate field, all on the basis of cellular patterns analyzed in median longitudinal section of the root apex. Inferring growth distribution from meristem sections is possible also with the aid of GT-based models, but with the assumptions that growth is steady in time and the GT field is axisymmetric. The pattern of periclines and anticlines, i.e., the natural coordinate system, has to be recognized in the median longitudinal meristem section, and displacement velocity has to be known either along the meristem axis (like the data provided by Van der Weele et al. 2003) or for points on the meristem surface. Such input data are enough to compute PDGs for the whole meristem (Hejnowicz et al. 1984a, b, 1988; Nakielski 2008).

The required empirical data are more challenging if growth of the meristem is unsteady even in the short time frame, as is often the case in the SAM. One solution is sequential in vivo observations followed by computational protocols. Such sequential observations and data collection may be performed with the aid of confocal laser scanning microscopy (CLSM) of individual living meristems (Grandjean et al. 2004; Reddy et al. 2004; Campilho et al. 2006) or scanning electron microscopy (SEM) in which epoxy resin replicas of living meristems are analyzed (Williams and Green 1988). This imaging methods enable 3D reconstruction of the SAM surface with the aid of adequate computational protocols (Dumais and Kwiatkowska 2002; Barbier de Reuille et al. 2005; Routier-Kierzkowska and Kwiatkowska 2008). The major limitation is that only the meristem surface can be studied. The reconstructed 3D surfaces of the individual meristem at consecutive time intervals are further used for computation of growth variables like PDGs (or strain rate crosses) (Dumais and Kwiatkowska 2002). An alternative to sequential observation of the meristem has been developed by Rolland-Lagan et al. (2005) and applied to developing petal of *Antirhinum majus* though it may be possible to adapt the method for apical meristems. Rolland-Lagan and collaborators method is based on clonal analysis and a growth model that accounts for growth anisotropy of the studied surface. The clones are recognized as cell sectors, in this particular case obtained in a temperature-sensitive flower pigment mutant. For growth characteristics, four types of regional parameters proposed by Coen et al. (2004) are employed: growth rate (referring to areal or volumetric growth rate), anisotropy, direction, and rotation. This description is equivalent to the description based on the GT. Anisotropy is a measure of how different is growth in PDGs. Direction of Coen and collaborators refers to PDG_{max}. The rotation of an examined organ portion results from growth taking place in other portions of the same organ, thus it is not its intrinsic property.

It is well documented that cell division frequency and thus presumably also growth rates are the lowest in distal SAM and RAM portions (Clowes 1961), i.e., the central zone and the QC, respectively. In the case of the SAM, the slowest growing part of the central zone is most likely the organizing center (discussed in Kwiatkowska 2004a). Computation of growth rate and PDGs at the surface of the central zone shows that it is characterized also by nearly isotropic growth (Kwiatkowska and Dumais 2003; Kwiatkowska 2004b). In species producing relatively big primordia, like in *Anagallis* or *Arabidopsis*, the growth rate distribution on the peripheral zone surface is not uniform. Areal growth rates, growth anisotropy, and the orientation of PDG_{max} are all related to the developmental stage of the neighboring primordium. The fastest and the most anisotropic growth, with the PDG_{max} in the meridional SAM direction, is characteristic for the regions where the SAM portion used up for the primordium formation is being rebuilt. Regions adjacent to primordia that are not yet separated from the SAM by a saddle-shaped boundary grow slower, less anisotropically, and with the PDG_{max} along the latitudinal direction.

Detailed growth assessment for the root apex is provided by GT models. Figure 5e, f shows the displacement velocity field and spatial variation of $RERG_l$ suggested to occur in the root apex with QC by GT-based modeling (Hejnowicz and Karczewski 1993) and computed using a new 3D approach (Nakielski and Lipowczan unpublished). The $\boldsymbol{V}$ vectors change within the organ except for the QC where velocity is zero (Fig. 5e). Everywhere the magnitude of $\boldsymbol{V}$ vectors increases with the distance from the QC. The highest increase is in the root cap columella (zone 3 in Fig. 5a). Also the direction (inclination with respect to the root axis) of the vectors changes in some places, the most significant changes are in the lateral part of the root cap (zone 4 in Fig. 5a). Because of the symplastic growth, all the changes are coordinated between successive points within the apex. The $RERG_l$ anisotropy is also changing in space (Fig. 5f). The $RERG_l$ values change with the position and depend strongly on a direction (except for a small region in the central part of the root cap where growth is isotropic). In the root proper (excluding the epidermis), the direction p is the PDG_{max}, while in the root cap the directional preferences are not uniform. Interestingly, the epidermis is distinguished from the adjacent tissues located on its two sides (i.e., the cortex and lateral parts of the root cap) by a higher growth rate in the anticlinal direction (a). Near the root cap surface where cells are shed off, the contraction along the a direction occurs. Such contraction has been observed also in margins of a leaf lamina (Erickson 1966; Wiese et al. 2007). The variation of $RERG_l$ reflects a distribution of the volumetric growth rate ($RERG_{vol}$ equals the sum of $RERG_l$ along any three mutually orthogonal directions, in particular those representing PDGs). In the root apex represented in Fig. 5, the $RERG_{vol}$ values are within the range 1.0–2.0 h^{-1} (Hejnowicz and Karczewski 1993) with maximum in the zone 4, in the region corresponding to the epidermis.

5.2 Unique Position of the Centers Creating the Stem Cell Niche in SAM and RAM

The organizing center of the SAM and the QC of the RAM are necessary to create a so-called stem cell niche in the meristems. The term stem cell niche, originating from animal developmental biology, means a cellular microenvironment necessary for the stem cell maintenance (Scheres 2007). The niche is therefore necessary for the meristem self-perpetuation. The gene network regulating the self-perpetuation in the SAM is rather well recognized unlike in the RAM (Stahl et al. 2009). The molecular basis of stem cell maintenance has nevertheless been shown to be to a certain extent similar in the SAM and RAM (Miwa et al. 2009, Stahl et al. 2009). In both, cell-to-cell communication is involved in this maintenance and orthologous genes participating in the process are known for the two types of meristems.

Strikingly, the position occupied by the organizing center and by the QC within the SAM or RAM, respectively, is unique in mechanical terms. This is the region located around a singular point of the growth and presumably also of stress tensor fields. The region surrounding a singular point in the tensor field is the point to which the PD trajectories converge. It may be unique in the sense postulated by Lintilhac (1974b) for the stresses in a cotton ovule. In this case compressive stresses generated during growth of the nucellus do not lead to collapse of the embryo sac but pass around it, leaving the embryo sac as a compression-free space within the growing ovule.

To discuss the singular point in the putative stress tensor field let us return once again to the Lynch and Lintilhac (1997) experiment with mechanical load applied to the agarose block. There is an interesting resemblance between the pattern of the PDS trajectories in the agarose block and the one expected to occur in the root proper portion of the root apex (Nakielski 2008). A special aspect is the similarity between the shape of the "interface contact" place, through which the force was applied (Lynch and Lintilhac 1997), and the QC boundary on its proximal side. The region corresponding to QC includes the singular point, and to this point all the stress trajectories converge. Thus, QC may be unique in terms of stress. It may for example be the stress-free space, postulated by Lintilhac (1974b), but in the case of the RAM it would be related to tension.

Therefore, a putative common mechanism of stem cell maintenance in the SAM and RAM may be related to the unique features of the singular point of stress and/or GT field. Clonal analysis complemented with a tensor-based simulation model for growth and cell divisions has shown that the singular point may occur also in growing petals (Nakielski unpublished).

6 Cambium

6.1 Growth and Cell Divisions in Cambium

As explained earlier, symplastically growing cells "observe rules" of continuum mechanics and tensorial approach applies to description of symplastic growth. However, continuum mechanics rules do not strictly apply to cambium growth. This is because in cambium besides the symplastic growth also intrusive growth takes place (Romberger et al. 1993). The intrusive growth occurs only at cell edges or tips and is directed toward a middle lamella between two neighboring cells. The middle lamella is believed to be softened (digested) though the evidence on this is only indirect (Lev-Yadun 2001; Siedlecka et al. 2008). The intruding edge is growing only afterwards (Hejnowicz 1997). Since the intrusive growth is restricted to cell edges and can be regarded as a local modification of generally symplastic growth of the cambium one can generalize and speculate on directions of extreme growth rates also in the case of this meristem.

Cambium surrounds a solid cylinder of differentiated secondary xylem. Thus, the symplastic growth of cambium cells in the longitudinal direction (along the stem or root axis) is virtually completely restricted, though fusiform cells may increase their dimension in the longitudinal direction due to the intrusive growth (Romberger et al. 1993). The direction of maximal symplastic growth in cambium is therefore in the plane normal to the longitudinal direction, i.e., the plane including the radial and tangential directions. It may be assumed that growth in this plane is not isotropic. Presumably either radial or tangential direction (Fig. 6a) is the direction of maximal growth, and the difference between the radial and tangential growth rates is changing with the distance from the stem axis (Karczewska et al. 2009). It may thus be speculated that the pattern of periclines and anticlines in the cambium would be similar to the pattern shown in Fig. 6a. In such a pattern, the anticlinal direction in cambium is often referred to as the radial direction, the periclinal transverse direction – as tangential, while the periclinal longitudinal – as longitudinal. It should be kept in mind, however, that such pattern is a simplification and may be not adequate for cambium forming spiral grain wood (Romberger et al. 1993; Hejnowicz 2002) where, for example, anticlinal longitudinal surfaces would be "wavy."

The majority of divisions of fusiform cells are in disagreement with the Errera's rule. The new walls formed in periclinal divisions are the largest of walls that halve the mother cells (Fig. 6b). These walls, however, are in planes normal to the radial direction, which is quite likely the direction of maximal growth rate (Romberger et al. 1993). Anticlinal divisions of fusiform cells are most often oblique (Fig. 6b),

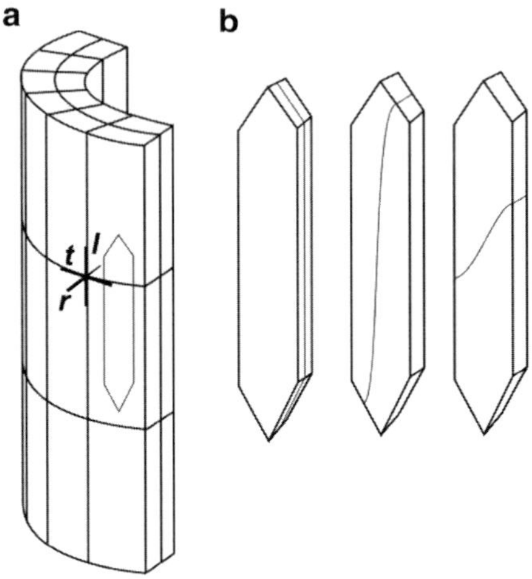

Fig. 6 The putative system of periclines and anticlines describing the cell wall system in cambium and schematic representations of various division planes typical for fusiform cambial cells. (**a**) Scheme of a one-third of the cambium cylinder. The *pointed directions* represent: *r* – the radial direction equivalent to anticlinal; *t* – the tangential direction, i.e. periclinal transverse; *l* – the longitudinal, i.e. periclinal longitudinal. (**b**) Three types of divisions (*from left to right*): periclinal, nearly longitudinal anticlinal; and nearly transverse anticlinal

i.e., the new walls also do not follow the shortest path. In storied cambium, the anticlinal divisions are usually nearly longitudinal while in nonstoried cambium the new wall orientation is nearly transverse. It can nevertheless be speculated that the central portion of new wall formed during anticlinal, especially nearly longitudinal, division is perpendicular to the tangential direction that is one of the putative directions of extreme growth. The remaining, apical and basal, marginal portions of the new wall have different orientation, but they in turn "obey" the Sachs' rule.

Direct measurements of cambium cell growth are obviously obstructed by the presence of bark, but also by the requirements of specific mechanical-stress environment for the proper cambium function (Lintilhac and Vesecky 1984). However, cambium is unique in that its growth and especially cell divisions history are preserved over years in the arrangement of some xylem cells. This phenomenon enables the reconstruction of anticlinal divisions and growth, in particular intrusive, of cambium initial cells even for hundreds of years (Romberger et al. 1993).

6.2 Specific Distribution of Stresses in Cambium

Forces that produce stresses in cambium come not only from turgor within the cambium cells themselves but also from the radial growth of differentiating secondary xylem and from bark imposed constraints (Hejnowicz 1980, 1997). In particular, the growing and differentiating xylem pushes cambium outward.

These forces coupled with the constraints coming from bark that generates forces opposite in direction in cambium result in compressive stress in radial direction (direction *r* in Fig. 6a), i.e., across the tangential cambial cell wall lamellae. However, these stresses are not uniform along the cambium circumference. It is manifested by variation in opening of gaps formed after differently oriented cambium cuts and may be one of factors involved in differentiation of cambial derivatives (Hejnowicz 1980). Presumably, during the early phase of growing season, especially in trees forming ring-porous wood, the radial stress in some cambium regions can be tensile, not compressive. This tensile stress may be a result of a collapse of phloem cells that were active during the preceding season. Therefore, it is expected to be a transient phenomenon and to depend on the synchronization of the collapse (Hejnowicz 1997). In general then, the radial stress in cambium is compressive.

Because the growing and differentiating xylem pushes cambium outward, the cambium circumference is increasing and tensile stresses are generated in tangential direction (*t* in Fig. 6a), i.e., across radial middle lamellae. As described earlier, the same forces imposed by xylem coupled with the constraints coming from bark result in cambium compression in radial direction. This compression is transformed into compression across the radial middle lamellae, i.e., the stress opposite (i.e., compressive vs. tensile) to the one generated by growing xylem itself. However, the experiments show that in mitotically active cambium the resultant stress in the tangential direction (across the radial wall lamellae) is tensile (Hejnowicz 1980).

7 Mechanical Factors in Plant Morphogenesis

The evidence on the role of mechanical factors in morphogenesis comes from both plants and animals. Moreover, the number of candidate mechanosensors and receptors is increasing (Telewski 2006; Humphrey et al. 2007). In plants these candidates include mechanosensing ion channels located in the plasma membrane, various families of receptor kinases, leucine-rich extensins, and other cell wall proteins (see chapter "Mechanics of the cytoskeleton"). For both plants and animals, two cases of the mechanical factor influence on morphogenesis can be considered. The first one is the influence of mechanical factors generated within the developing structure and resulting from its growth, geometry, or changing mechanical properties that accompany development. This type of interactions has been discussed in previous sections. The second one, perhaps somewhat simpler to elucidate, is the influence of external mechanical factors.

Classical examples on how external mechanical factors affect morphogenesis come from experiments on young sunflower, *Helianthus annuus*, capitulum. Physical constraints applied to the capitulum cause it to grow into oval and lead to alterations in both the pattern of floret primordia (phyllotaxis) and the identity of some primordia (Hernández and Green 1993). The applied constraints most likely generate compressive stress in the capitulum surface. Resulting differences in stress distribution between the treated and untreated capitula may be responsible for the observed anomalies. Similar conclusions can be drawn from Palmer and Marc (1982) experiments in which capitulum wounding, at the developmental stage when primordia have not yet been initiated, led to formation of involucral bracts and ray florets instead of disk florets in the wound area. The wound edges are the places of local stress concentration. The authors propose that this phenomenon may explain the observed abnormalities. Since in both the experiments the treatment led to a change in typical stress distribution it may be concluded that mechanical stress is a factor playing the role in capitulum morphogenesis.

Analogous examples come from animals. Elegant experiments with mechanical manipulation of in vitro cultured sheets of mammalian cells show the existence of feedback between tissue form and its proliferation and growth pattern (Nelson et al. 2005). Within the sheets of cells cultured on the adhesive surface, mechanical stresses arise from tension generated by actomyosin cytoskeleton and are transmitted between cells through cadherin-mediated intercellular adhesion. Both mechanical model predictions and stress measurements show that the proliferation and growth of cells is related to the stress pattern in the cultured sheets, which in turn depends on the sheet geometry (Nelson et al. 2005). Other studies (reviewed by Patwari and Lee 2008) show a critical role that forces generated by actomyosin cytoskeleton play in embryogenesis. A classical example is the cytoskeleton generated contraction of the apical cell side of epithelial cells that takes place in gastrulation of *Drosophila* embryo. A number of transcription factors involved in morphogenesis, which gene expression is induced by mechanical stimuli, are also known in animals (Brouzés and Farge 2004).

7.1 The SAM Surface Buckling as Phyllotaxis Mechanism

An attractive mechanical phenomenon that has been proposed as a mechanism generating biological patterns in plants is buckling (Green 1999; see also Martynov 1975). This is a deformation taking place under excessive compression (and may be a significant problem for the "construction" of tree trunks). In the case of Eulerian buckling, the deformation is elastic and distributed all along the element length (Romberger et al. 1993). It causes formation of a discrete number of undulations (folds) without any prepatterning. This observation led Green et al. (1996) to postulate the buckling mechanism of phyllotaxis. According to this postulate, the pattern of primordia is defined by a periodic undulation pattern that can be modeled as arising due to the SAM surface buckling. The presence of compressive stress in the surface of organogenic region of the SAM is prerequisite for such mechanism (Green et al. 1996).

The buckling mechanism of phyllotaxis is a continuation of Schüepp (1926) postulates that an excessive growth and self-folding of the SAM tunica is the basis of shoot patterning. Another explanation on how compressive stresses arise in the SAM surface is based on the parallel between the meristem and a pressurized vessel

(Selker et al. 1992; Dumais and Steele 2000; Steele 2000). The support for this parallel comes from the observations that outer tunica walls are the thickest of all the meristem walls and tunica (protodermis) may therefore be growth limiting similar to epidermis of differentiating shoot organs (Lyndon 1994; Savaldi-Goldstein and Chory 2008). The pressurized vessel model predicts that although most of the SAM surface should be under tensile stress, in a number of typical SAM geometry cases in some regions compressive stresses can be expected (Dumais and Steele 2000). This is true especially for the common SAM shape of a shallow dome where circumferential compression is expected at the periphery. Dumais and Steele (2000) show that in the young sunflower capitulum the predicted region of compressive circumferential stress overlaps with the organogenic region where the buckling would be expected to contribute to floret primordia patterning. They provide also empirical evidence on compression that is based on the gaping pattern of meristem surface cuts (see Dumais and Steele 2000 for a discussion of their own and previous results of similar experiments).

The significance of buckling mechanism of phyllotaxis has been questioned when the special role has been shown for polar auxin transport that is nearly sufficient to model generation of various phyllotactic patterns (e.g., Smith et al. 2006). Nevertheless, the role of mechanics in pattern generation does not have to be excluded and modeling of a combination of the two mechanisms, mechanical and biochemical, has been shown to generate phyllotactic patterns (Newell et al. 2008).

7.2 *Influence of Mechanical Properties of the Cell Wall on Morphogenesis at the SAM*

The clear evidence on the role of cell wall mechanics in morphogenesis comes from relationships between primordium initiation and function of expansins. The expansins have been referred to as a wall loosening factor (Cosgrove 2000). The actual influence of expansin on mechanical parameters has been analyzed by Wang et al. (2008) for single suspension culture cells of tomato (*Lycopersicon esculentum*). This is the first investigation in which the parameter changes due to expansin action have been measured. Wang et al. (2008) have shown that addition of expansin to the cell suspension culture medium led to increase in the cell wall elasticity (i.e., decrease of the mean Young's modulus). Using a model where the cell is treated as a liquid-filled sphere with thin elastic walls (Wang et al. 2004), they have estimated also stresses and strains at cell wall failure (from the measurements of the force leading to cell bursting). Such estimated stress and strain values are increased in the presence of expansin.

The above described quantitative results nicely complement a series of experiments performed by Kuhlemeier and Fleming groups. They have shown that sites of leaf primordium formation at the tomato apex are the sites of increased expansin gene activity even at the stage when the mitotic activity is not yet increased

(Reinhardt et al. 1998). A local application of expansin on the periphery of tomato SAM leads to induction of leaf primordium formation (Fleming et al. 1997). The induced primordia, however, do not realize a complete organogenesis (Fleming et al. 1999). The incomplete organogenesis could be explained by a quantitative discrepancy between expansin-induced growth and other systems involved in organogenesis or by the need for some biochemical determinants (Green 1997). Possibly, the former explanation holds as shown in the experiments where expression of expansin gene has been induced in SAM of transgenic tobacco. Local induction of expansin in several cell layers at the SAM periphery, not only in the surface layer as it has been done previously, led to formation of a typical leaf (Pien et al. 2001). This provides a clear example on how local manipulation leading to increase of cell wall elasticity can trigger the whole program of leaf development.

The earliest stages of primordia formation, in this case flower primordia in *Arabidopsis* inflorescence SAM, are related also to local demethyl-esterification of cell wall pectins in which pectin methyl-esterases are involved (Peaucelle et al. 2008). The SAM does not form primordia in transgenic plants exhibiting pectin hypermethyl-esterification. Since pectin esterification level is known to affect growth of other organs, e.g., hypocotyl cell elongation (Derbyshire et al. 2007), it may be expected that the effect observed in the SAM is due to changes in cell wall mechanics. Yet another example on the role of cell wall mechanics in primordia formation comes from the effect of cellulose synthesis inhibitor (2,6-dichloro-benzonitrile) on the number of flower organs in *Anagallis* (Green 1992). In this case, the treatment with the drug leads to decrease in flower primordium size. The consequence of this is a decrease in the number of sepals, petals, and stamens.

Summarizing, modifications of cell wall mechanical properties (biophysical system) accompanying the formation of primordia may be an important player in feedback loops regulating primordia spacing, i.e., phyllotaxis, in which auxin (biochemical system) has been shown to play a major role (Nakayama and Kuhlemeier 2008). These two systems, biophysical and biochemical, are believed to complement each other (Lyndon 1994).

7.3 The Role of Stress in Cambium Maintenance and in Intrusive Growth

The cambium provides clear evidence of the role and importance of mechanical stress for the proper meristem function. Cambial cells require specific stresses both for maintaining the shape via specific planes of cell divisions and for the intrusive growth. As described earlier because of a special localization in the plant body, this meristem is unique in terms of stress distribution.

As already mentioned, Lintilhac and Vesecky (1981, 1984) postulate that initial disorganized growth of explanted plant tissues is due to the loss of the in vivo stress-mechanical environment of growth that cannot be restored by nutritional or hormonal manipulation in vitro. In the case of cambium, the unique feature of this stress-mechanical environment is the existence of compressive stresses, acting along an axis normal to the plane of cambium, that control the planes of cell divisions. Brown and Sax (1962) showed that if longitudinal bark strips are released from the tree trunk and isolated from the wood by polyethylene plastic film, such exposed cambium cells proliferate and form initially unorganized callus tissue. On the contrary, in the similar strips that are returned to initial position and put under relatively low external pressure, the cambium continues its normal function producing virtually normal xylem and phloem. The applied pressure results in compressive stresses in the cambium of the directionality similar to the stresses in situ. This observation provides a strong support for the Lintilhac and Vesecky (1984) postulate. Similar support comes from observations of the isolated cambial explants cultured in vitro in which external pressure can to a certain extent restore the typical cell behavior (Brown 1964). It has also been shown that after large-scale superficial wounding of a tree trunk the wound cambium forms only after the wound periderm formation has been completed, possibly because only then the pressure (compressive stress) exerted on the cambium by the outer tissues is to some extent restored (Stobbe et al. 2002).

The tensile stress in turn is a prerequisite condition for intrusive growth at edges of fusiform cambial cells (Hejnowicz 1997). Prior to local expansion of the cell wall at the close vicinity of the edge the middle lamella is softened. The lamella softening coupled with tensile stress across the lamella allows the gap formation. The intrusively growing cell edge fills this gap. This is true both for radial edges of fusiform cells, in the case of "elongating" fusiform initials as well as mother cells of wood fibers, and for longitudinal edges of mother cells of xylem vessel members that in this way increase their diameter (Hejnowicz 1997). Of course the directions of tensile stress have to be different in these two cases: tangential for "elongating" cells, and radial for future vessel members. As described earlier (Sect. 6.2), both are present at least temporary in some places.

8 Conclusions

In order to get more insight into the meristem regulation by mechanical factors one needs first to understand the meristem mechanics or, in other words, characterize meristems as mechanical objects. The need for quantitative approach to plant development, its mechanical aspects among others, has for long been well recognized by plant biologists.

Empirical data and computer-based simulations describing growth of meristems are increasing in number and we get more insight in the growth accompanying plant morphogenesis, at least at apical meristems. The assessment of stress distribution in plant tissues is, however, technically and theoretically challenging because of the mechanical anisotropy of cell walls and the fact that while attempting to measure the stress we most likely disturb its distribution. In case of meristems, apical meristems, or cambium, the problem is further complicated by their minute dimensions and/or difficult access. Therefore, the available empirical data on stress distribution in the meristems are mainly qualitative. The local stress is estimated indirectly – the differentiation between compressive and tensile stresses is made on the basis of opening of cuts made in the meristem surface. Other data come from mechanical models, where in case of apical meristems the geometry plays an important role. One may hope that this state of knowledge will be improved once new methods, like for example the atomic force microscopy, are introduced to developmental plant biology.

Nevertheless, the available empirical data strongly support the postulate that plant cells perceive the directional signals of tensorial nature and employ them while setting the division planes. It is worth to realize that principal directions of tensorial quantities, like stress or growth, point to three mutually orthogonal directions but not to three arrows along these directions. This is unlike a direction pointed by a vector, e.g., gradient. The difference is not only in that a vector defines a single direction, but also in that it defines one of two possible arrows along this direction. Gradients can therefore be thought of as giving a signal for polarization while the principal directions may provide a directional information on anisotropy, in particular pointing to three mutually orthogonal directions or planes. In case of the PDSs, these three planes are unique in terms of shear.

Dedication and Acknowledgements We dedicate this chapter to Professor Zygmunt Hejnowicz on the occasion of his 80th birthday and also his nearly as long amazement and fascination with nature, and devotion to plant biology.

We would like to thank colleagues from the Department of Biophysics and Morphogenesis of Plants, University of Silesia, for all the inseminating discussions, Dr. Marcin Lipowczan for the help in preparation of the figures presenting the ***V*** field and GT indicatrices, and Dr. Agata Burian for the comments of this manuscript. Work in our lab is partly supported by a bilateral grant from the Ministry of Science and Higher Education, Poland and the Institut National de la Recherche Agronomique, France.

References

Aida M, Tasaka M (2006) Morphogenesis and patterning at the organ boundaries in the higher plant shoot apex. Plant Mol Biol 60:915–928

Anastasiou E, Kenz S, Gerstung M, MacLean D, Timmer J, Fleck Ch, Lenhard M (2007) Control of plant organ size by KLUH/CYP78A5-dependent intercellular signaling. Dev Cell 13:843–856

Baluška F, Šamaj J, Wojtaszek P, Volkmann D, Menzel D (2003) Cytoskeleton-plasma membrane-cell wall continuum in plants. Emerging links revisited. Plant Physiol 133:482–491

Barbier de Reuille P, Bohn-Courseau I, Godin Ch, Traas J (2005) A protocol to analyse cellular dynamics during plant development. Plant J 44:1045–1053

Barlow PW, Carr DJ (1984) Positional controls in plant development. Cambridge University Press, Cambridge

Baskin TJ (2005) Anisotropic expansion of the plant cell wall. Annu Rev Cell Dev Biol 21:203–222

Breuil-Broyer S, Morel P, de Almeida-Engler J, Coustham V, Negrutiu I, Trehin C (2004) High-resolution boundary analysis during *Arabidopsis thaliana* flower development. Plant J 38:182–192
Brouzés E, Farge E (2004) Interplay of mechanical deformation and patterned gene expression in developing embryos. Curr Opin Gen Dev 14:367–374
Brown CL (1964) The influence of external pressure on the differentiation of cells and tissues cultured in vitro. In: Zimmermann MH (ed) The formation of wood in forest trees. Academic, New York, pp 389–404
Brown CL, Sax K (1962) The influence of pressure on the differentiation of secondary tissues. Am J Bot 49:683–691
Campilho A, Garcia B, vd Toorn H, Wijk H v, Campilho A, Scheres B (2006) Time-lapse analysis of stem-cell divisions in the *Arabidopsis thaliana* root meristem. Plant J 48:619–627
Clowes FAL (1959) Apical meristems of roots. Biol Rev 34:501–529
Clowes FAL (1961) Apical meristems. Blackwell Scientific, Oxford
Coen E, Rolland-Lagan A-G, Matthews M, Bangham A, Prusinkiewicz S (2004) The genetics of geometry. Proc Natl Acad Sci USA 101:4728–4735
Cosgrove DJ (2000) Loosening of plant cell walls by expansins. Nature 407:321–326
Crowell EF, Bischoff V, Desprez T, Rolland A, Stierhof Y-D, Schumacher K, Gonneau M, Höfte H, Vernhettes S (2009) Pausing of Golgi bodies on microtubules regulates secretion of cellulose synthase complexes in *Arabidopsis*. Plant Cell 21:1141–1154
Day SJ, Lawrence PA (2000) Measuring dimensions: the regulation of size and shape. Development 127:2977–2987
Derbyshire P, McCann MC, Roberts K (2007) Restricted cell elongation in Arabidopsis hypocotyls is associated with a reduced average pectin esterification level. BMC Plant Biol 7:31
Dumais J (2009) Plant morphogenesis: a role for mechanical signals. Curr Biol 19:207–208
Dumais J, Kwiatkowska D (2002) Analysis of surface growth in shoot apices. Plant J 31:229–241
Dumais J, Steele CS (2000) New evidence for the role of mechanical forces in the shoot apical meristem. J Plant Growth Regul 19:7–18
Dumais J, Shaw SL, Steele ChR, Long SR, Ray PM (2006) An anisotropic-viscoplastic model of plant cell morphogenesis by tip growth. Int J Dev Biol 50:209–222
Erickson RO (1966) Relative elemental rates and anisotropy of growth in area: a computer programme. J Exp Bot 51:390–403
Erickson RO (1986) Symplastic growth and symplasmic transport. Plant Physiol 82:1153
Erickson RO, Mitchelini FJ (1957) The plastochron index. Am J Bot 44:297–305
Erickson RO, Sax KB (1956) Rates of cell division and cell elongation in the growth of the primary root of *Zea mays*. Proc Am Phil Soc 100:499–515
Feldman LJ, Torrey JG (1976) The isolation and culture in vitro of the quiescent center of *Zea mays*. Am J Bot 63:345–355
Fiorani F, Beemster GTS (2006) Quantitative analyses of cell division in plants. Plant Mol Biol 60:963–979
Fisher DD, Cyr RJ (1998) Extending the microtubule/microfibril paradigm. Plant Physiol 116:1043–1051
Flanders DJ, Rawlins DJ, Shaw PJ, Lloyd CW (1990) Nucleus-associated microtubules help determine the division plane of plant epidermal cells: avoidance of four-way junctions and role of cell geometry. J Cell Biol 110:1111–1122
Fleming AJ (2006) The co-ordination of cell division, differentiation and morphogenesis in the shoot apical meristem: a perspective. J Exp Bot 57:25–32
Fleming AJ, McQueen-Mason S, Mandel T, Kuhlemeier C (1997) Induction of leaf primordia by the cell wall protein expansin. Science 276:1415–1418
Fleming AJ, Caderas D, Wehrli E, McQueen-Mason S, Kuhlemeier C (1999) Analysis of expansin-induced morphogenesis on the apical meristem of tomato. Planta 208:166–174
Fosket DE (1994) Plant growth and development. A molecular approach. Academic, San Diego

Foster AS (1939) Problems of structure, growth and evolution in the shoot apex of seed plants. Bot Rev 5:454–470

Foster AS (1943) Zonal structure and growth of the shoot apex in *Microcycas calocoma* (Miq.) A. DC. Am J Bot 30:56–73

Fung YC (1984) Biomechanics. Springer, New York

Gandar PW (1980) Analysis of growth and cell production in root apices. Bot Gaz 144:1–10

Gandar PW (1983a) Growth in root apices I. The kinematic description of growth. Bot Gaz 144:1–10

Gandar PW (1983b) Growth in root apices II. Deformation and rate of deformation. Bot Gaz 144:11–19

Grandjean O, Vernoux T, Laufs P, Belcram K, Mizukami Y, Traas J (2004) *In vivo* analysis of cell division, cell growth, and differentiation at the shoot apical meristem in *Arabidopsis*. Plant Cell 16:74–87

Green PB (1962) Mechanism for plant cellular morphogenesis. Science 138:1404–1405

Green PB (1992) Pattern formation in shoots: a likely role for minimal energy configurations of the tunica. Int J Plant Sci 153(Suppl):59–75

Green PB (1997) Expansin and morphology: a role for biophysics. Trends Plant Sci 2:365–366

Green PB (1999) Expression of pattern in plants: combining molecular and calculus-based biophysical paradigms. Am J Bot 86:1059–1076

Green PB, Poethig RS (1982) Biophysics of the extension and initiation of plant organs. In: Subtelny S, Green PB (eds) Developmental order: its origin and regulation. Alan R. Liss, New York, pp 485–509

Green PB, Selker JML (1991) Mutual alignments of cell walls, cellulose, and cytoskeletons: their role in meristems. In: Lloyd CW (ed) The cytoskeletal basis of plant growth and form. Academic, London, pp 303–322

Green PB, Erickson RO, Richmond PA (1970) On the physical basis of cell morphogenesis. Annu NY Acad Sci 175:712–731

Green PB, Steele CS, Rennich SC (1996) Phyllotactic patterns: a biophysical mechanism for their origin. Ann Bot 77:515–527

Gunning BES, Hughes JE, Hardham AR (1978) Formative and proliferative cell divisions, cell differentiation, and developmental changes in the meristem of *Azolla* roots. Planta 143:121–144

Hamant O, Traas J (2009) The mechanics behind plant development. New Phytol 185:369–385

Hamant O, Heisler MG, Jönsson H, Krupiński P, Uyttewaal M, Bokov P, Corson F, Sahlin P, Boudaoud A, Meyerowitz M, Couder Y, Traas J (2008) Developmental pattering by mechanical signals in *Arabidopsis*. Science 322:1650–1655

Hejnowicz Z (1980) Tensional stress in the cambium and its developmental significance. Am J Bot 67:1–5

Hejnowicz Z (1984) Trajectories of principal directions of growth, natural coordinate system in growing plant organ. Acta Soc Bot Pol 53:301–316

Hejnowicz Z (1989) Differential growth resulting in the specification of different types of cellular architecture in root meristems. Environ Exp Bot 29:85–93

Hejnowicz Z (1997) Graviresponses in herbs and trees: a major role for the redistribution of tissue and growth stresses. Planta 203:S136–S146

Hejnowicz Z (2002) Anatomy and histogenesis of vascular plants [Anatomia i histogeneza roślin naczyniowych]. Wydawnictwo Naukowe PWN, Warszawa (in Polish)

Hejnowicz Z, Karczewski J (1993) Modeling of meristematic growth of root apices in a natural coordinate system. Am J Bot 80:309–315

Hejnowicz Z, Romberger JA (1984) Growth tensor of plant organs. J Theor Biol 110:93–114

Hejnowicz Z, Sievers A (1995) Tissue stresses in organs of herbaceous plants II. Determination in three dimensions in the hypocotyl of sunflower. J Exp Bot 46:1045–2053

Hejnowicz Z, Sievers A (1996) Acid-induced elongation of *Reynoutria* stems requires tissue stresses. Physiol Plant 98:345–348

Hejnowicz Z, Nakielski J, Hejnowicz K (1984a) Modeling of spatial variations of growth within apical domes by means of growth tensor I. Growth specified on dome axis. Acta Soc Bot Pol 53:17–28

Hejnowicz Z, Nakielski J, Hejnowicz K (1984b) Modeling of spatial variations of growth within apical domes by means of growth tensor II. Growth specified on dome surface. Acta Soc Bot Pol 53:301–316

Hejnowicz Z, Nakielski J, Włoch W, Bełtowski M (1988) Growth and development of the shoot apex of barley III. Study of growth rate variation by means of the growth tensor. Acta Soc Bot Pol 57:31–50

Hejnowicz Z, Rusin A, Rusin T (2000) Tensile tissue stress affects the orientation of cortical microtubules in the epidermis of sunflower hypocotyls. J Plant Growth Regul 19:31–44

Hernández LF, Green PB (1993) Transductions for the expression of structural pattern: analysis in sunflower. Plant Cell 5:1725–1738

Humphrey TV, Bonetta DT, Goring DR (2007) Sentinels at the wall: cell wall receptors and sensors. New Phytol 176:7–21

Ingber DE (2003) Tensegrity I Cell structure and hierarchical systems biology. J Cell Sci 116:1157–1173

Jaeger J, Irons D, Monk N (2008) Regulative feedback in pattern formation: towards a general relativistic theory of positional information. Development 135:3175–3183

Jarvis MC, Briggs SPH, Knox JP (2003) Intercellular adhesion and cell separation in plants. Plant Cell Environ 26:977–989

Jiang K, Feldman LJ (2005) Regulation of root apical meristem development. Annu Rev Cell Dev Biol 21:485–509

Kaplan DR (1992) The relationship of cells to organisms in plants: problem and implications of an organismal perspective. Int J Plant Sci 153:28–37

Karczewska D, Karczewski J, Włoch W, Jura-Morawiec J, Kojs P, Iqbal M, Krawczyszyn J (2009) Mathematical modeling of intrusive growth of fusiform initials in relation to radial growth and expanding cambial circumference in *Pinus sylvestris* L. Acta Biotheor 57:331–348

Kerstens S, Decraemer WF, Verbelen J-P (2001) Cell wall at plant surface behave mechanically like fiber-reinforced composite materials. Plant Physiol 127:381–385

Kidner C, Sundaresan V, Roberts K, Dolan L (2000) Clonal analysis of the *Arabidopsis* root confirms that position, not lineage, determines cell fate. Planta 211:191–199

Kwiatkowska D (2004a) Structural integration at the shoot apical meristem: models, measurements, and experiments. Am J Bot 91:1277–1293

Kwiatkowska D (2004b) Surface growth at the reproductive shoot apex of *Arabidopsis thaliana – pin-formed 1* and wild type. J Exp Bot 55:1021–1032

Kwiatkowska D (2008) Flowering and apical meristem growth dynamics. J Exp Bot 59:187–201

Kwiatkowska D, Dumais J (2003) Growth and morphogenesis at the vegetative shoot apex of *Anagallis arvensis* L. J Exp Bot 54:1585–1595

Kwiatkowska D, Routier-Kierzkowska A-L (2009) Morphogenesis at the inflorescence shoot apex of *Anagallis arvensis*: surface geometry and growth in comparison with the vegetative shoot. J Exp Bot 60:3407–3418

Lev-Yadun S (2001) Intrusive growth – the plant analog of dendrite and axon growth in animals. New Phytol 150:508–512

Lintilhac PM (1974a) Differentiation, organogenesis, and the tectonics of cell wall orientation III. Theoretical consideration of cell wall mechanics. Am J Bot 61:230–237

Lintilhac PM (1974b) Differentiation, organogenesis, and the tectonics of cell wall orientation II. Separation of stresses in a two-dimensional model. Am J Bot 61:135–140

Lintilhac PM, Vesecky TB (1980) Mechanical stress and cell wall orientation in plant. I. Photoelastic derivation of principal stresses; with a discussion of a concept of axillarity and the significance of the "arcuate shell zone". Am J Bot 67:1477–1483

Lintilhac PM, Vesecky TB (1981) Mechanical stress and cell wall orientation in plants. II. The application of controlled directional stress to growing plants; with a discussion on the nature of the wound reaction. Am J Bot 68:1222–1230

Lintilhac PM, Vesecky TB (1984) Stress-induced alignment of division plane in plant tissues grown *in vitro*. Nature 307:363–364

Lloyd CW (1991a) How does the cytoskeleton read the laws of geometry in aligning the division plane of plant cells? Development (Suppl) 1:55–65

Lloyd CW (1991b) Cytoskeletal elements of the phragmosome establish the division plane in vacuolated higher plant cells. In: Lloyd L (ed) The cytoskeletal basis of plant growth and form. Academic, London, pp 245–257

Lynch TM, Lintilhac PM (1997) Mechanical signals in plant development: a new method for single cell studies. Dev Biol 181:246–256

Lyndon RF (1968) Changes in volume and cell number in the different regions of shoot apex of *Pisum* during a single plastochron. Ann Bot 32:371–390

Lyndon RF (1994) Control of organogenesis at the shot apex. New Phytol 128:1–18

Martynov LA (1975) A morphogenetic mechanism involving instability of initial form. J Theor Biol 52:471–480

Miwa H, Kinoshita A, Fukuda H, Sawa S (2009) Plant meristems: *CLAVATA3/ESR*-related signaling in the shoot apical meristem and the root apical meristem. J Plant Res 122:31–39

Nakayama N, Kuhlemeier C (2008) Leaf development: untangling the spirals. Curr Biol 19:71–74

Nakielski J (1987) Spatial variations of growth within domes having different patterns of principal growth directions. Acta Soc Bot Pol 56:611–623

Nakielski J (1991) Distribution of linear growth rates in different directions in root apical meristems. Acta Soc Bot Pol 60:77–86

Nakielski J (1992) Regeneration in the root apex: modelling study by means of the growth tensor. In: Karalis TK (ed) Mechanics of swelling. NATO ASI Series H 64, pp 179–191

Nakielski J (2008) The tensor-based model for growth and cell divisions of the root apex I. The significance of principal directions. Planta 228:179–189

Nakielski J, Barlow PW (1995) Principal direction of growth and the generation of cell patterns in wild-type and *gib-1* mutant roots of tomato (*Lycopersicon esculentum* Mill.) grown in vitro. Planta 196:30–39

Nakielski J, Hejnowicz Z (2003) The description of growth of plant organs: a continuous approach based on the growth tensor. In: Nation J et al. (eds) Formal description of developing systems, NATO science series II, vol 121. Kluwer Academic, Dordrecht, pp 119–136

Nelson CM, Jean RP, Tan JL, Liu WF, Sniadecki NJ, Spector AA, Chen CS (2005) Emergent patterns of growth controlled by multicellular form and mechanics. Proc Natl Acad Sci USA 102:11594–11599

Newell AC, Shipman PD, Sun Z (2008) Phyllotaxis: cooperation and competition between mechanical and biochemical processes. J Theor Biol 251:421–439

Niklas KJ, Paolillo DJ (1998) Preferential states of longitudinal tension in the outer tissues of *Taraxacum officinale* (Asteraceae) peduncles. Am J Bot 85:1068–1081

Palmer JH, Marc J (1982) Wound-induced initiation of involucral bracts and florets in the developing sunflower inflorescence. Plant Cell Physiol 23:1401–1409

Paredez AR, Somerville CR, Ehrhardt DW (2006) Visualization of cellulose synthase demonstrates functional association with microtubules. Science 312:1491–1495

Paredez AR, Persson S, Ehrhardt DW, Somerville CR (2008) Genetic evidence that cellulose synthase activity influences microtubule cortical array organization. Plant Physiol 147: 1723–1734

Patwari P, Lee RT (2008) Mechanical control of tissue morphogenesis. Circ Res 103:234–243

Peaucelle A, Louvet R, Johansen JN, Höfte H, Laufs P, Pelloux J, Mouille G (2008) Arabidopsis phyllotaxis is controlled by the methyl-esterification status of cell-wall pectins. Curr Biol 18:1943–1948

Peters WS, Tomos AD (1996) The history of tissues tension. Ann Bot 77:657–665

Peters WS, Hagemann W, Tomos AD (2000) What makes plants different? Principles of extracellular matrix function in 'soft' plant tissues. Comp Biochem Physiol A 125:151–167
Pien S, Wyrzykowska J, McQueen-Mason S, Smart C, Fleming A (2001) Local expression of expansin induces the entire process of leaf development and modifies leaf shape. Proc Natl Acad Sci USA 98:11812–11817
Preston RD (1964) Structural and mechanical aspects of plant cell walls with particular reference to synthesis and growth. In: Zimmermann MH (ed) The formation of wood in forest trees. Academic, New York, pp 169–188
Priestley JH (1930) Studies in the physiology of cambial activity II. The concept of sliding growth. New Phytol 29:96–140
Probine MC, Barber NF (1966) The structure and plastic properties of the cell wall of *Nitella* in relation to extension growth. Aust J Biol Sci 19:439–457
Reddy GV, Heisler MG, Ehrhardt DW, Meyerowitz EM (2004) Real-time lineage analysis reveals oriented cell divisions associated with morphogenesis at the shoot apex of *Arabidopsis thaliana*. Development 131:4225–4237
Reinhardt D, Wittwer F, Mandel T, Kuhlemeier C (1998) Localized upregulation of a new expansin gene predicts the site of leaf formation in the tomato meristem. Plant Cell 10:1427–1437
Richards OW, Kavanagh AJ (1943) The analysis of the relative growth gradients and changing form of growing organisms illustrated by tobacco leaf. Am Nat 77:385–399
Rolland-Lagan A-G, Coen E, Impey SJ, Bangham JA (2005) A computational method for inferring growth parameters and shape changes during development based on clonal analysis. J Theor Biol 232:157–177
Romberger JA, Hejnowicz Z, Hill JF (1993) Plant structure: function and development. Springer, Berlin
Routier-Kierzkowska A-L, Kwiatkowska D (2008) New stereoscopic reconstruction protocol for scanning electron microscope images and its application to *in vivo* replicas of the shoot apical meristem. Funct Plant Biol 35:1034–1046
Sachs J (1887) Lecture XXVII: Relations between growth and cell-division in the embryonic tissues. In: Lectures in plant physiology. Clarendon, Oxford, pp 431–459
Savaldi-Goldstein S, Chory J (2008) Growth coordination and the shoot epidermis. Curr Opin Plant Biol 11:42–48
Scheres B (2007) Stem-cell niches: nursery rhymes across kingdoms. Nat Cell Biol 8:345–354
Schopfer P (2006) Biomechanics of plant growth. Am J Bot 93:1415–1425
Schüepp O (1926) Meristeme. Handbuch der Pflanzenanatomie Ab. I. Teil 2: Histologie. Verlag von Gebrüder Borntraeger, Berlin
Selker JML, Steucek GL, Green PB (1992) Biophysical mechanisms for morphogenetic progressions at the shoot apex. Dev Biol 153:29–43
Siedlecka A, Wiklund S, Péronne M-A, Micheli F, Leśniewska J, Sethson I, Edlund U, Richard L, Sundberg B, Mellerowicz EJ (2008) Pectin methyl esterase inhibits intrusive and symplastic cell growth in developing wood cells of *Populus*. Plant Physiol 146:554–565
Silk WK (1984) Quantitative descriptions of development. Annu Rev Plant Physiol 35:479–518
Silk WK, Lord EM, Eckard KJ (1989) Growth patterns inferred from anatomical records: empirical tests using longisections of roots of *Zea mays* L. Plant Physiol 90:708–713
Sinnott EW (1960) Plant morphogenesis. McGraw-Hill, New York
Sinnott EW, Bloch R (1941) The relative position of cell walls in developing plant tissues. Am J Bot 28:607–617
Skalak RG, Dasgupta M, Moss E, Otten P, Dullemeijer Vilmann H (1982) Analytical description of growth. J Theor Biol 94:555–577
Smith RS, Guyomarc'h S, Mandel T, Reinhard D, Kuhlemeier C (2006) A plausible model of phyllotaxis. Proc Natl Acad Sci USA 103:1301–1306
Stahl Y, Wink RH, Ingram GC, Simon R (2009) A signaling module controlling the stem cell niche in *Arabidopsis* root meristems. Curr Biol 19:1–6

Steele CR (2000) Shell stability related to pattern formation in plants. J Appl Mech 67:237–247
Stobbe H, Schmitt U, Eckstein D, Dujesiefken D (2002) Developmental stages and fine structure of surface callus formed after debarking of living lime trees (*Tilia* sp.). Ann Bot 89:773–782
Struik DL (1988) Lectures on classical differential geometry. Dover, New York
Szymanski DB, Cosgrove DJ (2009) Dynamic coordination of cytoskeletal and cell wall systems during plant cell morphogenesis. Curr Biol 19:R800–R811
Telewski FW (2006) A unified hypothesis of mechanoperception in plants. Am J Bot 93: 1466–1476
Théry M, Racine V, Pépin A, Piel M, Chen Y, Sibarita J-B, Bornens M (2005) The extracellular matrix guides the orientation of the cell division axis. Nat Cell Biol 7:947–953
Thompson D'Arcy W (1942) On growth and form: a new edition. Cambridge University Press, Cambridge
Traas J, Doonan JH (2001) Cellular basis of shoot apical meristem development. Int Rev Cytol 208:161–206
Uyttewaal M, Traas J, Hamant O (2009) Integrating physical stress, growth, and development. Curr Opin Plant Biol 13:1–7
Van der Weele CM, Jiang HS, Palaniappan KK, Ivanov VB, Palaniappan K, Baskin TI (2003) A new algorithm for computational image analysis of deformable motion at high spatial and temporal resolution applied to root growth. Roughly uniform elongation in the meristem and also, after an abrupt acceleration, in the elongation zone. Plant Physiol 132:1138–1148
Wang CX, Wang L, Thomas CR (2004) Modelling the mechanical properties of single suspension-cultured tomato cells. Ann Bot 93:443–453
Wang CX, Wang L, McQueen-Mason SJ, Pritchard J, Thomas CR (2008) pH and expansin action on single suspension-cultured tomato (*Lycopersicon esculentum*) cells. J Plant Res 121: 527–534
Wasteneys GO (2004) Progress in understanding the role of microtubules in plant cells. Curr Opin Plant Biol 7:651–660
Wei C, Lintilhac PM (2003) Loss of stability – a new model for stress relaxation in plant cell walls. J Theor Biol 224:305–312
Wei C, Lintilhac LS, Lintilhac PM (2006) Loss of stability, pH, and the anisotropic extensibility of *Chara* cell walls. Planta 223:1058–1067
Wiese A, Christ MM, Virnich O, Schurr U, Walter A (2007) Spatio-temporal leaf growth patterns of *Arabidopsis thaliana* and evidence for sugar control of the diel leaf growth cycle. New Phytol 174:752–761
Williams MH, Green PB (1988) Sequential scanning electron microscopy of a growing plant meristem. Protoplasma 147:77–79
Wolpert L (1994) Positional information and pattern formation in development. Dev Genet 15:485–490
Wyrzykowska J, Fleming AJ (2003) Cell division pattern influences gene expression in the shoot apical meristem. Proc Natl Acad Sci USA 100:5561–5566
Wyrzykowska J, Pien S, Shen WH, Fleming AJ (2002) Manipulation of leaf shape by modulation of cell division. Development 129:957–964
Zhou J, Wang B, Li Y, Wang Y, Zhu L (2006) Responses of chrysanthemum cells to mechanical stimulation require intact microtubules and plasma membrane-cell wall adhesion. J Plant Growth Regul 26:55–68
Zhou J, Wang B, Zhu L, Li Y, Wang Y (2007) A system for studying the effect of mechanical stress on the elongation behavior of immobilized plant cells. Colloids Surf B Biointerfaces 49:165–174
Zonia L, Munnik T (2007) Life under pressure: hydrostatic pressure in cell growth and function. Trends Plant Sci 12:90–97

Mechanical Force Responses of Plant Cells and Plants

E. Wassim Chehab, Yichuan Wang, and Janet Braam

Abstract Although plants lack animal-like sensory organs, they still perceive and react to stresses imposed on them by their surrounding environments. Of the various stimuli encountered, mechanical stresses, such as wind, can be detrimental if plants had not evolved mechano-stimulus-induced responses to acclimate to such stimuli. Many of these reactions have long been overlooked but over the past 40 years great progress has been made to identify the components involved in sensing and responding to such stresses. In this chapter, we briefly review some of the mechano-stimulus-induced responses of specialized plants and then discuss the various morphological and molecular responses to mechano-stimuli observed in nonspecialized plants.

1 Introduction

Living organisms have evolved to sense and respond to various environmental stresses, such as mechanical stimuli, to survive. Primitive responses to mechano-stimuli are observed in protozoa. Although these unicellular eukaryotic marine organisms lack a nervous system, nevertheless upon colliding with an object they sense the contact and reorient themselves away from the neighboring structure (Naitoh 1984). Animals are also able to respond to mechano-stimuli, often by relocating to less stressful environments. Plants are comparatively immobile. To cope with such an apparent disadvantage, plants have evolved sensitive perception

E.W. Chehab (✉) and J. Braam
Biochemistry and Cell Biology, Rice University, 6100 Main St, Houston, TX 77005, USA
e-mail: braam@rice.edu; ewchehab@rice.edu

Y. Wang
Biochemistry and Cell Biology, Rice University, 6100 Main St, Houston, TX 77005, USA
and
Chongqing University, College of Bioengineering, Chongqing 400044, People's Republic of China
e-mail: yichuan.wang@rice.edu

P. Wojtaszek (ed.), *Mechanical Integration of Plant Cells and Plants*, Signaling and Communication in Plants 9, DOI 10.1007/978-3-642-19091-9_7,

© Springer-Verlag Berlin Heidelberg 2011

and response mechanisms enabling them to acclimate to subtle environmental signals, even the seemingly innocuous stimulus of touch.

In 1881, Charles Darwin provided a detailed report on mechano-stimulus-induced plant responses. In "The Power of Movement in Plants," Darwin reported that roots of a wide variety of species change growth direction away from a touch stimulus (Darwin and Darwin 1881). In 1893 (Darwin 1893), Darwin expressed amazement of the touch-induced response of the carnivorous plants, Sundew *Drosera rotundifolia* and Venus' Flytrap *Dionaea muscipula*. The latter he referred to as "one of the most wonderful in the world." These fast-responding sensitive plants use highly specialized machinery that leads to a rapid leaf folding response (Braam 2005). Other plants, such as *Arabidopsis*, do not display rapid touch responses. However, they do sense and react to mechano-stimulation (Braam 2005). Their mechano-stimulus responses are often not recognized because the responses occur slowly over time; these responses are nonetheless quite dramatic.

A key question to address is how does a mechanical force, such as touch, get perceived and subsequently converted into biochemical signals leading to morphological responses? *Arabidopsis* and other plant species have been found to undergo various mechano-stimulus-induced physiological changes including alterations in membrane potential and turgor pressure (Shepherd et al. 2001; Shimmen 2006), rapid changes in gene transcription (Braam and Davis 1990; Braam 1992b; Lee et al. 2005), fluctuations in phytohormone as well as calcium (Ca^{2+}) levels (Mitchell and Myers 1995; Polisensky and Braam 1996; Walley et al. 2007; Tretner et al. 2008), reorientation of the cytoskeleton (Hush and Overall 1991; Cleary and Hardham 1993; Wymer et al. 1996; Komis et al. 2002; Hamant et al. 2008), in addition to modifications of the cell wall (Antosiewicz et al. 1997; Saidi et al. 2009). In this chapter, we discuss these responses and their potential implications in plant acclimation to mechanical perturbations.

2 Fast and Catchy Mechano-Stimulus Responses

Some plants respond to mechanical perturbations with impressive rapidity. Such mechano-induced responses are exhibited by the insect-catching plants the Venus' Flytrap (*D. muscipula*) and the Sundew (*D. rotundifolia*) in addition to the Sensitive Plant (*Mimosa pudica*). These plant species are examples of touch-sensitive plants that respond within seconds to mechano-stimulation.

2.1 Venus' Flytrap (D. muscipula)

The Venus' Flytrap, like other plants, absorbs inorganic nutrients from its surroundings. However, Venus' Flytrap usually grows in marsh-like areas and wet savannahs where soil is often nitrogen poor (Givnish et al. 1984). To survive such a deficit in an important nutrient, Venus' Flytrap has evolved a mechanism to supplement its diet by catching and feeding on insects. Such a behavior is made possible by an intriguing mechano-stimulus response of the plant's specially modified leaves. Each leaf is bilobed with tines or thin prong-like extensions lining the leaf edges. This specialized leaf is usually open, exposing three to five small sensitive hairs essential for triggering the shutting of the trap (Curtis 1834). The lobed leaf contains anthocyanin and its edges secrete shiny mucilage that attracts insects. When an insect crawls onto the leaf and touches the trigger hairs, an intracellular signal is generated that subsequently induces differential enlargement of leaf cells. This causes the trap to close within 0.1 s, locking in large insects. This fast closure of the trap results from a snap-buckling instability, the onset of which is controlled by the plant itself (Forterre et al. 2005). Small prey can escape through the tines; this selection for larger prey may be a consequence of the energy cost of triggering the full closure of the trap. The trapped prey struggles between the leaf lobes and stimulates further closure and tightening of the trap. The captive creature is then digested and the nutrients absorbed (Burdon-Sanderson 1873; Jacobs 1954; Jacobson 1965; Robins and Juniper 1980; Simons 1981; Fagerberg and Allain 1991). A triggered trap will reopen within 10–12 h (Darwin 1893). The requirement for multiple stimulations of the sensory hairs within a short time frame to elicit response likely enables the plant to differentiate between mechanical perturbations resulting from stimuli such as rain or wind from nutrient-rich insect prey.

2.2 Sundew (*D. rotundifolia*)

The Sundew plant is another type of touch-sensitive carnivorous plant examined by Darwin in 1893. Similar to the Venus' Flytrap, Sundew also grows in moist environments of severe nitrogen deficiency. Therefore, it has evolved to supplement itself with nitrogen provided by insectivorous behavior made possible by its specialized leaves, also known as laminae. Each lamina is covered with more than 100 red glandular hairs which secrete gluey sugary mucilage that shines in the sunlight; this appearance is likely responsible for its common name, Sundew. Insects are attracted to the leaf's red color as well as its sugary secretions and get stuck in the mucilage on the glandular tentacles. As the insect struggles to try to free itself, its movements cause a touch-induced response of the neighboring glandular tentacles that bend toward the struggling insect, creating a cup-like structure to enclose the prey (Darwin 1893; Lloyd 1942). Subsequently, the plant releases digestive enzymes to dissolve the insect and absorb the released nutrients. The mechano-sensitivity of the Sundew is amazing in its selectivity. Similar to the Venus' Flytrap, it can differentiate between a mechano-stimulation arising from living prey as compared to those arising from other environmental stimuli, such as rain (Darwin and Darwin 1881; Darwin 1893).

2.3 *Sensitive Plant (M. pudica)*

Some plants evolved touch responses for purposes other than carnivory. One example is the Sensitive Plant, also known as *M. pudica*, which has a touch-induced behavior believed to be protective more than aggressive. Upon touch, *Mimosa*'s compound leaves fold inward and, depending on the intensity of the stimulus, the leaf petiole may also droop (Fig. 1a). The leaf and petiole movements occur as a result of loss in turgor pressure in specialized cells of the pulvini, an organ located at the base of the petiole and leaflet (Esau 1965). The stimulus can also be transmitted to neighboring as well as distant leaflets on other petioles (Simons 1981; Malone 1994; Braam 2005). Since 1916, many signaling mechanisms have been implicated in carrying out such touch-induced responses. Chemical compounds have been suggested to play a role in transducing the signal to local and distal leaflets. One of these factors is Ricca's factor (Ricca 1916) and more recently turgorins (Schildknecht and Meier-Augenstein 1990). Action potentials and changes in hydraulic pressure are also believed to play a role in carrying the stimulus information within the plant allowing nontouched leaves to respond (Simons 1981; Fromm and Eschrich 1988; Fleurat-Lessard et al. 1993; Malone 1994; Stahlberg and Cosgrove 1997). The advantage of this trait to *Mimosa* is not fully understood. However, one possibility is that the leaf folding may scare away herbivorous predators and exposes protective thorns (Eisner 1981).

Fig. 1 Plant responses to mechano-stimulation. (**a**) *M. pudica* plant before and after a single mechano-stimulation. (**b**) Repetitive touch stimulated *Arabidopsis* plants exhibit thigmomorphogenetic responses characterized by stunted growth, inhibition of inflorescence elongation and a delay in the timing of the transition to flowering as compared to nonstimulated plants

3 Thigmomorphogenesis: Slow Yet Robust Mechano-Stimulus Response

Not all plants have specialized sensory organs and rapid responses to touch. Nevertheless, when these "less fortunate" organisms are subjected to repetitive mechano-stimulation, such as wind or touch, they respond with morphological alterations. These responses occur gradually and generally include a decrease in elongation and an increase in radial growth, resulting in shorter and stockier phenotypes. As pointed out by Jaffe and Forbes (1993), these touch-induced growth changes have been recognized at least since the third century B.C. in the book "*De Causis Plantarum*," written by Aristotle's student and successor Theophrastus (300 b.c.e.). In 1973, Mark Jaffe introduced the term "thigmomorphogenesis" to describe such mechano-stimulus-induced plant growth responses (Jaffe 1973). Thigmomorphogenesis is observed in diverse plant species ranging from herbaceous ones (Biddington 1986; Braam and Davis 1990) to woody trees (Telewski and Pruyn 1998; Coutand et al. 2009; Porter et al. 2009). These morphological alterations are thought to allow plants to resist and acclimate to environmental mechanical perturbations of their potentially hostile surroundings (Rodriguez et al. 2008).

3.1 Morphological Changes

Thigmomorphogenetic responses vary among different plant species. Generally, however, morphological changes occur slowly over time. An example of thigmomorphogenesis displayed by *Arabidopsis* is shown in Fig. 1b. The most common thigmomorphogenetic alterations reported include changes in reduction in shoot elongation and increase in radial expansion. Thigmomorphogenesis correlates with changes in hormone levels, chlorophyll content, stomata distribution, and cell wall structure, leaf size and wax composition, in addition to inhibition in phloem transport, and early senescence (Whitehead and Luti 1962; Jaffe and Biro 1979; Biddington 1986; Boyer et al. 1979; Jaeger et al. 1988; Jaffe and Forbes 1993; Saidi et al. 2009). For example, repeatedly touching the leaves of young cocklebur plants causes a 30% elongation growth inhibition as well as an increase in the rate of leaf senescence (Salisbury 1963). Bean plants subjected to mechano-stimulation exhibit decreased axial cell elongation and increased radial cell division (Biro et al. 1980). Tomato plants subjected to mechanical elicitations show an enhancement of cell wall lignification. However, the newly synthesized lignin in response to such stresses has a distinct composition with more syringyl units as compared to the lignin synthesized under nonstimulated growth (Saidi et al. 2009). Interestingly, although many thigmomorphogenetic responses have been documented to date, novel ones are still being reported. For example, mechanically stimulated *Carica papaya* plants display thigmomorphogenetic responses characterized by a decrease in anthocyanin levels, in addition to tissue outgrowths on the abaxial side of petioles

(Porter et al. 2009). Usually plants react to stress by increasing their levels of anthocyanins. However, the thigmomorphogenetic characteristics displayed by *Carica* suggests that each plant species evolved mechano-stimulus-induced responses that may be species-specific and perhaps more suitable for the environments in which distinct species grow.

Plants displaying thigmomorphogenetic characteristics also have decreased susceptibility to various biotic and abiotic stresses as compared to nonstimulated controls (Grace 1977; Glidden 1982; Jaffe et al. 1984; Biddington 1986; Telewski and Jaffe 1986; Mitchell and Myers 1995; Telewski 1995; Telewski and Pruyn 1998; Saidi et al. 2009). For example, leaves from mechanically stimulated *Zea mays* are thicker and have more but smaller stomata as compared to control plants. Furthermore, the surface of the stimulated leaves have more wax as compared to those from untreated plants (Whitehead and Luti 1962). Similar mechano-stimulus-induced changes are also reported in beans and rice (Whitehead and Luti 1962; Weyers and Hillman 1979; Jaffe and Forbes 1993). Such characteristics are advantageous with respect to plant water economy since waxier leaves and smaller stomata cause less water loss, and thus the plants are more able to withstand drought as compared to plants that did not undergo such alterations (Whitehead and Luti 1962; Weyers and Hillman 1979). As previously indicated, mechano-perturbations can also increase plant resistance to biotic stress (Cipollini 1997, 1998; Wang et al. 2006; Zhao et al. 2005a, b). For example, mechanically stimulated bean plants are more resistant to herbivorous pests as compared to nontreated plants (Cipollini 1997, 1998). This is attributed to a more rigid cell wall due to increased accumulation of lignin as a result of the mechano-stimulation (Cipollini 1997). The findings that touch-induced genes have diverse functions suggest that mechano-stimulus responses may provide diverse physiological effects. Genes up-regulated by mechano-stimulation have been implicated in functioning in biotic resistance (Ma et al. 2008), ion stress response (Delk et al. 2005), regulation in the transition to flowering (Tsai et al. 2007), cell wall biogenesis (Xu et al. 1995), temperature stress responses (Braam 1992a; Polisensky and Braam 1996), and phytohormone synthesis (Biro and Jaffe 1984; Botella et al. 1995; Arteca and Arteca 1999). Therefore, these mechano-stimulus-induced changes may be adaptive, making plants that sense and respond to mechanical perturbations able to withstand not only future mechanical stresses but also other types of stimuli (Jaffe et al. 1984; Biddington 1986; Telewski and Jaffe 1986; Coutand et al. 2000; McCormack et al. 2005; Braam 2005; Chehab et al. 2009).

3.2 Cellular Signaling

Upon mechano-stimulation, many of the above-mentioned morphological changes occur simultaneously as part of the plant's thigmomorphogenetic response. Therefore, it is very difficult to isolate their individual effects on plant growth. To dissect transduction of each of the diverse responses and subsequently identify what role(s) each plays in the mechano-stimulus-induced morphological changes, the different biochemical and molecular responses associated with thigmomorphogenesis have been investigated. Research is focused on changes in membrane potential as well as the involvement of various chemicals such as phytohormones, Ca^{2+}, reactive oxygen species (ROS), and nitric oxide. These cellular signals may have roles in regulating gene expression that subsequently results in the observed mechano-stimulus-induced morphological changes.

3.2.1 Membrane Potential

Changes in membrane potential have been suggested to play an early role in the response to touch stimuli (Shimmen 2006). Ion flux can lead to many cellular effects including changes in plasma membrane curvature (Zhang and Sachs 2001), modifications in the composition and conformation of membrane proteins (Gilroy and Trewavas 2001), as well as the initiation of action potentials possibly causing alterations in the gating of voltage-sensitive ion channels, such as those permeable to Ca^{2+} (Iijima and Sibaoka 1981; Hodick and Sievers 1988; Schroeder and Keller 1992; Grabov and Blatt 1998, 1999; Shimmen 2006). Shepherd et al. (2001) investigated the involvement of membrane potentials in the thigmo-perception/response mechanism of the green algae *Chara corallina*. They reported that upon strong mechano-stimulation, mechano-sensitive Ca^{2+} channels appear to be responsible for the increase in intracellular Ca^{2+} levels. This, in turn, activates Ca^{2+}-sensitive chloride channels. The inflow of Ca^{2+} and outflow of chloride cause the depolarization of the membrane in the mechano-stimulated region. If the depolarization reaches a certain threshold, neighboring voltage-regulated ion channels are recruited to propagate the wave of depolarization to the rest of the cell as well as to neighboring cells. It is hypothesized that this wave of depolarization along the plasma membrane is sensed by other voltage-gated ion channels, such as those permeable to Ca^{2+}. The influx of these ions potentially triggers the signal transduction machinery leading to physiological changes associated with the onset of thigmomorphogenesis. Nevertheless, the exact components involved in this pathway are yet to be deciphered.

Plants respond differently depending on the intensity of the stimulus (Coutand et al. 2000). For example, Hunt and Jaffe (1980) showed that thigmomorphogenetic responses increase almost linearly with increased mechanical stimulations. Interestingly, very weak mechano-perturbations can cause the plant cell to become hyperpolarized (Monshausen and Sievers 1998). Such a difference in membrane potential between strong mechano-stresses and weak ones may explain how cells are able to sense differences among different mechano-stimuli.

3.2.2 Hormones

Over the years, many investigations suggested the involvement of phytohormones in thigmomorphogenetic responses. Ethylene has been thought to be a major hormone-regulating aspect of mechano-stimulus responses and thus is the most long-studied signal with respect to its involvement in thigmomorphogenesis (Mitchell and Myers 1995). Exogenous application of this hormone results in changes that mimic those observed in thigmomorphogenesis (Goeschl et al. 1966; Brown and Leopold 1972; Jaffe and Biro 1979; Erner and Jaffe 1983; de Jaegher et al. 1987; Telewski 1995). For example, mechano-stimulating tomato plants result in a thigmomorphogenetic response characterized by stunted stem growth. A similar morphological change is observed upon ethylene exposure (Jaffe 1973; Abeles et al. 1992). Therefore, it was widely believed that such a response is at least partly mediated through ethylene. However, mechano-stimulated *Arabidopsis* mutants defective in ethylene responses still exhibit a decrease in stem elongation (Johnson et al. 1998; Coutand et al. 2000). These results indicate that a fully functional ethylene pathway is not required for aspects of thigmomorphogenesis.

Auxin is another possible thigmomorphogenetic candidate hormone involved in mechano-stimulus responses. This phytohormone promotes shoot elongation, a morphological change reversed upon mechanical stimulation (Victor and Vanderhoef 1975; Mitchell 1977). Interestingly, the levels of auxin in different plant species, such as *Bryonia dioica* and *Liquidambar styraciflua*, decrease upon mechano-perturbations (Hofinger et al. 1979; Neel and Harris 1971). Furthermore, the decrease in growth rate observed post mechano-stimulation can be restored to normal upon exogenous auxin application (Neel and Harris 1971). Altogether, these data suggest that the thigmomorphogenetic response of stunted shoot growth might be mediated by auxin.

Abscisic acid (ABA) is another phytohormone involved in regulating various developmental as well as stress responses in plants (Giraudat 1995). In particular, ABA negatively regulates growth, a process observed in plants following mechano-stimulation (Erner and Jaffe 1982). Therefore, ABA might be playing a role in mechano-stimulus responses. This hypothesis is corroborated by many findings showing that mechanically stimulated plants, such as sunflower, beans, and rice, exhibit a touch-induced increase in ABA levels (Whitehead and Luti 1962; Weyers and Hillman 1979; Beyl and Mitchell 1983). Such a surge in the phytohormone might contribute to the observed growth retardation (Jeong and Ota 1980; Erner and Jaffe 1982; Beyl and Mitchell 1983). Similar mechano-stimulus-induced morphological responses are also observed subsequent to exogenous ABA application (Erner and Jaffe 1982). The touch-inducible *CML24/TCH2* gene has been shown to be necessary for ABA responses, including an inhibition of germination and seedling growth (Delk et al. 2005), potentially linking ABA and touch responses. *Arabidopsis* mutants that are either deficient in ABA production, such as *aba-1* (Koornneef et al. 1982), or are insensitive to this phytohormone, such as *abi-1* (Koornneef et al. 1984), are invaluable tools to confirm and possibly dissect roles this phytohormone might play in thigmomorphogenesis.

Cytokinins are a group of phytohormones that influence cell division and shoot formation. They also help delay senescence, mediate auxin transport, and affect internodal length and leaf growth (Skoog 1973). All these processes regulated by cytokinins overlap with morphology alterations in plants exhibiting thigmomorphogenesis. Whether cytokinins are involved in mechano-stimulus-induced responses is not known. Further investigations are necessary to test such a link.

The oxylipin jasmonates (JAs), in particular jasmonic acid (JA), its methyl ester (MeJA), and its precursor 12-Oxo-10,15-phytodienoic acid (12-OPDA) have been implicated in mechano-stimulus responses (Staswick 1992; Weiler et al. 1993, 1994; Stelmach et al. 1998; Berger 2002; Yan et al. 2007; Tretner et al. 2008). *B. dioica* is a wild tendril-climbing cucumber. The endogenous MeJA levels of its tendrils increase dramatically upon encountering an obstacle after which they begin to coil. Noncoiling tendrils have basal levels of MeJA. Interestingly, exogenous MeJA or 12-OPDA applied to *B. dioica* tendrils induces coiling even if they have not experienced touch stimulation (Weiler et al. 1993). Mechanically stimulating *P. vulgaris* results in thigmomorphogenetic responses that are also coupled to increases in 12-OPDA levels. Similar morphological changes are also observed when coronatine, an OPDA analog, is applied to *P. vulgaris* (Stelmach et al. 1998). *Arabidopsis cev1* mutant plants, producing constitutively high levels of JA and 12-OPDA (Ellis et al. 2002), exhibit morphological alterations reminiscent of thigmomorphogenesis. Touch-stimulated *Medicago trancatula* plants have elevated levels of JA coupled with increases in chlorophyll content and inhibition of shoot growth as compared to untreated plants (Tretner et al. 2008). These data suggest that the phytohormone JA and/or perhaps its precursor and derived metabolites play roles in thigmomorphogenesis. The availability and use of the JA-insensitive *Arabidopsis coi1* mutant as well as mutants defective in the biosynthesis of JAs, such as *aos* and *opr3*, are crucial to understand and unravel the true role(s), if any, JAs play in thigmomorphogenesis.

3.2.3 Calcium and Calcium Sensors

Ca^{2+} is an important ubiquitous second messenger in eukaryotic cell signaling. When present at high cytosolic concentrations, it is toxic to cells. Subjecting soybean plants to the Ca^{2+} ionophore A23187 coupled with the exogenous application of extraordinarily high Ca^{2+} concentrations (10 mM) causes plants to have stunted growth in addition to other thigmomorphogenetic-like responses (Jones and Mitchell 1989). Upon subtle mechanical stimulation, cells show rapid and transient fluctuations of intracellular Ca^{2+} (Trewavas and Knight 1994; Allen et al. 1999; Monshausen et al. 2009). The amplitude and duration of such brief oscillations depend on the stimulated tissue as well as the type and intensity of the mechano-stimulus (Knight et al. 1991; Haley et al. 1995; Legue et al. 1997; Allen et al. 1999). For example, mimicking the pathogen entry of a cell by triggering a localized deformation of a root cell surface causes a monophasic Ca^{2+} increase which initiates at the stress site and spreads to the cytoplasm (Monshausen et al. 2009). On the other hand, bending the root to create tensile stress on some cells but compressive stress on others results in the former cells exhibiting a biphasic Ca^{2+} concentration surge; whereas cells under compression show only weak and transient intracellular Ca^{2+} elevation (Monshausen et al. 2009). The mechanism(s) by which intracellular Ca^{2+} levels increase upon mechano-stimulation remain unclear. One possible explanation is briefly discussed in a previous section of this chapter. Upon changes in plasma membrane potential, voltage-regulated Ca^{2+} channels are activated allowing the flow of extracellular Ca^{2+} to the cytosol (Shepherd et al. 2001). These initial spikes in Ca^{2+} levels might mediate further increases in intracellular Ca^{2+} through the activation of Ca^{2+}-induced Ca^{2+} release channels present on vacuolar membranes (Ward and Schroeder 1994). Interestingly, mechanically stimulating transgenic *Arabidopsis* expressing the Ca^{2+} sensor aequorin confirms that increases in intracellular Ca^{2+} originate mainly from internal stores (Knight et al. 1991, 1992). The vacuole and the endoplasmic reticulum are the major sources for stored intracellular Ca^{2+}. More recently, Kikuyama and Tazawa (2001) used *Nitella flexis* as a model system to propose that increases in cytosolic Ca^{2+} also occur through another mechanism. They showed that mechano-stimulations result in membrane stretching including those of intracellular organelles which ultimately causes the release of Ca^{2+} from internal stores. Plants are documented to have stretch-activated Ca^{2+} channels (Gens et al. 2000). Therefore, one hypothesis is that membrane stretching causes elevated Ca^{2+} levels possibly through the activation of such stretch-activated Ca^{2+} channels. Intriguingly, a similar mechanism is proposed to regulate the surge in intracellular Ca^{2+} levels in animal cells following mechano-stimulation (Sachs 1986). However, how external mechano-stimuli cause membrane stretching is still a mystery. A prime suspect in relaying these perturbations is the cytoskeleton.

Fluctuations in intracellular Ca^{2+} levels observed upon mechano-stimulations suggest that this ion may have a role in mediating signal transduction linking mechano-stimulus perception to mechano-stimulus responses. Calmodulin (*CaM*) and calmodulin-like (*CML*) genes were among the first touch-inducible genes identified (Braam and Davis 1990; Sistrunk et al. 1994). The encoded products of these genes are potential Ca^{2+} sensors (McCormack and Braam 2003), with possible roles in transducing Ca^{2+} signals into cellular responses through Ca^{2+}-dependent changes in target protein activities. Interestingly, CaM inhibitors suppress mechano-stimulation-induced growth inhibition (Jones and Mitchell 1989). These data further strengthen the hypothesis that Ca^{2+}, through CaM and/or CMLs, may mediate responses to mechanical perturbations.

CaMs and *CMLs* constitute a large multigene family in plants (Zielinski 2002; McCormack et al. 2005). The encoded proteins are predicted to bind Ca^{2+} with conserved motifs called EF-hands. The highly conserved CaMs have four EF hands. The number of Ca^{2+}-binding motifs is variable among the CMLs. For example, in *Arabidopsis*, CML1 has one identifiable EF-hand whereas CML12 (TCH3) has six (McCormack et al. 2005). Modifications in such important domains of the protein may reflect possible alterations in the Ca^{2+}-binding characteristics. Therefore, fluctuations in intracellular Ca^{2+} levels following mechano-stimulation could theoretically selectively activate specific CaMs and/or CMLs. When bound to Ca^{2+}, CaMs, and at least a subset of CMLs, undergo conformational changes (Babu et al. 1985; Wriggers et al. 1998) and expose hydrophobic domains through which they interact with target proteins. Although CaMs and CMLs are highly conserved, they are also likely to have different tissue, organ, and subcellular localization (Antosiewicz et al. 1997; McCormack et al. 2005). Therefore, variations in localization in addition to the differences in Ca^{2+}-binding behaviors may underlie potential differential roles for the Ca^{2+} sensor isoforms in Ca^{2+}-mediated touch response pathways.

3.2.4 Reactive Oxygen Species and Nitric Oxide

Reactive oxygen species (ROS) and nitric oxide (NO) are important signaling molecules mediating plant responses to various stimuli (Mori and Schroeder 2004; Shapiro 2005; Qiao and Fan 2008). Changes in ROS levels following mechano-stimulations have implicated ROS in thigmomorphogenesis (Legendre et al. 1993; Yahraus et al. 1995; Legue et al. 1997; Gus-Mayer et al. 1998; Bergey et al. 1999; Bolwell 1999; Orozco-Cardenas and Ryan 1999; Jih et al. 2003; Slesak et al. 2008). For example, mechanically perturbing soybean culture cells result in elevated ROS levels (Yahraus et al. 1995). Furthermore, the mechanical stimulation of *Mesembryanthemum crystallinum* leaves causes an increase in the *in planta* levels of ROS (Slesak et al. 2008). Such increases suggest that cells may utilize ROS molecules as signaling compounds involved in regulating the expression of genes causing thigmomorphogenesis (Van Breusegem et al. 2001). Interestingly, Mori and Shroeder (2004) reported that ROS might be involved in regulating the levels of intracellular Ca^{2+} through controlling the activity of Ca^{2+} channels. NO levels also increase in response to mechanical stimulation (Garces et al. 2001). *CML24*, a strong touch-inducible gene (Braam and Davis 1990; Delk et al. 2005; Lee et al. 2005), is necessary for proper NO accumulation (Tsai et al. 2007; Ma et al. 2008). The concurrent variations in the levels of ROS, NO, and Ca^{2+} following mechanical perturbations suggest that these signals might act in concert to regulate mechano-stimulus responses in plants.

3.2.5 Gene Expression

The first identified touch-inducible plant genes were the *Arabidopsis TCH* genes (Braam and Davis 1990). Three *TCH* genes code for potential Ca^{2+} sensors. *TCH1* encodes CaM2 (Braam and Davis 1990; Lee et al. 2005). *TCH2* and *TCH3* encode CML24 and CML12 (Braam and Davis 1990; Sistrunk et al. 1994; Khan et al. 1997; McCormack and Braam 2003). Lee et al. (2005) performed microarray analysis as a genome-wide screen for touch-inducible genes in *Arabidopsis*. They reported a 3.3-fold enrichment of Ca^{2+}-binding protein genes among the genes up-regulated at least twofold by mechano-stimulation (Lee et al. 2005). *Arabidopsis* is not the only plant showing induction in the expression of *CaM*s and *CML*s following touch; for example, potato and mung bean also show *CaM* and *CML* gene induction upon mechanical stimulation (Ling et al. 1991; Perera and Zielinski 1992; Gawienowski et al. 1993; Botella and Arteca 1994; Botella et al. 1996; Oh et al. 1996). Therefore, the touch-inducible induction of genes encoding potential Ca^{2+} sensors is conserved among diverse plants and may therefore be a feature of all plant cells.

Many plant morphological changes in response to mechano-stimulation likely require cell wall modifications. Structural characteristics of the cell wall in addition to its magnitude of extensibility play important roles in affecting plant size, form, and mechanical properties (McNeil et al. 1984; Varner and Lin 1989; Carpita and Gibeaut 1993). Interestingly, some touch-induced genes encode cell wall modifying proteins, including xyloglucan endotransglucosylase/hydrolases (XTHs) (Braam and Davis 1990; Xu et al. 1995; Lee et al. 2005). XTHs are predicted to have a role in cell wall modification by incorporating xyloglucan to reinforce the walls and/or facilitate cell growth (Campbell and Braam 1998; Osato et al. 2006; Shin et al. 2006; Liu et al. 2007; Maris et al. 2009). *TCH4*, also known as *XTH22*, is one of the several *TCH* genes first identified by Braam and Davis (1990) to be up-regulated in expression following mechano-stimulation. Furthermore, Lee et al. (2005) identified cell wall modifying protein-encoding genes, including 12 out of 33 *XTHs*, as one of the most represented classes of the touch up-regulated genes.

The first handful of touch-inducible genes discovered proved to be only the tip of the iceberg. In addition to the *CAMs*/*CMLs* and *XTHs*, many other genes, indeed approximately 2.5% of the *Arabidopsis* genome (Lee et al. 2005), are regulated following mechano-stimulation. Many of these genes identified by Lee et al. (2005) encode numerous protein kinases, transcription factors, and disease-resistance proteins. Interestingly, the latter set of genes constitutes the third most represented functional class in the *Arabidopsis* touch-regulated genes (Lee et al. 2005). Many of those defense-relevant genes are regulated by JAs. Therefore, changes in their expression following mechano-stimulation might be explained by the surge in JAs following similar stimuli (Tretner et al. 2008).

The original *TCH* genes are regulated not only by mechano-stimulation, but also by hormonal, developmental, and environmental stimuli (Braam and Davis 1990; Braam 1992a, b; Sistrunk et al. 1994; Antosiewicz et al. 1995; Xu et al. 1995; Polisensky and Braam 1996; Lee et al. 2005). Although these stimuli do not appear to share mechanical properties similar to those of touch or wounding, these stimuli induce expression of similar sets of genes. Lee et al. (2005) reported that ~53% of the touch-inducible genes showing at least twofold increase in response to touch are also up-regulated by darkness. For example, *TCH2* (*CML24*) transcripts increase in abundance by touch, heat, and cold (Braam and Davis 1990; Braam 1992b; Polisensky and Braam 1996). The finding that distinct stimuli result in shared gene expression responses suggests that different stimuli utilize independent signal transduction pathways that eventually converge to regulate touch-inducible genes, or that these distinct stimuli result in mechanical perturbations leading to the activation of a signal transduction mechanism leading to the regulation of *TCH* gene expression.

4 Microtubules in Mechano-Stimulus Perception

The cytoskeleton in animal cells plays important roles in mechano-scaffolding as well as in the perception and response to mechanical stimuli (Ingber 2003). In plants, mechano-scaffolding has been mainly attributed to the cell wall. However, in the last two decades, investigation in several plant species has provided evidence for the involvement of the plant cytoskeleton in the perception and response to mechano-perturbations (Hush and Overall 1991; Wymer et al. 1996; Hamant et al. 2008).

Plant cell microtubules have been observed to adopt a predominant intracellular orientation (Ledbetter and Porter 1963; Marc et al. 1998; Ueda et al. 1999; Nakamura et al. 2004). These cytoskeletal structures can depolymerize and/or reorient in response to mechano-stimulation. This behavior may suggest a role for microtubules in plant mechano-perception and/or response. Hush and Overall (1991) observed the reorientation of microtubules in pea root cells following mechano-perturbation. Specifically, these cytoskeletal structures assume a direction that is perpendicular to that of the applied mechanical stress (Hush and Overall 1991). A similar mechano-stress-induced reorientation of the microtubules is observed in the *Arabidopsis* shoot meristem (Hamant et al. 2008). Interestingly, mechanical stresses causing such a cytoskeletal response range from subtle touch to more robust stresses, such as centrifugation (Cleary and Hardham 1993; Wymer et al. 1996; Komis et al. 2002). Recently, *CML24*, a touch-inducible gene has been shown to be required for appropriate mechano-responses of roots, possibly through the regulation of cortical microtubule orientation (Wang et al. 2011). In addition to these descriptive observations of mechano-stimulus-induced cytoskeletal reorientation, pharmacological studies have demonstrated a requirement of intact microtubules for appropriate plant cell responses to mechano-stresses. For example, agarose-embedded plant cells can set up a preferential axis of elongation following mechano-stimuli (Wymer et al. 1996; Zhou et al. 2007). This behavior is believed to be part of the cellular response to withstand such mechano-stresses. Mechanically stimulated cells that are pretreated with microtubule depolymerizing drugs do not change their axis of elongation (Wymer et al. 1996; Zhou et al. 2007), whereas treatment with actin-depolymerizing drugs do not cause an observable effect on the cells' response to mechano-stimulation (Zhou et al. 2007). Therefore, microtubules may be the primary cytoskeletal component with a potential role in mediating plant cell responses to mechanical perturbations.

The intracellular microtubules rearrangement post mechano-perturbations may underlie plant thigmomorphogenetic responses to mechanical stress. The morphology assumed by the plant following mechano-stimulation is established by changes in the growth direction and shape of the cells, which in turn is attributed to the orientation and mechanical properties of the cellulose microfibrils. Microtubules guide the deposition of cellulose microfibrils (Paredez et al. 2006; Gutierrez et al. 2009). Therefore, it is hypothesized that mechano-stimulus-induced reorganization of microtubules may enable the deposition of cellulose microfibrils in patterns that counter the mechano-stimulus-derived strain.

Another effect of mechano-stimulus-induced alterations in the microtubule network may be that microtubule depolymerization could potentially cause membrane stretching, which in turn may facilitate the opening of stretch-activated ion channels (Nick 2008). As there are likely wall–cytoskeleton connections through the plasma membrane (Baluška et al. 2003), mechanical stress may first be received by wall, which in turn transmits the stimulus into the protoplast. Several reports have shown that removal of wall connections inhibit plant cell response to mechanical stress (Jaffe et al. 2002; Zhou et al. 2007).

5 Conclusions

The diverse effects of mechanical stimulation on plants have long been appreciated by ancient philosophers and pioneering biologists. However, it was not until the early 1970s that Mark Jaffe initiated intensive investigations to unravel physiological and biochemical aspects of thigmomorphogenesis. The 1990s were the start of identification of touch-inducible genes. Approximately two decades later, there are significant advancements in the knowledge of the diversity of plant and plant cell responses to mechanical stimulation. Progress has also been made on comprehending how the macrothigmomorphogenetic characteristics observed at the whole plant level can be attributed to micro changes at the cellular level. These intracellular modifications include the regulation of various signaling components such as Ca^{2+}, phytohormones, membrane potential, and reactive oxygen species. Such signals may regulate expression of genes that lead to the observed mechano-stimulus responses. However, what specific role(s) each of these factors contributes to the mechano-stimulus perception/response pathway remains to be elucidated. Finally, the mechanism and machinery involved in the perception of mechanical stimulation remain largely uncertain.

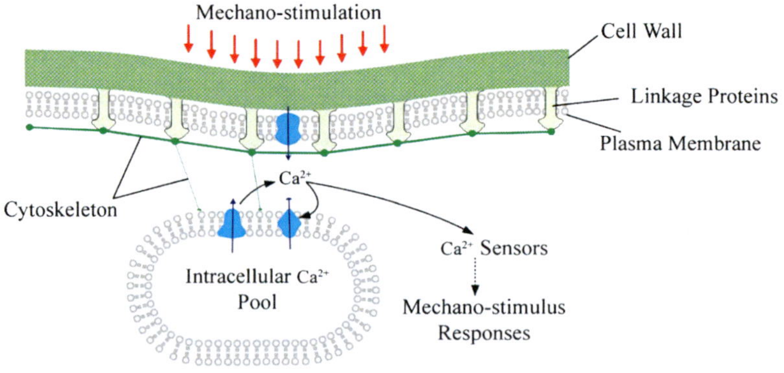

Fig. 2 A proposed model in which calcium and its sensors mediate mechano-stimulations to mechano-responses

Here, we propose a model in which the cytoskeleton plays an important role in perception and consequent induction of plant responses to mechanical perturbations (Fig. 2). Upon such stresses, the cell wall, which is the first to receive the force, transmits directly to the cytoskeleton through the linker protein complexes spanning the plasma membrane. This causes stretching of the plasma membrane as well as intracellular lipid bilayers which, in turn, results in activation of stretch-induced Ca^{2+} channels. The flow of Ca^{2+} into the cytosol has two main functions. First, it may initiate a signal that could potentially spread to neighboring cells. The second is to induce the release of more Ca^{2+} from internal stores through Ca^{2+}-activated Ca^{2+} channels. Such Ca^{2+} fluctuations are likely sensed by Ca^{2+} sensors such as CaMs and CMLs, which bind their target proteins to regulate downstream physiological effects leading to the observed mechano-stimulus-induced phenotypic changes. Microtubules may also reorient and guide the biogenesis of new cell wall to withstand further mechano-stimuli.

We are still far from unraveling the mechanism(s) through which plants sense and appropriately respond to mechano-stimuli. Over the years, many components have been implicated in taking part in such a mechanism. However, much remains to be understood. The use of modern genetic and biochemical approaches coupled with the employment of mathematical and systems modeling will be necessary to solve the intriguing puzzle of plant mechano-stimulus responses.

Acknowledgments Our work in this research area at Rice University is supported by the National Science Foundation (MCB 0817976) and the Rice University Shell Center for Sustainability. Y.W. is also supported by the National Natural Science of China (10872223).

References

Abeles FB, Morgan PW, Saltveit ME (1992) Ethylene in plant biology, 2nd edn. Academic, San Diego

Allen GJ, Kwak JM, Chu SP, Llopis J, Tsien RY, Harper JF, Schroeder JI (1999) Cameleon calcium indicator reports cytoplasmic calcium dynamics in *Arabidopsis* guard cells. Plant J 19:735–747

Antosiewicz DM, Polisensky DH, Braam J (1995) Cellular localization of the Ca^{2+} binding TCH3 protein of *Arabidopsis*. Plant J 8:623–636

Antosiewicz DM, Purugganan MM, Polisensky DH, Braam J (1997) Cellular localization of *Arabidopsis* xyloglucan endotransglycosylase-related proteins during development and after wind stimulation. Plant Physiol 115:1319–1328

Arteca JM, Arteca RN (1999) A multi-responsive gene encoding 1-aminocyclopropane-1-carboxylate synthase (ACS6) in mature *Arabidopsis* leaves. Plant Mol Biol 39:209–219

Babu YS, Sack JS, Greenhough TJ, Bugg CE, Means AR, Cook WJ (1985) Three-dimensional structure of calmodulin. Nature 315:37–40

Baluška F, Šamaj J, Wojtaszek P, Volkmann D, Menzel D (2003) Cytoskeleton-plasma membrane-cell wall continuum in plants. Emerging links revisited. Plant Physiol 133:482–491

Berger S (2002) Jasmonate-related mutants of Arabidopsis as tools for studying stress signaling. Planta 214:497–504

Bergey DR, Orozco-Cardenas M, de Moura DS, Ryan CA (1999) A wound- and systemin-inducible polygalacturonase in tomato leaves. Proc Natl Acad Sci USA 96:1756–1760
Beyl CA, Mitchell CA (1983) Alteration of growth, exudation rate, and endogenous hormone profiles in mechanically-dwarfed sunflower. J Am Soc Hortic Sci 108:257–262
Biddington NL (1986) The effects of mechanically-induced stress in plants – a review. Plant Growth Regul 4:103–123
Biro R, Jaffe MJ (1984) Thigmomorphogenesis: ethylene evolution and its role in the changes observed in mechanically perturbed bean plants. Physiol Plant 62:289–296
Biro RL, Hunt ERJ, Erner Y, Jaffe MJ (1980) Thigmomorphogenesis: changes in cell division and elongation in the internodes of mechanically-perturbed or ethrel-treated bean plants. Ann Bot 45:655–664
Bolwell GP (1999) Role of active oxygen species and NO in plant defence responses. Curr Opin Plant Biol 2:287–294
Botella JR, Arteca RN (1994) Differential expression of two calmodulin genes in response to physical and chemical stimuli. Plant Mol Biol 24:757–766
Botella JR, Arteca RN, Frangos JA (1995) A mechanical strain-induced 1-aminocyclopropane-1-carboxylic acid synthase gene. Proc Natl Acad Sci USA 92:1595–1598
Botella JR, Arteca JM, Somodevilla M, Arteca RN (1996) Calcium-dependent protein kinase gene expression in response to physical and chemical stimuli in mungbean (*Vigna radiata*). Plant Mol Biol 30:1129–1137
Boyer N, Chapelle B, Gaspar T (1979) Lithium inhibition of the thigmomorphogenetic response in *Bryonia diocia*. Plant Physiol 63:1215–1216
Braam J (1992a) Regulated expression of the calmodulin-related *TCH* genes in cultured *Arabidopsis* cells: induction by calcium and heat shock. Proc Natl Acad Sci USA 89:3213–3216
Braam J (1992b) Regulation of expression of calmodulin and calmodulin-related genes by environmental stimuli in plants. Cell Calcium 13:457–463
Braam J (2005) In touch: plant responses to mechanical stimuli. New Phytol 165:373–389
Braam J, Davis RW (1990) Rain-, wind-, and touch-induced expression of calmodulin and calmodulin-related genes in *Arabidopsis*. Cell 60:357–364
Brown KM, Leopold AC (1972) Ethylene and the regulation of growth in pine. Can J For Res 3:143–145
Burdon-Sanderson J (1873) Note on the electrical phenomena which accompany stimulation of leaf of *Dionaea muscipula*. Proc Philos Trans R Soc London 21:495–496
Campbell P, Braam J (1998) Co- and/or post-translational modifications are critical for TCH4 XET activity. Plant J 15:553–561
Carpita NC, Gibeaut DM (1993) Structural models of primary cell walls in flowering plants: consistency of molecular structure with the physical properties of the walls during growth. Plant J 3:1–30
Chehab EW, Eich E, Braam J (2009) Thigmomorphogenesis: a complex plant response to mechano-stimulation. J Exp Bot 60:43–56
Cipollini DFJ (1997) Wind-induced mechanical stimulation increases pest resistance in common bean. Oecologia 11:84–90
Cipollini DFJ (1998) The induction of soluble peroxidase activity in bean leaves by wind-induced mechanical perturbation. Am J Bot 85:1586–1591
Cleary AL, Hardham AR (1993) Pressure induced reorientation of cortical microtubules in epidermal cells of *Lolium rigidum* leaves. Plant Cell Physiol 34:1003–1008
Coutand C, Julien JL, Moulia B, Mauget JC, Guitard D (2000) Biomechanical study of the effect of a controlled bending on tomato stem elongation: global mechanical analysis. J Exp Bot 51:1813–1824
Coutand C, Martin L, Leblanc-Fournier N, Decourteix M, Julien JL, Moulia B (2009) Strain mechanosensing quantitatively controls diameter growth and *PtaZFP2* gene expression in poplar. Plant Physiol 151:223–232
Curtis M (1834) Enumeration of plants around Wilmington, NC. Boston J Nat Histol 1:123–237

Darwin C (1893) Insectivorous plants. London
Darwin C, Darwin F (1881) The power of movement in plants. D. Appleton, New York
de Jaegher G, Boyer N, Bon MC, Gaspar T (1987) Thigmomorphogenesis in *Bryonia dioica*: early events in ethylene biosynthesis pathway. Biochem Physiol Pflanz 182:49–56
Delk NA, Johnson KA, Chowdhury NI, Braam J (2005) CML24, regulated in expression by diverse stimuli, encodes a potential Ca^{2+} sensor that functions in responses to abscisic acid, daylength, and ion stress. Plant Physiol 139:240–253
Eisner T (1981) Leaf folding in a sensitive plant: a defensive thorn-exposure mechanism? Proc Natl Acad Sci USA 78:402–404
Ellis C, Karafyllidis I, Wasternack C, Turner JG (2002) The *Arabidopsis* mutant *cev1* links cell wall signaling to jasmonate and ethylene responses. Plant Cell 14:1557–1566
Erner Y, Jaffe MJ (1982) Thigmomorphogenesis: the involvement of auxin and abscisic acid in growth retardation due to mechanical perturbation. Plant Cell Physiol 23:935–941
Erner Y, Jaffe MJ (1983) Thigmomorphogenesis: membrane lipid and protein changes in bean plants as affected by mechanical perturbation and ethrel. Physiol Plant 58:197–203
Esau K (1965) Plant anatomy. Wiley, New York
Fagerberg W, Allain D (1991) A quantitative study of tissue dynamics during closure in the traps of Venus's flytrap *Dionaea muscipula* (Ellis). Am J Bot 78:647–657
Fleurat-Lessard P, Roblin G, Bonmort J, Besse C (1993) Effects of colchicine, vinblastine, cytochalasin B and phalloidin on the seismonastic movement of *Mimosa pudica* leaf and on motor cell ultrastructure. J Exp Bot 39:209–221
Forterre Y, Skotheim JM, Dumais J, Mahadevan L (2005) How the Venus flytrap snaps. Nature 433:421–425
Fromm J, Eschrich W (1988) Transport processes in stimulated and nonstimulated leaves of *Mimosa pudica*. II. Energesis and transmission of seismic stimulations. Trees 2:18–24
Garces H, Durzan D, Pedroso MC (2001) Mechanical stress elicits nitric oxide formation and DNA fragmentation in *Arabidopsis thaliana*. Ann Bot 87:567–574
Gawienowski MC, Szymanski D, Perera IY, Zielinski RE (1993) Calmodulin isoforms in *Arabidopsis* encoded by multiple divergent mRNAs. Plant Mol Biol 22:215–225
Gens JS, Fujiki M, Pickard BG (2000) Arabinoglactan protein and wall-associated kinases in a plasmalemmal reticulum with specialized vertices. Protoplasma 212:115–134
Gilroy S, Trewavas A (2001) Signal processing and transduction in plant cells: the end of the beginning? Nat Rev Mol Cell Biol 2:307–314
Giraudat J (1995) Abscisic acid signaling. Curr Opin Cell Biol 7:232–238
Givnish T, Burkhardt E, Happel R, Weintraub J (1984) Carnivory in the bromeliad *Brocchinia reducta*, with a cost/benefit model for the general restriction of carnivorous plants to sunny, moist, nutrient-poor habitats. Am Nat 124:479–497
Glidden DE (1982) Winter wind studies in Rocky Mountain National Park. Rocky Mountain Nature Association, Estes Park, CO
Goeschl JD, Rappaport L, Pratt HK (1966) Ethylene as a factor regulating the growth of pea epicotyls subjected to physical stress. Plant Physiol 41:877–844
Grabov A, Blatt MR (1998) Membrane voltage initiates Ca^{2+} waves and potentiates Ca^{2+} increases with abscisic acid in stomatal guard cells. Proc Natl Acad Sci USA 95:4778–4783
Grabov A, Blatt MR (1999) A steep dependence of inward-rectifying potassium channels on cytosolic free calcium concentration increase evoked by hyperpolarization in guard cells. Plant Physiol 119:277–288
Grace J (1977) Plant response to wind. Academic, London
Gus-Mayer S, Naton B, Hahlbrock K, Schmelzer E (1998) Local mechanical stimulation induces components of the pathogen defense response in parsley. Proc Natl Acad Sci USA 95:8398–8403
Gutierrez R, Lindeboom JJ, Paredez AR, Emons AMC, Ehrhardt DW (2009) *Arabidopsis* cortical microtubules position cellulose synthase delivery to the plasma membrane and interact with cellulose synthase trafficking compartments. Nat Cell Biol 11:797–806

Haley A, Russell AJ, Wood N, Allan AC, Knight M, Campbell AK, Trewavas AJ (1995) Effects of mechanical signaling on plant cell cytosolic calcium. Proc Natl Acad Sci USA 92:4124–4128
Hamant O, Heisler MG, Jönsson H, Krupinski P, Uyttewaal M, Bokov P, Corson F, Sahlin P, Boudaoud A, Meyerowitz EM, Couder Y, Traas J (2008) Developmental patterning by mechanical signals in *Arabidopsis*. Science 322:1650–1655
Hodick D, Sievers A (1988) The action potential of *Dionaea muscipula* Ellis. Planta 174:8–18
Hofinger M, Chapelle B, Boyer N, Gaspar T (1979) GC-MS identification and titration of IAA in mechanically perturbed *Bryonia dioica*. Plant Physiol 63:52
Hunt ERJ, Jaffe MJ (1980) Thigmomorphogenesis: the interaction of wind and temperature in the field on the growth of *Phaseolus vulgaris* L. Ann Bot 45:665–672
Hush J, Overall RL (1991) Electrical and mechanical fields orient cortical microtubules in higher plant tissue. Cell Biol Int Rep 15:551–560
Iijima T, Sibaoka T (1981) Actions potentials in the trap-lobes of *Aldrovanda vesiculosa*. Plant Cell Physiol 22:1595–1601
Ingber DE (2003) Tensegrity I. Cell structure and hierarchical systems biology. J Cell Sci 116:1157–1173
Jacobs M (1954) The effect of wind sway on the form and development of *Pinus radiata*. Aust J Bot 2:35–51
Jacobson S (1965) Receptor response in the Venus's flytrap. J Gen Physiol 49:117–129
Jaeger CH, Goeschl JD, Magnuson CE, Fares Y, Strain BR (1988) Short-term responses of phloem transport to mechanical perturbation. Physiol Plant 72:588–594
Jaffe MJ (1973) Thigmomorphogenesis: the response of plant growth and development to mechanical stimulation. Planta 114:143–157
Jaffe MJ, Biro R (1979) Thigmomorphogenesis: the effect of mechanical perturbation on the growth of plants, with special reference to anatomical changes, the role of ethylene, and interaction with other environmental stresses. In: Mussel H, Staples RC (eds) Stress physiology of crop plants. Wiley, New York, pp 25–29
Jaffe MJ, Forbes S (1993) Thigmomorphogenesis: the effect of mechanical perturbation on plants. Plant Growth Regul 12:313–324
Jaffe MJ, Telewski FW, Cooke PW (1984) Thigmomorphogenesis: on the mechanical properties of mechanically perturbed bean plants. Physiol Plant 62:73–78
Jaffe MJ, Leopold AC, Staples RA (2002) Thigmo responses in plants and fungi. Am J Bot 89:375–382
Jeong Y, Ota Y (1980) A relationship between growth inhibition and abscisic acid content by mechanical stimulation in rice plant. Jpn J Crop Sci 49:615–616
Jih PJ, Chen YC, Jeng ST (2003) Involvement of hydrogen peroxide and nitric oxide in expression of the ipomoelin gene from sweet potato. Plant Physiol 132:381–389
Johnson KA, Sistrunk ML, Polisensky DH, Braam J (1998) *Arabidopsis thaliana* responses to mechanical stimulation do not require ETR1 or EIN2. Plant Physiol 116:643–649
Jones RS, Mitchell CA (1989) Calcium ion involvement in growth inhibition of mechanically stressed soybean (Glycine max) seedlings. Physiol Plant 76:598–602
Khan AR, Johnson KA, Braam J, James MN (1997) Comparative modeling of the three-dimensional structure of the calmodulin-related TCH2 protein from *Arabidopsis*. Proteins 27:144–153
Kikuyama M, Tazawa M (2001) Mechanosensitive Ca^{2+} release from intracellular stores in *Nitella flexilis*. Plant Cell Physiol 42:358–365
Knight MR, Campbell AK, Smith SM, Trewavas AJ (1991) Transgenic plant aequorin reports the effects of touch and cold-shock and elicitors on cytoplasmic calcium. Nature 352:524–526
Knight MR, Smith SM, Trewavas AJ (1992) Wind-induced plant motion immediately increases cytosolic calcium. Proc Natl Acad Sci USA 89:4967–4971
Komis G, Apostolakos P, Galatis B (2002) Hyperosmotic stress induces formation of tubulin macrotubules in root-tip cells of triticum turgidum: their probable involvement in protoplast volume control. Plant Cell Physiol 43:911–922

Koornneef M, Jorna ML, Brinkhorst-van der Swan DLC, Karssen CM (1982) The isolation of abscisic acid (ABA)-deficient mutants by selection of induced revertants in non-germinating gibberellin sensitive lines of *Arabidopsis thaliana* (L.) Heynh. Theor Appl Genet 61:385–393

Koornneef M, Reuling G, Karssen CM (1984) The isolation and characterization of abscisic acid-insensitive mutants of *Arabidopsis thaliana*. Physiol Plant 61:377–383

Ledbetter MC, Porter KR (1963) A "microtubule" in plant cell fine structure. J Cell Biol 19:239–250

Lee D, Polisensky DH, Braam J (2005) Genome-wide identification of touch- and darkness-regulated *Arabidopsis* genes: a focus on calmodulin-like and XTH genes. New Phytol 165:429–444

Legendre L, Rueter S, Heinstein PF, Low PS (1993) Characterization of the oligogalacturonide-induced oxidative burst in cultured soybean (*Glucine max*) cells. Plant Physiol 102:133–140

Legue V, Blancaflor E, Wymer C, Perbal G, Fantin D, Gilroy S (1997) Cytoplasmic free Ca^{2+} in *Arabidopsis* roots changes in response to touch but not gravity. Plant Physiol 114:789–800

Ling V, Perera I, Zielinski RE (1991) Primary structures of *Arabidopsis* calmodulin isoforms deduced from the sequences of cDNA clones. Plant Physiol 96:1196–1202

Liu YB, Lu SM, Zhang JF, Liu S, Lu YT (2007) A xyloglucan endotransglucosylase/hydrolase involves in growth of primary root and alters the deposition of cellulose in *Arabidopsis*. Planta 226:1547–1560

Lloyd F (1942) The carnivorous plants. Chronica Botanica, Watham

Ma W, Smigel A, Tsai YC, Braam J, Berkowitz GA (2008) Innate immunity signaling: cytosolic Ca^{2+} elevation is linked to downstream nitric oxide generation through the action of calmodulin or a calmodulin-like protein. Plant Physiol 148:818–828

Malone M (1994) Wound-induced hydraulic signals and stimulus transmission in *Mimosa pudica* L. New Phytol 128:49–56

Marc J, Granger CL, Brincat J, Fisher DD, Kao T, McCubbin AG, Cyr RJ (1998) A GFP–MAP4 reporter gene for visualizing cortical microtubule rearrangements in living epidermal cells. Plant Cell 10:1927–1939

Maris A, Suslov D, Fry SC, Verbelen JP, Vissenberg K (2009) Enzymic characterization of two recombinant xyloglucan endotransglucosylase/hydrolase (XTH) proteins of *Arabidopsis* and their effect on root growth and cell wall extension. J Exp Bot 60:3959–3972

McCormack E, Braam J (2003) Calmodulins and related potential calcium sensors of *Arabidopsis*. New Phytol 159:585–598

McCormack E, Tsai YC, Braam J (2005) Handling calcium signaling: *Arabidopsis* CaMs and CMLs. Trends Plant Sci 10:383–389

McNeil M, Darvill AG, Fry SC, Albersheim P (1984) Structure and function of the primary cell walls of plants. Annu Rev Biochem 53:625–663

Mitchell CA (1977) Influence of mechanical stress on auxin-stimulated growth on excised pea stem sections. Physiol Plant 41:412–424

Mitchell CA, Myers PN (1995) Mechanical stress regulation of plant growth and development. Hortic Rev (Am Soc Hortic Sci) 17:1–42

Monshausen GB, Sievers A (1998) Weak mechanical stimulation causes hyperpolarisation in root cells of *Lepidium*. Bot Acta 111:303–306

Monshausen GB, Bibikova TN, Weisenseel MH, Gilroy S (2009) Ca^{2+} regulates reactive oxygen species production and pH during mechanosensing in *Arabidopsis* roots. Plant Cell 21:2341–2356

Mori IC, Schroeder JI (2004) Reactive oxygen species activation of plant Ca^{2+} channels. A signaling mechanism in polar growth, hormone transduction, stress signaling, and hypothetically mechanotransduction. Plant Physiol 135:702–708

Naitoh Y (1984) Mechanosensory transduction in Protozoa. In membranes and sensory transduction. Plenum, New York

Nakamura M, Naoi K, Shoji T, Hashimoto T (2004) Low concentrations of propyzamide and oryzalin alter microtubule dynamics in *Arabidopsis* epidermal cells. Plant Cell Physiol 45:1330–1334
Neel PL, Harris RW (1971) Motion-induced inhibition of elongation and induction of dormancy in liquidambar. Science 173:58–59
Nick P (2008) Plant microtubules: development and flexibility. In: Plant cell monographs, vol 11. Springer, New York
Oh SA, Kwak JM, Kwun IC, Nam HG (1996) Rapid and transient induction of calmodulin-encoding gene(s) of *Brassica napus* by a touch stimulus. Plant Cell Rep 15:586–590
Orozco-Cardenas M, Ryan CA (1999) Hydrogen peroxide is generated systemically in plant leaves by wounding and systemin via the octadecanoid pathway. Proc Natl Acad Sci USA 96:6553–6557
Osato Y, Yokoyama R, Nishitani K (2006) A principal role for AtXTH18 in *Arabidopsis thaliana* root growth: a functional analysis using RNAi plants. J Plant Res 119:153–162
Paredez AR, Somerville CR, Ehrhardt DW (2006) Visualization of cellulose synthase demonstrates functional association with microtubules. Science 312:1491–1495
Perera IY, Zielinski RE (1992) Structure and expression of the *Arabidopsis CaM3* calmodulin gene. Plant Mol Biol 19:649–664
Polisensky DH, Braam J (1996) Cold-shock regulation of the *Arabidopsis TCH* genes and the effects of modulating intracellular calcium levels. Plant Physiol 111:1271–1279
Porter BW, Zhu YJ, Webb DT, Christopher DA (2009) Novel thigmomorphogenetic responses in *Carica papaya*: touch decreases anthocyanin levels and stimulates petiole cork outgrowths. Ann Bot 103:847–858
Qiao W, Fan LM (2008) Nitric oxide signaling in plant responses to abiotic stresses. J Integr Plant Biol 50:1238–1246
Ricca J (1916) Soluzione d'un problema di fisiologia. La propagazione di stimulo nella Mimosa. Nuove Giornale Botanico Italiano 23:51–170
Robins R, Juniper B (1980) The secretory cycle of *Dionaea muscipula* Ellis. V. The absorption of nutrients. New Phytol 86:413–422
Rodriguez M, De Langre E, Moulia B (2008) A scaling law for the effects of architecture and allometry on tree vibration modes suggests a biological tuning to modal compartmentalization. Am J Bot 95:1523–1537
Sachs F (1986) Mechanotransducing ion channels. In: Latorre R (ed) Ionic channels in cells and model systems. Plenum, New York, pp xxiii, 437 p
Saidi I, Ammar S, Demont-Cauleta N, Thévenina J, Lapierrec C, Bouzid S, Jouanin L (2009) Thigmomorphogenesis in *Solanum lycopersicum*: morphological and biochemical responses in stem after mechanical stimulation. Plant Sci 177:1–6
Salisbury FB (1963) The flowering process. Macmillan, New York
Schildknecht H, Meier-Augenstein W (1990) Role of turgorins in leaf movement. In: The pulvinus: motor organ for leaf movement. American Society of Plant Physiology, Rockville
Schroeder JI, Keller BU (1992) Two types of anion channel currents in guard cells with distinct voltage regulation. Proc Natl Acad Sci USA 89:5025–5029
Shapiro AD (2005) Nitric oxide signaling in plants. Vitam Horm 72:339–398
Shepherd VA, Shimmen T, Beilby MJ (2001) Mechanosensory ion channels in *Chara*: the influence of cell turgor pressure on touch-activated receptor potentials and action potentials. J Plant Physiol 28:551–566
Shimmen T (2006) Electrical perception of "Death Message" in *Chara*: characterization of K^+-induced depolarization. Plant Cell Physiol 47:559–562
Shin YK, Yum H, Kim ES, Cho H, Gothandam KM, Hyun J, Chung YY (2006) BcXTH1, a *Brassica campestris* homologue of *Arabidopsis* XTH9, is associated with cell expansion. Planta 224:32–41
Simons P (1981) The role of electricity in plant movements. New Phytol 87:11–37

Sistrunk ML, Antosiewicz DM, Purugganan MM, Braam J (1994) *Arabidopsis TCH3* encodes a novel Ca^{2+} binding protein and shows environmentally induced and tissue-specific regulation. Plant Cell 6:1553–1565

Skoog F (1973) Cytokinins in regulation of plant growth. Basic Life Sci 2:147–184

Slesak I, Slesak H, Libik M, Miszalski Z (2008) Antioxidant response system in the short-term post-wounding effect in *Mesembryanthemum crystallinum* leaves. J Plant Physiol 165:127–137

Stahlberg R, Cosgrove D (1997) The propagation of slow wave potentials in pea epicotyls. Plant Physiol 113:209–217

Staswick PE (1992) Jasmonate, genes, and fragrant signals. Plant Physiol 99:804–807

Stelmach BA, Muller A, Hennig P, Laudert D, Andert L, Weiler EW (1998) Quantitation of the octadecanoid 12-oxo-phytodienoic acid, a signalling compound in plant mechanotransduction. Phytochemistry 47:539–546

Telewski FW (1995) Wind induced physiological and developmental responses in trees. In: Coutts MP, Grace J (eds) Wind and trees. Cambridge University Press, Cambridge, New York, pp 237–263

Telewski FW, Jaffe MJ (1986) Thigmomorphogenesis: anatomical, morphological and mechanical analysis of genetically different sibs of *Pinus taeda* in response to mechanical perturbation. Physiol Plant 66:219–226

Telewski FW, Pruyn ML (1998) Thigmomorphogenesis: a dose response to flexing in Ulmus americana seedlings. Tree Physiol 18:65–68

Theophrastus (300 B.C.E.) De Causis Plantarum, 3 vols (English Translation). Harvard University Press, Cambridge, MA

Tretner C, Huth U, Hause B (2008) Mechanostimulation of *Medicago truncatula* leads to enhanced levels of jasmonic acid. J Exp Bot 59:2847–2856

Trewavas A, Knight M (1994) Mechanical signalling, calcium and plant form. Plant Mol Biol 26:1329–1341

Tsai YC, Delk NA, Chowdhury NI, Braam J (2007) Arabidopsis potential calcium sensors regulate nitric oxide levels and the transition to flowering. Plant Signal Behav 6:446–454

Ueda K, Matsuyama T, Hashimoto T (1999) Visualization of microtubules in living cells of transgenic Arabidopsis thaliana. Protoplasma 206:201–206

Van Breusegem F, Vranova E, Dat JF, Inze D (2001) The role of active oxygen species in plant signal transduction. Plant Sci 161:405–414

Varner JE, Lin LS (1989) Plant cell wall architecture. Cell 56:231–239

Victor TS, Vanderhoef LN (1975) Mechanical inhibition of hypocotyl elongation induces radial enlargement: implications for cytokinin action. Plant Physiol 56:845–846

Walley JW, Coughlan S, Hudson ME, Covington MF, Kaspi R, Banu G, Harmer SL, Dehesh K (2007) Mechanical stress induces biotic and abiotic stress responses via a novel *cis*-element. PLoS Genet 3:1800–1812

Wang BC, Wang JB, Zhao HC, Zhao H (2006) Stress induced plant resistance and enzyme activity varying in cucumber. Colloid Surf B 48:138–142

Wang Y, Wang B, Gilroy S, Chehab EW, Braam J (2011) CML24 is involved in root mechanoresponses and cortical microtubule orientation in *Arabidopsis*. Plant Growth Regul. In Press

Ward JM, Schroeder JI (1994) Calcium-activated K^+ channels and calcium-induced calcium release by slow vacuolar ion channels in guard cell vacuoles implicated in the control of stomatal closure. Plant Cell 6:669–683

Weiler EW, Albrecht T, Groth B, Xia ZQ, Luxem M, Lib H, Andert L, Spengler P (1993) Evidence for the involvement of jasmonates and their octadecanoid precursors in the tendril coiling response of *Bryonia dioica*. Phytochemistry 32:591–600

Weiler EW, Kutchan TM, Gorba T, Brodschelm W, Niesel U, Bublitz F (1994) The *Pseudomonas* phytotoxin coronatine mimics octadecanoid signalling molecules of higher plants. FEBS Lett 345:9–13

Weyers JDB, Hillman JR (1979) Sensitivity of *Commelina* stomata to abscisic acid. Planta 146:623–628

Whitehead H, Luti R (1962) Experimental studies of the effect of wind on plant growth and anatomy. I. Zea mays. New Phytol 61:56–58
Wriggers W, Mehler E, Pitici F, Weinstein H, Schulten K (1998) Structure and dynamics of calmodulin in solution. Biophys J 74:1622–1639
Wymer CL, Wymer SA, Cosgrove DJ, Cyr RJ (1996) Plant cell growth responds to external forces and the response requires intact microtubules. Plant Physiol 110:425–430
Xu W, Purugganan MM, Polisensky DH, Antosiewicz DM, Fry SC, Braam J (1995) *Arabidopsis TCH4*, regulated by hormones and the environment, encodes a xyloglucan endotransglycosylase. Plant Cell 7:1555–1567
Yahraus T, Chandra S, Legendre L, Low PS (1995) Evidence for a mechanically induced oxidative burst. Plant Physiol 109:1259–1266
Yan Y, Stolz S, Chetelat A, Reymond P, Pagni M, Dubugnon L, Farmer EE (2007) A downstream mediator in the growth repression limb of the jasmonate pathway. Plant Cell 19:2470–2483
Zhang PC, Sachs F (2001) Voltage-induced membrane movement. Nature 413:428–431
Zhao H, Wang BC, Zhao HC, Wang JB (2005a) Stress stimulus induced resistance to *Cladosporium cucumerinum* in cucumber seeding. Colloids Surf B 44:36–40
Zhao HC, Zhao H, Wang JB, Wang BC, Wang YN (2005b) Stress stimulation induced resistance of plant. Colloids Surf B 43:174–178
Zhou J, Wang B, Li Y, Wang Y, Zhu L (2007) Responses of chrysanthemum cells to mechanical stimulation require intact microtubules and plasma membrane-cell wall adhesion. J Plant Growth Regul 26:55–68
Zielinski RE (2002) Characterization of three new members of the *Arabidopsis thaliana* calmodulin gene family: conserved and highly diverged members of the gene family functionally complement a yeast calmodulin null. Planta 214:446–455

Mechanical Aspects of Gravity-Controlled Growth, Development and Morphogenesis

František Baluška and Dieter Volkmann

Motto: *Physical Forces Shape Life*

Abstract Plants are "addicted" to gravity. Gravity is the most constant physical force and acts on all organisms in the same way. Plants need this vectorial physical force for their roots to grow down into soil to search for water and mineral nutrients, and for their shoots to grow up to allow optimal exposure of their leaves to light. The plant body form and plant morphogenesis depend on gravity, which acts as a reference force for plant development. Gravity-controlled morphogenesis of plants seems to rely on multiple gravity-sensing mechanisms. Despite more than 100 years of study, plant gravisensing is still as mysterious as it was in 1900, when, for the first time, the statolith theory was proposed by Bohumil Nemec and Gottlieb Haberlandt. Some 10 years after its initial acceptance, this theory was put aside for some 50 years. In the 1960s, it was resuscitated and presently dominates our scientific thinking. Nevertheless, how gravity is perceived and what the gravity transduction pathways are remain enigmatic. In this chapter, we first discuss the physical properties of the cytoplasm. Then, we provide an overview the starch-based amyloplasts and some other possible statolith candidates for sedimentation to the physical bottom. We list several unconventional processes and structures, both at the subcellular and at the supracellular level, which emerge to play a role in plant gravisensing. Finally, we point out surprising differences in sensing of gravity between roots and shoots. Although there are several common themes, such as amyloplast sedimentation, relevance of endomembranes and endocytic vesicle recycling, as well as of the actin cytoskeleton, and the polar transport of auxin, there are very profound differences between root and shoot sensing of gravity. Obviously, plant gravisensing will keep plant scientists busy in the future.

F. Baluška (✉) and D. Volkmann
Institut für Zelluläre und Molekulare Botanik, Friedrich-Wilhelms-Universität Bonn, Kirschallee 1, 53115 Bonn, Germany
e-mail: baluska@uni-bonn.de; unb110@uni-bonn.de

P. Wojtaszek (ed.), *Mechanical Integration of Plant Cells and Plants*, Signaling and Communication in Plants 9, DOI 10.1007/978-3-642-19091-9_8,
© Springer-Verlag Berlin Heidelberg 2011

1 Gravity Is a Unique Force with Respect to Life

Morphogenesis of plants is under the control of physicochemical factors, in particular the gravitational force of the earth and the electromagnetic spectrum of the sun. In recent years, clear examples for mechano-induced cell fate transitions, development and morphogenesis in plants have been obtained for mechanically induced organogenesis both in shoots and in roots (Fleming et al. 1997; Reinhardt et al. 1998; Baluška et al. 2000a; Pien et al. 2001; Cho and Cosgrove 2002; Laskowski et al. 2008; Ditengou et al. 2008; Braybrook and Kuhlemeier 2010).

Ever since plants moved out of the oceans to develop a terrestrial mode of their lives, they have been under the strong physical stress of gravity (Trewavas and Knight 1994; Volkmann and Baluška 2006; for a more general review see Morey-Holton 2003). To survive this new physical situation, immobile plants adapting to land life harmonized their cell biological processes, physiological processes, and plant body organization with this physical force. The mechanical stimulus of gravity has to be transformed (graviperception) first into an electrical response and subsequently into a biochemical signal. As a consequence, physical stimuli are transduced into biological "language", resulting in gravity-controlled growth (gravitropism) and gravity-controlled morphogenesis (gravimorphosis), both being inherent features of terrestrial plant life (for recent reviews see Boudaoud 2009; Chehab et al. 2009; Geitmann and Ortega 2009). However, gravity is a unique physical force with respect to life as it acts on all tissues, cells and their molecules in the same way throughout evolution (Morey-Holton 2003; Volkmann and Baluška 2006). In contrast to gravity, light, mechanical stresses induced both from the outside (touch and wind) and from the inside (tissue stress) and volatile chemicals act via gradients that are detectable by cells and organisms (Chehab et al. 2009; Baldwin 2010). Last but not least, the gravity-induced effects are initiated from within the cells (Trewavas and Knight 1994) owing to the gravity-induced acceleration vector acting on all structures having sufficiently high mass, density and/or flexibility.

In spite of diverse experimental approaches by electrophysiologists, as well as cell and molecular biologists, both in ground laboratories and under microgravity conditions, the initial step of graviperception is still a matter of debate. At the cellular level, there are two controversial lines of argumentation related to perception–transduction of the mechanical stimulus: firstly, the gravity-controlled sedimentation of amyloplasts, the more than 100-year-old starch-statolith hypothesis (Němec 1900; Haberlandt 1900; Darwin 1903, 1907; reviewed by Sack 1997); secondly, mechanical deformation of large, continuous and flexible subcellular structures such as membranes, cytoskeletal elements and cell wall components. This so-called gravitational pressure hypothesis (Pfeffer 1894; Czapek 1898; Staves 1997; Staves et al. 1997a, b) is closely related to the tensegrity concept of Donald Ingber (1997, 1999, 2003, 2008).

2 Sedimentation of Subcellular Particles on the Basis of Their Masses and Density Differences

2.1 *Viscosity of Cytoplasm and Its Dependence on the Actin Cytoskeleton*

Under the nearly constant acceleration of Earth's gravitational force (9.81 m s^{-2}), the process of sedimentation of subcellular particles and its velocity depend on the diameter and density of the particles and on the solvent viscosity. Cytoplasm as a non-Newtonian fluid is characterized by anisotropy and specific viscosity (Luby-Phelps 2000; Leterrier 2001; Shepherd 2006), mainly resulting from polymerization and depolymerization processes of cytoskeletal elements such as actin filaments and microtubules (Leterrier 2001; Liverpool 2006; Wagner et al. 2006; Janmey and McCulloch 2007; Lieleg et al. 2009). Spatiotemporal variations of the cytoskeleton and dynamic assembly/disassembly of its elements contribute to inhomogeneities of the cytoplasmic structure and nature (Luby-Phelps 2000; Janmey and McCulloch 2007). The viscoelastic properties of the cytoplasmic cross-linked actin networks are determined via cross-linker off rates, binding energies, and bond lengths between F-actin and actin-binding proteins (Wagner et al. 2006; Schmoller et al. 2009; Lieleg et al. 2009). Besides the viscosity status, the physical properties of the cytoplasm are also determined by molecular crowding, which is rather high (Cueno et al. 1992; Burg 2000; Bancaud et al. 2009; Zhou et al. 2009) and is often ignored by biologists. High molecular crowding affects the polymerization status of the cytoskeleton, as well as diffusion and binding of proteins and structures, acting also as a cell volume sensor (Cueno et al. 1992; Burg 2000; Bancaud et al. 2009). Importantly, crowded cytoplasmic space resists better compression but results in a weakening of cytoskeletal networks, and the compression-induced weakening of cytoskeletal networks is overwhelmed by crowding-induced stiffening of the cytoplasm (Zhou et al. 2009). Under the gravity-induced settlement of the protoplasm to the physical bottom, one can expect weakening of the cytoskeleton and increased macromolecular crowding at the physical bottom. This is because the mechanical properties of a cytoplasm domain under high compression are dominated by crowded macromolecular structures and colloidal glass transitions (Zhou et al. 2009).

Indirect evidence for inherent intracellular inhomogeneities is provided by measuring the speed and observing the behaviour of sedimenting particles such as starch-filled plastids (amyloplasts) in putative gravity-perceiving cells (Yoder et al. 2001; Gaina et al. 2003; Švegždienė et al. 2010). The sedimentation speed is highest at the very beginning of the gravity-induced sedimentation process and then it decreases remarkably. The speed depends on the position of the particles within the cell. It is highest at the cell centre, whereas it is much lower at the cell periphery (Yoder et al. 2001), where abundant endoplasmic reticulum (ER) membranes are located (Volkmann 1974; Yoder et al. 2001) and hamper the sedimentation process.

Viscosity measurements in *Chara* rhizoids (Scherp and Hasenstein 2007), the earliest gravicompetent cells related to plant evolution, demonstrate the high anisotropic viscosity of the cytoplasm. With use of supraparamagnetic particles, the highest viscosity was detected particularly near small vacuoles (or endosomes?) filled with barium sulphate, which function as "statoliths" in these cells (Braun and Limbach 2006). Depending on the direction of the measurements, the viscosity η varied from 78 to 139 mPa s. After application of latrunculin B, cytoplasmic viscosity decreased strongly and rapidly, except in the immediate surroundings of nuclei (Scherp and Hasenstein 2007). Lowering of the cytoplasmic viscosity after depolymerization of actin filaments with both latrunculin B and cytochalasin D stimulates gravisensitivity of roots and shoots (Staves et al. 1997a; Yamamoto and Kiss 2002; Blancaflor et al. 2003; Hou et al. 2003, 2004; Mancuso et al. 2006; Morita 2010).

Experiments under microgravity conditions, i.e. on sounding rockets and Space Shuttle flights (Volkmann et al. 1991; Driss-Ecole et al. 2000; Perbal et al. 2004), brought important insight into these complex sedimentation processes. Control experiments in ground laboratories and in $1 \times g$ centrifuges in orbit demonstrated that sedimentable particles do not sediment completely to the cell bottom, as they should do in isotropic Newtonian fluids. On the other hand, under microgravity, particles move in the opposite direction to the former-acting gravity-induced acceleration vector. These data indicate that the position of particles is a result of balance between two forces, the external-acting gravitational force and an internal force exerted by the cytoskeleton on these particles (Volkmann et al. 1991; Driss-Ecole et al. 2000; Perbal et al. 2004).

2.2 *Possible Candidates for Sedimentation: Statoliths in Algae and Terrestrial Plants*

Besides solvent viscosity, the diameter and the density of particles are the relevant physical parameters influencing the speed of sedimentation. With respect to these aspects, organelles such as mitochondria, starch-filled plastids (amyloplasts) and chloroplasts as well as the nucleus are possible candidates. In some specific cases, crystals enclosed by membranes (vacuoles or endosomes?) show abilities for sedimentation, e.g. vacuoles filled with barium sulphate in rhizoids of the green alga *Chara* (see earlier) having a density of 4.5 g cm^{-3}. The density of mitochondria and chloroplasts varies between 1.08 and 1.10 g cm^{-3} and they would need at least a few minutes to travel over a distance equivalent to their own diameter even under ideal conditions. Starch-filled plastids (amyloplasts) having a density of 1.5 g cm^{-3} sediment at a rate of 5.5 μm min^{-1}, i.e. they need just 30 s to travel a distance equivalent to their own diameter (Sack et al. 1986). This situation is true just for specific cells, presumptive gravity-perceiving cells known as statocytes (Sack 1997). Root cap statocytes are distinguished in particular by viscosity probably in the range of blood viscosity, which is approximately 3 mPa s, and sedimenting particles in these cells are called statoliths.

Another top candidate is the nucleus, which has been seldom discussed (but see Kordyum and Guikema 2001). After application of cytochalasins or latrunculin B, nuclei of root cap statocytes sediment even faster than amyloplasts (Lorenzi and Perbal 1990), and this indicates a density greater than 1.5 g cm^{-3}. The density of isolated DNA from plant cell cultures is 1.8 g cm^{-3} (Willmitzer and Wagner 1981). Nuclear shape is linked to mechanotransduction in animal cells (Dahl et al. 2008; Wang et al. 2009). Concerning nuclei as presumptive statoliths, it is important to take into consideration that nuclei are not isolated structures, but are inherently connected to the endomembrane system by the nuclear envelope (Volkmann 1974). The nuclear envelope supports microtubules (Stoppin et al. 1994; Baluška et al. 1996; Nakayama et al. 2008) and presumably also the actin polymerization. Finally, actin filaments form a cage around nuclei in suspension cells (Seagull et al. 1987) and in root cells organized into tissues (Baluška et al. 1997b, 2000a, b, c). In plant cells the whole nuclear surface acts as a microtubule-organizing centre, and the nucleus with associated microtubules and other cytoskeletal and membraneous elements (Baluška et al. 1997c, 1998) behaves as the smallest autonomous living unit of the eukaryotic life domain. We first termed this smallest unit the "cell body", according to the original proposal of Daniel Mazia (Baluška et al. 1997c, 1998, 2000c, 2004a, b). Realizing that an early version of this concept had been proposed by Julius Sachs (1892), we renamed the cell body as "*Energide*" (Baluška et al. 2006a; Agnati et al. 2009).

Obviously, there are many organelles which might by their density contribute to the pressure exerted on the plasma membrane of cells, in particular of presumptive statocytes showing low viscosity. Lowering of the cytoplasmic viscosity after depolymerization of actin filaments with latrunculin B and cytochalasin D stimulates gravisensitivity of both roots and shoots (Staves et al. 1997a; Yamamoto and Kiss 2002; Blancaflor et al. 2003; Hou et al. 2003, 2004; Mancuso et al. 2006; Morita 2010). As roots and shoots differ in the cellular and tissue nature of cells representing statocytes, as well as in the molecules involved in gravity sensing and signalling (Tasaka et al. 1999; Silady et al. 2004, 2008; Morita and Tasaka 2004; Perrin et al. 2005; Tamura et al. 2007; Yoshihara and Iino 2007; Hála et al. 2010; Morita 2010), we propose that the lowering of the cytoplasmic viscosity by F-actin drugs (Wagner et al. 2006; Schmoller et al. 2009; Lieleg et al. 2009) increases the gravisensitivity of plant cells devoid of F-actin in the roots and shoots. Interestingly in this respect, the absence of prominent F-actin elements is a characteristic feature of root cap statocytes (Baluška et al. 1997b; Baluška and Hasenstein 1997), but not of the shoot statocytes (reviewed by Morita 2010).

2.3 Are There Specific Mechanoreceptors and Gravireceptors?

Barium sulphate filled compartments in *Chara* rhizoids (Schröter et al. 1973, 1975) and starch-filled plastids (amyloplasts) in statocytes of land-living plants are most widely accepted as statoliths (for reviews see Sack 1997; Volkmann and Baluška 2006). Nevertheless, only a signal-amplifying nature of these structures has never been excluded. In the further search for structures involved in gravisensing, in particular regarding the idea of putative sensor–receptor-based processes between the outer envelope of statoliths and other membranes (Limbach et al. 2005; Braun and Limbach 2006), it is important to know more about these envelopes. In *Chara* rhizoids, the envelope of statoliths originates from aggregating endomembranes generating a vacuole-like compartment (Schröter et al. 1973, 1975). Nothing is known about the molecular composition of the surrounding membrane. Concerning land plants, there is some information available for the outer envelope of chloroplasts (Ferro et al. 2003), but almost nothing is known about the outer envelopes of amyloplasts (Guikema et al. 1993). Thus, without knowledge of the proteome of amyloplast envelopes, there is no need to speculate about the plant gravisensing accomplished via activation of putative but elusive gravireceptors (Braun and Limbach 2006). In fact, animal cell mechanosensing of gravity via sedimenting structures, and mechanosensing in general, is also not known to be mediated via specific well-defined receptors (Kernan et al. 1994; Kung 2005; O'Hagan et al. 2005; Chalfie 2009). This is in stark contrast with the situation with photosensing, where numerous receptors have been discovered and characterized both in animals and in plants.

Recently, components of the translocon of outer membrane of chloroplasts (TOC) complex have been reported to interact with the ARG1 protein essential for root gravitropism (Stanga et al. 2009a, b). Interestingly in this respect, the TOC159 component of the TOC complex has been reported to interact with actin on the cytosolic side of the outer envelope membrane of chloroplasts (Jouhet and Gray 2009). It is important to find out if this interaction is relevant for statolith-based amyloplasts. Nevertheless, as the complete depolymerization of the actin cytoskeleton increases, rather than compromises, root gravitropism, it is difficult to reconcile these new data with the classical cytoskeleton-based gravisensing model.

3 Deformation of Subcellular Structures

3.1 The Cytoskeleton–Plasma Membrane–Cell Wall Continuum

Plant cells are known to use their cytoskeleton–plasma membrane–cell wall (WMC) continuum for both sensory and signalling purposes (Baluška et al. 2003a; Szymanski and Cosgrove 2009). In parallel to Ingber's concept of tensegrity (Ingber 1997, 1999, 2003, 2008), as long ago as 1993 Barbara Pickard discussed for plants the WMC continuum as a possible mechanosensing system including the system of gravisensing (Pickard and Ding 1993; Pickard 2007). In this model, mechanical tension of and/or pressure on the cell wall is transmitted via membrane domains and mechanosensitive membrane components, for instance mechanosensitive channels, to the cytoskeleton. These first steps of stimulus transduction should finally result in regulatory processes downstream of the physical stimulus to this continuum. The main problem with Pickard's model was the general lack of knowledge of plants related to adhesion domains, integrin-like molecules, mechanosensitive channels and even cytoskeletal proteins such as actin, myosin and profilin. More importantly, however, the major reason for non-acceptance of Pickard's vision was the nearly 100-year-old dogma of "statolith theory" (Němec 1900; Haberlandt 1900). This "theory", which is still nothing more than a working hypothesis (for the relevant literature see Wayne and Staves 1996), is based on correlations between sedimenting amyloplast-based statoliths and the gravitropic growth of roots and shoots. Nevertheless, convincing evidence for statoliths as the sole prerequisite for gravitropic growth is lacking. There is just one experimental approach taking physical rules related to buoyancy into consideration (Wayne and Staves 1996; Staves et al. 1997a, b). The latter authors used two different systems with respect of the growth behaviour after a change in the positions within the gravity vector: rice roots (Staves et al. 1997a, b) and *Chara* internodal cells (Wayne et al. 1990, 1992; Staves 1997). After the density of the external medium had been increased to a density greater than that of the cytoplasm, rice roots did not show gravity-dependent bending in spite of sedimenting statoliths (Staves et al. 1997b), and the gravity-dependent polarity of plasma streaming in *Chara* changed its direction (Staves 1997). Both these observations are not compatible with the statolith theory (Wayne and Staves 1996). As a consequence, these authors introduced, on the basis of the old concept originally proposed by Friedrich Czapek and Wilhelm Pfeffer (Pfeffer 1894; Czapek 1898; Darwin 1899, 1907; Haberlandt 1908), as well as later ideas of Pickard (Pickard and Ding 1993; Pickard 2007), the gravitational pressure model for gravity sensing. This model proposes that gravity causes cells to settle within the confines of the extracellular matrix, resulting in a tension between the plasma membrane and the cell wall at the top of the cell and a compression between the plasma membrane and the cell wall at the bottom of the statocyte. In this model, sedimenting statoliths of gravicompetent cells (statocytes) may act not as sensors but rather as a signal-amplifying system.

3.2 When "Outside" Is "Inside": Mechanosensitive Signalling Endosomes?

Our discovery of endocytic internalization of cell wall pectins and xyloglucans into endosomes of root apex cells (Baluška et al. 2002, 2005b) changes completely our perspective of the WMC continuum. As the internal space of the endosome corresponds topologically to the outside space, with the outer leaflet of the plasma membrane facing the endosomal interior, it is perhaps not so surprising to have these endosomes filled with the cell wall molecules (Baluška et al. 2002; Dhonukshe et al. 2006). This inverted outside–inside topology means that the extracellular space is protruding deeply into plant cells, making the WMC sensory interface much more extensive, surely having an impact on mechanosensing too. It is well known that endosomes have a complex sensory and signalling nature both in animal and in plant cells (Teis and Huber 2003; Šamaj et al. 2004; Geldner and Robatzek 2008; Bar and Avni 2009; Ivanov and Gaude 2009; Birtwistle and Kholodenko 2009; Sorkin and von Zastrow 2009; Drakakaki et al. 2009; Murphy et al. 2009; Scita and di Fiore 2010). This unexpected twist in our understanding of signal transduction will dominate future studies on gravisensing in plants. It is interesting that the outer plastid membrane is also topologically inverted owing to the endosymbiotic origin of plastids. Should the barium sulphate filled vesicles of *Chara* rhizoids turn out to be endosomes, then all known plant statoliths would have the inverted topology of their outer (limiting) membrane.

3.3 Mechanosensitive Channels

Mechanosensitive channels, in particular mechanosensory calcium-selective cation channels (Ding and Pickard 1993a, b; Pickard and Ding 1993; Ding et al. 1993) play an important role in the gravitational pressure hypothesis. Meanwhile, these channels are well established for mechanosensing processes in plants in relation to touch, wind, physical barriers and predation (reviewed by Chehab et al. 2009). It is difficult to measure precisely cytosolic calcium concentrations in relation to gravitropic signalling and growth responses and, therefore, conclusions drawn from such measurements are a matter of fierce debate (Legué et al. 1997; Plieth and Trewavas 2002; Perrin et al. 2005; Plieth 2005; Pickard 2007). Transient changes of cytosolic calcium concentrations in the range of minutes after stimulation might have a role in gravity-controlled growth. Moreover, calcium signatures emerge as the key feature of signal transduction networks related to environmental sensing and signalling (Luan 2009; Dodd et al. 2010; Kudla et al. 2010).

Mechanosensitive channels of small conductance (MscSs) are mechanosensitive channels that protect *Escherichia coli* from osmotic shocks (Levina et al. 1999). In *Arabidopsis*, ten MscS-like (MSL) proteins are present, of which MSL2 and MSL3 are involved in the control of plastid size and shape (Haswell and Meyerowitz 2006). Moreover, MSL9 and MSL10 were reported to be localized at the plasma membranes of the end poles and cytokinetic cell plates (Haswell et al. 2008; Peyronnet et al. 2008), known to be generated from endosomes (Baluška et al. 2002, 2005a; Dhonuskhe et al. 2006). Root apices are extremely mechanosensitive, especially at their transition zone, to mechanostimulation (Massa and Gilroy 2003; Ishikawa and Evans 1992), which induces coherent transcellular calcium waves spanning the whole root apex (see Fig. 3D in Fasano et al. 2002). Moreover, the same root zone was reported to accomplish touch-induced ATP release from cells into the extracellular space (Weerasinghe et al. 2009). These phenomena are relevant for the obstacle-avoidance behaviour of growing root apices (Fasano et al. 2002; Massa and Gilroy 2003; Weerasinghe et al. 2009). In addition, extracellular ATP does not only act as a signalling molecule in plants (Demidchik et al. 2003), it also inhibits both polar auxin transport and root gravitropism (Tang et al. 2003). Moreover, extracellular ATP induces production of nitric oxide and reactive oxygen species in the cell wall too (Song et al. 2006; Foresi et al. 2007). Aluminium inhibits mechanosensory calcium-selective cation channels (Ding et al. 1993), basipetal auxin transport (Kollmeier et al. 2000; Shen et al. 2008) and PIN2 endocytosis, increasing the abundance of PIN2 at the plasma membrane (Shen et al. 2008).

4 Old and New Arguments for the Tension/Pressure Hypothesis

4.1 Time Constants

Dose–response curves and the underlying time constants are important data for judging the steps of stimulus transformation. The minimum stimulation time for plant roots is about 20 s, a value which was discussed earlier on the basis of clinostat experiments (Johnsson and Pickard 1979; Volkmann and Sievers 1979). Most convincing in this respect, however, are data obtained from experiments performed under microgravity conditions, i.e. under more realistic conditions of plant sensory physiological processes (Volkmann and Tewinkel 1996; Perbal et al. 1997, 2004). These data, which indicate 25 s as the minimum time for stimulation, are in agreement with the former theoretical estimations. For this time constant, a statolith would travel a distance less than that equivalent to its diameter, a distance which is not far enough to reach presumptive sites of gravity transduction, e.g. the plasma membrane or underlying ER membranes. Time constants for mechanosensing of any other organism are extremely small, e.g. in the range of milliseconds, indicating a very fast physical nature of the stimulus transformation (Kernan 2007).

4.2 Buoyancy Experiments

The classical physical approach to identify statoliths as possible candidates for transformation of the gravitational force is experiments related to buoyancy. Results from those experiments (Staves et al. 1997a, b) strongly support the gravitational pressure hypothesis and contradict the strong version of the statolith hypothesis. They indicate fast and immediate transformation of the gravity stimulus. Importantly, lowering of the cytoplasmic viscosity after depolymerization of actin filaments with latrunculin B and cytochalasin D strongly stimulates gravisensitivity (Staves et al. 1997a; Yamamoto and Kiss 2002; Blancaflor et al. 2003; Hou et al. 2003, 2004; Mancuso et al. 2006; Morita 2010). This unexpected effect was reported for both roots and shoots, despite the fact that the molecular basis of gravity sensing in these two plant organs is different (see the discussion in Sect. 6 and Perrin et al. 2005; Morita 2010).

4.3 New Fast Signals

Fast signals, about 10 s after gravitational stimulation, were recorded in maize pulvini for the phospholipase C mediated generation of the second-messenger molecule inositol 1,4,5-trisphosphate (Perera et al. 1999, 2001). With use of a non-invasive electrophysiological approach, changes in oxygen fluxes in maize roots have been found within less than 30 s after horizontal stimulation (Stefano Mancuso et al., unpublished results). Within less than 60 s after stimulation, changes in cytosolic pH are evident in root cap cells of *Arabidopsis* (Scott and Allen 1999; Fasano et al. 2001). Transient changes in cytosolic calcium concentration have been measured 90 s after reorientation of *Arabidopsis* seedlings (Plieth and Trewavas 2002). Finally, very rapid (within less than 120 s), but transient, transcriptional regulation occurs in root apices of *Arabidopsis* (Kimbrough et al. 2004).

Skilful surgery experiments by Juniper et al. (1966) brought the old Nemec–Haberlandt statolith theory back into general acceptance. Since this paper was published, the root cap columella cells harbouring sedimentable amyloplasts have been widely accepted as the major site of gravity sensing (for experimental confirmation in the *Arabidopsis* root cap by laser ablation see Blancaflor et al. 1998). Nevertheless, evidence has accumulated for a site of stimulus perception and transduction outside the root cap. Using the sophisticated ROTATO device, Mullen et al. (2000) analysed the kinetics of gravity-controlled root growth under constant vertical orientation of the root cap. This is the more sophisticated version of the original experimental approach of Czapek in the Pfeffer laboratory using bent capillaries (Pfeffer 1894; Czapek 1898; Darwin 1907; Haberlandt 1908). These new data confirmed the old findings that besides the root cap statocytes, there is gravity sensing outside the cap (Wolverton et al. 2002a, b; Perrin et al. 2005; Mancuso et al. 2006). As discussed in more detail later, electrical responses in cells in the transition zone (Baluška et al. 2010; Verbelen et al. 2006) occur almost immediately after repositioning the roots in the gravity vector (Iwabuchi et al. 1989; Collings et al. 1992; Ishikawa and Evans 1990a, 1994). Recently, we have also observed instant changes in oxygen uptake and reactive oxygen species generation in these mechanosensitive cells after gravistimulation (Stefano Mancuso et al., unpublished data).

5 Tension/Pressure Hypothesis: Root Gravisensing Outside the Root Cap

5.1 Subcellular Candidates

Strong support for the tension/pressure hypothesis of gravisensing in plants came from studies reporting gravity perception outside the root cap. The concept of dual sensors and dual motors of root graviresponse, as proposed by Mike Evans and his colleagues, is based on results showing the ability of postmitotic (transition zone) and early elongating cells to show detectable gravity-induced electric changes within some 30 s of gravistimulation (Ishikawa and Evans 1990a, 1994; Evans and Ishikawa 1997). These rapid electrical responses, which were reported also by Collings et al. (1992), are much quicker than the auxin transport polarization in the root cap. In root caps, the earliest changes in the auxin asymmetry have been documented as occurring only after 30 min of gravistimulation in maize (Young et al. 1990) and after 15 min in *Arabidopsis* (Ottenschläger et al. 2003). These times are too slow to determine the bending in postmitotic root cells, which is detectable after only 10 min of gravistimulation in roots of *Arabidopsis* (Mullen et al. 2000). As there are no obvious sedimenting particles/organelles in the transition zone cells, the gravisensing of these cells relies perhaps on pressure differentials within cells and/or exerted by the whole protoplast weight on some sensitive structures at the physically lower portion of these gravity-sensitive root cells. The WMC structural continuum (Baluška et al. 2003a; Szymanski and Cosgrove 2009) is one of the best candidates for this sensory superstructure interlinking the nucleus, via cytoskeletal and membranous elements, with supracellular cell wall in the sense of Ingber's tensegrity ideas (Ingber 1997, 1999, 2003, 2008).

5.1.1 Gravisensing Outside the Root Cap: Myosin VIII as an Endocytic and Gravisensing Motor

Plants are unique with respect to myosins. As the only eukaryotic organism, plants do not have any myosins from classes I and II, which are represented in any other eukaryotic organisms (Thompson and Langford 2002; Foth et al. 2006). On the other hand, plants have two classes of myosins which are plant-specific and these plant myosins are not found in any other eukaryotic organism (Thompson and Langford 2002; Foth et al. 2006). Class XI myosin is localized to diverse organelles and is involved in their intracellular transport, whereas class VIII myosin is localized preferentially at the plasma membrane domains specialized for endocytosis and endocytic vesicle recycling (Baluška et al. 2003b, 2004b; Volkmann et al. 2003). Myosin VIII acts as an endocytic motor in plants (Baluška et al. 2004b; Volkmann et al. 2003; Golomb et al. 2008; Sattarzadeh et al. 2008). Interestingly, myosin VIII is an evolutionarily more recent addition to the plant cytoskeleton and it evolved from the more ancient myosin XI, which is closely related to myosin V (Thompson and Langford 2002). All this suggests that the subcellular domains based on myosin VIII (Baluška et al. 2000b, c, 2001, 2004b; Wojtaszek et al. 2005, 2007) represent relatively novel evolutionary innovation of higher plants. Both features, the evolutionary conservation and proliferation of the two plant-specific classes of myosins in plants, point to the functional importance of these myosins for plant biological processes.

In roots cells, myosin VIII localizes to end poles (synapses), where it anchors F-actin bundles (Baluška et al. 1997a, 2000b) and also supports endocytosis as the only endocytic myosin motor in plants (which lack myosins from classes I and II).

Myosin VIII might be an important player in the plant-specific WMC structural continuum (Baluška et al. 2003a). Similarly in *Vallisneria gigantea* leaf cells and *Chara corallina* internodal cells, F-actin bundles are organized in parallel with the cell axis and are anchored at the end walls (Wayne et al. 1992; Ryu et al. 1995, 1997; Hayashi and Takagi 2003). It can be expected that also in these systems, myosin VIII, or a similar myosin is responsible for the anchoring longitudinal bundles of F-actin and organizing cytoplasmic streaming, which is sensitive to gravity and light (Ewart 1902; Wayne et al. 1990). These end poles of *Chara* cells, having a very active and physiologically relevant WMC continuum, are mechanosensitive and initiate membrane depolarization and action potentials after subtle mechanical changes in the plasma membrane tension (Shimmen 2001, 2002, 2005). They seem to be optimized to sense mechanical changes and transduce these pressure signals into electrical signals (Shimmen 2003). This resembles closely the situation known in sensory neurons of animal cells specialized to mechanosensing (Kernan 2007; Kernan et al. 1994; Kung 2005; O'Hagan et al. 2005; Chalfie 2009).

5.1.2 Gravisensing Outside the Root Cap: Actomyosin-Based Forces as a Negative Player

The root cap statocytes act as the principal gravisensory organ of roots. However, there are numerous data – both old and more recent – convincingly documenting that the postmitotic cells of the transition zone (Baluška et al. 2010) are also capable of independent gravisensing (Pfeffer 1894; Czapek 1898; Darwin 1907; Haberlandt 1908; Wolverton et al. 2000, 2002a, b; Mullen et al. 2000; Mancuso et al. 2006). We reported that not only the depolymerization of F-actin but also general inhibition of myosin motors increased the ability of decapped roots to accomplish gravitropism (Mancuso et al. 2006).

These findings suggest that either the gravisensing outside the root cap is enhanced by the inhibited actomyosin forces or the signal transduction cascade interlinking gravisensing with the motoric gravibending response is more effective if the actomyosin-based forces are inhibited. In the first scenario, a dense meshwork of F-actin bundles would prevent mechanically any sedimentation of heavy particles/organelles. Importantly in this respect, root cap statocytes are known to be devoid of a dense meshwork of F-actin bundles (Baluška et al. 1997b), a feature which allows them to act as plant statocytes. Similarly, myosins anchor vesicles/organelles to the F-actin bundles, precluding their free mobility within the cytoplasm, which is the requirement for the free gravity-mediated sedimentation of putative statoliths. Therefore, we can expect that depolymerization of F-actin as well as inhibition of myosin forces will increase gravisensitivity of plant cells. The second scenario is rather improbable as the intact networks of actin filaments are important for signal transduction in diverse other sensory signal transduction pathways. Nevertheless, as there are no obvious sedimentable structures in these cells, the first scenario is also not able to explain the perplexing ability of the transition zone cells to perceive gravity and to perform gravity-induced motoric responses if compromised by actomyosin forces (Mancuso et al. 2006). The decapped roots undergo significant gravitropism if they are exposed to F-actin and myosin inhibitors.

This suggests that some processes are activated when there is an inactive actomyosin system, which would then allow gravisensing and sensory–motoric coupling. A study using rat kidney cells suggests, rather surprisingly, that this process might be mechanoinjury-induced endocytosis (Idone et al. 2008a). Injury of cells using the bacterial toxin stroptolysin O or via scape woundings initiated a novel type of endocytosis which removed the injured plasma membrane patches, allowing the cells to recover from this mechanical stress (Idone et al. 2008a, b). Depolymerization of F-actin via cytochalasin D and latrunculin A enhanced this endocytosis and enhanced membrane repair (Idone et al. 2008a). An older article reported that this resealing repair of the plasma membrane is accomplished via vesicular mechanisms resembling neurotransmitter release (Steinhardt et al. 1994; discussed in Idone et al. 2008b).

Interestingly, the actomyosin-based cytoplasmic streaming in large Characean cells is sensitive to gravity when the downward streaming is faster than the upward streaming of the cytoplasm (Wayne et al. 1990). As early as 1902, Alfred Ewart (1902) wrote a book entitled *On the Physics and Physiology of Protoplasmic Streaming in Plants* in which he reported that the cytoplasmic streaming in plants is sensitive to both gravity and light. Could it be that preventing the gravisensitive cytoplasmic streaming increases effectivity of some other, evolutionarily more recent gravisensing mechanisms allowing effective gravitropism of the decapped maize roots?

5.1.3 Gravisensing Outside the Root Cap: Mechanosensitive *Energides?*

On the subcellular level, the largest structure is the nucleus with its associated cytoskeleton–membrane assembly via perinuclear microtubules, which we discussed first as the *cell body* (Baluška et al. 2004a, b), as a legacy from Daniel Mazia, and later as the *Energide* in the sense of the concept originally proposed by Julius Sachs in 1892 (Sachs 1892; Baluška et al. 2006a; Agnati et al. 2009). It is still unknown what keeps the *Energide* in the central position in cells of the transition zone of the root apex (Baluška et al. 2000a, c, 2010). Our preliminary data indicate that besides the dynamic cytoskeleton interacting with the ER membranes, spanning from the nuclear surface up to the cell periphery and protruding into plasmodesmata, also endocytic vesicle recycling (synaptic activity) is critical for the maintenance of *Energides* in their central cellular positions (Baluška and Hlavacka 2005, Baluška et al. 2009a, unpublished data). After exposure of roots to the inhibitor of endocytic vesicle recycling brefeldin A (BFA), nuclei are shifted basally (shootward, towards the shoot apex – see Baskin et al. 2010 for the new terminology) in cells in the transition zone (Baluška and Hlavacka 2005). The *Energide* integrates structurally DNA molecules with the cytoskeleton–plasma membrane–extracellular matrix (Baluška et al. 2003a, b, for animal cells, see Ingber

1997, 1999, 2003, 2008; Dahl et al. 2008; Wang et al. 2009) and this superstructure has several features, such as the central cellular position with radiating microtubules (Baluška et al. 2001, 2006a), implicating its suitability to act as a mechanosensitive superstructure. It can be easily envisaged that the *Energide* position in the cellular centre, which requires the cytoskeletal and membranous activities of the cell to keep the central position of *Energides*, is a dynamic one owing to a noise in the system manifesting itself as perpetual deviations from its centrality. It can be expected that repositioning within the gravity field will result in immediate shifting out of the *Energide*'s position from the cell's centre. Interestingly, the *Energides* are effectively attracted to weak cell periphery sites as evidenced in trichoblasts during root hair initiation via local expansin-induced cell wall weakening (Baluška et al. 2000a, b; Cho and Cosgrove 2002). That the nucleus, being an inherent part of the *Energide* (Baluška et al. 1997a, b, c, 1998, 2006a, b), is a suitable structure for mechanosensing and mechanotransduction (Kordyum and Guikema 2001) is also obvious from studies on animal cells (Dahl et al. 2008; Wang et al. 2009). Interestingly in this respect, amyloplasts encircle closely centrally positioned nuclei in the transition zone cells of maize root apices (see Fig. 7 in Baluška et al. 1990 and Fig. 6 in Baluška et al. 1993). Recently, very strong support for a mechanosensitive *Energide* has been obtained in animal cells when the nucleus with perinuclear microtubules was shown to act as a force sensor (Minc et al. 2011).

5.2 *Supracellular Candidates*

5.2.1 Gravisensing Outside the Root Cap: Mechanosensitive Plasmodesmata?

Although seldom discussed in this respect, plasmodesmata (plant-specific cell–cell channels) represent one of the best candidates for gravisensing in plants (Juniper 1976). Plasmodesmata are known to be mechanosensitive (Cote et al. 1987; Oparka and Prior 1992; Oparka 1993). Pressure gradients generated between adjacent cells using different mannitol concentrations increased electrical resistance and inhibited transport via their plasmodesmata (Trebacz and Fensom 1989; Reid and Overall 1992). Our in vivo analysis of root cap statocytes of *Arabidopsis* revealed that these are symplastically closed in vertical roots but are selectively gated (opened) in gravistimulated roots (Šamaj et al. 2006). These findings suggest that plasmodesmata, or some structures/molecules associated with them, act as one of the still elusive gravisensors of plants. Interestingly in this respect, F-actin and myosin VIII are associated with the plasmodesmata and are relevant for their gating (Volkmann et al. 2003; Šamaj et al. 2006) as well as for their maintenance (Baluška et al. 1997b; Volkmann et al. 2003, Volkmann and Baluška 2008). In root tissues which are depleted of myosin VIII (Šamaj et al. 2006; Baluška et al. 2004c) or lose myosin VIII and F-actin from their plasmodesmata, such as the middle cortex in maize roots (Baluška et al. 2000a), their cell–cell contacts also disintegrate (Wojtaszek et al. 2005). As myosin VIII is a good candidate for a plant mechanosensing molecule (Volkmann et al. 2003; Volkmann and Baluška 2008; Baluška et al. 2003a, 2004c; Wojtaszek et al. 2005), enrichment of myosin VIII in plasmodesmata (Baluška et al. 2000b; Wojtaszek et al. 2004c, 2005, 2007) might be relevant for the putative gravisensing via plasmodesmata. In the root cap, this pathway towards tissue disintegration is prominent and leads to the complete release of these secretory cells from the root cap periphery. Plasmodesmata are often grouped into large myosin VIII enriched pit fields (Baluška et al. 2000b, 2001, 2004b) which might act, similarly to the diverse adhesions of animal epithelium cells (Bershadsky et al. 2006; Schwarz et al. 2006; Geiger et al. 2009), as mechanosensitive domains. In fact, besides recent advances in the study of domain-based and polarity-related organization of plant cells (Baluška et al. 2000c; Zárský et al. 2009; Zárský and Potocký 2010), as discussed later, root endodermis (Alassimone et al. 2010) and epidermis (Łangowski et al. 2010) show features and properties resembling those of animal epithelium tissues.

5.2.2 Gravisensing Outside the Root Cap: Mechanosensitive Synapses?

In 2005, we proposed that cellular end poles of the transition zone, which show the highest rates of endocytosis of endocytic vesicle recycling in the whole root apex (Baluška et al. 2002, 2005a, 2009a, b, 2010) – and it seems also in the whole plant body – act as plant synapses (Baluška et al. 2003b, 2005b, 2007). This synaptic concept for plant gravisensing is based on the gravity-induced mechanical asymmetry at the plant synapses: F-actin/myosin VIII enriched cell–cell adhesion domains specialized for cell–cell communication (Baluška et al. 2005b, 2007, 2009a, 2010). The physically upper (shootward – towards the shoot apex) pole of root cells (Baskin et al. 2010) has very low tension stress on the plasma membrane, which is proposed to be relieved by a higher rate of endocytosis and a lower rate of exocytosis (Baluška et al. 2005b). On the other hand, the physically lower (rootward – towards the root apex, see Baskin et al. 2010) pole of root cells has very high tension stress on the plasma membrane, which is proposed to be relieved by a higher rate of exocytosis and a lower rate of endocytosis. Owing to this mechanical asymmetry, there is obvious polarity of cells active in the endocytic vesicle recycling at their end poles, which is highest in the transition zone of root apices (Baluška et al. 2005b, 2007, 2009a, b, 2010). There is polarity in the endocytic vesicle recycling system, when the apical (physically lower) exocytic synaptic domain is controlled by the GNOM-based endosome, whereas the basal (physically upper) endocytic synaptic domain is maintained by the SNX1-based endosome (Baluška et al. 2010). Moreover, although both recycling domains need an intact actin cytoskeleton, the apical domain is also dependent on microtubules (Kleine-Vehn et al. 2008).

In support of these links between the plant synapses, plasma membrane tension, myosins and mechanosensitive vesicle recycling, there are several relevant data obtained on animal cells. As already mentioned, mechanosensitive cells in animals and humans are typically neurons (Kernan et al. 1994; Kung 2005; O'Hagan et al. 2005; Chalfie 2009). For the neuromuscular presynaptic terminal, actomyosin-based mechanical tension was shown to control clustering of synaptic vesicles (Siechen et al. 2009). Similarly to what is emerging for plant-specific myosin VIII, myosin I was also reported to control the membrane tension in animal cells (Nambiar et al. 2009; McConnell and Tyska 2010). Moreover, myosin VI, which is another class of unconventional myosins and an endocytic motor of animal cells (Buss et al. 2001; Hasson 2003), is required for the maturation and function of inner hair cell ribbon synapses (Roux et al. 2009). These auditory synapses of the mammalian hearing organ cochlea mediate mechanotransduction via the C2-domain transmembrane protein otoferlin (Roux et al. 2006). Otoferlin belongs to the ferlin group of proteins, resembling synaptotagmin-like (SYT-like) proteins (discussed in Roux et al. 2006), involved in the plasma membrane repair via calcium-mediated exocytosis (for plants see Schapire et al. 2008, 2009; Yamazaki et al. 2008). Interestingly in this respect, *Arabidopsis* SYT1 protein localizes to the plant synapses in the transition zone of the root apex (Schapire et al. 2008) and supports plant endocytosis (Lewis and Lazarowitz 2009).

The endodermis, which is the major gravisensitive tissue in the shoot, has been shown to resemble animal epithelium tissues (Alassimone et al. 2010). Similarly, root epidermis shows polarity properties resembling those of animal epithelia (Łangowski et al. 2010). In the case of the endodermis, its special epithelium-like polarity is established in early embryogenesis and requires endocytic vesicle recycling. The adhesive Casparian bands of the endodermis, having very tight connections between the plasma membrane and the cell wall, resemble closely the tight junctions of animal epithelia. Interestingly in this respect, epithelial tight and/or adherens junctions are enriched with synaptic molecules (Tang 2006) and are mechanosensitive (Yonemura et al. 2010; Lecuit 2010), suggesting that these junctions emerge as epithelial synapses for cell–cell communication (Tang 2006; Yamada and Nelson 2007). The plant synapses concept, which is closely linked with the polar transport of auxin in plants (Baluška et al. 2003a, b, 2005a, 2007, 2009a, b, 2010), as well as plant and animal gravisensing too.

5.2.3 Gravisensing Outside the Root Cap: Cell Wall in Flux?

Plant cell walls are defined as a mechanically robust superstructure, composed of diverse homopolymer and heteropolymer polysaccharides and proteoglycans. Cell wall components are partially assembled in the endomembranes but the final form is acquired outside the plasma membrane. Cellulose microfibrils polymerize in an ordered manner at the outer leaflet of the plasma membrane via the directional movements of cellulose synthase complexes (Lindeboom et al. 2008; Geisler et al.

2008). Matrix polysaccharides originate in the Golgi apparatus and are secreted within vesicles from the cytoplasm (Geisler et al. 2008; Sandhu et al. 2009). This one-way view of the plant cell wall assembly was challenged in 2002 when we published data documenting endocytic internalization of cell wall pectins, cross-linked with calcium and boron, into the cytoplasm of maize root apex cells (Baluška et al. 2002). Our later studies revealed that these endosomes loaded with cell wall pectins are used for building the cell plate (primordial cell wall) during plant cytokinesis, both in maize and in *Arabidopsis* (Baluška et al. 2005a, 2006b; Dhonukshe et al. 2006). Viotti et al. (2010) criticized our data by stating that the cell wall specificity of the pectin antibodies used is not sufficient to conclude that cell wall pectins are internalized via endocytosis. However, our data are not based only on the cell wall specificity of these antibodies. We also documented using these well-characterized antibodies that the overall cell wall pectin signal is greatly decreased in BFA-exposed roots only in those cells which accumulate these cross-linked pectins within BFA-induced compartments (Baluška et al. 2002, 2005a, b; Dhonukshe et al. 2006). When plant endocytosis is generally inhibited by low temperature, cell wall pectins remain in the cell walls of both control and BFA-exposed roots (Baluška et al. 2002). The size of these compartments depends on the synaptic activity (the rate of endocytic vesicle recycling), showing the highest activity in the transition zone (Baluška et al. 2002, 2009a, 2010). Finally, there are close similarities between the composition of synapses and that of BFA-induced compartments (Baluška et al. 2003a; Hause et al. 2006).

5.2.4 Gravisensing Outside the Root Cap: Gravisensitive Electric Fields?

Growing roots generate electric fields with a characteristic pattern around their growth zones (Scott 1967; Iwabuchi et al. 1989; Ishikawa and Evans 1990a, 1994; Collings et al. 1992). Importantly, growing roots not only generate electric fields around themselves, they are also sensitive to them by inducing root electrotropism (Iwabuchi et al. 1989; Ishikawa and Evans 1990b; Wolverton et al. 2000) or by changing the root morphogenesis (Wawrzecki and Zagorska-Marek 2007). Whereas the exposure to electric fields does not inhibit root growth under $1 \times g$ on Earth, space experiments revealed that an electric field rapidly and strongly inhibits root growth in the microgravity situation (Wolverton et al. 1999). Moreover, gravistimulation of roots in ground experiments rapidly induces changes to these electric fields (Collings et al. 1992), strongly suggesting that these, or the physiological processes which generate them, are rapidly affected by repositioning of roots in the gravity field. Interestingly in this respect, imposing nanosecond pulsed electric fields on plant cells results in disintegration of the cytoskeleton in the cell cortex, followed by contraction of actin filaments towards the nucleus and alterations to the nuclear envelope (Berghöfer et al. 2009). The most recent electrophysiological measurements (Masi et al. 2009) by multi-electrode-array apparatus support the occurrence of fast electrical signals as a feature of plants in general for integration of diverse stimuli from the environment.

6 Profound Differences Between Root and Shoot Gravisensing: Evo–Devo Mystery?

Finally, it is a small evolutionary mystery that shoot and root gravisensing differ not only in the tissue identity of cells which are specialized for gravisensing, but also in the cellular nature of statocytes and, particularly, in the proteins underlying their gravisensing. Despite the crucial role of sedimenting amyloplasts both in roots and in shoots, there are profound differences between the root and shoot amyloplast-based statocytes. Whereas in the root apex the root cap statocytes are rich in the cytoplasm but devoid of any visible F-actin elements, the shoot statocytes are endodermis cells which are highly vacuolated and equipped with prominent F-actin bundles (Baluška and Hasenstein 1997; Baluška et al. 1997b; Tasaka et al. 1999; Morita and Tasaka 2004). Surprisingly, the root endodermis, although reaching up to the very root apex, is not relevant for root apex gravisensing. Moreover, although both shoot gravisensing and root gravisensing are related to the endosomal vesicle recycling, and general inhibition of endocytic vesicle recycling via BFA, *N*-1-naphthylphthalamic acid and 2,3,5-triiodobenzoic acid inhibits gravitropism of shoots and roots, the molecules involved are different (for roots see Perrin et al. 2005; Harrison and Masson 2008a, b; Stanga et al. 2009a, b). Diverse mutants affecting shoot sensitivity to gravity do not compromise root gravisensing, and the same is true in the opposite way (Silady et al. 2004, 2008; Perrin et al. 2005; Tamura et al. 2007; Yoshihara and Iino 2007; Harrison and Masson 2008a, b; Stanga et al. 2009a, b; Hála et al. 2010; Morita 2010).

In the transition zone, the gravisensing is not based on sedimenting amyloplasts and involves mechanisms such as rapid electrical responses and gravisensitive auxin-secreting plant synapses (Baluška et al. 2005b, 2007, 2009a, b, 2010). As life on Earth has evolved under the invariant $1 \times g$ gravitational force (Morey-Holton 2003; Volkmann and Baluška 2006), it is to expect that multiple mechanisms have been selected, which are based on different sensory mechanisms coupled to a distinct stimulus – response pathways of gravitropism (Barlow 1995, 1996; Volkmann and Baluška 2006). One can envisage that ancient gravity-sensing mechanisms, which were optimized for the low gravity pressure of the high-buoyancy life in oceans of ancient plants, have been retained in present-day plants. But they were overlaid by evolutionarily more recent gravity-sensing mechanisms implemented during transfer of ancient plants from the low-gravity-loaded water of the oceans to the high-gravity-loaded terrestrial environment. Under specific situations, such as when all actin filaments are effectively depolymerized, the ancient gravity-sensing mechanisms are activated so strongly that, counterintuitively, the sensitivity to gravity increases.

References

Agnati LF, Fuxe K, Baluška F, Guidolin D (2009) Implications of the 'Energide' concept for communication and information handling in the central nervous system. J Neural Transm 11:1037–1052

Alassimone J, Naseer S, Geldner N (2010) A developmental framework for endodermal differentiation and polarity. Proc Natl Acad Sci USA 107:5214–5219

Baldwin IT (2010) Plant volatiles. Curr Biol 20:R392–R397

Baluška F, Hasenstein KH (1997) Root cytoskeleton: its role in perception of and response to gravity. Planta 203:S69–S78

Baluška F, Hlavacka A (2005) Plant formins come to age: something special about cross-walls. New Phytol 168:499–503

Baluška F, Kubica Š, Hauskrecht M (1990) Postmitotic 'isodiametric' cell growth in the maize root apex. Planta 181:269–274

Baluška F, Parker JS, Barlow PW (1993) A role for gibberellic acid in orienting microtubules and regulating cell growth polarity in the maize root cortex. Planta 191:149–157

Baluška F, Barlow PW, Parker JS, Volkmann D (1996) Symmetric reorganizations of radiating microtubules around pre-mitotic and post-mitotic nuclei of dividing cells organized within intact root meristems. J Plant Physiol 149:119–128

Baluška F, Vitha S, Barlow PW, Volkmann D (1997a) Rearrangements of F-actin arrays in growing cells of intact maize root apex tissues: a major developmental switch occurs in the postmitotic transition region. Eur J Cell Biol 72:113–121

Baluška F, Kreibaum A, Vitha S, Parker JS, Barlow PW, Sievers A (1997b) Central root cap cells are depleted of endoplasmic microtubules and actin microfilament bundles: implications for their role as gravity-sensing statocytes. Protoplasma 196:212–223

Baluška F, Volkmann D, Barlow PW (1997c) Nuclear components with microtubule organizing properties in multicellular eukaryotes: functional and evolutionary considerations. Int Rev Cytol 175:91–135

Baluška F, Lichtscheidl IK, Volkmann D, Barlow PW (1998) The plant cell body: a cytoskeletal tool for cellular development and morphogenesis. Protoplasma 202:1–10

Baluška F, Salaj J, Mathur J, Braun M, Jasper F, Šamaj J, Chua N-H, Barlow PW, Volkmann D (2000a) Root hair formation: F-actin-dependent tip growth is initiated by local assembly of profilin-supported F-actin meshworks accumulated within expansin-enriched bulges. Dev Biol 227:618–632

Baluška F, Barlow PW, Volkmann D (2000b) Actin and myosin VIII in developing root cells. In: Staiger CJ, Baluška F, Volkmann D, Barlow PW (eds) Actin: a dynamic framework for multiple plant cell functions. Kluwer, Dordrecht, pp 457–476

Baluška F, Volkmann D, Barlow PW (2000c) Actin-based domains of the 'cell periphery complex' and their associations with polarized 'cell bodies' in higher plants. Plant Biol 2:253–267

Baluška F, Cvrčková F, Kendrick-Jones J, Volkmann D (2001) Sink plasmodesmata as gateways for phloem unloading. Myosin VIII and calreticulin as molecular determinants of sink strength? Plant Physiol 126:39–46

Baluška F, Hlavačka A, Šamaj J, Palme K, Robinson DG, Matoh T, McCurdy DW, Menzel D, Volkmann D (2002) F-actin-dependent endocytosis of cell wall pectins in meristematic root cells: insights from brefeldin A-induced compartments. Plant Physiol 130:422–431

Baluška F, Šamaj J, Wojtaszek P, Volkmann D, Menzel D (2003a) Cytoskeleton – plasma membrane – cell wall continuum in plants: emerging links revisited. Plant Physiol 133:482–491

Baluška F, Šamaj J, Menzel D (2003b) Polar transport of auxin: carrier-mediated flux across the plasma membrane or neurotransmitter-like secretion? Trends Cell Biol 13:282–285

Baluška F, Volkmann D, Barlow PW (2004a) Cell bodies in a cage. Nature 428:371

Baluška F, Volkmann D, Barlow PW (2004b) Eukaryotic cells and their *cell bodies*: cell theory revisited. Ann Bot 94:9–32

Baluška F et al (2004c) Myosin VIII and F-actin enriched plasmodesmata in maize root inner cortex cells accomplish fluid-phase endocytosis via an actomyosin-dependent process. J Exp Bot 55:463–473

Baluška F, Volkmann D, Menzel D (2005a) Plant synapses: actin-based adhesion domains for cell-to-cell communication. Trends Plant Sci 10:106–111

Baluška F, Liners F, Hlavačka A, Schlicht M, Van Cutsem P, McCurdy D, Menzel D (2005b) Cell wall pectins and xyloglucans are internalized into dividing root cells and accumulate within cell plates during cytokinesis. Protoplasma 225:141–155

Baluška F, Volkmann D, Barlow PW (2006a) Cell-cell channels and their implications for cell theory. In: Baluška F, Volkmann D, Barlow PW (eds) Cell-cell channels. Landes Bioscience, Austin, pp 1–18

Baluška F, Menzel D, Barlow PW (2006b) Cytokinesis in plant and animal cells: endosomes 'shut the door'. Dev Biol 294:1–10

Baluška F, Barlow PW, Volkmann D, Mancuso S (2007) Gravity related paradoxes in plants: plant neurobiology provides the means for their resolution. In: Witzany G (ed) Biosemiotics in transdisciplinary context. Proceedings of the gathering in Biosemiotics 6, Salzburg 2006. Umweb, Helsinki

Baluška F, Schlicht M, Wan Y-L, Burbach C, Volkmann D (2009a) Intracellular domains and polarity in root apices: from synaptic domains to plant neurobiology. Nova Acta Leopold 96:103–122

Baluška F, Mancuso S, Volkmann D, Barlow PW (2009b) The 'root-brain' hypothesis of Charles and Francis Darwin: revival after more than 125 years. Plant Signal Behav 4:1121–1127

Baluška F, Mancuso S, Volkmann D, Barlow PW (2010) Root apex transition zone: a signalling – response nexus in the root. Trends Plant Sci 15:402–408

Bancaud A, Huet S, Daigle N, Mozziconacci J, Beaudouin J, Ellenberg J (2009) Molecular crowding affects diffusion and binding of nuclear proteins in heterochromatin and reveals the fractal organization of chromatin. EMBO J 28:3785–3798

Bar M, Avni A (2009) EHD2 inhibits signaling of leucine rich repeat receptor-like proteins. Plant Signal Behav 4:682–684

Barlow PW (1995) Gravity perception in plants: a multiplicity of systems derived by evolution? Plant Cell Environ 18:951–962

Barlow PW (1996) An introduction to gravity perception in plants and fungi: a multiplicity of mechanisms. Adv Space Res 17:69–72

Baskin TI, Peret B, Baluška F, Benfey P, Bennett M, Forde BG, Gilroy S, Helariutta Y, Hepler PK, Leyser O, Masson PH, Muday GK, Poethig S, Roberts K, Scheres B, Sharp RE, Somerville C (2010) Shootward and rootward: peak terminology for plant polarity. Trends Plant Sci 15:593–594

Berghöfer T, Eing C, Flickinger B, Hohenberger P, Wegner LH, Frey W, Nick P (2009) Nanosecond electric pulses trigger actin responses in plant cells. Biochem Biophys Res Commun 387:590–595

Bershadsky A, Kozlov M, Geiger B (2006) Adhesion-mediated mechanosensitivity: a time to experiment and a time to theorize. Curr Opin Cell Biol 18:472–481

Birtwistle MR, Kholodenko BN (2009) Endocytosis and signalling: a meeting with mathematics. Mol Oncol 3:308–320

Blancaflor EB, Fasano JM, Gilroy S (1998) Mapping the functional roles of cap cells in the response of Arabidopsis primary roots to gravity. Plant Physiol 116:213–222

Blancaflor EB, Hou GC, Mohamalawari DR (2003) The promotive effect of latrunculin B on maize root gravitropism is concentration dependent. Adv Space Res 31:2215–2220

Boudaoud A (2009) An introduction to the mechanics of morphogenesis for plant biologists. Trends Plant Sci 14:467–478

Braun M, Limbach C (2006) Rhizoids and protonemata of characean algae: model cells for research on polarized growth and plant gravity sensing. Protoplasma 229:133–142

Braybrook SA, Kuhlemeier C (2010) How a plant builds leaves. Plant Cell 22:1006–1018

Burg MB (2000) Macromolecular crowding as a cell volume sensor. Cell Physiol Biochem 10:251–256

Buss F, Arden SD, Lindsay M, Luzio JP, Kendrick-Jones J (2001) Myosin VI isoform localized to clathrin-coated vesicles with a role in clathrin-mediated endocytosis. EMBO J 20:3676–3684

Chalfie M (2009) Neurosensory mechanotransduction. Nat Rev Mol Cell Biol 10:45–52

Chehab EW, Eich E, Braam J (2009) Thigmomorphogenesis: a complex plant response to mechano-stimulation. J Exp Bot 60:43–56

Cho HT, Cosgrove DJ (2002) Regulation of root hair initiation and expansin gene expression in Arabidopsis. Plant Cell 14:3237–3253

Collings DA, White RG, Overall RL (1992) Ionic current changes associated with the gravity-induced bending response in roots of *Zea mays* L. Plant Physiol 100:1417–1426

Cote R, Thain JF, Fensom DS (1987) Increase in electrical resistance of plasmodesmata of Chara induced by an applied pressure gradient across nodes. Can J Bot 65:509–511

Cueno P, Magri E, Verzola A, Grazi E (1992) Macromolecular crowding is a primary factor in the organization of the cytoskeleton. Biochem J 281:507–512

Czapek F (1898) Weitere Beiträge zur Kentniss der geotropischen Reizbewegungen. Jahrb Wiss Bot 32:175–308

Dahl KN, Ribeiro AJS, Lammerding J (2008) Nuclear shape, mechanics, and mechanotransduction. Circ Res 102:1307–1318

Darwin F (1899) On geotropism and the localization of the sensitive region. Ann Bot 13:567–574

Darwin F (1903) The statolith theory of geotropism. Nature 67:571–572

Darwin F (1907) Lectures on the physiology of movement in plants. IV. The localisation of perception. New Phytol 6:35–42

Demidchik V, Nichols C, Oliynyk M, Dark A, Glover BJ, Davies JM (2003) Is ATP a signaling agent in plants? Plant Physiol 133:456–461

Dhonukshe P, Baluška F, Schlicht M, Hlavačka A, Šamaj J, Friml J, Gadella TWJ Jr (2006) Endocytosis of cell surface material mediates cell plate formation during plant cytokinesis. Dev Cell 10:137–150

Ding JP, Pickard BG (1993a) Mechanosensory calcium-selective cation channels in epidermal cells. Plant J 3:83–110

Ding JP, Pickard BG (1993b) Modulation of mechanosensitive calcium-selective cation channels by temperature. Plant J 3:713–720

Ding JP, Badot P-M, Pickard BG (1993) Aluminium and hydrogen ions inhibit a mechanosensory calcium-selective cation channel. Aust J Plant Physiol 20:771–778

Ditengou FA, Teale WD, Kochersperger P, Flittner KA, Kneuper I, van der Graaff E, Nziengui H, Pinosa F, Li X, Nitschke R, Laux T, Palme K (2008) Mechanical induction of lateral root initiation in *Arabidopsis thaliana*. Proc Natl Acad Sci USA 105:18818–18823

Dodd AN, Kudla J, Sanders D (2010) The language of calcium signaling. Annu Rev Plant Biol 61:593–620

Drakakaki G, Robert S, Raikhel NV, Hicks GR (2009) Chemical dissection of endosomal pathways. Plant Signal Behav 4:57–62

Driss-Ecole D, Jeune B, Prouteau M, Julianus P, Perbal G (2000) Lentil root statoliths reach a stable state in microgravity. Planta 211:396–405

Evans ML, Ishikawa H (1997) Cellular specificity of the gravitropic motor response in roots. Planta 203(Suppl 1):S115–S122

Ewart AJ (1902) On the physics and physiology of protoplasmic streaming in plants. Clarendon, Oxford

Fasano JM, Swanson SJ, Blancaflor EB, Dowd PE, Kao TH, Gilroy S (2001) Changes in root cap pH are required for the gravity response of the Arabidopsis root. Plant Cell 13:907–921

Fasano JM, Massa GD, Gilroy S (2002) Ionic signaling in plant responses to gravity and touch. J Plant Growth Regul 21:71–88

Ferro M, Salvi D, Brugière S, Miras S, Kowalski S, Louwagie M, Garin J, Joyard J, Rolland N (2003) Proteomics of the chloroplast envelope membranes from *Arabidopsis thaliana*. Mol Cell Proteomics 2:325–345

Fleming AJ, McQueen-Mason S, Mandel T, Kuhlemeier C (1997) Induction of leaf primordia by the cell wall protein expansin. Science 276:1415–1418

Foresi NP, Laxalt AM, Tonon CV, Casalongue CA, Lamattina L (2007) Extracellular ATP induces nitric oxide production in tomato cell suspensions. Plant Physiol 145:589–592

Foth BJ, Goedecke MC, Soldati D (2006) New insights into myosin evolution and classification. Proc Natl Acad Sci USA 103:3681–3686

Gaina V, Svegzdiene D, Rakleviciene D, Koryzniene D, Staneviciene R, Laurinavicius R (2003) Kinetics of amyloplast movement in cress root statocytes under different gravitational loads. Adv Space Res 31:2275–2281

Geiger B, Spatz JP, Bershadsky AD (2009) Environmental sensing through focal adhesions. Nat Rev Mol Cell Biol 10:21–33

Geisler DA, Sampathkumar A, Mutwil M, Persson S (2008) Laying down the bricks: logistic aspects of cell wall biosynthesis. Curr Opin Plant Biol 11:647–652

Geitmann A, Ortega JKE (2009) Mechanics and modeling of plant cell growth. Trends Plant Sci 14:467–478

Geldner N, Robatzek S (2008) Plant receptors go endosomal: a moving view on signal transduction. Plant Physiol 147:1565–1574

Golomb L, Abu-Abied M, Belausov E, Sadot E (2008) Different subcellular localizations and functions of Arabidopsis myosin VIII. BMC Plant Biol 8:3

Guikema JA, Hilaire E, Odom WR (1993) Cytoskeleton-amyloplast interactions in sweet clover. Trans KS Acad Sci 96:13–19

Haberlandt G (1900) Über die Perzeption des geotropischen Reizes. Ber Dtsch Bot Ges 18:261–272

Haberlandt G (1908) Über die Verteilung der geotropischen Sensibilität in der Wurzel. Jahrb Wiss Bot 45:575–600

Hála M, Soukupová H, Synek L, Zárský V (2010) Arabidopsis RAB geranylgeranyl transferase beta-subunit mutant is constitutively photomorphogenic, and has shoot growth and gravitropic defects. Plant J 62:615–627

Harrison BR, Masson PH (2008a) ARL2, ARG1 and PIN3 define a gravity signal transduction pathway in root statocytes. Plant J 53:380–392

Harrison B, Masson PH (2008b) ARG1 and ARL2 form an actin-based gravity-signaling chaperone complex in root statocytes? Plant Signal Behav 3:650–653

Hasson T (2003) Myosin VI: two distinct roles in endocytosis. J Cell Sci 116:3453–3461

Haswell ES, Meyerowitz EM (2006) MscS-like proteins control plastid size and shape in *Arabidopsis thaliana*. Curr Biol 16:1–11

Haswell ES, Peyronnet R, Barbier-Brygoo H, Meyerowitz EM, Frachisse J-M (2008) Two MscS homologs provide mechanosensitive channel activities in the Arabidopsis root. Curr Biol 18:730–734

Hause G, Šamaj J, Menzel D, Baluška F (2006) Fine structural analysis of brefeldin A-induced compartment formation after high-pressure freeze fixation of maize root epidermis: compound exocytosis resembling cell plate formation during cytokinesis. Plant Signal Behav 1:134–139

Hayashi T, Takagi S (2003) Ca^{2+}-dependent cessation of cytoplasmic streaming induced by hypertonic treatment in *Vallisneria mesophyll* cells: possible role of cell wall–plasma membrane adhesion. Plant Cell Physiol 44:1027–1036

Hou G, Mohamalawari DR, Blancaflor EB (2003) Enhanced gravitropism of roots with a disrupted cap actin cytoskeleton. Plant Physiol 131:1360–1373

Hou G, Kramer VL, Wang YS, Chen R, Perbal G, Gilroy S, Blancaflor EB (2004) The promotion of gravitropism in Arabidopsis roots upon actin disruption is coupled with the extended alkalinization of the columella cytoplasm and a persistent lateral auxin gradient. Plant J 39:113–125

Idone V, Tam C, Goss JW, Toomre D, Pypaert M, Andrews NW (2008a) Repair of injured plasma membrane by rapid Ca^{2+}-dependent endocytosis. J Cell Biol 180:905–914

Idone V, Tam C, Andrews NW (2008b) Two-way traffic on the road to plasma membrane repair. Trends Cell Biol 18:552–559

Ingber DE (1997) Tensegrity: the architectural basis of cellular mechanotransduction. Annu Rev Physiol 59:575–599
Ingber D (1999) How cells (might) sense microgravity. FASEB J 13(Suppl):S3–S15
Ingber DE (2003) Tensegrity II. How structural networks influence cellular information processing networks. J Cell Sci 116:1397–1408
Ingber DE (2008) Tensegrity-based mechanosensing from macro to micro. Prog Biophys Mol Biol 97:163–179
Ishikawa H, Evans ML (1990a) Gravity-induced changes in intracellular potentials in elongating cortical cells of mung bean roots. Plant Cell Physiol 31:457–462
Ishikawa H, Evans ML (1990b) Electrotropism of maize roots. Role of the root cap and relationship to gravitropism. Plant Physiol 94:913–918
Ishikawa H, Evans ML (1992) Induction of curvature in maize roots by calcium or by thigmostimulation: role of the postmitotic isodiametric growth zone. Plant Physiol 100:762–768
Ishikawa H, Evans ML (1994) Correlations between changes in electrical parameters and changes in cell elongation rates in gravistimulated roots. Adv Space Res 14:125–133
Ivanov R, Gaude T (2009) Endocytosis and endosomal regulation of the S-receptor kinase during the self-incompatibility response in *Brassica oleracea*. Plant Cell 21:2107–2117
Iwabuchi A, Yano M, Shimizu H (1989) Development of extracellular electric pattern around Lepidium roots: its possible role in root growth and gravitropism. Protoplasma 148:98–100
Janmey PA, McCulloch CA (2007) Cell mechanics: integrating cell responses to mechanical stimuli. Annu Rev Biomed Eng 9:1–34
Johnsson A, Pickard BG (1979) The threshold stimulus for geotropism. Physiol Plant 45: 315–319
Jouhet J, Gray JC (2009) Interaction of actin and the chloroplast protein import apparatus. J Biol Chem 284:19132–19141
Juniper BE (1976) Geotropism. Annu Rev Plant Physiol 27:385–406
Juniper BE, Groves S, Landau-Schachar B, Audus LJ (1966) Root cap and the perception of gravity. Nature 209:93–94
Kernan MJ (2007) Mechanotransduction and auditory transduction in Drosophila. Pflugers Arch 454:703–720
Kernan M, Cowan D, Zuker C (1994) Genetic dissection of mechanosensory transduction: mechanoreception-defective mutations of Drosophila. Neuron 12:1195–1206
Kimbrough JM, Salinas-Mondragon R, Boss WF, Brown CS, Sederoff HW (2004) The fast and transient transcriptional network of gravity and mechanical stimulation in the Arabidopsis root apex. Plant Physiol 136:2790–2805
Kleine-Vehn J, Langowski L, Wisniewska J, Dhonukshe P, Brewer PB, Friml J (2008) Cellular and molecular requirements for polar PIN targeting and transcytosis in plants. Mol Plant 1:1056–1066
Kollmeier M, Felle HH, Horst WJ (2000) Genotypical differences in aluminum resistance of maize are expressed in the distal part of the transition zone. Is reduced basipetal auxin flow involved in inhibition of root elongation by aluminum? Plant Physiol 122:945–956
Kordyum E, Guikema J (2001) An active role of the amyloplasts and nuclei of root statocytes in graviperception. Adv Space Res 27:951–956
Kudla J, Batistic O, Hashimoto K (2010) Calcium signals: the lead currency of plant information processing. Plant Cell 22:541–563
Kung C (2005) A possible unifying principle for mechanosensation. Nature 436:647–654
Łangowski Ł, Růžička K, Naramoto S, Kleine-Vehn J, Friml J (2010) Trafficking to the outer polar domain defines the root-soil interface. Curr Biol 10:904–908
Laskowski M, Grieneisen VA, Hofhuis H, Hove CA, Hogeweg P, Marée AF, Scheres B (2008) Root system architecture from coupling cell shape to auxin transport. PloS Biol 6:e307
Lecuit T (2010) α-Catenin mechanosensing for adherens junctions. Nat Cell Biol 12:522–524
Legué V, Blancaflor E, Wymer C, Perbal G, Fantin D, Gilroy S (1997) Cytoplasmic free Ca^{2+} in Arabidopsis roots changes in response to touch but not gravity. Plant Physiol 114:789–800

Leterrier JF (2001) Water and the cytoskeleton. Cell Mol Biol 47:901–923
Levina N, Totemeyer S, Stokes NR, Louis P, Jones MA, Booth IR (1999) Protection of *Escherichia coli* cells against extreme turgor by activation of MscS and MscL mechanosensitive channels: identification of genes required for MscS activity. EMBO J 18:1730–1737
Lewis JD, Lazarowitz SG (2009) Arabidopsis synaptotagmin SYTA regulates endocytosis and virus movement protein cell-to-cell transport. Proc Natl Acad Sci USA 107:2491–2496
Lieleg O, Schmoller KM, Claessens MM, Bausch AR (2009) Cytoskeletal polymer networks: viscoelastic properties are determined by the microscopic interaction potential of cross-links. Biophys J 96:4725–4732
Limbach C, Hauslage J, Schäfer C, Braun M (2005) How to activate a plant gravireceptor. Early mechanisms of gravity sensing studied in characean rhizoids during parabolic flights. Plant Physiol 139:1030–1040
Lindeboom J, Mulder BM, Vos JW, Ketelaar T, Emons AM (2008) Cellulose microfibril deposition: coordinated activity at the plant plasma membrane. J Microsc 231:192–200
Liverpool TB (2006) Active gels: where polymer physics meets cytoskeletal dynamics. Philos Trans A Math Phys Eng Sci 364:3335–3355
Lorenzi G, Perbal G (1990) Actin filaments responsible for the location of the nucleus in the lentil statocyte are sensitive to gravity. Biol Cell 68:259–263
Luan S (2009) The CBL-CIPK network in plant calcium signaling. Trends Plant Sci 14:37–42
Luby-Phelps K (2000) Cytoarchitecture and physical properties of cytoplasm: volume, viscosity, diffusion, intracellular surface area. Int Rev Cytol 192:189–221
Mancuso S, Barlow PW, Volkmann D, Baluška F (2006) Actin turnover-mediated gravity response in maize root apices: gravitropism of decapped roots implicates gravisensing outside of the root cap. Plant Signal Behav 1:52–58
Masi E, Ciszak M, Stefano G, Renna L, Azzarello E, Pandolfi C, Mugnai S, Baluška F, Arecchi FT, Mancuso S (2009) Spatio-temporal dynamics of the electrical network activity in the root apex. Proc Natl Acad Sci USA 106:4048–4053
Massa GD, Gilroy S (2003) Touch modulates gravity sensing to regulate the growth of primary roots of *Arabidopsis thaliana*. Plant J 33:435–445
McConnell RE, Tyska MJ (2010) Leveraging the membrane – cytoskeleton interface with myosin-1. Trends Cell Biol 20:418–426
Minc N, Burgess D, Chang F (2011) Influence of cell geometry on division-plane positioning. Cell 144:414–426
Morey-Holton ER (2003) The impact of gravity on life. In: Rothschild LJ, Lister AM (eds) Evolution on planet Earth: impact of the physical environment. Academic, New York, pp 143–160
Morita MT (2010) Directional gravity sensing in gravitropism. Annu Rev Plant Biol 61:705–720
Morita MT, Tasaka M (2004) Gravity sensing and signaling. Curr Opin Plant Biol 7:712–718
Mullen JL, Wolverton C, Ishikawa H, Evans ML (2000) Kinetics of constant gravitropic stimulus responses in Arabidopsis roots using a feedback system. Plant Physiol 123:665–670
Murphy JE, Padilla BE, Hasdemir B, Cottrell GS, Bunnett NW (2009) Endosomes: a legitimate platform for the signaling train. Proc Natl Acad Sci USA 106:17615–17622
Nakayama T, Ishii T, Hotta T, Mizuno K (2008) Radial microtubule organization by histone H1 on nuclei of cultured tobacco BY-2 cells. J Biol Chem 283:16632–16640
Nambiar R, McConnell RE, Tyska MJ (2009) Control of cell membrane tension by myosin-I. Proc Natl Acad Sci USA 106:11972–11977
Němec B (1900) Über die Art der Wahrnehmung des Schwerkraftreizes bei den Pflanzen. Ber Dtsch Bot Ges 18:241–245
O'Hagan R, Chalfie M, Goodman MB (2005) The MEC-4 DEG/ENaC channel of *Caenorhabditis elegans* touch receptor neurons transduces mechanical signals. Nat Neurosci 8:43–50
Oparka KJ (1993) Signalling via plasmodesmata – the neglected pathway. Sem Cell Biol 4:131–138
Oparka KJ, Prior DAM (1992) Direct evidence for pressure-generated closure of plasmodesmata. Plant J 2:741–750

Ottenschläger I, Wolff P, Wolverton C, Bhalerao RP, Sandberg G, Ishikawa H, Evans M, Palme K (2003) Gravity-regulated differential auxin transport from columella to lateral root cap cells. Proc Natl Acad Sci USA 100:2987–2991

Perbal G, Driss-Ecole D, Tewinkel M, Volkmann D (1997) Statocyte polarity and gravisensitivity in seedling roots grown in microgravity. Planta 203(Suppl 1):S57–S62

Perbal G, Lefranc A, Jeune B, Driss-Ecole D (2004) Mechanotransduction in root gravity sensing cells. Physiol Plant 120:303–311

Perera IY, Heilmann I, Boss WF (1999) Transient and sustained increases in inositol 1,4,5-trisphosphate precede the differential growth response in gravistimulated maize pulvini. Proc Natl Acad Sci USA 96:5838–5843

Perera IY, Heilmann I, Chang SC, Boss WF, Kaufman PB (2001) A role for inositol 1,4,5-trisphosphate in gravitropic signaling and the retention of cold-perceived gravistimulation of oat shoot pulvini. Plant Physiol 125:1499–1507

Perrin RM, Young LS, Murthy UMN, Harrison BR, Wang Y, Will JL, Masson PH (2005) Gravity signal transduction in primary roots. Ann Bot 96:737–743

Peyronnet R, Haswell ES, Barbier-Brygoo H, Frachisse J-M (2008) AtMSL9 and AtMSL10. Sensors of plasma membrane tension in Arabidopsis roots. Plant Signal Behav 3:726–729

Pfeffer W (1894) Geotropic sensitiveness of the root-tip. Ann Bot 8:317–320

Pickard BG (2007) Delivering force and amplifying signals in plant mechanosensing. Curr Top Membr 58:361–392

Pickard BG, Ding JP (1993) The mechanosensory calcium-selective ion channel: key component of a plasmalemmal control centre? Aust J Plant Physiol 20:439–459

Pien S, Wyrzykowska J, McQueen-Mason S, Smart C, Fleming A (2001) Local expression of expansin induces the entire process of leaf development and modifies leaf shape. Proc Natl Acad Sci USA 98:11812–11817

Plieth C (2005) Calcium: just another regulator in the machinery of life? Ann Bot 96:1–8

Plieth C, Trewavas AJ (2002) Reorientation of seedlings in the earth's gravitational field induces cytosolic calcium transients. Plant Physiol 129:786–796

Reid RJ, Overall RL (1992) Intercellular communication in Chara: factors affecting transnodal electrical resistance and solute fluxes. Plant Cell Environ 15:507–517

Reinhardt D, Wittwer F, Mandel T, Kuhlemeier C (1998) Localized upregulation of a new expansin gene predicts the site of leaf formation in the tomato meristem. Plant Cell 10:1427–1437

Roux I, Safieddine S, Nouvian R, Grati M et al (2006) Otoferlin, defective in a human deafness form, is essential for exocytosis at the auditory ribbon synapse. Cell 127:277–289

Roux I, Hosie S, Johnson SL, Bahloul A, Cayet N, Nouaille S, Kros CJ, Petit C, Safieddine S (2009) Myosin VI is required for the proper maturation and function of inner hair cell ribbon synapses. Hum Mol Genet 18:4615–4628

Ryu J-H, Takagi S, Nagai R (1995) Stationary organization of the actin cytoskeleton in Vallisneria: the role of stable microfilaments at the end walls. J Cell Sci 108:1531–1539

Ryu J-H, Mizuno K, Takagi S, Nagai R (1997) Extracellular components implicated in the stationary organization of the actin cytoskeleton in mesophyll cells of Vallisneria. Plant Cell Physiol 38:420–432

Sachs J (1892) Beiträge zur Zellentheorie. Energiden und Zellen. Flora 75:57–67

Sack FD (1997) Plastids and gravitropic sensing. Planta 203:S63–S68

Sack FD, Suyemoto MM, Leopold AC (1986) Amyloplast sedimentation and organelle saltation in living corn columella cells. Am J Bot 73:1692–1698

Šamaj J, Baluška F, Hirt H (2004) From signal to cell polarity: mitogen-activated protein kinases as sensors and effectors of dynamic cytoskeleton and signalling endosomes. J Exp Bot 55:189–198

Šamaj J, Chaffey NJ, Tirlapur U, Jasik J, Volkmann D, Menzel D, Baluška F (2006) Actin and myosin VIII in plasmodesmata cell-cell channels. In: Baluška F, Volkmann D, Barlow PW (eds) Cell-cell channels. Landes Bioscience, Austin, pp 119–134

Sandhu AP, Randhawa GS, Dhugga KS (2009) Plant cell wall matrix polysaccharide biosynthesis. Mol Plant 2:840–850

Sattarzadeh A, Franzen R, Schmelzer E (2008) The Arabidopsis class VIII myosin ATM2 is involved in endocytosis. Cell Motil Cytoskelet 65:457–468

Schapire AL, Voigt B, Jasik J, Rosado A, Lopez-Cobollo R, Menzel D, Salinas J, Mancuso S, Valpuesta V, Baluška F, Botella MA (2008) Arabidopsis synaptotagmin-like protein 1 is essential for plasma membrane viability and survival under abiotic stress. Plant Cell 20:3374–3388

Schapire AL, Valpuesta V, Botella MA (2009) Plasma membrane repair in plants. Trends Plant Sci 14:645–653

Scherp P, Hasenstein KH (2007) Anisotropic viscosity of the Chara (Characeae) rhizoid cytoplasm. Am J Bot 94:1930–1934

Schmoller KM, Lieleg O, Bausch AR (2009) Structural and viscoelastic properties of actin/filamin networks: cross-linked versus bundled networks. Biophys J 97:83–89

Schröter K, Rodriguez-Garcia M, Sievers A (1973) Die Rolle des endoplasmatischen Retikulums bei der Genese der Chara-Statolithen. Protoplasma 76:435–442

Schröter K, Läuchli A, Sievers A (1975) Mikroanalytische Identifikation von Bariumsulfat-Kristallen in den Statolithen der Rhizoide von *Chara fragilis* Desv. Planta 122:213–225

Schwarz US, Erdmann T, Bischofs IB (2006) Focal adhesions as mechanosensors: the two-spring model. Biosystems 83:225–232

Scita G, Di Fiore PP (2010) The endocytic matrix. Nature 463:464–473

Scott BIH (1967) Electric fields in plants. Annu Rev Plant Physiol 18:409–418

Scott AC, Allen NS (1999) Changes in cytosolic pH within Arabidopsis root columella cells play a key role in the early signaling pathway for root gravitropism. Plant Physiol 121:1291–1298

Seagull RW, Falconer MM, Weerdenburg CA (1987) Microfilaments: dynamic arrays in higher plant cells. J Cell Biol 104:995–1004

Shen H, Hou NY, Schlicht M, Wan Y, Mancuso S, Baluška F (2008) Aluminium toxicity targets PIN2 in Arabidopsis root apices: effects on PIN2 endocytosis, vesicular recycling, and polar auxin transport. Chin Sci Bull 53:2480–2487

Shepherd VA (2006) The cytomatrix as a cooperative system of macromolecular and water networks. Curr Top Dev Biol 75:171–223

Shimmen T (2001) Electrical perception of "death message" in Chara: involvement of turgor pressure. Plant Cell Physiol 42:366–373

Shimmen T (2002) Electrical perception of "death message" in Chara: analysis of rapid component and ionic process. Plant Cell Physiol 43:1575–1584

Shimmen T (2003) Studies on mechano-perception in the Characeae: transduction of pressure signals into electrical signals. Plant Cell Physiol 44:1215–1224

Shimmen T (2005) Studies on mechano-perception in the Characeae: analysis of K^+-sensitive depolarization. Plant Cell Physiol 46:1839–1847

Siechen S, Yang S, Chiba A, Saif T (2009) Mechanical tension contributes to clustering of neurotransmitter vesicles at presynaptic terminals. Proc Natl Acad Sci USA 106:12611–12616

Silady RA, Kato T, Lukowitz W, Sieber P, Tasaka M, Somerville CR (2004) The gravitropism defective 2 mutants of Arabidopsis are deficient in a protein implicated in endocytosis in *Caenorhabditis elegans*. Plant Physiol 136:3095–3103

Silady RA, Ehrhardt DW, Jackson K, Faulkner C, Oparka K, Somerville CR (2008) The GRV2/RME-8 protein of Arabidopsis functions in the late endocytic pathway and is required for vacuolar membrane flow. Plant J 53:29–41

Song SJ, Steinebrunner I, Wang X, Stout SC, Roux SJ (2006) Extracellular ATP induces the accumulation of superoxide via NADPH oxidases in Arabidopsis. Plant Physiol 140:1222–1232

Sorkin A, von Zastrow M (2009) Endocytosis and signalling: intertwining molecular networks. Nat Rev Mol Cell Biol 10:609–622

Stanga JP, Boonsirichai K, Sedbrook JC, Otegui MS, Masson PH (2009a) A role for the TOC complex in Arabidopsis root gravitropism. Plant Physiol 149:1896–1905

Stanga J, Baldwin K, Masson PH (2009b) Joining forces: the interface of gravitropism and plastid protein import. Plant Signal Behav 4:933–941

Staves MP (1997) Cytoplasmic streaming and gravity sensing in Chara internodal cells. Planta 203:S79–S84

Staves MP, Wayne R, Leopold AC (1997a) Cytochalasin D does not inhibit gravitropism in roots. Am J Bot 84:1530–1535

Staves MP, Wayne R, Leopold AC (1997b) The effect of the external medium on the gravitropic curvature of rice (*Oryza sativa*, Poaceae) roots. Am J Bot 84:1522–1529

Steinhardt RA, Guoqiang B, Alderton JM (1994) Cell membrane resealing by a vesicular mechanism similar to neurotransmitter release. Science 263:390–393

Stoppin V, Vantard M, Schmit AC, Lambert AM (1994) Isolated plant nuclei nucleate microtubule assembly: the nuclear surface in higher plants has centrosome-like activity. Plant Cell 6:1099–1106

Švegždienė D, Koryznienė D, Raklevičienė D (2010) Comparison study of gravity-dependent displacement of amyloplasts in statocytes of cress roots and hypocotyls. Microgravity Sci Technol 23:235–241

Szymanski DB, Cosgrove DJ (2009) Dynamic coordination of cytoskeletal and cell wall systems during plant cell morphogenesis. Curr Biol 19:R800–R811

Tamura K, Takahashi H, Kunieda T, Fuji K, Shimada T, Hara-Nishimura I (2007) Arabidopsis KAM2/GRV2 is required for proper endosome formation and functions in vacuolar sorting and determination of the embryo growth axis. Plant Cell 19:320–332

Tang VW (2006) Proteomic and bioinformatic analysis of epithelial tight junction reveals an unexpected cluster of synaptic molecules. Biol Direct 1:37

Tang W, Brady SR, Sun Y, Muday GK, Roux SJ (2003) Extracellular ATP inhibits root gravitropism at concentrations that inhibit polar auxin transport. Plant Physiol 131:147–154

Tasaka M, Kato T, Fukaki H (1999) The endodermis and shoot gravitropism. Trends Plant Sci 4:103–106

Teis D, Huber LA (2003) The odd couple: signal transduction and endocytosis. Cell Mol Life Sci 60:2020–2033

Thompson RF, Langford GM (2002) Myosin superfamily evolutionary history. Anat Rec 268:276–289

Trebacz K, Fensom DS (1989) The uptake and transport of ^{14}C in cells of *Conocephallum conicum* in light. J Exp Bot 40:1089–1092

Trewavas A, Knight M (1994) Mechanical signalling, calcium and plant form. Plant Mol Biol 26:1329–1341

Verbelen J-P, De Cnodder T, Le J, Vissenberg K, Baluška F (2006) The root apex of *Arabidopsis thaliana* consists of four distinct zones of cellular activities: meristematic zone, transition zone, fast elongation zone, and growth terminating zone. Plant Signal Behav 1:296–304

Viotti C, Bubeck J, Stierhof YD, Krebs M, Langhans M, van den Berg W, van Dongen W, Richter S, Geldner N, Takano J, Jürgens G, de Vries SC, Robinson DG, Schumacher K (2010) Endocytic and secretory traffic in Arabidopsis merge in the trans-Golgi network/early endosome, an independent and highly dynamic organelle. Plant Cell 22:1344–1357

Volkmann D (1974) Amyloplasten und Endomembranen: das Geoperzeptionssystem der Primärwurzel. Protoplasma 79:159–183

Volkmann D, Sievers A (1979) Graviperception in multicellular organs. In: Haupt W, Feinleib ME (eds) Encyclopedia of Plant Physiology. Vol. 7 - Physiology of Movements, Springer-Verlag, Berlin, pp 573–609

Volkmann D, Baluška F (2006) Gravity: one of the driving forces of evolution. Protoplasma 229:143–148

Volkmann D, Baluška F (2008) Plant myosins: do they have roles in gravi- and mechano-sensing? In: Blume YB, Baird WV, Yemets AI, Breviario D (eds) The plant cytoskeleton: a key tool for agro-biotechnology. Springer, Berlin, pp 161–172

Volkmann D, Tewinkel M (1996) Gravisensitivity of cress roots: investigations of threshold values under specific conditions of sensor physiology in microgravity. Plant Cell Environ 19:1195–1202

Volkmann D, Buchen B, Hejnowicz Z, Tewinkel M, Sievers A (1991) Oriented movement of statoliths studied in a reduced gravitational field during parabolic flights of rockets. Planta 185:153–161

Volkmann D, Mori T, Tirlapur UK, König K, Fujiwara T, Kendrick-Jones J, Baluška F (2003) Unconventional myosins of the plant-specific class VIII: endocytosis, cytokinesis, plasmodesmata – pit-fields, and cell-to-cell coupling. Cell Biol Int 27:289–291

Wagner B, Tharmann R, Haase I, Fischer M, Bausch AR (2006) Cytoskeletal polymer networks: the molecular structure of cross-linkers determines macroscopic properties. Proc Natl Acad Sci USA 103:13974–13978

Wang N, Tytell JD, Ingber DE (2009) Mechanotransduction at a distance: mechanically coupling the extracellular matrix with the nucleus. Nat Rev Mol Cell Biol 10:75–82

Wawrzecki W, Zagorska-Marek B (2007) Influence of a weak DC electric field on root meristem architecture. Ann Bot 100:791–796

Wayne R, Staves MP (1996) A down to earth model of gravisensing, or Newton's law of gravitation from the apple's perspective. Physiol Plant 98:917–921

Wayne R, Staves MP, Leoplod AC (1990) Gravity-dependent polarity of cytoplasmic streaming in *Nitellopsis*. Protoplasma 155:43–57

Wayne R, Staves MP, Leopold AC (1992) The contribution of the extracellular matrix to gravisensing in characean cells. J Cell Sci 101:611–623

Weerasinghe RR, Swanson SJ, Okada SF, Garrett MB, Kim SY, Stacey G, Boucher RC, Gilroy S, Jones AM (2009) Touch induces ATP release in Arabidopsis roots that is modulated by the heterotrimeric G-protein complex. FEBS Lett 583:2521–2526

Willmitzer L, Wagner KG (1981) The isolation of nuclei from tissue-cultured plant cells. Exp Cell Res 135:69–77

Wojtaszek P, Anielska-Mazur A, Gabryś H, Baluška F, Volkmann D (2005) Recruitments of myosin VIII towards plastid surfaces are root cap-specific and provide the evidence for actomyosin involvement in root osmosensing. Funct Plant Biol 32:721–736

Wojtaszek P, Baluška F, Kasprowicz A, Łuczak M, Volkmann D (2007) Domain-specific mechanosensory transmission of osmotic and enzymatic cell wall disturbances to the actin cytoskeleton. Protoplasma 230:217–230

Wolverton C, Mullen JL, Aizawa S, Yoshizaki I, Kamigaichi S, Mukai C, Shimazu T, Fukui K, Evans ML, Ishikawa H (1999) Inhibition of root elongation in microgravity by an applied electric field. J Plant Res 112:493–496

Wolverton C, Mullen JL, Ishikawa H, Evans ML (2000) Two distinct regions of response drive differential growth in Vigna root electrotropism. Plant Cell Environ 23:1275–1280

Wolverton C, Ishikawa H, Evans ML (2002a) The kinetics of root gravitropism: dual motors and sensors. J Plant Growth Regul 21:102–112

Wolverton C, Mullen JL, Ishikawa H, Evans ML (2002b) Root gravitropism in response to a signal originating outside of the cap. Planta 215:153–157

Yamada S, Nelson WJ (2007) Synapses: sites of cell recognition, adhesion, and functional specification. Annu Rev Biochem 76:267–294

Yamamoto K, Kiss JZ (2002) Disruption of the actin cytoskeleton results in the promotion of gravitropism in inflorescence stems and hypocotyls of Arabidopsis. Plant Physiol 128: 669–681

Yamazaki T, Kawamura Y, Minami A, Uemura M (2008) Calcium-dependent freezing tolerance in Arabidopsis involves membrane resealing via synaptotagmin SYT1. Plant Cell 20:3389–3404·

Yoder TL, Zheng HQ, Todd P, Staehelin LA (2001) Amyloplast sedimentation dynamics in maize columella cells support a new model for the gravity-sensing apparatus of roots. Plant Physiol 125:1045–1060

Yonemura S, Wada Y, Watanabe T, Nagafuchi A, Shibata M (2010) α-Catenin as a tension transducer that induces adherens junction development. Nat Cell Biol 12:533–542

Yoshihara T, Iino M (2007) Identification of the gravitropism-related rice gene LAZY1 and elucidation of LAZY1-dependent and -independent gravity signaling pathways. Plant Cell Physiol 48:678–688

Young LM, Evans ML, Hertel R (1990) Correlations between gravitropic curvature and auxin movement across gravistimulated roots of *Zea mays*. Plant Physiol 92:792–796

Zárský V, Potocký M (2010) Recycling domains in plant cell morphogenesis: small GTPase effectors, plasma membrane signalling and the exocyst. Biochem Soc Trans 38:723–738

Zárský V, Cvrcková F, Potocký M, Hála M (2009) Exocytosis and cell polarity in plants – exocyst and recycling domains. New Phytol 183:255–272

Zhou EH, Trepat X, Park CY, Lenormand G, Oliver MN, Mijailovich SM, Hardin C, Weitz DA, Butler JP, Fredberg JJ (2009) Universal behavior of the osmotically compressed cell and its analogy to the colloidal glass transition. Proc Natl Acad Sci USA 106:10632–10637

Osmosensing

Anna Kasprowicz

Abstract As plants are not able to escape from changing environmental conditions, e.g., osmotic ones, their survival depends on their ability to react quickly and efficiently to such external stimuli. To achieve this they have developed and optimized many different signal perception and transduction pathways. Because plant stress sensors are poorly characterized and most of the signaling intermediates have not been identified, there is little definitive information regarding specificity of osmosensing pathway and the cross-talk between different stress signal sensing and transduction pathways in plants. This chapter aims to summarize the current understanding of osmosensing events in plants by focusing on converging contributions from biochemistry, molecular biology, and biophysics. Highlighted is the role of cell walls and several plasma membrane proteins as primary sensors of disturbances in osmotic pressure as well as the possible downstream signaling cascades and the influence of osmotic conditions on proper plant growth and development.

1 Plant Life Under Constant Osmotic Stress Threat

All organisms possess mechanisms that provide them with the flexibility to survive and grow under hypo- or hyperosmotic conditions. Osmosensing targets are found in plants, microbes, and mammals, and include receptors, signaling molecules, and ion transport mechanisms. Changes in hydrostatic pressure, evoked by disturbances in cellular water conditions, belong to fundamental physical properties, determining shape and functioning of plant cells. In plants, due to their sessile nature, special mechanisms have evolved, allowing for the adjustment of metabolism to the variable osmotic conditions and due to this, making possible the tolerance and survival in severe conditions (Walbot 1996). Shape of plant cells is determined by the shape of cell walls exposed to multidirectional stress. Therefore, plant cell

A. Kasprowicz
Faculty of Biology, Department of Molecular and Cellular Biology, Adam Mickiewicz University, Umultowska 89, 61-614 Poznań, Poland
e-mail: akas@amu.edu.pl

P. Wojtaszek (ed.), *Mechanical Integration of Plant Cells and Plants*,
Signaling and Communication in Plants 9, DOI 10.1007/978-3-642-19091-9_9,

© Springer-Verlag Berlin Heidelberg 2011

shape is influenced by plastic deformations of cell walls in response to isotropic turgor pressure. The magnitude of turgor pressure is dependent on osmotic equilibrium across plasma membrane and the mechanical properties of cell walls (Peters et al. 2000). Regulation of the osmotic pressure at the cellular level is a key process for proper growth and development. Plant cells have evolved many adaptations allowing monitoring of changes in osmotic potential outside the cell. Osmotic stress, causing disturbances in distribution of forces acting on plasma membrane from both inside and outside, can be considered as special case of mechanical stress (see also chapter "Plants as Mechano-osmotic Transducers"). Disturbances in water relations within and outside the cell cause osmotic stress: hypoosmotic stress is evoked by the rise of water potential outside the cell and promotes water inflow and swelling of the protoplast. On the other hand, hyperosmotic stress is evoked by the reduction of osmotic potential outside the cell, and this leads to the turgor decrease, metabolic disturbances, cessation of minerals uptake from soil and, in consequence, can even cause cell death, because water uptake from soil is impeded due to the presence of osmotic agents. These symptoms are specific for physiological drought. Plant cells are evolutionary adjusted to survive in variable environment. Classical example could be the reversible phenomenon of plasmolysis – adaptive response of the cell to the changes in water content in immediate surrounding environment (Lang-Pauluzzi and Gunning 2000). NaCl is a specific stress agent that causes both osmotic stress (due to Na^+ accumulation in the soil) and ionic stress (salt stress, due to perturbations in ionic metabolism of the cell), launching different mechanisms of cellular response (Ueda et al. 2003). It is thought that in the early phase of salt stress, cells respond similarly as upon osmotic stress, and this has been documented during analysis of expression profiles of genes involved in cellular reaction to salt and osmotic stress (Ueda et al. 2004).

2 Plant Complex Responses to Osmotic Stress

Despite the growing knowledge about particular elements of signaling pathways involved in stress signal transduction and activation of proper genes in response to stress, the question arises how plants detect individual stress signals. Various stress types (e.g. osmotic, salt, cold, drought) could be recognized independently by specific sensors that could initiate downstream cascades of secondary messengers transmitting the stress signal in the cellular interior (Xiong and Zhu 2002). The type of stress agent and intensity of its action are important for the initiation of plant response. For example, different receptors are activated by hypo- and hyperosmotic stress evoked by NaCl applied in different concentrations to the culture media (Munnik and Meijer 2001). It was proposed that in response to mild stress caused by disturbances in water conditions, changes in contacts between cell wall and plasma membrane are leading to the activation of receptor-like kinases (RLK), cytoskeletal elements sensitive to mechanical stimuli, and pathways dependent on redox potential balance (Kacperska 2004). Hydrostatic pressure can generate compressing or decompressing forces inside the cell, and activate ion channels dependent on mechanical stimulation (Hayashi et al. 2006; Roberts 2006). This mechanism is also proposed to be one of the most evolutionary ancient signal transduction pathways (Zonia and Munnik 2007). Mild stress conditions initiate activation of metabolic pathways that allow the plants to grow in the presence of stress agent. Sudden, drastic change in the gradient of osmotic potential across the plasma membrane evokes primary response that results in ion flow across plasma membrane leading to ionic balance (Schroeder and Hagiwara 1989; Schroeder and Hedrich 1989; Ward et al. 1995; Teodoro et al. 1998; Liu and Luan 1998; Barbier-Brygoo et al. 2000; Blatt 2000; Shabala et al. 2000; Ivashikina et al. 2001; Schroeder et al. 2001). High concentration of stress agent causes depolarization of plasma membrane due to disturbances in ion transport. This is sensed by the cell as destabilization of plasma membrane and initiates signal cascade dependent on phospholipids, like phospholipase C or phosphoinositols (Blatt et al. 1990; Gilroy et al. 1990; Lemtiri-Chlieh et al. 2000; Zonia et al. 2002; Staxen et al. 1999; Drøbak and Watkins 2000; DeWald et al. 2001; Munnik and Meijer 2001; Takahashi et al. 2001). Within a few minutes after stimulation with stress agent, activation of signaling cascade dependent of mitogen-activated protein serine/threonine kinases (MAPK) can be noticed (Cazale et al. 1999; Munnik et al. 1999; Felix et al. 2000; Munnik and Meijer 2001). This leads, in consequence, to enhanced production of reactive oxygen species, accumulation of hydrogen peroxide, lipid peroxidation, and enhanced synthesis of hormones, like jasmonic acid and ethylene (Assmann and Wang 2001; Schroeder et al. 2001; Seo and Koshiba 2002). Accumulation and distribution within the cell of compatible compounds playing the role of osmoprotectants (Borowitzka and Brown 1974; Hare et al. 1998, 1999; Hasegawa et al. 2000; Lefévre et al. 2001; Raymond and Smirnoff 2002) allows for decline of the water potential in the cell and prevents further loss of water. Compatible solutes appear in high amounts (up to 20% total dry mass) upon osmotic adjustment. These include, e.g., proline, glycinobetaine, and sorbitol. In contrast to ions, they do not inhibit cytosolic enzymes, and thus do not negatively affecting the cellular metabolism, whereas their accumulation contributes to enhanced stress tolerance (Gorham et al. 1985; Delauney and Verma 1985; Kishor et al. 1995). Interestingly, they are also involved in stabilization of biological membranes under stress conditions (Hincha and Hagemann 2004). It was shown, that sucrose, trehalose, and sorbitol protect liposomes from leakage of solutes and from membrane fusion during drying and rehydration. Exposition of plants to the hyperosmotic conditions induces specific gene expression and synthesis of proteins involved in specific cellular stress response pathways (Zhu et al. 1998; Hasegawa et al. 2000; Zhu 2000). Finally, under conditions of severe stress, the resulting damage of the cell can be observed. Depending on the properties of the stress agent, when its activity is diminishing, cellular reconstruction or cell death would be the final result. Recently, due to progress in comparative transcriptomics and proteomics, many genes and proteins involved in osmotic and salt stress response have been identified (Kawasaki et al. 2001; Seki et al. 2002; Takahashi et al. 2004; Ueda et al. 2004; Yan et al. 2005; Wang et al. 2003; Parker et al. 2006).

3 Plant Cell Walls as Osmosensor

The plant cell wall serves not only as a mechanical support for the plant cell, but also for the growth in and adaptation to hostile environments (Wojtaszek 2000). Due to an isotropic action of turgor pressure, plant cells depleted of cell walls – protoplasts – would preferentially develop spherical shape. This simple observation demonstrates that joint action of cell walls and the turgor pressure is a key mechanism of cell's shape and growth mode determination (e.g. elongation, tip growth) (Kutschera 2000, 2001). Cell walls and the steep gradient of hydrostatic pressure across the plasma membrane, exceeding 2 MPa, could be used for mechanical stabilization of plant cells and plants (Peters et al. 2000). Cell walls act like corset, squeezing protoplasts and shaping them up. Protoplast's turgor pressure keeps cell walls under constant tension and this defines local biophysical conditions around individual cells, and, subsequently, around neighboring cells (Carpita and Gibeaut 1993). Some data indicate that monitoring of wall integrity in yeasts might be tightly coupled to osmosensing (Hohmann 2002) and this is also postulated for plant cells (Marshall and Dumbroff 1999; Nakagawa and Sakurai 2001). Yeast cell wall integrity sensing pathway involves a family of plasma membrane proteins with periplasmic ectodomain functioning as the cell surface sensor that detects and transmits signals to small G-protein Rho1, as well as a set of Rho1-activated effectors. This pathway forms a part of signaling network regulating the architecture of actin cytoskeleton, correlating and coordinating in this way such cellular processes as cell wall biosynthesis and cell polarization (Levin 2005). THESEUS1 (THE1), a RLK expressed in elongating cells and in the vascular tissue, has recently emerged as a putative cell wall integrity sensor in plants. This protein mediates the response of growing plant cells to the perturbation of cellulose synthesis. It also controls ectopic lignin accumulation in cellulose-deficient mutant backgrounds (Hématy et al. 2007; Hématy and Höfte 2008).

The structure of cell walls allows them to specifically react to pressure and extension. In response to turgor pressure the walls act as a viscoelastic material, as upon external forces the walls can undergo elastic (reversible) and plastic (irreversible) deformations. Cell growth is determined by plastic deformations and interactions (Brett and Waldron 1996). Therefore, disturbances in the turgor pressure upon osmotic stress, due to specific properties of cell walls, can modify growth of plant cells. Associations between particular polymers building the walls are precisely controlled, and the composition and spatial distribution of cell wall compounds are different and highly specific to particular cell wall domains, cell types, developmental stages, and different plant species (Somerville et al. 2004; Wojtaszek et al. 2007). Water loss from cell walls influence polymer positioning in the wall. Polymers separated in hydrated walls change their position and adhere in walls collapsing upon dehydration. In this situation hydrogen bounds between polysaccharide chains could be enhanced, and this would change biophysical properties of the walls, affecting finally plant growth (Moorea et al. 2008). Water loss from the cell directly modifies extensibility of cell walls.

It was documented that cells react to turgor changes by relaxation or fortification of cell walls. In tissues indispensible for plant growth the loosening of cell walls was observed, whereas in the others hardening. Thus, even upon lowered water potential, some growth phenomena, like tip growth of root hairs, are still possible (Wu and Cosgrove 2000).

The growth of plant cells is made possible thanks to the enlargement of total area of plasma membrane and cell wall. The extensibility of plasma membrane is relatively small, and estimated to be in the range of 2% (Wolfe and Steponkus 1981). In growing cells the walls do not become thinner, what indicates that cell growth depends on constant delivery of building materials to plasma membrane and cell wall (Miller et al. 1997). Expansins affect the extensibility of the cell wall due to their ability to disrupt hydrogen bonds between cellulose and the surrounding polymers, mostly hemicelluloses (McQueen-Mason et al. 1992). One of the first acclimation responses to drought/osmotic stress is the decrease in leaf growth, allowing for the maintenance of cell turgor and reducing the transpiration area (Mathews et al. 1984; Neumann 1995). Under mild drought, *Arabidopsis* leaves compensate for low expansion rate by increasing the duration of the expansion period (Aguirrezabal et al. 2006). It was discovered that in *Arabidopsis* mild osmotic stress induces expression of expansin genes (Skirycz et al. 2010). A study on the resurrection plant (*Craterostigma plantagineum*) showed an increase in expansin expression and activity at an early stage of dehydration, resulting in a flexible cell wall and allowing for adaptation to dehydration (Jones and McQueen-Mason 2004). Overexpression of expansin genes leads to enhanced growth in rice and high sensitivity to hormones and salt stress in *Arabidopsis* (Choi et al. 2003; Kwon et al. 2008). Expansins are also important for localized events during plant development when local loosening of wall structure could initiate e.g. formation of leaf primordia (Fleming et al. 1997) or cellular bulging leading to the development of root hairs (Baluška et al. 2000).

The changes in cell wall composition observed during development and in response to stress, suggest the role of directed transport via exo- and endocytosis of polysaccharides and proteins to and from strictly defined regions of cell wall and plasma membrane (Meckel et al. 2004; Dhonukshe et al. 2006; Johansen et al. 2006; Reichardt et al. 2007). Research on tip-growing cells, like pollen tubes or root hairs, have shown that turgor pressure propelled by dynamic changes in water conditions and interacting with the cell wall domains, characterized by specific mechanical properties, is the driving force of cell elongation. Cell growth is initiated in the weakest regions of cell walls (Baluška et al. 2000; Zonia et al. 2002, 2006; Zonia and Munnik 2004, 2007). Local osmotic pressure changes in certain areas of the cell influence the intensity of endo- and exocytosis. In hypoosmotic conditions, the swelling of the protoplast stimulates exocytosis, inhibits endocytosis and initiates cell growth, whereas in hyperosmotic conditions, the shrinking of the protoplast leads subsequently to the stimulation of endocytosis and cessation of exocytosis (Zonia and Munnik 2008). Increase in the osmotic pressure leads to growth retardation and decrease of pectin ratio in pollen tube cell wall (Parre and Geitmann 2005). It was documented that increase in the turgor pressure promotes transport of wall components to the extracellular space, whereas the turgor decrease acts in the opposite way (Proseus and Boyer 2005, 2006). Similar observations have been made in stomata where transport via endo- and exocytosis is strictly related to changes in the cellular volume (Shope et al. 2003; Meckel et al. 2005). It seems therefore, that constitutive internal membrane flow enables the cell the quick reaction to changes in osmotic pressure and due to this regulates tension of plasma membrane and cell wall. Moreover, it is suggested that endo- and exocytosis, ensuring the directional transport of cell wall compounds, enable the control of mechanical properties of cell walls and help to balance high-pressure gradient across plasma membrane (Zonia and Munnik 2007).

It is thought, that structural and functional continuum between cell *w*all, plasma *m*embrane, and *c*ytoskeleton (WMC) is involved in sensing and transduction of signals from external environment to the inside of the cell (Wyatt and Carpita 1993; see also chapter "Introduction: Tensegral World of Plants"). Plasma membrane proteins with ectodomains embedded in cell wall or strongly interacting with the cell wall components could serve as linkers within the WMC continuum. These ectodomains, connected with transmembrane domains, and, if present, with cytoplasmic domains of individual proteins could interact with the cytoskeleton at the intracellular side of the membrane either directly or via cytoskeleton-binding proteins. There could be different scenarios of signal transduction, depending on the linker protein features. Based on animal models, it is thought that mechanical stimuli could be propagated as conformational changes within a chain of interacting proteins or transduced into electric signal due to the activity of ion channels. Another possibility includes the transformation of mechanical signal into a chemical one due to kinase activity of sensor protein (Ingber 2003a, b; Orr et al. 2006). Plant cells lack integrins that are major molecules linking external environment with plasma membrane and cytoskeleton, but there are many other molecules suggested to play this role (Baluška et al. 2003), e.g. wall-associated kinases (WAK) (Anderson et al. 2001; Kohorn 2001; Verica and He 2002), formins (Favery et al. 2004; Baluška and Hlavacka 2005; Deeks et al. 2005), lectin receptor kinases (LRKs) (Gouget et al. 2006), glycoproteins, and arabinogalactan proteins (AGPs) in particular (Fincher et al. 1983; Pont-Lezica et al. 1993), proteins attached to plasma membrane with galactosylphosphatidylinositol (GPI) anchor (Borner et al. 2003), as well as plasma membrane-located cellulose and callose synthase complexes (Delmer 1999). With respect to cell wall sensing, a very interesting observation is that WAK2 interacts with pectins at the cell surface and probably via MAPK3 (mitogen-activated protein kinase) signal transduction pathway activates a vacuolar invertase, leading to the production of glucose and fructose, which increases the turgor pressure and hence cell expansion (Kohorn et al. 2006, 2009). It was reported that on medium with no sucrose, *wak2* mutant showed a reduced cell elongation rate and the phenotype could be compensated for by expressing a maize (*Zea mays*) sucrose phosphate synthase (Kohorn et al. 2006). The low level of glucose and fructose in the vacuole could reduce turgor pressure in the cell and affect cell elongation (Martin et al. 2001), and in *wak2* mutant the vacuolar invertase activity is indeed reduced (Kohorn et al. 2006). LRKs have been identified as proteins that could serve as a direct link to cell wall polysaccharides due to their ability to bind carbohydrates and proteins (André et al. 2005; Bouwmeester and Govers 2009). Overexpression of a tobacco transmembrane protein NtC7, most similar to RLKs, confers tolerance to osmotic stress evoked by 500 mM mannitol. Interestingly, this tolerance appears to be stress-specific as seeds were not able to germinate on media containing high salts (Tamura et al. 2003). It was also proposed that NtC7 could form dimers whose conformation may be sensitive to changes in the membrane architecture in response to hyperosmolarity. NtC7 lacks the kinase catalytic domain, specific for RLK proteins, and this suggests that NtC7 transduces the osmotic signal into the cell through plasma membrane by interacting with another molecules through its C-terminal tail region (Boudsocq and Lauriére 2005). Such phenomenon of stress-specific responses has also been demonstrated earlier for other steps in osmotic signaling (Zonia and Munnik 2004).

4 Potential Plant Membrane Osmosensors

The sensing and signaling systems for osmotic conditions occur in all groups of organisms. Relatively little is known about membranous osmosensors in plants (Grefen and Harter 2004), but several plasma membrane proteins have been postulated to play this role. Plants have specialized receptors that may sense and transduce diverse stimuli, such as salinity, drought, and cold (Urao et al. 1999, 2000). Plasma membrane histidine kinase AtHK1 has been identified in *Arabidopsis* and shown to be a homologue of yeast osmosensory two-component histidine kinase SLN1 (Urao et al. 1999). Interestingly, one of the cytokinin receptors – plant histidine kinase cytokinin response 1 (CRE1) – is also regulated by changes in turgor pressure in a manner identical to that of SLN1, in the presence of cytokinin (Reiser et al. 2003). A close homologue of another cytokinin receptor (AtHK3) from *Medicago sativa* has been shown to be transcriptionally activated in response to high salinity, which suggests that MsHK1 can also function as an osmosensor (Coba de la Peña et al. 2008). Involvement of AtHK1 *in planta* in water stress responses has been demonstrated (Wohlbach et al. 2008). In total, in *Arabidopsis* genome there are eight genes coding for two-component histidine kinases, among which are also potential receptors for cytokinins and ethylene (Grefen and Harter 2004). The Arabidopsis histidine kinase AtHK1 is more abundant in roots than other tissues under normal growth conditions and accumulates under conditions of high or low osmolarity. AtHK1 gene is transcriptionally up-regulated in response to changes in external osmolarity what implies that it acts as an osmosensor and transmits the stress signal to a downstream MAPK cascade. Also worth mentioning is that AtHK1-defective *Arabidopsis* mutant plants are unable to withstand water stress (Wohlbach et al. 2008). Interestingly, it was discovered in heterological system (*Xenopus laevis* oocytes), that the two AtHKT1 homologs from *Eucalyptus camaldulensis* EcHKT1 and EcHKT2, thought to function in plasma membrane in the roots, possess intrinsic capability to sense changes in the osmolarity of the external medium in contrast to homologs from wheat and *Arabidopsis* (Liu et al. 2001). *E. camaldulensis* grows in dry river beds that are periodically flooded after high rainfalls (Williams and Brooker 1997). It might indicate that in long lifetime trees exposed to temporary drought and high salinity stresses, adaptive osmosensing mechanism based on K^+/Na^+ channels has evolved making possible the survival upon extremely variable environmental conditions (Zubrinich et al. 2000). This suggests different mechanisms allowing the adaptation to harsh environmental conditions in plants of short and long lifetime (Liu et al. 2001). Osmosensitivity was observed only in response to hypoosmotic solutions and was not found in the wheat HKT1. The enhanced ion uptake was not accompanied by increased water flux (Liu et al. 2001). Under hypotonic conditions plant cells uptake K^+ and this response is more rapid than under hypertonic stress (Felix et al. 2000). Therefore, it is speculated that the increased K^+ uptake via EcHKT1 and EcHKT2 may provide the plant with a mechanism for maintaining K^+ homeostasis when soil K^+ concentrations are diluted by floods. EcHKT may either provide the plant with a source of K^+ or may help to maintain cellular homeostasis by taking up K^+ that is lost from root cells during stress (Liu et al. 2001).

Changes in the hydrostatic pressure across the plasma membrane generate stretch and compression forces that induce rapid responses in plant cells. In animal and yeast models proteins belonging to the transient receptor potential (TRP) channels (and the yeast TRPY homolog) (Christensen and Corey 2007; Folgering et al. 2008), the DEG/ENaC voltage-independent Na^+ channel family (Christensen and Corey 2007; Bianchi 2007), and the TREK K^+ channel family (Folgering et al. 2008) are described as promising candidates for mechano- and osmosensing pathways, but there are no clear homologs of these mechanosensitive (MS) channels in any sequenced plant genomes (Monshausen and Gilroy 2010). Only the bacterial MS channels have been so far proven to be a useful model for potential plant mechanoperception. The bacterial mechanosensitive channels of small (MscS) and large (MscL) conductance open when tension of the plasma membrane rises upon hyperosmotic stress, allowing for water efflux and preventing bursting of the cell (Corry and Martinac 2008). Although in *Arabidopsis* genome ten MscS-like (MSL) genes were identified (Haswell and Meyerowitz 2006), only knockouts of two genes (*msl2* and *msl3* mutants) gave obvious phenotype related to plastid division (Haswell et al. 2008). In quintuple mutant of all root-expressed MSL genes (*msl4/msl5/msl6/msl9/msl10*), all the MS channel activity was disturbed and MSL9 and MSL10 proteins are thought to form the core of a multimeric Cl^- permeable channel consistent with the known subunits of iris-like structure of bacterial MscS (Corry and Martinac 2008; Haswell et al. 2008). However, this mutant shows no obvious developmental perturbations and no changes in the phenotype are visible upon osmotic, salt, mechanical, dehydration, and rehydration stresses (Haswell et al. 2008), which suggests that there are also other mechanosensors in plants. Based on protein structure and localization, at least two classes of MS channels could be distinguished in plants (Haswell 2007). Proteins belonging to the first class are involved in sensing the tension within biological membranes (Martinac 2004), whereas proteins belonging to the second group are able to sense mechanical distortions in either cytoskeleton or cell walls. Some of MS channels from first class are involved in regulation of the cellular volume (Roberts 2006), and gating of Ca^{2+} influx, e.g. the MCA1 protein from *Arabidopsis* roots (Dutta and Robinson 2004; Nakagawa et al. 2007). A knockout mutant of MCA1 showed disturbances in root growth and response to mechanical stress, and constitutive MCA1 overexpressing lines exhibited strong defects in the development with short stems, small rosettes, no petals and shrunken seed pods. MCA1 overexpressing lines showed also an increased basal Ca^{2+} uptake and an elevated cytosolic Ca^{2+} level in response to osmotic shock. Interestingly, when MCA1 was heterologously expressed in Chinese hamster ovary cells, a novel Ca^{2+} increase was observed upon stretching the cells, which suggests that MCA1 could link the Ca^{2+} fluxes and mechanical response in *Arabidopsis* (Nakagawa et al. 2007). MS channels from the second group are involved in the WMC continuum formation as, due to the presence of transmembrane domains, they are embedded within cell wall from one side, whereas on another one interact directly or indirectly with cytoskeletal elements. A potassium channel KAT1, located in plasma membrane and probably associated with the surrounding cell walls belongs to this group of MS channels (Homann et al. 2007). However, it was not shown so far experimentally if it transmits mechanical stress signals into the cell. Stretch-activated MS Ca^{2+} channel located in tobacco plasma membrane was shown to be associated with WAKs and AGPs that are also the putative members of WMC continuum (Gens et al. 2000). Changes in osmotic conditions as e.g. hypertonicity introduce tension into WMC continuum and these channels are crucial for sensing of both hypotonic and hypertonic conditions, whereas the WMC continuum itself – only in sensing of hypertonic environment (Hayashi et al. 2006). It was demonstrated that both hypertonic and hypotonic treatments induce a transient and rapid increase in cytosolic Ca^{2+} concentration and that stretch-activated Ca^{2+} channels are involved in both responses. On the other hand, application of cellulase which disturbs connections between plasma membrane and cell wall showed that functional adhesion between those two cell domains is necessary for the Ca^{2+} transients mediated by stretch-activated Ca^{2+} channels, in the response to hypertonic treatment (Hayashi et al. 2006).

Plant ion transporters are also involved in response to changes in osmolarity. Outward and inward potassium channels in guard cells regulate the movement of K^+ across the membrane of guard cells in response to ABA signals evoked by changes in external osmolarity (Schroeder et al. 2001; Nilson and Assmann 2007). The inward K^+ channels are activated by hypotonic solutions, whereas outward K^+ channels are activated by hypertonic solutions (Liu and Luan 1998). Under salt or drought stress, the outward channels are induced and inward channels are repressed, leading to a decrease in stomatal conductance. This serves as an avoidance response/strategy developed by plants to maintain turgor under limited water availability (Skirycz and Inze 2010).

5 Conclusion

The integration of molecular and cell biology with physics approaches is crucial to elucidate the osmosensing pathway and to understand the correlations between changes in external environment, signal transduction, gene expression, and plant growth and development. Despite the constant influence of osmotic environment on plant cells, the proteins mentioned above have been described only as potential osmosensors. Little is known about downstream signaling cascades and effectors induced by osmotic stimuli. Because plants, as sessile organisms had to evolve very specific pathways to cope with changing environment, the general comparison of plants with other systems such as animals, yeast and bacteria will not be applicable in many cases to identify possible candidate proteins. A good example here is an indication that the plant Msc channels resemble more the prokaryotic than the eukaryotic ones. More work is also required to determine whether the osmosensing events described herein are part of a specific osmotic stress response or of a general stress response.

References

Aguirrezabal L, Bouchier-Combaud S, Radziejowski A, Dauzat M, Cookson SJ, Granier C (2006) Plasticity to soil water deficit in *Arabidopsis thaliana*: dissection of leaf development into underlying growth dynamic and cellular variables reveals invisible phenotypes. Plant Cell Environ 29:2216–2227

Anderson C, Wagner T, Perret M, He Z, He D, Kohorn B (2001) WAKs: cell wall-associated kinases linking the cytoplasm to the extracellular matrix. Plant Mol Biol 47:197–206

André S, Siebert HC, Nishiguchi M, Tazaki K, Gabius HJ (2005) Evidence for lectin activity of a plant receptor-like protein kinase by application of neoglycoproteins and bioinformatic algorithms. Biochim Biophys Acta 1725:222–232

Assmann SM, Wang XQ (2001) From milliseconds to millions of years: guard cells and environmental responses. Curr Opin Plant Biol 4:421–428

Baluška F, Hlavacka A (2005) Plant formins come to age: something special about cross-walls. New Phytol 168:499–503

Baluška F, Salaj J, Mathur J, Braun M, Jasper F, Šamaj J, Chua N-H, Barlow PW, Volkmann D (2000) Root hair formation: F-actin-dependent tip growth is initiated by local assembly of profilin-supported F-actin meshworks accumulated within expansin-enriched bulges. Dev Biol 226:618–632

Baluška F, Šamaj J, Wojtaszek P, Volkmann D, Menzel D (2003) Cytoskeleton-plasma membrane-cell wall continuum in plants. Emerging links revisited. Plant Physiol 133:482–491

Barbier-Brygoo H, Vinauger M, Colcombet J, Ephritikhine G, Frachisse JM, Maurel C (2000) Anion channels in higher plants: functional characterization, molecular structure and physiological role. Biochim Biophys Acta 1465:199–218

Bianchi L (2007) Mechanotransduction: touch and feel at the molecular level as modeled in *Caenorhabditis elegans*. Mol Neurobiol 36:254–271

Blatt MR (2000) Ca^{2+} signaling and control of guard-cell volume in stomatal movements. Curr Opin Plant Biol 3:196–204

Blatt MR, Thiel G, Trentham DR (1990) Reversible inactivation of K^+ channels of *Vicia* stomatal guard cells following the photolysis of caged inositol 1,4,5-*tris*phosphate. Nature 346:766–769

Borner GHH, Lilley KS, Stevens TJ, Dupree P (2003) Identification of glycosylphosphatidylinositol-anchored proteins in *Arabidopsis*: a proteomic and genomic analysis. Plant Physiol 132:568–577

Borowitzka LJ, Brown AD (1974) The salt relations of marine and halophilic species of unicellular green alga. Dunaliella. The role of glycerol as a compatible solute. Arch Microbiol 96:37–52

Boudsocq M, Lauriére C (2005) Osmotic signaling in plants. Multiple pathways mediated by emerging kinase families. Plant Physiol 138:1185–1194

Bouwmeester K, Govers F (2009) Arabidopsis L-type lectin receptor kinases: phylogeny, classification, and expression profiles. J Exp Bot 60:4383–4396

Brett C, Waldron K (1996) The physiology and biochemistry of the plant cell wall. Chapman and Hall, London

Carpita NC, Gibeaut DM (1993) Structural models of primary cell walls in flowering plants: consistency of molecular structure with the physical properties of the walls during growth. Plant J 3:1–30

Cazale AC, Droillard MJ, Wilson C, Heberle-Bors E, Barbier-Brygoo H, Lauriere C (1999) MAP kinase activation by hypoosmotic stress of tobacco cell suspensions: towards the oxidative burst response? Plant J 19:297–307

Choi D, Lee Y, Cho HT, Kende H (2003) Regulation of expansin gene expression affects growth and development in transgenic rice plants. Plant Cell 15:1386–1398

Christensen AP, Corey DP (2007) TRP channels in mechanosensation: direct or indirect activation? Nat Rev Neurosci 8:510–521

Coba de la Peña T, Cárcamo CB, Almonacid L, Zaballos A, Lucas MM, Balomenos D, Pueyo JJ (2008) A salt stress-responsive cytokinin receptor homologue isolated from Medicago sativa nodules. Planta 227:769–779

Corry B, Martinac P (2008) Bacterial mechanosensitive channels: experiment and theory. Biochim Biophys Acta 1778:1859–1870

Deeks MJ, Cvrcková F, Machesky LM, Mikitová V, Ketelaar T, Zársky V, Davies B, Hussey PJ (2005) *Arabidopsis* group Ie formins localize to specific cell membrane domains, interact with actin-binding proteins and cause defects in cell expansion upon aberrant expression. New Phytol 168:529–540

Delauney AJ, Verma DP (1985) Proline biosynthesis and osmoregulation in plants. Plant J 4:215–223

Delmer DP (1999) Cellulose biosynthesis: exciting times for a difficult field of study. Annu Rev Plant Physiol Plant Mol Biol 50:245–276

DeWald DB, Torabinejad J, Jones CA, Shope JC, Cangelosi AR, Thompson JE, Prestwich GD, Hama H (2001) Rapid accumulation of phosphatidylinositol 4,5-*bis*phosphate and inositol 1,4,5-*tris*phosphate correlates with calcium mobilization in salt-stressed *Arabidopsis*. Plant Physiol 126:759–769

Dhonukshe P, Baluška F, Schlicht M, Hlavacka A, Samaj J, Friml J, Gadella TWJ (2006) Endocytosis of cell surface material mediates cell plate formation during plant cytokinesis. Dev Cell 10:37–50

Drøbak BK, Watkins PAC (2000) Inositol (1,4,5)*tris*phosphate production in plant cells: an early response to salinity and hyperosmotic stress. FEBS Lett 481:240–244

Dutta R, Robinson KR (2004) Identification and characterization of stretch-activated ion channels in pollen protoplasts. Plant Physiol 135:1–9

Favery B, Chelysheva LA, Lebris M, Jammes F, Marmagne A, Almeida-Engler J, Lecomte P, Vaury C, Arkowitz RA, Abada P (2004) *Arabidopsis* formin AtFH6 is a plasma membrane-associated protein upregulated in giant cells induced by parasitic nematodes. Plant Cell 16:2529–2540

Felix G, Regenass M, Boller T (2000) Sensing of osmotic pressure changes in tomato cells. Plant Physiol 124:1169–1179

Fincher GB, Stone BA, Clarke AE (1983) Arabinogalactan-proteins: structure, biosynthesis, and function. Annu Rev Plant Physiol 34:47–70

Fleming AJ, McQueen-Mason S, Mandel T, Kuhlemeier C (1997) Local expression of expansin induces the entire process of leaf development and modifies leaf shape. Science 276: 1415–1418

Folgering JH, Sharif-Naeini R, Dedman A, Patel A, Delmas P, Honoré E (2008) Molecular basis of the mammalian pressure-sensitive ion channels: focus on vascular mechanotransduction. Prog Biophys Mol Biol 97:180–195

Gens JS, Fujiki M, Pickard BG (2000) Arabinogalactan protein and wall-associated kinase in a plasmalemmal reticulum with specialized vertices. Protoplasma 212:115–134

Gilroy S, Read ND, Trewavas AJ (1990) Elevation of cytoplasmic calcium by caged calcium or caged inositol *tris*phosphate initiates stomatal closure. Nature 346:769–771

Gorham J, Jones RGW, McDonnell E (1985) Some mechanisms of salt tolerance in crop plants. Plant Soil 89:15–40

Gouget A, Senchou V, Govers F, Sanson A, Barre A, Rouge P, Pont-Lezica R, Canut H (2006) Lectin receptor kinases participate in protein-protein interactions to mediate plasma membrane-cell wall adhesions in *Arabidopsis*. Plant Physiol 140:81–90

Grefen C, Harter K (2004) Plant two-component systems: principles, functions, complexity and cross talk. Planta 219:733–742

Hare PD, Cress WA, van Staden J (1998) Dissecting the roles of osmolyte accumulation during stress. Plant Cell Environ 21:535–553

Hare PD, Cress WA, van Staden J (1999) Proline synthesis and degradation: a model system for elucidating stress-related signal transduction. J Exp Bot 50:413–434

Hasegawa PM, Bressan RA, Zhu J-K, Bohnert HJ (2000) Plant cellular and molecular responses to high salinity. Annu Rev Plant Physiol Plant Mol Biol 51:463–499

Haswell ES (2007) MscS-like proteins in plants. Curr Top Membr 58:329–359

Haswell ES, Meyerowitz EM (2006) MscS-like proteins control plastid size and shape in *Arabidopsis thaliana*. Curr Biol 16:1–11

Haswell ES, Peyronnet R, Barbier-Brygoo H, Meyerowitz EM, Frachisse JM (2008) Two MscS homologs provide mechanosensitive channel activities in the *Arabidopsis* root. Curr Biol 18:730–734

Hayashi T, Harada A, Sakai T, Takagi A (2006) Ca^{2+} transient induced by extracellular changes in osmotic pressure in *Arabidopsis* leaves: differential involvement of cell wall-plasma membrane adhesion. Plant Cell Environ 29:661–672

Hématy K, Höfte H (2008) Novel receptor kinases involved in growth regulation. Curr Opin Plant Biol 11:321–328

Hématy K, Sado P-E, Van Tuinen A, Rochange S, Desnos T, Balzergue S, Pelletier S, Renou J-P, Höfte H (2007) A receptor-like kinase mediates the response of *Arabidopsis* cells to the inhibition of cellulose synthesis. Curr Biol 17:922–931

Hincha DK, Hagemann M (2004) Stabilization of model membranes during drying by compatible solutes involved in the stress tolerance of plants and microorganisms. Biochem J 383:277–283

Hohmann S (2002) Osmotic stress signaling and osmoadaptation in yeast. Microbiol Mol Rev 66:300–372

Homann U, Meckell T, Hewing J, Hütt M-T, Hurst AC (2007) Distinct fluorescent pattern of KAT1::GFP in the plasma membrane of Vicia faba guard cells. Eur J Cell Biol 86:489–500

Ingber D (2003a) Tensegrity I. Cell structure and hierarchical systems biology. J Cell Sci 116:1157–1173

Ingber D (2003b) Tensegrity II. How structural networks influence cellular information processing networks. J Cell Sci 116:1397–1408

Ivashikina N, Becker D, Ache P, Meyerhoff O, Felle HH, Hedrich R (2001) K^+ channel profile and electrical properties of *Arabidopsis* root hairs. FEBS Lett 508:463–469

Johansen JN, Vernhettes S, Hofte H (2006) The ins and outs of plant cell walls. Curr Opin Plant Biol 9:616–620

Jones L, McQueen-Mason S (2004) A role for expansins in dehydration and rehydration of the resurrection plant Craterostigma plantagineum. FEBS Lett 559:61–65

Kacperska A (2004) Sensor types in signal transduction pathways in plant cells responding to abiotic stressors: do they depend on stress intensity? Physiol Plant 122:159–168

Kawasaki S, Borchert C, Deyholos M, Wang H, Brazille S, Kawai K, Galbraith D, Bohnert HJ (2001) Gene expression profiles during the initial phase of salt stress in rice. Plant Cell 13:889–905

Kishor P, Hong Z, Miao GH, Hu C, Verma DP (1995) Overexpression of (delta)-pyrroline-5-carboxylate synthetase increases proline production and confers osmotolerance in transgenic plants. Plant Physiol 108:1387–1394

Kohorn BD (2001) WAKs; cell wall associated kinases. Curr Opin Cell Biol 13:529–533

Kohorn BD, Kobayashi M, Johansen S, Riese J, Huang LF, Koch K, Fu S, Dotson A, Byers N (2006) An *Arabidopsis* cell wall-associated kinase required for invertase activity and cell growth. Plant J 46:307–316

Kohorn BD, Johansen S, Shishido A, Todorova T, Martinez R, Defeo E, Obregon P (2009) Pectin activation of MAP kinase and gene expression is WAK2 dependent. Plant J 60:974–982

Kutschera U (2000) Cell expansion in plant development. Rev Bras Fisiol Veg 12:65–95

Kutschera U (2001) Stem elongation and cell wall proteins in flowering plants. Plant Biol 3:466–480

Kwon YR, Lee HJ, Kim KH, Hong SW, Lee SJ, Lee H (2008) Ectopic expression of Expansin3 or Expansinbeta1 causes enhanced hormone and salt stress sensitivity in *Arabidopsis*. Biotechnol Lett 30:1281–1288

Lang-Pauluzzi I, Gunning BES (2000) A plasmolytic cycle: the fate of cytoskeletal elements. Protoplasma 212:174–185

Lefévre I, Gratia E, Lutts S (2001) Discrimination between the ionic and osmotic components of salt stress in relation to free polyamine level in rice (*Oryza sativa*). Plant Sci 161:943–952

Lemtiri-Chlieh F, MacRobbie EAC, Brearley CA (2000) Inositol hexakisphosphate is a physiological signal regulating the K-inward rectifying conductance in guard cells. Proc Natl Acad Sci USA 97:8687–8692

Levin DE (2005) Cell wall integrity signaling in Saccharomyces cerevisiae. Microbiol Mol Biol Rev 69:262–291

Liu K, Luan S (1998) Voltage-dependent K^+ channels as targets of osmosensing in guard cells. Plant Cell 10:1957–1970

Liu W, Fairbairn DJ, Reid RJ, Schachtman DP (2001) Characterization of two HKT1 homologues from Eucalyptus camaldulensis that display intrinsic osmosensing capability. Plant Physiol 127:283–294

Marshall JG, Dumbroff EB (1999) Turgor regulation via cell wall adjustment in white spruce. Plant Physiol 119:313–319

Martin C, Bhatt K, Baumann K (2001) Shaping in plant cells. Curr Opin Plant Biol 4:540–549

Martinac B (2004) Mechanosensitive ion channels: molecules of mechanotransduction. J Cell Sci 117:2449–2460

Mathews M, Volkenburgh E, Boyer J (1984) Acclimation of leaf growth to low water potentials in sunflower. Plant Cell Environ 7:199–206

McQueen-Mason S, Durachko DM, Cosgrove DJ (1992) Two endogenous proteins that induce cell wall extension in plants. Plant Cell 4:1425–1433

Meckel T, Hurst AC, Thiel G, Homann U (2004) Endocytosis against high turgor: intact guard cells of Vicia faba constitutively endocytose fluorescently labelled plasma membrane and GFP-tagged K-channel KAT1. Plant J 39:182–193

Meckel T, Hurst AC, Thiel G, Homann U (2005) Guard cells undergo constitutive and pressure-driven membrane turnover. Protoplasma 226:23–29

Miller DD, De Ruijter NCA, Emons AMC (1997) From signal to form: aspects of the cytoskeleton-plasma membrane-cell wall continuum in root hairs. J Exp Bot 48:1881–1896

Monshausen G, Gilroy S (2010) Feeling green: mechanosensing in plants. Trends Cell Biol 19:228–235

Moorea JP, Vicre-Gibouinb M, Farrantc JM, Driouich A (2008) Adaptations of higher plant cell walls to water loss: drought vs desiccation. Physiol Plant 134:237–245
Munnik T, Meijer HJG (2001) Osmotic stress activates distinct lipid and MAPK signalling pathways in plants. FEBS Lett 498:172–178
Munnik T, Ligterink W, Meskiene I, Calderini O, Beyerly J, Musgrave A, Hirt H (1999) Distinct osmo-sensing protein kinase pathways are involved in signaling moderate and severe hyper-osmotic stress. Plant J 20:381–388
Nakagawa N, Sakurai N (2001) Cell wall integrity controls expression of endoxyloglucan transferase in tobacco BY2 cells. Plant Cell Physiol 42:240–244
Nakagawa Y, Katagiri T, Shinozaki K, Qi Z, Tatsumi H, Furuichi T, Kishigami A, Sokabe M, Kojima I, Sato S, Kato T, Tabata S, Iida K, Terashima A, Nakano M, Ikeda M, Yamanaka T, Iida H (2007) *Arabidopsis* plasma membrane protein crucial for Ca^{2+} influx and touch sensing in roots. Proc Natl Acad Sci USA 104:3639–3644
Neumann P (1995) The role of cell wall adjustment in plant resistance to water deficits. Crop Sci 35:1258–1266
Nilson SE, Assmann SM (2007) The control of transpiration. Insights from Arabidopsis. Plant Physiol 143:19–27
Orr AW, Helmke BP, Blackman BR, Schwartz MA (2006) Mechanisms of mechanotransduction. Dev Cell 10:11–20
Parker R, Flowers TJ, Moore AL, Harpham NV (2006) An accurate and reproducible method for proteome profiling of the effects of salt stress in the rice leaf lamina. J Exp Bot 57:1109–1118
Parre E, Geitmann A (2005) Pectin and the role of the physical properties of the cell wall in pollen tube growth of Solanum chacoense. Planta 220:582–592
Peters WS, Hagemann W, Tomos DA (2000) What makes plants different? Principles of extracellular matrix function in 'soft' plant tissues. Comp Biochem Physiol 125:151–167
Pont-Lezica RF, McNally JG, Pickard BG (1993) Wall-to-membrane linkers in onion epidermis: some hypotheses. Plant Cell Environ 16:111–123
Proseus TE, Boyer JS (2005) Turgor pressure moves polysaccharides into growing cell walls of *Chara corallina*. Ann Bot 95:967–979
Proseus TE, Boyer JS (2006) Periplasm turgor pressure controls wall deposition and assembly in growing *Chara corallina* cells. Ann Bot 98:93–105
Raymond MJ, Smirnoff N (2002) Proline metabolism and transport in maize seedlings at low water potential. Ann Bot 89:813–823
Reichardt I, Stierhof YD, Mayer U, Richter S, Schwarz H, Schumacher K, Jűrgens G (2007) Plant cytokinesis requires de novo secretory trafficking but not endocytosis. Curr Biol 17:2047–2053
Reiser V, Raitt DC, Saito H (2003) Yeast osmosensor Sln1 and plant cytokinin receptor Cre1 respond to changes in turgor pressure. J Cell Biol 161:1035–1040
Roberts SK (2006) Plasma membrane anion channels in higher plants and their putative functions in roots. New Phytol 169:647–666
Schroeder JI, Hagiwara S (1989) Cytosolic calcium regulates ion channels in the plasma membrane of *Vicia faba* guard cells. Nature 338:427–430
Schroeder JI, Hedrich R (1989) Involvement of ion channels and active transport in osmoregulation and signaling of higher plant cells. Trends Biochem Sci 14:187–192
Schroeder JI, Allen GJ, Hugouvieux V, Kwak JM, Warner D (2001) Guard cell signal transduction. Annu Rev Plant Physiol Plant Mol Biol 52:627–658
Seki M, Narusaka M, Ishida J, Nanjo T, Fujita M, Oono Y, Kamiya A, Nakajima M, Enju A, Sakurai T, Satou M, Akiyama K, Taji T, Yamaguchi-Shinozaki K, Carninci P, Kawai J, Hayashizaki Y, Shinozaki K (2002) Monitoring the expression profiles of 7000 *Arabidopsis* genes under drought, cold, and high-salinity stresses using a full-length cDNA microarray. Plant J 31:279–292
Seo M, Koshiba T (2002) Complex regulation of ABA biosynthesis in plants. Trends Plant Sci 7:41–48

Shabala S, Babourina O, Newman I (2000) Ion-specific mechanisms of osmoregulation in bean mesophyll cells. J Exp Bot 51:1243–1253

Shope JC, DeWald GB, Mott KA (2003) Changes in surface area of intact guard cells are correlated with membrane internalization. Plant Physiol 133:1314–1321

Skirycz A, Inze D (2010) More from less: plant growth under limited water. Curr Opin Biotechnol 21:1–7

Skirycz A, De Bodt S, Obata T, De Clercq I, Claeys H, De Rycke R, Andriankaja M, Van Aken O, Van Breusegem F, Fernie AR, Inze D (2010) Developmental stage specificity and the role of mitochondrial metabolism in the response of *Arabidopsis* leaves to prolonged mild osmotic stress. Plant Physiol 152:226–244

Somerville C, Bauer S, Brininstool G, Facette M, Hamann T, Milne J, Osborne E, Paredez A, Persson S, Raab T, Vorwerk S, Youngs H (2004) Toward a system approach to understanding plant cell walls. Science 306:2206–2211

Staxen I, Pical C, Montgomery LT, Gray JE, Hetherington AM, McAinsh MR (1999) Abscisic acid induces oscillations in guard-cell cytosolic free calcium that involve phosphoinositide-specific phospholipase C. Proc Natl Acad Sci USA 96:1779–1784

Takahashi S, Katagiri T, Hirayama T, Yamaguchi-Shinozaki K, Shinozaki K (2001) Hyperosmotic stress induces a rapid and transient increase in inositol 1,4,5-*tris*phosphate independent of abscisic acid in *Arabidopsis* cell culture. Plant Cell Physiol 42:214–222

Takahashi S, Seki M, Ishida J, Satou M, Sakurai T, Narusaka M, Kamiya A, Nakajima M, Enju A, Akiyama K, Yamaguchi-Shinozaki K, Shinozaki K (2004) Monitoring the expression profiles of genes induced by hyperosmotic, high salinity, and oxidative stress and abscisic acid treatment in *Arabidopsis* cell culture using a full-length cDNA microarray. Plant Mol Biol 56:29–55

Tamura T, Hara K, Yamaguchi Y, Koizumi N, Sano H (2003) Osmotic stress tolerance of transgenic tobacco expressing a gene encoding a membrane-located receptor-like protein from tobacco plants. Plant Physiol 131:454–462

Teodoro AE, Zingarelli L, Lado P (1998) Early changes in Cl^- efflux and H^+ extrusion induced by osmotic stress in *Arabidopsis thaliana* cells. Physiol Plant 102:29–37

Ueda A, Kanechi M, Uno Y, Inagaki N (2003) Photosynthetic limitations of a halophyte sea aster (Aster tripolium L.) under water stress and NaCl stress. J Plant Res 116:65–70

Ueda A, Kathiresan A, Inada M, Narita Y, Nakamura T, Shi W, Takabe T, Bennet J (2004) Osmotic stress in barley regulates expression of a different set of genes than salt stress does. J Exp Bot 55:2213–2218

Urao T, Yakubov B, Satoh R, Yamaguchi-Shinozaki K, Seki M, Hirayama T, Shinozaki K (1999) A transmembrane hybrid-type kinase in *Arabidopsis* functions as an osmosensor. Plant Cell 11:1743–1754

Urao T, Yamaguchi-Shinozaki K, Shinozaki K (2000) Two-component system in plant signal transduction. Trends Plant Sci 5:67–74

Verica JA, He Z-H (2002) The cell wall-associated kinase (WAK) and WAK-like kinase gene family. Plant Physiol 129:455–459

Walbot V (1996) Sources and consequences of phenotypic and genotypic plasticity in flowering plants. Trends Plant Sci 1:27–32

Wang H, Miyazaki S, Kawai K, Deyholos M, Galbraith DW, Bohnert HJ (2003) Temporal progression of gene expression responses to salt shock in maize roots. Plant Mol Biol 52:873–891

Ward JM, Pei Z-M, Schroeder JI (1995) Roles of ion channels in initiation of signal transduction in higher plants. Plant Cell 7:833–844

Williams JE, Brooker IH (1997) Eucalypts: an introduction. In: Woinarski J, Williams J (eds) Eucalypt ecology. Cambridge University Press, Cambridge, pp 1–15

Wohlbach DJ, Quirino BF, Sussman MR (2008) Analysis of the *Arabidopsis* histidine kinase ATHK1 reveals a connection between vegetative osmotic stress sensing and seed maturation. Plant Cell 20:1101–1117

Wojtaszek P (2000) Genes and plant cell walls: a difficult relationship. Biol Rev 75:437–475
Wojtaszek P, Baluška F, Kasprowicz A, Łuczak M, Volkmann D (2007) Domain-specific mechanosensory transmission of osmotic and enzymatic cell wall disturbances to the actin cytoskeleton. Protoplasma 230:217–230
Wolfe J, Steponkus PL (1981) The stress-strain relation of the plasma membrane of isolated protoplasts. Biochim Biophys Acta 643:663–668
Wu Y, Cosgrove DJ (2000) Adaptation of roots to low water potentials by changes in cell wall extensibility and cell wall proteins. J Exp Bot 51:1543–1554
Wyatt SE, Carpita NC (1993) The plant cytoskeleton-cell-wall continuum. Trends Cell Biol 3:413–417
Xiong L, Zhu J-K (2002) Molecular and genetic aspects of plant responses to osmotic stress. Plant Cell Environ 25:131–139
Yan S, Tang Z, Su W, Sun W (2005) Proteomic analysis of salt stress-responsive proteins in rice root. Proteomics 5:235–244
Zhu JK (2000) Genetic analysis of plant salt tolerance using *Arabidopsis*. Plant Physiol 124:941–948
Zhu JK, Liu J, Xiong L (1998) Genetic analysis of salt tolerance in *Arabidopsis*: evidence for a critical role of potassium nutrition. Plant Cell 10:1181–1191
Zonia L, Munnik T (2004) Osmotically induced cell swelling versus cell shrinking elicits specific changes in phospholipid signals in tobacco pollen tubes. Plant Physiol 134:813–823
Zonia L, Munnik T (2007) Life under pressure: hydrostatic pressure in cell growth and function. Trends Plant Sci 12:90–97
Zonia L, Munnik T (2008) Vesicle trafficking dynamics and visualization of zones of exocytosis and endocytosis in tobacco pollen tubes. J Exp Bot 59:861–873
Zonia L, Cordeiro S, Tupy J, Feijo JA (2002) Oscillatory chloride efflux at the pollen tube apex has a role in growth and osmoregulation and is targeted by inositol 3,4,5,6-tetra*kis*phosphate. Plant Cell 14:2233–2249
Zonia L, Müller M, Munnik T (2006) Hydrodynamics and cell volume oscillations in the pollen tube apical region are integral components of the biomechanics of *Nicotiana tabacum* pollen tube growth. Cell Biochem Biophys 46:209–232
Zubrinich TM, Loveys B, Gallasch S, Seekamp JV, Tyerman SD (2000) Tolerance of salinized floodplain conditions in a naturally occurring Eucalyptus hybrid related to lowered plant water potential. Tree Physiol 20:953–963

Plants as Mechano-Osmotic Transducers

Zygmunt Hejnowicz

Abstract Supportive function and expansion of a soft tissue are considered from the energetic perspective. The minimum of the potential energy in a cell determines the cell shape. The minimum of the potential energy in a cylindrical organ composed of turgid tissues, which differ in their elasticity moduli, predicts the occurrence of tissue stresses in the organ. The concept of turgor-driven cell wall extension is reexamined on the assumptions that (1) during stress relaxation the osmotic energy is transformed into strain energy of newly formed cell wall layers, and (2) only the outer cell wall layer undergoes the stress relaxation. This leads to an equation for a relative extension rate different from the rheological equation but also including a threshold turgor pressure. The cases of cell wall expansion that cannot be driven by turgor pressure (formation of intercalary gas spaces, expansion of convoluted anticlinal walls in leaf epidermis, expansion of cell wall invaginations in *Pinus* mesophyll) are described. A hypothesis is presented that in such cases the wall extension is driven by an increased swelling of the inner layer of the cell wall.

1 Introduction

Mature plant organs generally consist of a system of "stiff" scaffolding, mostly lignified, embedded in a "soft" tissue that, nevertheless, can resist deformation. Expanding organs are composed of the soft tissues only, yet are quite stable mechanically. In general, the mechanical significance of the soft tissue in organs decreases with organ age and size; however, even a few meters long leaf petioles in old *Amorphophallus gigas* or *A. titanum* plants, which resemble tree trunks, are composed entirely of the soft tissues, i.e., parenchyma, collenchyma, and epidermis (Hejnowicz and Barthlott 2005).

Z. Hejnowicz
Department of Biophysics and Morphogenesis of Plants, University of Silesia, Jagiellońska 28, 40-032 Katowice, Poland
e-mail: krystynahe@vp.pl

P. Wojtaszek (ed.), *Mechanical Integration of Plant Cells and Plants*, Signaling and Communication in Plants 9, DOI 10.1007/978-3-642-19091-9_10,

© Springer-Verlag Berlin Heidelberg 2011

In engineering terminology, a soft tissue cell is represented by the *hydrostat* – a thin-walled inflatable structure (like a tire or balloon) the mechanical stability of which depends on the internal pressure (P) that brings the wall into tensile stress (σ) approximately r/t higher than the P (r – hydrostat radius, t – thickness of its wall). The hydrostat wall is "soft" (flexible) but being under tension can support a load because any load, also a compressive one, increases only the tension to which the wall is highly resistant.

As the thickness of a hydrostat wall increases the tensile stress diminishes. Engineering theory precludes the hydrostat concept for the wall thickness higher than 20% of the radius (Niklas 1992). No doubt a thin-walled parenchyma cell represents a mechanically efficient hydrostat, but should collenchyma and/or epidermis cells be considered as hydrostats? The collenchyma is defined as a supportive tissue composed of elongated living cells with usually unevenly thickened nonlignified primary walls. The thickenings usually exceed 20% of the cell radius. However, it is rarely realized that the thickenings are mostly composed of water and after dehydration the collenchyma in a cross-section looks like a parenchyma. Even in the above-mentioned huge petioles of *Amorphophallus*, where the collenchyma is the only tissue specialized as a supportive one, the collenchyma cell walls when dehydrated can be hardly regarded as thicker than those of adjoining parenchyma cells (Figs. 19–22 in Hejnowicz and Barthlott 2005). Obviously, water does not transmit the tensile stress in the cell wall and therefore precluding of collenchyma cells as hydrostats on the basis of wall thickness, is incorrect. That such cells are hydrostats follows from the author's own observations (unpublished) that isolated collenchyma strands immersed in water are stiff while they are floppy when immersed in a hypertonic solution. However, the hydrostatic nature of collenchyma functioning within an organ may be unimportant (Spatz et al. 1998; Speck et al. 1998) because it may be in tensile stress due to the tissue stresses (see Sect. 3), independent of its own turgor pressure.

The epidermal cell with thick outer wall fulfills the hydrostat concept because the outer part of this wall is under a considerable turgor-derived tension as indicated by a reversible buckling of the inner layer observed in plasmolysis experiment (Hejnowicz and Borowska-Wykręt 2005; see Sect. 5). It should be also taken into account that the whole organ with primary structure may function as a hydrostat in which the epidermis represents a wall pressurized by the parenchyma (Hohl and Schopfer 1992a; Kutschera and Köhler 1992). This is a hydrostat at the organ level.

Hydrostats used by engineers are usually filled by air (gas). The air is compressible but its amount is constant in a hydrostat, while the solution in a cell is incompressible but the amount of water in the cell can change under pressure; water is taken up in normal *osmosis* and squeezed out in *reverse osmosis*. An inflated tire increases only little its pressure with an increasing load but its shape is changed so that the product of A (the area of the contact between the tire and the ground) and P (the tire pressure) equals the load supported by the tire. Also turgid cells react to an increasing load with a shape change. However, a significant increase in the hydrostatic pressure may occur in the cells squeezing water out of them. The pressure in a cell is an osmotic phenomenon, and an interchange between the mechanical and osmotic energies may occur in a cell if allowed by time, i.e., if the changed load persists for an adequately long time or the change proceeds slowly.

In any supportive system under a dynamic loading, the work done by external forces (load) in generating deformations is temporarily stored within the deformed structures or materials as *potential energy*, which is recoverable upon unloading. In technical elastic materials, the energy is stored as the *strain* energy, classified here as a mechanical one. In the soft tissue of plants, beside the mechanical energy in stretched cell walls, there is also a *potential energy of water*: *osmotic* (in solutions) and/or *swelling* (in meshwork type solids); that can both be classified as *osmotic* (in a wide meaning) energy. If allowed by time, one form of the potential energy may be transformed into another. This is the *mechano-osmotic transduction* referred to in the title of this chapter. The energy transformations are often overlooked, though they are significant for supportive functioning of the soft tissues and for the cell wall expansion. This chapter aims at changing such an attitude. It shows that energetic perspective facilitates better understanding of the soft tissues in their supportive functioning and growth. Attention is paid to the hypothetical role of swelling energy performed in these cases of cell wall expansion that cannot be driven by turgor pressure. Typically, energetic approach needs mathematical formalism, however, this chapter is written so as to render it intuitively understandable.

The terms *osmotic* and *swelling energies* are rather rarely used; however, at least the first term has become more familiar, because the difference in osmotic energy between freshwater from a river and sea water, which attains 28 bars per volume unit, can be used to drive a turbine and generate electricity.

2 General Characteristics of the Potential Energy in Cells

2.1 Energy, Pressure, and Stress

Potential energy due to a work done on an object is stored within the object. This means that there is a factor (force) that tends to "pull" the object back toward some lower energy level.

The *first object* of our interest is water. It can move due to the presence of solutes or materials capable of swelling, due to a hydrostatic pressure, and due to gravity. Thus we can distinguish the following forms of potential energy of water: osmotic (in the narrow meaning of this term, i.e., pertaining to the solution); swelling; hydrostatic; and gravitational. The latter may be disregarded when considering a plant cell, tissue, or a small organ because the position in gravitational field is nearly the same for all its parts. However, when considering a tree, the gravitational effect must be taken into account.

Energy per unit volume is the corresponding pressure or stress (an important relationship, often not appreciated by biologists). We thus have: the osmotic pressure; the swelling stress; and the hydrostatic pressure. To estimate the energy stored in a particular volume of water, the corresponding pressure (stress) should be multiplied by the volume, e.g., osmotic energy of a cell (due to an osmotic pressure of a solution) equals the product of the osmotic pressure and the cell volume.

Osmotic and hydrostatic pressures are components of *water potential*, which is an extremely useful term in many considerations concerning the direction of water movement. However, in this chapter, the two pressures are considered separately, so this term is not applicable.

The *second object* of our interest is the cell wall. The wall moves when stretched. An elastic strain then appears in the wall and tends to bring it back to its original (not stretched) position. The corresponding energy is the *strain energy*. It is calculated per unit volume and has the pressure dimension but is still called the energy.

2.2 Forms of Potential Energy per Unit of Volume (Pressure, Stress) in Soft Tissues

Osmotic pressure (π) is that determined by the Van der Hook equation:

$$\pi = RTc, \tag{1}$$

where c is osmole concentration, T is temperature, and R is the universal gas constant. It should be remembered that this equation gives only an approximation of the osmotic pressure defined in thermodynamics as a quantity proportional to logarithm of a mole fraction of water in the solution (see Nobel 1974).

Hydrostatic pressure in a cell is its *turgor pressure* (P), which is the counter-pressure exerted by the cell wall on the protoplast. In a steady state, P is equal to the osmotic pressure. Here we shall consider the steady state and the processes tending to achieve this state, assuming that the cell membrane is not permeable for solutes, and that there is pure water on the outer side of the membrane (in cell wall apoplast). In the steady state, in the case of nongrowing tissue, the osmotic pressure and turgor pressure are equal, i.e.:

$$P = \pi = RTc. \tag{2}$$

If $P < \pi$ water is taken up by the cell until the steady state is achieved. During this process osmotic energy decreases, thus the process can run spontaneously. If the hydrostatic pressure is increased to above π (for instance due to loading of the cell), the solution must change its concentration to achieve a new steady state. Differentiating (2) we obtain:

$$\mathrm{d}P = \mathrm{d}\pi = RT\mathrm{d}c.$$

For positive $\mathrm{d}P$, $\mathrm{d}c$ must be positive, i.e., the concentration must increase, which means that water must be squeezed out of the cell. Squeezing water out of the cell (protoplast) means a decrease of cell volume (V). It can be shown that $\Delta\pi/\pi \approx -\Delta V/V$. The situation when $P > \pi$ may take place when the cell is loaded. If such a loaded cell is turgid, water may be squeezed out of the cell; osmotic energy of the cell then increases storing a part of the work done by the load. This, however, will happen on the condition that the load affects P. As will be shown later, only a part of the load work is transformed into the cell osmotic energy (most of it is stored in a cell wall as the strain energy) and the process is rather slow because water must flow through the cell membrane.

Swelling stress is a function of water energy in the swelling material, which usually is a sort of an anisotropic meshwork (lattice) of solid. Such a material interacts with water through polar groups attached to internal surfaces. The resulting forces depend on direction, therefore the swelling "pressure" (stress) is a tensor, while the hydrostatic pressure is a scalar. No general equation for the swelling compressive stress can be formulated; however, for a particular material the stress can be measured in different directions, e.g., like in wood. When water is available and a swelling material is unsaturated, the material takes up water. This process can be prevented by imposing adequate compressive stress. In a steady state, there is a certain relationship between this stress and a water content. If the stress is increased above that in the steady state, water is squeezed out of the material. The process is reversible, i.e., in the swelling material, that is under stress, a certain amount of swelling potential energy in stored.

A strict analysis of swelling-induced stresses and strains needs advanced mathematics (see Singh et al. 2003; Huyghe and Bovendeerd 2004; Burgert et al. 2007); however, on such a computational level the swelling stress would not be a useful term in this chapter addressed to biologists.

One may ask whether it is possible that a compartment containing a swelling material, which might be under stress, occurs in plant cells. It is shown below that it may occur in the cell wall. Two of the several variants of such a situation are described:

Variant 1. Imagine a cell wall consisting of two layers, A and B, that differ in an ability of the matrix to swell (matrix in this chapter designates all components of a cell wall beyond the solid core within microfibrils). Let microfibrils be arranged mutually parallel at least in B, and the matrix in A be of a constant swelling ability, while in the B it undergoes a change that increases its swelling ability (e.g., an increase in a number of polar groups in the meshwork due to action of hydrolases). Let the wall be unstressed initially. If the swelling ability increases in B (and consequently the swelling takes places) a compressive stress appears in this layers while in A a tensile stress appears (similarly as in a bimetal with B warmer than A). The stress is perpendicular to the microfibril direction (tends to push them apart). When the compressive stress in B attains a threshold value, the hitherto existing stability is lost and buckling occurs in which the wall becomes bent with B on a convex side releasing the stresses.[1] It is hypothesized that such a case may occur

[1]Buckling is a reversible deformation due to pure compression resulting in bending when a threshold compressive stress is surpassed.

in a cell wall having B on the inner side (an evidence that such phenomenon may indeed take place is presented in Sect. 7.1). Then the wall bends inward. However, if the bending is prevented, the inner layer (B) is under compressive stress while the outer layer (A) is in tension, both stresses are perpendicular to the direction of the parallel microfibrils. This phenomenon can be easily visualized on the example of a piece of a board containing more compression wood on one side than on the other (compression wood swells longitudinally more than a normal wood). When the board has an access to water, it tends to bend. The bending may be prevented by pressing the board flat but then the normal wood is brought under tension that may be so high as to cause cracking of the wood.

In the case of the cell wall of the AB type, the bending may be prevented by the turgor pressure or by gluing of two cell walls as in a case of a double cell wall (BAAB) at the contact of two cells in a "soft" tissue. Then the compression in B and the tension in A may drive the expansion of the wall in the direction perpendicular to microfibrils. Because the B layers, being under compressive stress, tend to expand spontaneously, the A layers undergo a creep. Examples of the occurrence of both the above cases – bending and/or expansion – will be provided in the Sect. 7.

Variant 2. In a typical cell wall, patches of matrix are encased in a microfibrillar network as sketched in Fig. 1. When the network transmits a tensile stress, e.g., due to a loading, the patches are brought under compression. If the patches represent a swelled matrix, the compression may squeeze out water damping the effect of the loading. This explains why when a freshly-isolated strand of collenchyma is stretched in air, a film of water appears on its surface (own observations, unpublished).

Strain energy, U_{su}, is the energy stored (in the case considered here) in a unit volume of strained cell walls (the symbol U_s is reserved for the strain energy in the wall or walls of a particular object, e.g., of a cell). It is equal to the work performed while deforming the wall (its unit volume) from its unstrained state (less any energy dissipated during the deformation). In an ideal case $\sigma = \varepsilon E$; $U_{su} = \int \sigma d\varepsilon = E \int \varepsilon d\varepsilon$.

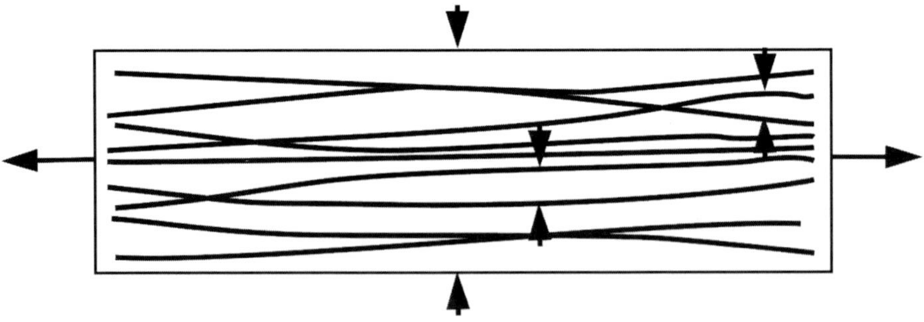

Fig. 1 Schematic diagram of cellulose microfibril network filled with a swellable matrix. *Longer arrows* symbolize the tensile forces, which stretch the network. Pairs of the *shorter arrows* symbolize compressive forces acting on the matrix

Hence:

$$U_{su} = \frac{\varepsilon^2 E}{2} = \frac{\sigma^2}{2E} \tag{3}$$

where σ is stress, ε strain, and E is Young modulus of the cell wall.

The energy in (3) is per unit volume, i.e., U_{su} is a kind of stress. The dimension of U_{su} is the same as that of σ and E (N m^{-2}). It should be mentioned that strain and stress are quantities at a tensor level that, as already mentioned, is not exploited in this chapter (but see chapter "Mechanics of the Meristems").

Since the work required for elastic expansion is stored in cell walls as U_{su}, this energy can be estimated by using the equation:

$$U_{su} = \frac{P\Delta V\rho}{2m} \tag{4}$$

where ΔV is the change in water volume of the cells from turgor-loss to turgid state, m is the mass of the cell wall, ρ the cell wall density. U_{su} is then the energy stored per unit volume. This volume unit can be, however, chosen in a different way, not necessarily as a volume unit of pure cell wall material (which multiplied by the volume of the wall of a single cell would give the strain energy in the cell). For instance, Marshall and Dumbroff (1999) used (4) for mesophyll cells in seedlings of white spruce, taking as a volume unit 100 ml of the mesophyll. They found that the osmotic work (in Joules per 100 ml) varied from ~6 to 64 in seedlings differing in drought tolerance; with a lower value in the case of the increased tolerance.

3 Consequences of the Principle of Minimum Potential Energy in Cells and Tissues. Tissue Stresses

Potential energy of a system always tends to attain the minimum value. This principle follows from the second law of thermodynamics and thus is of a fundamental importance. In a complex system, and such is a plant cell, tissue or organ, there are many local minima of the potential energy at which the system may reside for a long time. The potential energy of plant cells or tissues in an organ is the sum of (1) the osmotic energy, U_o, of the protoplasts; (2) the strain energy of cell walls, U_s; and (3) the swelling energy of the cell wall (for simplicity, this component is not considered in this section). In a steady state, this sum must attain a possible minimum. The minimum means "as much water as possible" in the case of U_o, and "as low stress in the wall and as low cell wall surface area as possible" in the case of U_s. The minimum of the area is a consequence of the fact that the product of the wall surface area and thickness gives the volume of the cell wall, which multiplied by U_{su} gives the total strain energy in the wall. Lower stress in the cell wall means lower turgor pressure, i.e., less water in the cell (at a given amount of solutes). Thus, the tendencies in U_o and U_s with regard to water uptake are opposite.

The osmotic energy is equal to the product of volume and π. The strain energy in turn is a complex function of the cell wall surface area, turgor pressure, and the pattern of elasticity moduli. The wall area and cell volume are related via the cell shape. Thus, the shape of a turgid cell is determined by the minimum of potential energy for a particular pattern of cell wall properties (volume enclosed, area and the pattern of elasticity moduli). Therefore, different properties at a given turgor pressure lead to different shapes (Dumais et al. 2006; Dumais 2007).

Our further consideration will concern simplified (idealized) cells by assuming (1) linear elasticity, i.e., $\sigma = E\varepsilon$; and (2) that the cell wall is characterized by a single E (isotropy and homogeny). This is far from reality but offers the first order approximations of the considered relationships.

For spherical cells under turgor pressure ($P = \pi$), the stress in the wall can be calculated as $\sigma = Pr/(2d)$ and used to estimate the strain energy. The osmotic energy is given by P multiplied by cell volume. It can be easily shown that the ratio of U_o to U_s is:

$$\frac{U_o}{U_s} \approx \frac{3Ed}{rP} \tag{5}$$

where r is radius and d wall thickness. Using the already mentioned data for mesophyll cells of white spruce seedlings (Marshall and Dumbroff 1999), the ratio can be estimated as about 3. It is seen from (5) that the ratio increases with E and decreases with r. Since the sphere has the highest volume at a given surface area, it can be expected that the ratio for nonspherical cells (but with similar parameters) will be rather lower than 3.

To illustrate the importance of the energetic approach, it will be shown that the principle of minimum potential energy applied to a cylindrical organ composed of soft tissues that differ in cell geometry (cell radius) and mechanical properties (elastic moduli) predicts the existence of longitudinal *tissue stresses* in the organ. However, first the relationship between the elastic modulus of a cell wall (E_w) and the modulus of a tissue (E_t) should be considered to enable using of E_t in the prediction.

To introduce the problem let us consider a single cylindrical cell under turgor pressure P. The longitudinal strain (ε) of the cell can be described in two ways (1) at the wall level $\varepsilon = \sigma/E_w$; (2) at the tissue (or cell) level $\varepsilon = P/E_t$. The longitudinal stress in the wall is $\sigma = Pr/(2d)$. Hence: in (1) $E_w = Pr/(2d\varepsilon)$; in (2) $E_t = P/\varepsilon$. It can be seen that the two moduli differ only in a combination of geometrical parameters. From this example, we can conclude that (1) the strain/stress relationship can be considered at the two levels; (2) the consideration at the tissue level renders technical simplification (no need for measurements of r, d); (3) (*a reminder*) at a given pressure or stress, the modulus is the ratio P/ε for the tissue modulus, or σ/ε for the wall modulus (knowing the orders of P or σ, the approximate value of the modulus can be easily estimated; the order of σ is r/d times higher than that of P).

It is known that a single tangent (elastic) modulus from a stress–strain curve is not sufficient to characterize the cell wall (Niklas 1992). Obviously, the same is true for a turgid plant tissue. Nevertheless, the idea of a single (mean) modulus and of a linear dependence between strain and stress facilitates logical analysis of cell wall (or tissue) mechanics.

Because at the tissue level there are two forms of the potential energy (strain and osmotic) and the reversibility of deformation at this level is rather complete (if enough time is allowed), it seems that the analysis of mechanical properties of plants may be simpler at the tissue level (in vivo) than at the cell wall level (in vitro). Yet, most research literature concerns the cell wall level.

The existence of tissue stresses is revealed upon dissection of a stem segment when the outer tissue becomes shorter and the inner tissue longer than the original length of the segment (Schopfer 2006). According to a concept rooted in the classical plant physiology, a tissue like epidermis or/and collenchyma may grow slower than the remaining tissues and is therefore brought under tension. This differential growth would lead to a *tissue tension* (Kutschera 1989; Peters and Tomos 1996; Hejnowicz 1997). Obviously, the tension in one layer must be compensated by compression in another (the epidermis and/or collenchyma are in tension, the parenchyma is compressed), so the plural form *tissue stresses* (TSs) should be used when considering an organ.

Studies of the sunflower hypocotyl (Hejnowicz and Sievers 1995a, b, 1996) showed that the TSs are an unavoidable physical consequence of the variation in structural characteristics of tissue layers or strands in turgid stems. An equation was obtained that related increments of TSs to increments of P and elasticity moduli of the outer and inner tissues in the organ. This equation applied to the sunflower hypocotyl enabled computation of forces involved in generation of TSs, which fitted well to the measured values. The conclusion of these studies is that the structural characteristics of tissues are sufficient to explain fully the TSs that exist in turgid stems and in many typical organs. By no means, however, is differential growth excluded from contributing to TSs but then such a growth can occur only transiently (when continued too long a catastrophic event must occur). Thus the differential growth can be expected in organs highly specialized. For instance, differential growth leads to the high TSs in traps of *Dionaea* (Forterre et al. 2005), in pulvini of *Mimosa*, or in fruits of *Impatiens* (unpublished data).

The occurrence of the structure-based TSs means that in an intact organ there are forces of opposite signs, which act on cross-sectional areas of different tissues. Occurrence of such forces in an intact organ (not as an artifact caused by dissection) is indicated by a characteristic sound (a "snap"), that can be heard when a collenchyma strand is cut within a turgid petiole of celery, and that is similar to the sound generated by a stretched string when cut.

A convincing support for the hypothesis, that the forces related to TSs originate as a consequence of turgor pressure and mechanical heterogeneity of tissues, can be obtained by applying the principle of minimum of potential energy to an organ displaying the heterogeneity. The argument is in form of a logical test: if we assume two forces of the same magnitude but of different signs (F and $-F$) that act on two tissue layers characterized by different E_t in a cylindrical organ, and consider the minimum of potential energy with respect to F, the resulting statement: F is different from zero will prove the hypothesis.

Let us consider a cylindrical organ composed of the dermal tissue (subscript d) and the inner ground tissue (g). Turgor pressure (P) is the same in both layers but the layers differ in tissue elastic modulus (E) so that the moduli are E_d and E_g, and $E_d > E_g$. The reason for the difference in E_{tissue} may be in different cell dimensions (in cross-section) and/or different E_{wall}.

For the logical test, we assume axially directed forces acting on the tissues (F_d, F_g). Static equilibrium makes that the forces differ only in sign, but which one is tensile is left to the test (by a comparison with P, which is known to generate tensile stresses).

The strain energies for the tissues per unit length are:

$$U_{sg} = \varepsilon_g^2 E_g A_g \quad U_{sd} = \varepsilon_d^2 E_d A_d.$$

The strains are:

$$\varepsilon_g = \frac{P}{E_g} + \frac{F}{E_g A_g} \quad \varepsilon_d = \frac{P}{E_d} - \frac{F}{E_d A_d}.$$

The total energy in the organ (per unit length) is:

$$U_s = U_{sg} + U_{sd}$$

$$U = \left(\frac{P}{E_g} + \frac{F}{E_g A_g}\right)^2 E_g A_g + \left(\frac{P}{E_d} - \frac{F}{E_d A_d}\right)^2 E_d A_d. \tag{6}$$

Differentiation with respect to F and comparing to zero (to obtain the minimum) gives:

$$\frac{dU}{dF} = \left(\frac{P}{E_g} + \frac{F}{E_g A_g}\right) - \left(\frac{P}{E_d} - \frac{F}{E_d A_d}\right) = 0.$$

This is fulfilled only when:

$$\frac{P}{E_g} + \frac{F}{E_g A_g} = \frac{P}{E_d} - \frac{F}{E_d A_d}, \tag{7}$$

which means that the strains ε_g and ε_d must be equal. Indeed, adjacent plant cells do not alter their position relative to each other (in contrast to animal organs where integrated growth is accompanied by cell movement), which results in the same longitudinal strain ε across the cylindrical organ (strain or growth symplasticity; Erickson 1986). Thus the (7) excludes differential growth (in the considered case). We obtain further:

$$F = \frac{(E_g - E_d)A_g A_d}{A_g E_g + A_d E_d} P. \tag{8}$$

This is the equation for F, which was assigned to the inner (ground) tissue. The denominator in (8) is always positive and different from zero. Because we assumed that $E_d > E_g$ the numerator is negative, which means that our force that acts on the inner tissue has opposite sign to that of P. Since P is tensile (causes tensile strain), the force that acts on the inner tissue is compressive. Thus the force that acts on the dermal tissue is tensile, as expected for TSs. We can conclude that (1) if a turgid organ is heterogenic with respect to mechanical properties, and is characterized by strain symplasticity, TSs will appear; (2) the tissue that is characterized by higher stiffness will be under tensile stress; (3) if the organ is homogenic, i.e., $E_d = E_g$ and is characterized by strain symplasticity, no structure-based TSs will appear.

It is worth mentioning that the equality in magnitude of the forces does not imply equality in magnitude of the compressive and tensile tissue stress because the stress depends also on the area. Tensile TS may attain very high value if the area of stiffer tissue is small.

The symplasticity implied by (7) concerns the elastic strain. However, it is obvious that if the strain has two components – elastic and plastic, as it is in the case of elongating (growing) organ, the symplastic growth (total strain is the same across the organ) allows for some difference in plastic strain that is compensated by an opposite difference in elastic strain across the organ. This would be the case of the TSs originating from differential growth. We see, however, that TSs may originate also without differential growth. Passioura and Boyer (2003) showed that TSs of that type conspire to uncouple the water potential of the epidermis from that of xylem in elongating plant stems. Because the uncoupling (independence of the epidermis water potential from that of xylem, when the water potential of xylem was changed) is a fact (Nonami et al. 1997) the mathematical logic of Passioura and Boyer study supports the concept that TSs may originate without differential growth.

The equation used in studies of TSs in sunflower hypocotyl (Hejnowicz and Sievers 1996) formulated for finite elements of strain, was similar to (8).

4 Potential Energy in Plant Cells or Tissues on Loading and Unloading

In general, a deformation in a turgid plant cell (tissue, organ) due to loading is fully reversible, i.e., the deformation disappears completely on the conditions that enough time is allowed for recovery and water is available. This is true for both growing and mature "soft" organs. The full reversibility is often not adequately appreciated because the two conditions mentioned above are often not fulfilled.

As already mentioned, any reversible deformation requires potential energy that can be used in the deformation reversal. This means that the work (W) of the force (load) that causes deformation characterized by the displacement (u), W = force $\cdot$ u, is transformed into U_s and U_o, i.e., at least one of them increases (if changes in swelling of cell wall matrix occur the corresponding energy may be considered as a part of U_s). Of course, $U_s + U_o \leq W$. Increase of U_s means an increase of cell wall tension, while increase of U_o means an increase in π, i.e., a decrease of the cell volume, hence squeezing water out of the cell.

The work connected with any very fast displacement caused by loading can be transformed only into U_s because the changing of U_o is a time-consuming process, since it necessitates water flow through the cell membrane. How much U_o will increase depends on: (1) how much the load increases turgor pressure (P); (2) water conductivity of cell membrane (L); and (3) time. When L and time are not limiting factors, the increase in P (with respect to that in a steady state) results in squeezing out water (reverse osmosis) and U_o increases.

In general, in the case of hydrostats, their shape on loading changes so that it allows for a decrease of volume, independently of whether the loading is compressive or tensile. However, this volume decrease is often relatively small.

The deformations in plants on loading and unloading are usually described in terms of viscoelasticity as a combination of the Hookean elastic elements (springs) and Newtonian viscous elements (dash-pots with irreversible movement of a fluid) (Finney et al. 1964; Dorrington 1980).

For the Hookean element the load (F) is proportional to displacement (u) ($F \propto u$) and does not depend on time, while for the Newtonian element the load is proportional to the rate of displacement ($F \propto \mathrm{d}u/\mathrm{d}t$). Hence, any combination of the elements gives the displacement as a function of time. If after an abrupt change of stress or strain, the strain is kept constant, the stress decreases with time. This is designated as a *stress relaxation*. When stress is kept constant, the strain increases – the process designated as a *creep*.

On loading of a turgid cell (tissue, organ) there is instantaneous increase in U_s followed by a slow decrease in U_s. This decrease has two components: *dissipation* of U_s related to an irreversible fluid flow (the dash-pot element); and *transformation* of U_s into U_o. The former always occurs in viscoelasticity (represented by hysteresis loop of stress/strain curves). The latter may occur in turgid cells and is related to an reversible flow of water out of the cell.

If the tissue is mounted on a tensiometer working in isotonic regime (constant stress during recording) and is loaded step-up in stress, the creep means a slow decrease in U_s; a corresponding decrease in stress is interpreted as the stress relaxation. Stress relaxation is typically considered as an irreversible process. However, it may include a transformation of U_s into U_o in terms of squeezing water out of cells that is a reversible process.

On unloading there is an instantaneous decrease of U_s and time-dependent decrease of U_o manifested by water uptake to make the reversal complete. However, for the latter, the cells must have an access to water. If not, the tendency to decrease U_o remains unfulfilled until water is provided. In case of a long delay in water provision, one has an impression that there is a long "memory" of the initial state in the tissue. Thus one of effects of the coexistence of osmotic and strain energies in plants is the memory.

As already mentioned, the change in U_o on loading depends on how much the load increases the turgor pressure. This is a complicated dependence. The change in P is the function of a load change that depends upon many factors: cell shape; cell wall structure; cell arrangement in a tissue; intercellular spaces; and the way of loading, e.g., tensile or compressive. It depends also on the rate of load change.

To illustrate the dependence of U_o change upon the cell wall structure, let us consider the case in which U_o does not change because the volume (*vol*) of cylindrical cells, with radius (r), remains constant on loading; the cells become wider but shorter. This is possible if helically arranged cellulose microfibrils are deformed by the load with no appreciable change in microfibril length (l). Let us consider a segment of the cylinder corresponding to one turn of the helix, and denote the length of the segment as h. The volume of this segment is the product of h and cross-sectional area of the cells. If the angle between longitudinal direction and the microfibrils is denoted as α, we can express h and r as function of α. We obtain:

$$vol = \frac{d^3}{4\pi} \sin^2\alpha \cos\alpha \tag{9}$$

By differentiating with respect to α, we find the extremum of the function given by (9), i.e., we find the α at which the volume remains approximately constant when the segment becomes slightly shorter but wider (or in other words such α that when its value slightly changes the volume remains approximately constant). The value of such an angle is $\alpha = 54^\circ 45'$. At this angle, the turgor pressure on loading remains approximately constant, i.e., the whole work of the load is stored as U_s. On the other hand, when the helix is very flat, the width of the cells on loading does not increase. However, the cell length decreases to a maximum extent, i.e., the cell volume is maximally decreased and turgor pressure is maximally increased. The work of the load is then stored in U_o.

In general, the turgor change may vary from negligible to such that should be taken into account when considering what happens with the work done by load. When a cylindrical block of parenchyma confined laterally is compressed longitudinally, the turgor change may be close to the pressure exerted by load. In such a case, the osmotic energy may be the main storage for the potential energy of deformation, but the load must act adequately long. Let us consider cells under compression on a concave side of stems and leaves, which are bent by a wind blowing from one direction. Initially, the work of bending is stored in the cell walls as U_s, then it is transformed partly into U_o. If after such a transformation the load (wind drag) is released, the elastic recovery is slow, because decrease in U_o requires water uptake. This slowness is desirable ecologically because the aerodynamic change in shoot shape caused by the wind is then allowed to “wait” for the next wind gust. However, the deformation is fully reversible.

If loading significantly increases U_o, the portion of water that is squeezed out of the tissue into apoplast can be used for other purposes (e.g., for transpiration) so that on unloading there may be not enough water for removal of the deformation.

We have thus to deal with viscoelasticity in which the reversibility is dependent on the access to water. It is proposed to designate this class of viscoelasticity as *osmoelasticity*. There is also another reason to distinguish osmoelasticity. While in typical viscoelasticity, the energy of deformation is stored in form of the strain energy, in osmoelasticity there is an additional form of the stored (potential) energy – the potential energy of water (osmotic and/or swelling). Specifically, osmoelasticity would be the class of viscoelasticity that concerns plant cells (tissues, organs).

The term *osmo-* in osmoelasticity is taken in its wider meaning pertaining to the potential energy of water. Thus osmoelasticity does not necessarily have to be connected with the presence of living cells whose cell membranes enable the osmosis. It is applicable also in the case of walls of dead cells, due to swelling of the wall matrix as sketched in Fig. 1. The longitudinal stress in the fibrillar network squeezes out water from the matrix, resulting in phenomena similar to those occurring on loading and unloading in living tissues. Thus the osmoelasticity would embrace all cases of elasticity in which water potential energy must be taken into account. Everybody knows that wet twigs break differently from dry twigs, especially when fast rates of breaking are considered. The same concerns rods of wood. It is convenient to consider the elasticity of dry and wet twigs (or wood) in two separate classes: visco- and osmo-elasticity.

5 Mechano-Osmotic Transductions at Turgor-Driven Cell Wall Expansion

Plant cell growth is defined as irreversible increase in cell volume and surface area of the cell wall, where volume increase means a corresponding uptake of water. The latter demands that $\pi - P > 0$. It is generally accepted that this is achieved by a decrease in P (Boyer 2001) caused by stress relaxation in the wall (see Schopfer 2006). In seed plants, the relaxation is due to chemical loosening of the wall in which process a special role is played by expansins as primary loosening agents, and by some enzymes as secondary agents (see Cosgrove 2000). The former are effective in loosening if pH is adequately low (acid growth hypothesis: Rayle and Cleland 1970; Lüthen et al. 1990; Kutschera 1994). In charophytes another mechanism for the relaxation emerges from the studies of Boyer's Group. The tensile strain distorts some of the wall pectate, weakening its calcium cross-links and causing the calcium to be preferentially lost to still not distorted pectate (Proseus and Boyer 2007).

At the stress relaxation, the pressure is lowered so that the osmotic pressure is not compensated by turgor pressure ($\pi - P = \Delta\pi > 0$). This difference drives water uptake while metabolic delivery of solutes at continued wall loosening and water uptake keeps the $\Delta\pi > 0$. Let us designate the above-sketched concept as *stress relaxation by wall loosening*. The relative rate of water uptake determines the relative rate of volumetric growth $dV/(Vdt)$. The growth rate can be thus considered as a function of $\Delta\pi$ and hydraulic conductance. On the other hand, it is generally accepted that growing cell wall yields under the turgor pressure in excess of a critical pressure (the yield threshold, P_y) according to the rheological equation of plastic flow, introduced to plant biology by Lockhart (1965):

$$\frac{dV}{Vdt} = m_v(P - P_y) \quad \text{if } P \geq P_y \tag{10a}$$

$$\frac{dV}{Vdt} = 0 \quad \text{if } P < P_y,$$

where m_v is designated as extensibility coefficient for volumetric growth.

An equivalent of (10a) for the relative rate of elongation is:

$$\frac{dl}{ldt} = m_l(P - P_{yl}) \quad \text{if } P \geq P_y \tag{10b}$$

$$\frac{dl}{ldt} = 0 \quad \text{if } P < P_y,$$

where m_l and P_{yl} refer to unidimensional expansion.

Since P is a requisite for tissue (organ) stiffness it can be expected that the stress relaxation accompanying growth will result in lowering the stiffness of a growing organ. Thus if the organ grows periodically, there should be a corresponding periodicity in its stiffness. The author's own observations (unpublished) do not support such an expectation. A more serious doubt whether the concept of *stress relaxation by wall loosening* is complete enough to describe growth arises when energetic aspect is considered. A question arises into what the osmotic energy is transformed when π is lowered due to water uptake? Obviously, a cell wall expansion is accompanied by the formation of new wall material, and the osmotic energy, which decreases when water is taken up, is transformed into U_s of the increasing strain of the cell walls. This is not explicitly indicated by the *stress relaxation by wall loosening* concept. It is so because the concept does not take into account (at least not fully) the two facts (1) that new layers of cell wall are formed during the wall expansion; and (2) that there is a gradient of stress across the wall.

The innermost cell wall layer, just deposited on the expanding wall, is obviously unstressed. The adjacent, earlier formed layer is already slightly stressed due to the strain that occurred after its formation. In general, the stress increases if one moves outward until it attains the value at which a loosening process starts. As an outer layer undergoes stress relaxation (and a new layer is deposited on the other side), water is taken up, and the remaining layers expand elastically keeping the stress gradient. Observe that this expansion is reversible. However, the sequentially deposited new layers are longer and longer (at instants of their deposition), which means that the cell grows in length irreversibly. This growth process cannot be described by visco-*plasticity* (rheology). The irreversible expansion of a growing cell wall pictured above is not a viscous or plastic effect like creep across the wall. It is mainly a physiological process connected with the formation of new wall layers, while the rheological process is limited to the outer layer of the wall. Particular layers become thinner and thinner when expanding and a collapse of such layers on the loosening may be of "the loss of stability" type (Wei and Lintilhac 2003, 2007). However, it should be stressed that the loss of stability concerns only the outer layer of the wall, not the whole wall.

The occurrence of stress gradient in cell walls of turgid cells is indicated by a pattern of transversely oriented folds visible on inner surface of the cells after loss of turgor (Hejnowicz and Borowska-Wykręt 2005). The appearance of the folds is a reversible phenomenon as demonstrated by plasmolysis and deplasmolysis. Electron microscopy discloses a gradient in the folds' amplitude across the cell wall: the highest amplitude is in the innermost layer, the outer layers are not folded. The folds are apparently caused by sinusoidal-type buckling of the inner part of the wall due to compressive stress exerted on this wall part by the outer part when it shrinks after loss of turgor.

The occurrence of the folds is a known phenomenon, in electron microscopy described as "the wash-board-like structured inner cell wall surface" (Kutschera et al. 1987) or "transverse ribs" (Bergfeld et al. 1987). However, being assumed to be a result of uneven deposition of wall material they did not deserve due attention. The occurrence of the folds ("Querstreifung," i.e. transverse striation) in epidermal cells has been noticed already in the nineteenth century by Correns (1892), who, however, interpreted them in the same way as the helical streaks ("Spiralstreifung") observed in fibers due to a variation in water content and in composition of the wall (fibrillar structure had not yet been known). However, Strasburger (1889) already had suggested that the "Querstreifung" may be a folding of wall lamellae resulting from the contraction of other parts. This hypothesis was proven by Heyn (1933), who showed that the folding of the inner surface of tangential and radial walls of epidermal cells takes place when the epidermis is detached from the coleoptile, and that the folds (crinkles) disappear when the dead epidermis is stretched. Heyn concluded that the outer layer has a greater elastic extension in the turgid cell, and becomes shorter than the inner layer when the tensile strain disappears.

The hypothesis of the existence of a stress gradient in a growing cell wall takes into account both the stress relaxation in the material already existing on one side of the cell wall, and the deposition of new wall material on the other side. The latter process is underestimated in *stress relaxation by wall loosening* concept, though several recent studies are challenging such a view (Proseus and Boyer 2006a, b; Refrégier et al. 2004). The hypothesis is thus a variant of the *stress relaxation by wall loosening* concept. Let us designate it as *stress gradient* variant (SG) of *stress relaxation by wall loosening*.

Both in the original variant and in the SG variant, the stress relaxation is considered but its localization is different. In the SG (1) there is no global stress relaxation across the wall but only the outer wall layer is loosened; (2) the longitudinal expansion of the existing cell wall is elastic but the successively formed new layers are longer and longer. It is the increasing length of newly formed layers that gives the irreversible longitudinal expansion of the whole wall.

Let us evaluate the rate of this expansion, $\mathrm{d}l/(l\mathrm{d}t)$, assuming (1) there is a gradient of stress from zero in the newly formed layer to σ_{los} in the layer in which loosening begins; (2) σ_{los} is a critical value that must be attained to bring about the chemical loosening of wall material; and (3) the Hookean law $\varepsilon = \sigma/E$ holds.

Each layer expands from l_{o} to $l_{\max}$ when it is displaced across the cell wall thickness, where l_{o} is its initial length and $l_{\max}$ is the length at the instant of the loosening. Let the duration of this displacement be τ. Thus the rate of the expansion is $(l_{\max} - l_{\mathrm{o}})/\tau$. This rate divided by l_{o} gives the relative rate of expansion in longitudinal direction, $\mathrm{d}l/(l\mathrm{d}t)$;

$$\frac{\mathrm{d}l}{l\mathrm{d}t} = \frac{l_{\max} - l_{\mathrm{o}}}{l_{\mathrm{o}}\tau} \tag{11}$$

But $l_{\max} - l_{\mathrm{o}}/l_{\mathrm{o}} = \varepsilon_{\max}$, where $\varepsilon_{\max}$ is the elastic strain of a wall layer undergoing loosening. Thus:

$$\frac{\mathrm{d}l}{l\mathrm{d}t} = \frac{\varepsilon_{\max}}{\tau} \tag{12}$$

From Hookean law we have:

$$\varepsilon_{\max} = \frac{\sigma_{\mathrm{los}}}{E} \tag{13}$$

Thus:

$$\frac{\mathrm{d}l}{l\mathrm{d}t} = \frac{\sigma_{\mathrm{los}}}{E\tau} \tag{14}$$

The wall stress (σ) is a linear function of turgor pressure, $\sigma = \kappa P$. The coefficient κ relates the stress (acting in cell walls in the longitudinal direction) to P. It depends on the cell geometry. We thus obtain:

$$\frac{\mathrm{d}l}{l\mathrm{d}t} = \frac{\kappa P_{\mathrm{los}}}{E\tau} = \frac{m P_{\mathrm{los}}}{E} \tag{15a}$$

where $m = \kappa/\tau$ may be designated as extensibility coefficient. It depends on the cell geometry and also on factors related to the duration (τ) of the period during which a layer is displaced across the wall.

The equation (15a) is valid when turgor pressure attains the value P_{los}. If $P < P_{\mathrm{los}}$ the stress relaxation does not occur, water is not taken up and the rate is zero. Thus (15a) should be rewritten in the form:

$$\frac{\mathrm{d}l}{l\mathrm{d}t} = \frac{m P_{\mathrm{los}}}{E} \quad \text{if } P = P_{\mathrm{los}}$$

$$\frac{dl}{l dt} = 0 \quad \text{if } P < P_{los}. \tag{15b}$$

Obviously, there must be a certain range of P_{los}. The P_{los} in (15a) would be the value at the upper limit of this range while the P_{los} in the inequality $P < P_{los}$ would be the value at the lower limit. Within this range, the growth rate would jump from zero to the normal value.

What would happen if P were increased above P_{los} assuming that the (15b) is valid (such an increase of P can be experimentally achieved by fluid injection into the cell from pressure probe)? If relations on which (15b) is based are linear, we can guess that increase of the growth rate (r) should be proportional to $\Delta P/P_{los}$, i.e., $\Delta r = r_o \Delta P/P_{los}$, at least transiently (where r_o is a normal growth rate). Since $\Delta P/P_{los}$ is a small value also Δr is small, which fits well to the data from several studies where increase in growth rate after increase of P was very small (Green et al. 1971; Zhu and Boyer 1992).

The equation (15b) can be compared with the Lockhart's equation (10b) that is basic in the original version of *stress relaxation by wall loosening* concept. Both are characterized by a threshold value for P, as it has been experimentally supported (see Taiz 1984; Cosgrove 1988; Hohl and Schopfer 1992a; Zhu and Boyer 1992; Proseus and Boyer 2006a, b). However, the SG variant explains better the cases where a very small reduction of turgor results in an immediate cessation of growth (Taiz 1984).

Many studies concerned growth rate (or strain rate in vitro) for P above the threshold (see Taiz 1984; Hohl and Schopfer 1992b; Proseus and Boyer 2006a, b). In the case of *stress relaxation by wall loosening,* the increase of rate should be proportional to $\Delta P/(P_o - P_y)$, where P_o is the turgor pressure at normal growth. That proportion is higher than $\Delta P/P_o$. It appears that the growth rate increases with P less than expected on the basis of *stress relaxation by wall loosening* concept. However, the interpretation of experimental data is difficult because (1) growth should be distinguished from elastic changes what is rarely done (Proseus et al. 1999); and (2) the proportion $\Delta P/(P_o - P_y)$ occurs only at constant m and P_y. Any deviation from the mentioned proportion can be explained by taking m or P_y as a function of P.

An intriguing feature of the SG variant is that the "driving force" is turgor pressure divided by elastic modulus: the lower the E, the lower the P to get the same growth rate. This generally seems to be correct though the profile of elastic properties does not correlate with the gradient of growth rate (Beusmans and Silk 1988; Edelmann 1995; Peters et al. 2001), which indicates that the rate or P_{los} do not depend only on E.

However, this chapter does not aim in evaluation of different growth concepts. The SG variant was presented here because it arose in answering the question into what the osmotic energy is transformed when π is lowered due to water uptake by growing cells. It was then necessary to take into account that newly formed layers of cell wall are initially unstrained.

Now we turn to these cases of cell wall expansion that cannot be driven by turgor pressure. A hypothesis will be presented that in such cases the swelling energy should be taken into account.

6 A Hypothesis that Swelling Stress May Drive the Cell Wall Expansion

In Sect. 2, a possible variant of the occurrence of swelling stress in a cell wall is presented in which two layers (A and B) of the wall differ in the ability to increase swelling and in microfibrillar structure. It is assumed that in the layer B the swelling of the matrix may be increased and microfibrils are mutually parallel, while in the layer A the swelling remains unchanged and microfibrils are unordered. It is hypothesized that such a case may occur in a cell wall having B on its inner side with microfibrils oriented transversely to the direction in which the wall is to expand. Then the layer A represents the outer wall portion with microfibrils disordered by a previous expansion. When the swelling increases in the inner layer, this portion is brought under a compressive stress in the direction perpendicular to the microfibrils. The static equilibrium condition demands that the sum of forces is zero. Hence the outer layer of the cell wall must be under tensile stress such that the forces related to the two opposite stresses compensate each other. When a threshold stress in the inner layer is surpassed, the wall buckles releasing the stresses. One possible mode of the buckling is bending inward (convex side toward cell lumen). Another mode is that of a sinusoidal-type occurring in isolated and not turgid (killed or plasmolysed) cells as described in Sect. 5. However, if the bending is prevented, the inner layer remains under compressive stress while the outer one is in tension, both stresses are perpendicular to the direction of the parallel microfibrils.

In the case of the double cell wall between two cells, e.g., the anticlinal wall in the epidermis, the structure AB becomes BAAB in which B is under compressive stress and A under tensile stress. This means a double gradient of stresses from a negative value in B to a positive value in A. The magnitude of the stresses cannot be high, however, since the elastic (Young) modulus has a relatively low value for low stress (Niklas 1992), the weak tensile stress in A may cause quite a significant strain in this portion of the wall. The bending is prevented as long as the middle lamella remains intact. It is hypothesized that the swelling-induced stresses in such a double wall may drive its expansion in the direction perpendicular to microfibrils in B. The B layer tends to expand spontaneously but A restrains this tendency. Consequently, the rate of this expansion is determined by the rate of creep in A. The expansion decreases the stresses and makes the wall thinner. The stress relaxation would be compensated by sustaining of the swelling, the thinning – by addition of new lamellae to B.

Similar mechanism of expansion may act in a single wall in a bent state, like that lining a gas space. The buckling that brought about bending released the stresses, but if the swelling of the inner layer (now on the convex side) is continued, the curvature increases or, if this process is restricted, the wall expands in the direction perpendicular to the microfibrils in B, with the rate determined by the creep rate in A, resulting in enlargement of the gas space. There may be different particular reasons why the swelling might increase (or might be continued) in B, but first of all it should be realized that B being the inner layer – by its very position – is the first target for the action of enzymes and other factors that are secreted by the protoplast and that may increase the number of polar groups in interstices of the cell wall matrix. Thus, if the cell wall contains substances in which the density of the polar groups may be increased, and the cell secretes factors, which can lead to this increase, a tendency to increased swelling in the inner portion of the wall is a natural consequence. The particular reasons would be a deposition of normal or special substances on the inner surface of the cell wall, and secretion of normal or special factors that increase the number of polar groups in these substances. The substances and factors that might be taken into consideration in relation to the presented hypothesis may be exemplified as: β-1,3 glucan (callose) and β-1,3 glucanase (Maltby et al. 1979; Jaffe and Leopold 1984; Parre and Geitmann 2005); xyloglucans and their transglycosylases and hydrolases (Rose et al. 2002; Verbelen et al. 2001); and expansins (Cosgrove 2000). Some microorganisms produce an expansin-like protein that causes swelling of fibers, named *swollenin* (Kenealy and Jeffries 2003). Maybe proteins specialized in swelling of cell wall substances are also produced by plants.

7 Examples of Cell-Wall Expansion that Cannot be Explained by Yielding Under Turgor-Born Tensile Stress: A Possible Role for Swelling Stress

7.1 Intercalary Gas Spaces

A common belief concerning the formation of intercellular gas spaces [gas instead of air is used because often the only gas in the space is CO_2 (Burström 1959)] is that they initiate by opening at T-shaped or X-shaped wall configurations (in 2D), at which three or four cell edges meet (in 3D). At such a place, the walls of neighboring cells tend to pull apart due to turgor-derived tensile stress rounding out the cells. The opening is quite a complicated process because the cell wall/gas interface is hydrophobic from the very beginning (Roland 1978). It is worth noting, however, that:

(1) A closer examination of newly formed gas spaces at the corners of cylindrical cells viewed in cross-sections through hydrated material indicates that cell walls lining the new spaces are concave toward the space, instead of the expected concavity toward cell lumen. In 3D, such gas spaces are in form of narrow but long cylindrical canals oriented perpendicularly to the section.
(2) The microfibrils in the inner layer of cell wall along the cell edge (corresponding to the corner in 2D) are longitudinally oriented with respect to the axis of cylindrical cells, though in the remaining part of the walls they are transverse (Wardrop and Cronshaw 1958).

(3) Gas spaces can be also initiated at *intercalary* position, which is remote from the corners in 2D (Hiller 1872; Weston and Cass 1973; Panteris et al. 1993). An intercalary initiated gas space has, at least on one side, a concave lining wall (concave toward the gas space). The arrangement of microfibrils in the walls lining the intercalary gas space, can be judged from studies of Panteris et al. (1993) on the development of undulated epidermis with intercalary gas spaces of *Vigna* leaves. They showed that the intercalary gas space, visible in inner paradermal sections through the epidermis, is lined by a wall in which the microfibrils run anticlinally, i.e., normal to the epidermis plane.

The formation of the concavity of the wall lining a gas space cannot be explained by a turgor-driven mechanism, but can be easily explained by the hypothesis proposed in Sect. 6.

In the case of the cylindrical gas space formed along cell edges, the (single) cell wall lining the space has longitudinal microfibrils at least in its inner layer. If the matrix between microfibrils in the inner layer swells, the resulting compressive stress brings about the buckling-type bending concave toward the gas space. The lateral expansion of the concave walls lining gas spaces could be driven by the swelling-born stresses according to the proposed hypothesis. In the case of the intercalary gas spaces, the buckling would be also the mechanism of opening (formation) such a space.

A special type of intercalary gas space formation is that of the pore between guard cells in a stoma (reviewed by Galatis and Apostolakos 2004). The simplest explanation of the pore formation, though not considered as yet in literature, is that the ventral wall of a young guard cell, which is usually connected by middle lamella to the sister cell, develops compressive stress in the inner wall portion. This stress causes wall bending with convexity directed to the cell lumen (concave toward the originating pore). This bending, accompanied by weakening of the middle lamella, would lead to the pore formation. The microfibrils in the ventral wall of a young guard cell run anticlinally thus allowing for the predominantly tangential compressive stress.

7.2 Undulated Epidermis

A characteristic feature of ordinary epidermal cells in leaves of many dicots, some monocots and almost all ferns, is the undulated contour of their anticlinal walls. The undulations occur either along the whole depth of the walls as in *Vigna sinensis* (Panteris et al. 1993) or only at their external part as in *Cyperus papillus* (Panteris et al. 1994). Different intermediate cases may also occur. During the development of undulated epidermis, wall thickenings, called pads, are deposited at least at the junctions of the external periclinal wall with the anticlinal ones, as in *Cyperus*, or being longer than the depth of the cell, they cover the anticlinal wall and extend beyond the opposite junction on the internal periclinal wall. The microfibrils in the pad on the anticlinal wall are mutually parallel and perpendicular to the epidermal surface (transverse to the length of the anticlinal wall). In the pad on periclinal wall, the microfibrils run radially (with respect to the pad location). The pads in one cell alternate with those of the neighboring cell. Restricting the growth along the microfibrils, the pads impose differential expansion of the periclinal walls, a consequence of which is an undulated outline of the periclinal cell wall and undulated anticlinal wall.

Let us consider the wall stresses and strains due to turgor pressure in an epidermal cell with undulated walls. No doubt, the periclinal walls (the outer and the inner) are under tensile stress in all tangential directions. The anticlinal walls are under tensile stress in anticlinal direction. The tensile stress in the outer periclinal walls and that in anticlinal direction in anticlinal walls is manifested by indentation of the outer surface above anticlinal walls. However, the anticlinal wall that is undulated at the whole depth (i.e., the wall is like an undulated band) is free of tensile stress in tangential (periclinal) direction, let us say along its *length* (as far as turgor-derived tension is considered), i.e., the wall is not stretched in the direction along which it should expand during leaf blade growth. Yet the undulated walls grow in length. The undulation becomes even more pronounced as the leaf blade expands (unpublished data of D. Kwiatkowska for *Arabidopsis* leaves), which means that the relative rate of an anticlinal wall elongation is not lower than the relative rate of growth of the periclinal walls in the direction approximating the anticlinal wall. The multinet concept (Refrégier et al. 2004) seems to be fully applicable to the undulated anticlinal walls; the microfibrils are arranged predominantly anticlinally on the inner side of the wall (transversely to its length) and randomly close to the middle lamella. On the other hand, the growth of the undulated walls in anticlinal direction (perpendicularly to epidermis surface) is very limited, in spite of the fact that in this direction the wall is stretched by turgor pressure.

An anticlinal wall is a double cell wall. If the outer and inner layers (portions) of a wall are denoted by A and B, respectively, the double anticlinal wall has the structure BAAB.

What may be the mechanism of the tangential (periclinal) expansion of the undulated walls? Since the wall is compressed between two epidermal cell protoplasts (turgor pressure) and presumably new lamellas are continuously added to the wall, the simplest mechanism could be a kind of a plastic expansion under compression (as during rolling out a dough). More realistic version of wall enlargement under compression would be the mechanism sketched by Proseus and Boyer (2005) on the basis of in vitro experiments with isolated walls of *Chara* internodes attached to glass capillary. Solution of dextran or suspension of gold colloids were pushed into the internode lumen. The large molecules or particles moved into the wall at the pressure value generally accepted to extend the wall for growth (a sort of intussusception). Such a mechanism probably lays at the very basis of undulated wall extension, however, it needs some additions pertinent to the stress–strain relationship in the extending wall.

Proceeding further, let us observe that in petals of some species, the anticlinal walls of the upper epidermis are not only undulated at the whole depth, but also the

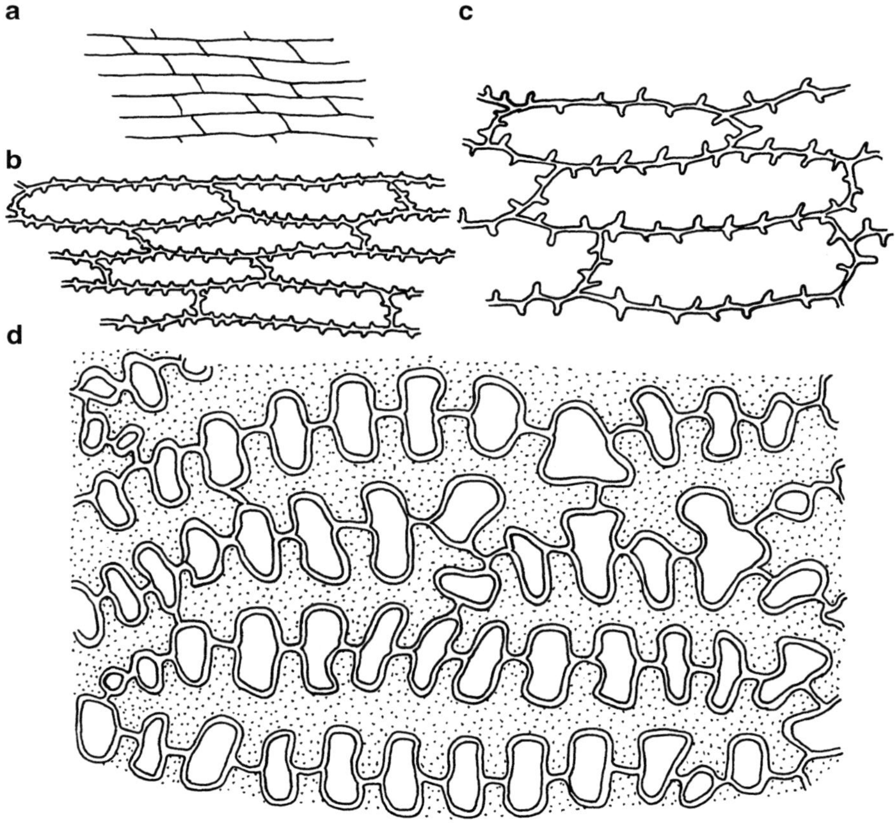

Fig. 2 Paradermal sections of the adaxial epidermis of *Linum usitatissimum* petals in four developmental stages. (**a**) A young bud. (**b**) An older bud. (**c**) Epidermis of the flower shortly before blossoming. (**d**) Epidermis of an open flower; cell lumens are dotted for easier recognition (based on Hiller 1872)

epidermis has numerous intercalary gas spaces, as exemplified by *Linum* (Fig. 2). The formation of these spaces can be explained by the hypothesis of swelling, as indicated in the previous section. It is thus appropriate to propose this hypothesis also for the expansion of the undulated anticlinal walls.

The anticlinal wall is fastened along its edges at the periclinal walls. As the periclinal walls extend in a turgor-driven way, the anticlinal wall would lengthen due to the creep in A. No loosening, or only little loosening of the wall in A layer (to adjust the creep rate to the expansion rate) would be required. Deposition of new lamellas followed by their swelling would sustain the stresses in B and A layers necessary for the wall expansion. In this process, the swelling energy would be transformed into the tensile strain energy of the A layer of the wall making the creep of this layer possible. The creep together with deposition of new lamellas – each longer than the previous one – would explain the irreversible expansion of the wall.

Water uptake necessary for cell growth would be connected with the stress relaxation in periclinal walls expanding in a turgor-driven manner.

7.3 Invaginations of Cell Walls

The mesophyll in needle leaves of *Pinus* consists of transversely oriented plates of cells alternating with similarly oriented air spaces. The mesophyll cells are characterized by a pattern of flat ingrowths of their longitudinal walls, designated as invaginations. They are in a form of shelves that run from one transverse wall to the other. The invaginations in adjacent (within a mesophyll plate) cells often lie opposite one another, or nearly so. In a cross-section through the leaf the invagination resembles a double cell wall with middle lamella – a wall folded back on itself (Campbell 1972; Harris 1971), which indicates that they grow intercalary and not at the edge. There may be a gas space instead of the lamella, locally.

The invaginations appear after separation of mesophyll precursor cells in a young leaf into the transverse plates. An invagination begins as a fold of the wall in its narrow region, previously covered by a band of microtubules, containing parallel oriented microfibrils (Hoss and Wernicke 1995). The fold extends then into the cell lumen. The microfibrils in an invagination are parallel to its free edge (the author's own observation using polarization microscopy). Surely, the invagination cannot extend being stretched by turgor pressure. However, the hypothesis presented in Sect. 6 for a double wall is fully applicable.

Dedication and Acknowledgments The author dedicates this chapter to Andreas Sievers. The author thanks Dorota Kwiatkowska for helpful discussion and comments, and Agata Burian and Anna Staroń for providing copies of literature issues.

References

Bergfeld R, Speth V, Schopfer P (1987) Reorientation of microfibrils and microtubules at the outer epidermal wall of maize coleoptiles during auxin-mediated growth. Bot Acta 101:31–41

Beusmans JMH, Silk WK (1988) Mechanical properties within the growth zone of corn roots investigated by bending experiments: II. Distribution of modulus and compliance in bending. Am J Bot 75:996–1002

Boyer JS (2001) Growth-induced water potentials originate from wall yielding during growth. J Exp Bot 52:1483–1488

Burgert I, Eder M, Gierlinger N, Fratzl P (2007) Tensile and compressive stresses in tracheids are induced by swelling based on geometrical constraints of the wood cell. Planta 226:981–987

Burström H (1959) Growth and formation of intercellularies in root meristems. Physiol Plant 12:371–385

Campbell R (1972) Electron microscopy of the development of needles of *Pinus nigra* var. *maritima*. Ann Bot 36:711–720

Correns C (1892) Zur Kenntnis der inneren Struktur der vegetabilischen Zellmembranen. Jahrb f wiss Bot 23:254–274

Cosgrove DJ (1988) Mechanism of rapid suppression of cell expansion in cucumber hypocotyl after blue-light irradiation. Planta 176:109–116

Cosgrove DJ (2000) Expansive growth of plant cell walls. Plant Physiol Biochem 38:109–124

Dorrington KL (1980) The theory of viscoelasticity in biomaterials. In: Vincent I et al. (eds) The mechanical properties of biological materials. Press Syndicate of the University of Cambridge, Cambridge, pp 289–314

Dumais J (2007) Can mechanics control pattern formation in plants? Curr Opin Plant Biol 10:58–62

Dumais J, Shaw SL, Steele CR, Long SR, Ray PM (2006) An anisotropic-viscoplastic model of plant cell morphogenesis by tip growth. Int J Dev Biol 50:209–222

Edelmann HG (1995) Wall extensibility during hypocotyl growth; a hypothesis to explain elastic induced wall loosening. Physiol Plant 95:296–303

Erickson RO (1986) Symplastic growth and symplasmic transport. Plant Physiol 82:1153

Finney EE, Hall CW, Mase GE (1964) Theory of linear viscoelasticity applied to potato. J Agric Eng Res 9:307–312

Forterre Y, Skotheim JM, Dumais J, Mahedevan L (2005) How the Venus flytrap snaps. Nature 433:421–425

Galatis B, Apostolakos P (2004) The role of the cytoskeleton in the morphogenesis and function of stomatal complexes. New Phytol 161:613–839

Green PB, Erickson RO, Buggy J (1971) Metabolic and physical control of cell elongation rate. In vivo studies in *Nitella*. Plant Physiol 47:423–430

Harris WM (1971) Ultrastructural observations on the mesophyll cells of pine leaves. Can J Bot 49:1107–1109

Hejnowicz Z (1997) Graviresponses in herbs and trees; a major role for the redistribution of tissue and growth stresses. Planta 203:S135–S146

Hejnowicz Z, Barthlott W (2005) Structural and mechanical peculiarities of the petioles of giant leaves of *Amorphophallus* (Araceae). Am J Bot 92:391–403

Hejnowicz Z, Borowska-Wykręt D (2005) Buckling of the inner cell wall layers after manipulations to reduce tensile stress: observations and interpretations for stress transmission. Planta 220:465–473

Hejnowicz Z, Sievers A (1995a) Tissue stresses in organs of herbaceous plants I. Poisson ratios and their role in determination of the stresses. J Exp Bot 46:1035–1044

Hejnowicz Z, Sievers A (1995b) Tissue stresses in organs of herbaceous plants II. Determination in three dimensions in the hypocotyl of sunflower. J Exp Bot 46:1045–1053

Hejnowicz Z, Sievers A (1996) Tissue stresses in organs of herbaceous plants III. Elastic properties of the tissues of sunflower hypocotyl and origin of tissue stresses. J Exp Bot 47:519–528

Heyn ANJ (1933) Further investigations on the mechanism of cell elongation and the properties of the cell wall in connection with elongation. I. The load extension relationship. Protoplasma 19:78–96

Hiller GH (1872) Untersuchungen über die Epidermis der Blüthenblätter. Jahrb f wiss Bot 15:411–452

Hohl M, Schopfer P (1992a) Growth at reduced turgor; irreversible and reversible cell-wall extension of maize coleoptiles and its implications for the theory of cell growth. Planta 187:209–217

Hohl M, Schopfer P (1992b) Physical extensibility of maize coleoptile cell walls: apparent plastic extensibility is due to elastic hysteresis. Planta 187:498–504

Hoss S, Wernicke W (1995) Microtubule and the establishment of apparent cell wall invaginations in mesophyll cells of *Pinus silvestris* L. J Plant Physiol 147:474–476

Huyghe JM, Bovendeerd PHM (2004) Swelling media: concepts and applications. In: Loret B, Huyghe JM (eds) Chemomechanical couplings in porous media – Geomechanics and biomechanics. Springer, Heidelberg, pp 57–124

Jaffe MJ, Leopold AC (1984) Callose deposition during gravitropism of *Zea mays* and *Pisum sativum* and its inhibition by 2-deoxy-D-glucose. Planta 161:20–26

Kenealy WR, Jeffries TW (2003) Enzyme processes for pulp and paper: a review of recent developments. In: Goodell B et al (eds) Wood deterioration and preservation. Oxford University Press, Oxford, pp 210–239
Kutschera U (1989) Tissue stresses in growing plant organs. Physiol Plant 77:157–163
Kutschera U (1994) The current status of the acid-growth hypothesis. New Phytol 126:349–369
Kutschera U, Köhler K (1992) Turgor and longitudinal tissue pressure in hypocotyls of *Helianthus annuus* L. J Exp Bot 43:1577–1581
Kutschera U, Bergfeld R, Schopfer P (1987) Cooperation of epidermis and inner tissues in auxin-mediated growth in maize coleoptiles. Planta 170:168–180
Lockhart JA (1965) An analysis of irreversible plant cell growth. J Theor Biol 8:264–375
Lüthen H, Bigdon M, Böttger M (1990) Reexamination of the acid growth theory of auxin action. Plant Physiol 93:931–939
Maltby D, Carpita NC, Montezinos D, Kulow C, Delmer DP (1979) β-1,3-Glucan in developing cotton fibers. Plant Physiol 63:1158–1164
Marshall JG, Dumbroff EB (1999) Turgor regulation via cell wall adjustment in white spruce. Plant Physiol 119:313–319
Niklas KJ (1992) Plant biomechanics. An engineering approach to plant form and function. University of Chicago Press, Chicago
Nobel PS (1974) Introduction to biophysical plant physiology. WH Freeman, San Francisco
Nonami H, Wu YJ, Boyer JS (1997) Decreased growth-induced water potentials; a primary cause of growth inhibition at low water potentials. Plant Physiol 114:501–509
Panteris E, Apostokalos P, Galatis B (1993) Microtubules and morphogenesis in ordinary epidermal cells of *Vigna sinensis* leaves. Protoplasma 174:91–100
Panteris E, Apostolakos P, Galatis B (1994) Sinuous ordinary epidermal cells; behind several patterns of waviness, a common morphogenetic mechanism. New Phytol 127:771–780
Parre E, Geitmann A (2005) More than a leak sealant. The mechanical properties of callose in pollen tubes. Plant Physiol 137:274–286
Passioura JB, Boyer JS (2003) Tissue stresses and resistance to water flow conspire to uncouple the water potential of the epidermis from that of the xylem in elongating plant stems. Funct Plant Biol 30:325–334
Peters WS, Tomos AD (1996) The history of tissue tension. Ann Bot 77:657–665
Peters WS, Farm MS, Kopf AJ (2001) Does growth correlate with turgor-induced elastic strain in stems? A re-evaluation of de Vries' classical experiments. Plant Physiol 125:2173–2179
Proseus TE, Boyer JS (2005) Turgor pressure moves polysaccharides into growing cell walls of *Chara corallina*. Ann Bot 95:967–979
Proseus TE, Boyer JS (2006a) Identifying cytoplasmic input to the cell wall of growing *Chara corallina*. J Exp Bot 57:3231–3242
Proseus TE, Boyer JS (2006b) Periplasm turgor pressure controls wall deposition and assembly in growing *Chara corallina* cells. Ann Bot 98:93–105
Proseus TE, Boyer JS (2007) Tension required for pectate chemistry to control growth in *Chara corallina*. J Exp Bot 58:4283–4292
Proseus TE, Ortega JKF, Boyer JS (1999) Separating growth from elastic deformation during cell enlargement. Plant Physiol 119:775–784
Rayle DL, Cleland RE (1970) Enhancement of wall loosening and elongation by acid solutions. Plant Physiol 46:250–253
Refrégier G, Pelletier S, Jaillard D, Höfte H (2004) Interaction between wall deposition and cell elongation in dark-grown hypocotyl cells in *Arabidopsis*. Plant Physiol 135:959–968
Roland JC (1978) Cell wall differentiation and stages involved with intercellular gas space opening. J Cell Sci 32:325–336
Rose JK, Braam J, Fry SC, Nishitani K (2002) The XTH family of enzymes involved in xyloglucan endotransglucosylation and endohydrolysis: current perspectives and a new unifying nomenclature. Plant Cell Physiol 43:1421–1435
Schopfer P (2006) Biomechanics of plant growth. Am J Bot 93:1415–1425

Singh PP, Cushman JN, Bennethum LS, Maier DE (2003) Thermodynamics of swelling biopolymeric systems. Transport Porous Media 53:1–24
Spatz H-Chr, Köhler L, Speck Th (1998) Biomechanics and functional anatomy of hollow-stemmed sphenopsids. I. *Equisetum giganteum* (Equisetaceae). Am J Bot 85:305–314
Speck Th, Speck O, Emanns A, H-Chr S (1998) Biomechanics and functional anatomy of hollow-stemmed sphenopsids. II. *Equisetum hyemale*. Bot Acta 111:366–376
Strasburger E (1889) Ueber das Wachsthum vegetabilischer Zellhäute. Verlag von Gustav Fischer, Jena
Taiz I (1984) Plant cell expansion: regulation of cell wall mechanical properties. Annu Rev Plant Physiol 35:585–657
Verbelen JP, Vissenberg K, Le J (2001) Cell expansion in the epidermis: microtubules, cellulose orientation and wall loosening enzymes. J Plant Physiol 158:537–543
Wardrop AB, Cronshaw J (1958) Changes in cell wall organization resulting from surface growth in parenchyma of oat coleoptiles. Aust J Bot 6:89–95
Wei C, Lintilhac PM (2003) Loss of stability – a new model for stress relaxation in plant cell walls. J Theor Biol 224:305–312
Wei C, Lintilhac PM (2007) Loss of stability: a new look at the physics of cell wall behavior during plant cell growth. Plant Physiol 145:763–772
Weston GD, Cass DD (1973) Observations on the development of the paraveinal mesophyll of soybean leaves. Bot Gaz 134:332–335
Zhu GL, Boyer JS (1992) Enlargement in *Chara* studied with a turgor clamp. Plant Physiol 100:2071–2080

Integrative Mechanobiology of Growth and Architectural Development in Changing Mechanical Environments

B. Moulia, C. Der Loughian, R. Bastien, O. Martin, M. Rodríguez, D. Gourcilleau, A. Barbacci, E. Badel, G. Franchel, C. Lenne, P. Roeckel-Drevet, J.M. Allain, J.M. Frachisse, E. de Langre, C. Coutand, N. Fournier-Leblanc, and J.L. Julien

Abstract Mechanosensitive control of plant growth is a major process shaping how terrestrial plants acclimate to the mechanical challenges set by wind, self-weight, and autostresses. Loads acting on the plant are distributed down to the tissues, following continuum mechanics. Mechanosensing, though, occurs within the cell, building up into integrated signals; yet the reviews on mechanosensing tend to address macroscopic and molecular responses, ignoring the biomechanical aspects of load distribution to tissues and reducing biological signal integration to a "mean plant cell." In this chapter, load distribution and biological signal integration are analyzed directly. The *Sum of Strain Sensing* model S^3m is then discussed as a

Authors Moulia, der Loughian and Bastien have equally contributed this work.

B. Moulia (✉), O. Martin, D. Gourcilleau, A. Barbacci, E. Badel, G. Franchel, C. Lenne, P. Roeckel-Drevet, C. Coutand, N. Fournier-Leblanc, and J.L. Julien
INRA, Université Blaise Pascal, UMR_A547 Physique et Physiologie Intégratives de l'Arbre Fruitier et Forestier, Domaine de Crouelle, 63039 Clermont Ferrand, France
e-mail: Bruno.Moulia@clermont.inra.fr

C. Der Loughian and M. Rodríguez
INRA, Université Blaise Pascal, UMR_A547 Physique et Physiologie Intégratives de l'Arbre Fruitier et Forestier, Domaine de Crouelle, 63039 Clermont Ferrand, France
and
LadHyX (Hydro-Dynamics), Ecole Polytechnique-CNRS, 91128 Palaiseau, France

R. Bastien
INRA, Université Blaise Pascal, UMR_A547 Physique et Physiologie Intégratives de l'Arbre Fruitier et Forestier, Domaine de Crouelle, 63039 Clermont Ferrand, France
and
Non-Linear Physics, Material and Complex Systems, CNRS UMR 7057, Université Paris 7, 10 rue Alice Domon et Léonie Duquet, 75205 Paris, cedex 13, France

J.M. Allain
LMS (Solid Mechanics), Ecole Polytechnique-CNRS, 91128 Palaiseau, France

J.M. Frachisse
Institut des Sciences du Végétal, CNRS, UPR 2355, Integrated ion transport, Avenue de la Terrasse, 91198 Gif-sur-Yvette Cedex, France

E. de Langre
LadHyX (Hydro-Dynamics), Ecole Polytechnique-CNRS, 91128 Palaiseau, France

P. Wojtaszek (ed.), *Mechanical Integration of Plant Cells and Plants*, 269
Signaling and Communication in Plants 9, DOI 10.1007/978-3-642-19091-9_11,
© Springer-Verlag Berlin Heidelberg 2011

synthesis of the state of the art in quantitative deterministic knowledge and as a template for the development of an integrative and system mechanobiology.

1 Introduction

From the instant cells first formed at the onset of life on Earth, they have been facing tough mechanical constraints linked to fluctuations in their osmotic environment (see chapter "Osmosensing"). Multicellularity and cell differentiation then added another source of mechanical stresses called tissue tensions or "autostresses" (see Hejnowicz 1997; Moulia 2000; Moulia and Fournier 2009, chapter "Mechanics of the Meristems"). This sparked the development of cellular systems for mechanosensing and subsequent control of mechanical cell properties very early on in the realm of evolution, and all the living cells that have been observed display mechanosensing systems (e.g., Haswell et al. 2008, see chapters "Introduction: Tensegral World of Plants" and "Mechanics of the Cytoskeleton").

However, land colonization and the evolution toward an erect habit under selective pressure for light competition and propagule dispersal has created three very challenging changes in the mechanical environment of plants (Niklas 1998).

First, the large mass density ratio between plant tissues and air (approximately 10^3) means land plants display very little buoyancy (compared to water plants) and thus get only negligible support from surrounding fluid. The vertical position can therefore become unstable over erected growth due to global Euler buckling. This sets a potential gravitational limit for self-supportive vertical growth on land (beyond that limit, additional growth in length or weight will actually reduce rather than increase the height of the stem (McMahon 1973; Jaouen et al. 2007; also see Moulia and Fournier-Djimbi 1997; Niklas and Spatz 2004 for critical reviews). Growth at low buoyancy sets a second instability in the erected form, as the continuous deposition of cell wall layers in growing cells fixes any long-lasting deformation, meaning deformation increases over time (see reviews in Moulia et al. 2006; Moulia and Fournier 2009; Almeras and Fournier 2009). This slow instability (compared to the instantaneous Euler buckling) is especially important for stems undergoing secondary growth (Fournier et al. 2006; Almeras and Fournier 2009).

The second change due to the mechanical in-land environment is the large fluctuations in the availability of water in the atmosphere (and in soils). As photosynthesis requirements have led to very high surface-to-volume ratios, the strong transpiration flows may induce losses of turgor and rigidity in hydrostatic tissues, despite the various control mechanisms plants have evolved (Niklas 1992; Tardieu 2003; also see chapter "Hydraulics of Vascular Water Transport"). Such changes induce broad variations in the mechanical stresses internal to plants (Chapters "Mechanics of the Cytoskeleton" and "Mechanical Force Responses of Plant Cells and Plants").

Last but not least, the very low kinematic viscosity of air and the existence of climatic air temperature and pressure gradients induce frequent and potentially strong winds (Stull 2007). Winds are major factors of land climate and thus of the physical environment of the plant. Therefore, any aerial organ in a terrestrial plant is submitted to fluctuating, intermittent but recurrent wind loads (de Langre 2008), apart from the tiny, well-protected shoot apical meristems. However, the extent of wind loads varies extensively over time and geographical space, depending on meteorology and local topography. Moreover, as plants grow erect, and display their foliage, wind drag and lever arms increase, producing a huge increase in wind-induced mechanical loads (Delus et al. 2004). Note that the vibrational excitability of a plant subjected to wind turbulence is also dependent on variables undergoing major changes during growth (Rodriguez et al. 2008).

It was been understood that the challenges set by wind drag and buckling instability may play an important role in the height growth of terrestrial plants (see Moulia and Fournier-Djimbi 1997 for a critical review), but the picture has become sharper over the last decade. It has become clear that growing plants have to contend with both static mechanical challenges (increased mean wind drag and risk of buckling instability, as well as the slow instability due to growth itself) and dynamic mechanical challenges related to the spectrum of their dynamic excitability in response to turbulent wind loads (Gardiner et al. 2008; Sellier et al. 2008; Rodriguez et al. 2008).

It has also become clear that these mechanical challenges depend very strongly on whether the plant is growing isolated or within a canopy. Plants growing isolated are submitted to a continuously increasing static and dynamic wind drag challenge. In contrast, plants growing within an existing canopy (like tree saplings) first experience a phase of low wind drag. But the low understory light resources force them to grow in height with little diametric growth, which generates a significant risk of buckling instability (Jaouen et al. 2007). As they get close to the canopy top, they become submitted to increased wind drag, but with a very intermittent gust regime produced by "honami" eddies peaking in a restricted waveband (Py et al. 2006; Dupont and Brunet 2008). This may induce resonant oscillations in plants that can be damaging, despite the lower mean static drag force compared to isolated plants (Sellier et al. 2008; Sellier and Fourcaud 2009; Gardiner et al. 2008). Finally, if the plant overreaches well beyond canopy height (as is the case with trees in savanna), it becomes an almost isolated plant.

It has been argued that the sheer diversity of mechanical challenges and the fact that they all change in quality and intensity during growth actually prevent a genetically fixed mechanical design from being competitive (except species colonizing a very specific niche; Fournier et al. 2006; Moulia et al. 2006). Therefore, a phenotypical developmental plasticity has been selected that includes (1) acclimation of the plant's load-bearing structure through thigmomorphogenesis, and (2) a process of active recovery and posture control through bending motors (Moulia et al. 2006; Barbacci et al. 2009). As the integrative mechanobiology of tropisms and postural controls has recently been reviewed in detail (e.g., Moulia et al. 2006; Moulia and Fournier 2009; Almeras and Fournier 2009), this chapter focuses on the integrative mechanobiology of thigmomorphogenetic acclimation.

Thigmomorphogenesis was first demonstrated by submitting plants to (artificial) mechanical bending. A syndrome of responses is then observed in a large number of species, involving (1) a reduction in longitudinal stem growth, (2) a stimulation of secondary radial stem growth (if a cambium is present), possibly with differentiation of a more flexible but stronger "flexure wood," and (3) a reallocation of biomass to the root system (see chapter "Mechanical Force Responses of Plant Cells and Plants," Telewski 2006; Moulia et al. 2006 and Coutand 2010 for reviews from complementary standpoints). This mechanosensitive control of growth allometries results in stunting and anchoring the shoots, while conserving most of the capacity for wind drag reduction through reconfiguration made possible by the more flexible wood. The thigmomorphogenetic syndrome thus seems to improve plant acclimation to the effects of static wind drag. Indeed, isolated plants submitted to high winds display a morphology that is reminiscent of the thigmomorphogenetic response (Jaffe et al. 2002). However, the fact that broad-scale thigmomorphogenetic responses were obtained with very small loads (e.g., one slight bending per day) cast doubt on the significance of thigmomorphogenesis in natural conditions, as plants responding as they did in laboratory experiments should not grow at all in natural windy settings. Furthermore, in some species and growth stages, thigmomorphogenesis seemed to be outweighed by photomorphogenesis (Holbrook and Putz 1989). However, studies on different dense stands, including a 15-m-high lodgepine forest (Meng et al. 2006) and a 0.7-m-high dense alfalfa stand (Moulia and Combes 2004) have shown that thigmomorphogenesis is a major response in many plant canopies. By comparing free-swaying stands to a control treatment with limited wind-induced motion, Moulia and Combes (2004) demonstrated that in the range of moderate, chronic winds ($U < 30$ km h^{-1}), in situ thigmomorphogenetic effects could range from a 40% decrease in stand height with a 65% decrease in aerial biomass down to no effect at all when little wind occurred. Thigmomorphogenesis was thus found to be a major process in the control of plant canopy growth in the ecological range of natural chronic winds. Further insight into the significance of thigmomorphogenesis for wind acclimation was recently provided by analysis of the dynamic excitability of trees (Rodriguez et al. 2008). It was found that throughout development, trees tend to tune their allometric growth to reduce changes in their resonance frequencies and compartmentalize their dynamic bending energy.

The question of how a coordinated thigmomorphogenetic syndrome of responses can be achieved through mechanoperception and signal integration within the plant is thus a major issue. Reviews on thigmomorphogenesis and mechanosensing (e.g., chapter "Mechanical Force Responses of Plant Cells and Plants," Braam 2005; Telewski 2006; Monshausen and Gilroy 2009) usually present the global thigmomorphogenetic syndrome and then zero in to detail mechano-induced genes in herbs (Lee et al. 2005) or trees (Leblanc-Fournier et al. 2008; Martin et al. 2009). However, there is little analysis of the link between the two levels of description (macroscopic growth responses and gene expression patterns in cells and tissues).

Mechanosensing occurs at cell level (see chapters "Introduction: Tensegral World of Plants," "Micromechanics of Cell Wall," "Mechanics of the Cytoskeleton,"

"Mechanics of the Meristems," and "Mechanical Force Responses of Plant Cells and Plants," for more in-depth analysis of mechanosensitive cell biology), yet mechanics stimulations apply mechanical loads at whole-plant level, acting either at its boundaries (as in the case of bending through wind drag) or across its full volume (as in the case of weight or inertial forces). Therefore, the links between the loads and the changes in mechanical state of tissue elements and cells that triggers cell mechanosensing depend not only on the load but also on the mechanical structure of the plant.

Analyzing and modeling the integrative biology of mechanosensing and thigmomorphogenetic response thus involves two phases (1) biomechanical analysis of how external mechanical loads on the plant are distributed over the constitutive plant tissue and cells, and (2) integrative mechanobiological modeling of local mechanosensing and how the plant integrates it. The aim of this chapter is to review the issue of plant integrative biomechanics and mechanobiology, from the load on the plant to tissue elements and cells, and then from mechanosensitive gene expression to global thigmomorphogenetic responses. The issue of the adaptive value of these responses will be then briefly discussed. Although the mechanical challenges of an erect habit on a terrestrial environment apply to any terrestrial taxa, our focus will mostly remain limited to the stems of terrestrial vascular seed plants. These phyla have evolved some of the largest and most perennial living erect structures on Earth – modern trees. A more general and complete view of the evolutionary aspects of mechanical design can be found in chapter 14.

2 From Whole-Plant Loads to Tissue Element Loads: Integrative Plant Biomechanics

2.1 *Some Basics of the Mechanics of Deformable Materials (fluids and solids), Including Biomaterials*

Before entering into biomechanics and mechanobiology, it is useful to first share a few concepts, principles, and tools of solid and fluid mechanics (definitions complementary to those given in chapter "Micromechanics of Cell Walls," see also Boudaoud 2010).

Mechanics is usually defined as the science that deals with the movements of bodies under the influence of forces (Mechanics 2010). For a given material domain within the body (characterized by its amount of matter, i.e., its mass), these movements can be split into (1) *global motion* of the domain, characterized by the motion of a central point – the center of mass – and possible rotations of the body around this point, and (2) the relative movement of the other parts of the domain with respect to this central point and altering its shape and/or size, called *deformation*. Motion can be measured by *velocity* (in meters per second) and deformation by the *strains* (relative change in lengths in all directions of a domain around a position within the body).

Any body compelled to change its global velocity of motion and/or to deform needs a *force* to do so (unit Newton, N). *Newton's second law* states that change of velocity (acceleration) is proportional to the force delivered and inversely proportional to the mass of the body. Reciprocally, any acceleration of a body of mass m (such as during wind-induced oscillations) will produce an additional force, called the *inertial force*.

A body cannot be completely at rest unless all forces and all moments acting on it are balanced. Once a complete balance exists, the body is said to be in *static equilibrium*. Note that the origin of some forces involved in static equilibrium is not directly observable. *Reaction forces* have to be taken into account. For example, when you are standing on the floor, the compressive force due to your weight is counterbalanced by a reaction of the floor. Although the mechanisms explaining the onset of such distant reaction forces are highly intuitive, they are sometimes misunderstood in the biological literature. These mechanisms are related to the requirements of internal mechanical equilibrium for all the domains within the body, and to the existence of physicochemical bonds within the materials. When subjected to the action of forces, a deformable body will strain. This deformation will stretch bonds and slide/shear internal elements, allowing internal reaction forces to set up until an internal and external mechanical static equilibrium is achieved (or quasi-static equilibrium if slow internal sliding occurs). A measure of the density of these internal reaction forces is given by the amount of internal force per unit area, called *stress* (unit: Pascal, Pa $=$ N/m^2). These internal reaction stresses are transmitted from place to place up to the sites of external constraints, where they generate external reaction forces. If the solid is *stiff* or *rigid*, then little strain will be necessary. However, if it is *compliant* or *flexible*, then large strains are necessary before a static equilibrium can be achieved with the applied loads. Note that there is a limit at which the internal stresses may overcome the *strength* of the material, leading to *fracture* (see Niklas 1992 and Moulia and Fournier-Djimbi 1997 for a more complete review on these topics). As stresses are a derived quantity, they are not directly observable and have to be inferred from kinematic measurements (strains and/or accelerations).

Finally, when there is no static equilibrium, accelerated movements take place. The notion of static equilibrium can then be extended to a *dynamic equilibrium* by taking into consideration inertial forces in addition to the static forces (d'Alembert's principle). The reaction forces, particularly stresses, may increase accordingly.

2.2 *Mechanical Modeling as an Integrative Structure–Function Tool for Plant Biomechanics*

It is virtually impossible for the human brain to directly and simultaneously process all the above-described effects from external loads through to internal bond

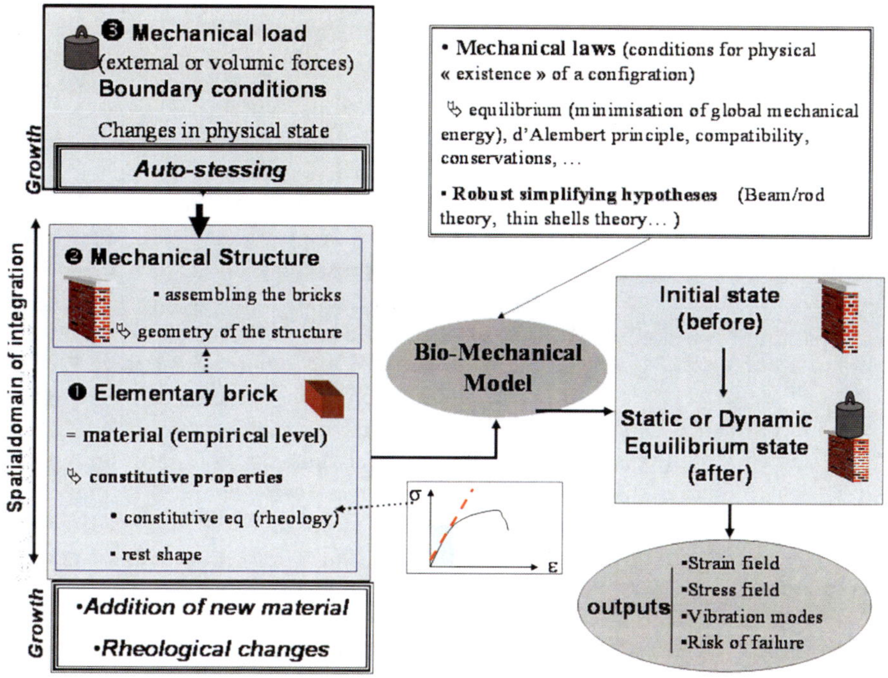

Fig. 1 The structure of a ISM Model for use in Plant Biomechanics. ISM models consider (at least) two scales in the system: a scale of phenomenological empiricism called the material scale, and a scale of mechanistic spatial integration, the mechanical structure. The internal and boundary loads (inputs) result in a change in mechanical state that can be calculated using mechanical principles and robust simplifying theories. ISM models can produce various outputs characterizing mechanical state or dynamics, such as strain (ε) and stress (σ) fields, vibration modes, or risk factors versus rupture

straining. We therefore employ dedicated tools called models, which can be produced through a scientific method called Integrative Structural Mechanics (ISM) modeling. It is useful to gain a general view of ISM models (Fig. 1) as (1) ISM modeling is central to understanding the link between external loads and the mechanical stimulus on cells (integrative plant biomechanics), and (2) it can serve as a paradigm for integrative mechanobiology (and more generally sensing biology), as we will see later.

Figure 1 sketches the general structure of an ISM model. The first step is to define the scale of the material, which can be viewed as the basic building block of the integrative model (the smallest level of organization taken into account). The mechanical properties of this material (its capacity for building stresses when reacting to a deformation of its "bonds," called its *rheology*) are defined by an empirical phenomenological equation, called the *constitutive equation*. This equation describes the relation between stresses (reaction) and strains or strain rates in each material element. For example, if the material can be modeled as *linear elastic*, the stresses will be proportional to the strains, and will vanish when the load disappears. But the constitutive equation also requires the definition of the at-rest configuration of the building block (its shape in a mechanically isolated state).

Next, the structure is defined by assembling the material elements together (i.e., defining the toplogy of the way they bind together and the resulting geometry of this assemblage). This structure defines the domain of integration of the model.

Finally, the load itself is modeled (how much, in what directions, where in the structure?). This includes external forces, for example, a wind-drag force distributed over the plant, but also boundary constraints giving rise to reaction forces. As mentioned in the Introduction section, multicellular plants also display significant autostresses (internal loads) linked to their growth and turgor or to cell wall differentiation (Hejnowicz 1997; Moulia 2000; Moulia and Fournier 2009).

Once the material structure and load have been defined, the model can be built, using the principles of mechanics (such as static or dynamic equilibrium, or the compatibility of strains between adjacent material elements), plus any simplifying hypotheses needed, such as those defining rods and beams theory. This model can then be solved either analytically (e.g., Jaouen et al. 2007) or numerically (e.g., Moulia et al. 1994; Rodriguez et al. 2008). In many cases, the model cannot be solved, making simulation studies over time (i.e., numerical experiments) the only achievable option (e.g., Dupont and Brunet 2008). Depending on the objectives of the model, many outputs can be computed, such as displacement velocity of the top of the plant, its bending rigidity (e.g., Speck et al. 1990; Moulia and Fournier 1997), the resonance modes of its dynamic excitability (Spatz and Speck 2002; Sellier et al. 2008; Rodriguez et al. 2008), the risk of global buckling instability (Jaouen et al. 2007), or even the full distribution of the stresses and strains (e.g., Coutand and Moulia 2000; Sellier and Fourcaud 2009).

It should be noted that plants are open systems in which amount of material and eventually rheology change through growth. The combination of these factors with autostressing has deep mechanical implications and forge the specificity of biomechanical models (see Fournier et al. 2006). Moreover, as we will see later, the growth and autostressing processes are under the control of mechanosensing (see also Moulia and Fournier 2009).

ISM models can be built to integrate several scales, from cell wall component to cell wall (as, for example, in cell growth models; see chapter "Intracellular Movement. Integration at the Cellular Level as Reflected in the Organization of Organelle Movements"), from cell wall element to tissue behavior, or from tissue through to whole plant level. But it is difficult and potentially misleading to span the whole range of scales in a plant. The choice of the relevant scales is an important aspect of the art of mechanical modeling. It is also important to remember that models are essentially a quantitative formalization of a set of hypotheses, which means they have to be validated before being used as a tool for analyzing plant biomechanics or mechanoperceptive experiments (e.g., Moulia and Fournier 1997). Once validated, though, they can then be exploited to analyze the distribution of stresses and strains in tissue elements in experimental conditions. Using controlled stem bending to study thigmomorphogenetic responses, Coutand and Moulia (2000) employed a validated composite-beam model of plant organs (Moulia and Fournier 1997) to assess the distribution of strains and stresses in the tissue elements. Other models can then be used to analyze the state of strains and stresses in a plant submitted to wind (e.g., Sellier et al. 2008), and thereby assess the ecophysiological range of mechanostimulation. Finally, models also make it possible to assess the functional performance of a given mechanical structure in terms of wind-induced fracture risks in plants (Gardiner et al. 2008) or buckling risks in saplings (Jaouen et al. 2007).

Extensive biomechanical modeling of plants has produced a small set of general features that can be usefully recalled here.

First of all, the intensity of stresses and strains within a plant organ is highly heterogeneous, both across the organ and across the component tissues. This heterogeneous spatial distribution does not depend solely on the location and intensity of mechanical loads. Geometry (length, diameter, tapering) and anatomy have major influences on stress and strain distribution across plant tissues within a plant organ (although acting in different combinations for stresses and strains; see Speck et al. 1990; Niklas 1992; Moulia and Fournier 1997; Coutand 2010; Boudaoud 2010 for details). This spatially heterogeneous distribution has profound consequences for the location of fracture risks in plants (Niklas 1992), as well for mechanosensing (Coutand and Moulia 2000, see Sect. 3). Note that the combination of load geometry and plant structure can lead to nonintuitive distributions. In pine trees under wind load, Ancelin et al. (2004) computed in many cases two maxima of bending stresses: one at the tree collar (as expected from lever arm effects) and one just below the crown. The ratio between the two peaks essentially depended on trunk taper.

Second, in a slender structure such as erect plant organs, both self-weight and wind loads acts mostly through organ bending. From a numerical example on a 38-m-high conifer tree (*Picea* sp.), Esser showed 65 years ago that static bending stresses overcome static compression stresses for a mean trunk inclination of only half a degree (!) or a mean wind velocity exceeding the low limit of 3.3 m s^{-1}, i.e., 12 km h^{-1} (see Moulia and Fournier 1997 for details on this calculation).

Focusing on plant morphology, the slenderness $S = L/R$ and the tapering of radius R along the length are probably the major mechanical characteristics of plant structures (Moulia and Fournier 1997; de Langre 2008; Sellier and Fourcaud 2009; Rodriguez et al. 2008; Jaouen et al. 2007; Almeras and Fournier 2009). For the same dynamic wind load, a 33% decrease in the trunk base diameter of an adult pine tree translates into a 60% higher stress at the stem periphery, whereas a change in height changes the distribution of stresses throughout the trunk (Sellier and Fourcaud 2009).

3 The Mechanosensing System in Plants: Integrative Sensing and Growth Control

As stated in the Introduction, very few studies have dealt with quantitative integrative mechanosensing. Moulia and coworkers (Coutand and Moulia 2000; Coutand et al. 2009) tackled the issue by producing the first integrative model of the mechanosensing system in plants, the Sum of Strains Sensing Model. We will study this model in more detail. This is not meant to lend it an absolute value, for as with all models they can be proved, improved, or disproved. However, this model does, in its current state, builds up a synthesis of our understanding of mechanosensitive integration, and it defines the minimal baseline modules that have to be considered in any model.

3.1 From Plant Loading to Cell Sensing and Gene Expression

The smallest complete unit of mechanosensing is a cell. Subcellular structures such as mechanosensitive channels (Haswell Peyronnet et al. 2008) and cytoskeleton (Baluška et al. 2003) are obviously involved, but the complete processing of mechanical signals can only be achieved at whole-cell level or, due to cell-to-cell coupling (through plasmodesmata for example), at the level of a small cluster of cells, called a *tissue element*. As detailed earlier, the links between loads and the changes in mechanical state of tissue elements triggering cell mechanosensing are dependent on the mechanical structure of the plant. As most plant organs are made of tissues with diverse mechanical stiffness, the distribution of stresses across and organ and within a population of organs does not parallel the distribution of strains (Moulia and Fournier 1997; Coutand and Moulia 2000), making it critical to identify which of the two mechanical state variables is involved in mechanosensing.

3.1.1 Stress or Strain Sensing?

Taking advantage on the natural variability of elastic tissue stiffness and stem diameters to decorrelate bending strains from bending stresses, Coutand and Moulia (2000) were able to show that neither stress-based, force-based, or energy-based criteria could explain the 1:10 variability in thigmomorphogenetic inhibition of longitudinal growth generated through controlled stem bending in the elastic (i.e., reversible) range. However, strain-based criteria could explain up to 76% of this variability. Strain sensing is also further supported by observations that animal tissues with stiff extracellular matrix also respond to local tissue strains (e.g., bones; see Ehrlich and Lanyon 2002; Moulia et al. 2006). Consistently with a central role for strain sensing in plants, transgenic tobacco plants with xylem of reduced stiffness undergo enhanced xylem development and attain overall stem stiffness and thus strain levels comparable to wild type (Hepworth and Vincent 1999). Therefore, local strain of the tissue element is the proper variable for assessing local mechanical stimuli, at least in tissues with stiff extracellular matrix such as plant tissues (Coutand and Moulia 2000, reviewed in Braam 2005; Telewski 2006; Baskin 2006; Moulia et al. 2006; Coutand 2010; Boudaoud 2010).

This point needs to be emphasized, and discussed carefully.

- First of all, the semantics of the concept of "stress" can generate epistemological confusion. Linguistically, the noun "stress" derives from distress. It has since been adopted in mechanics, biology, and psychology, probably vectored by the common human experience that distress causes somatic internal tensions (muscular autostresses). However, while stress then received a strict definition in mechanics with a clear dual opposition with strain, the definitional situation has not yet been cleared up in biology, as illustrated by the definition of stress as "any strain that disturbs the functioning of an organism" (Stress 2010). This situation has, for many biologists, fostered the implicit assumption that (mechanical) stress is the sensed variable.
- Second, the results by Coutand and Moulia (2000) demonstrate that mechanical strain is the local tissue variable making it possible to ascribe a local mechanical stimulus to a group of cells. This does not mean that stresses *within* the cell are not involved, but each cell is enveloped in a cell wall that is much stiffer than the cell itself, with the result that cell-wall stresses dominate the rheological behavior of the tissue element (see Niklas 1992 and even Hamant et al. 2008 on the thin-walled meristematic cells). If the mechanosensory system of the cell is unchanged but cell wall stiffness is doubled, then it will take twice as much force and wall stresses to bring the cell elements to the same level of stretching (strain) and internal reaction (stress *in* the cell microstructure).
- Reports of correlations between the intensity of the response and the applied force (and stresses) based on a large variation in the applied force and a low variation in the load-bearing mechanical structure (e.g., Jaffe et al. 1980; Mattheck and Bethge 1998) have failed to discuss the issue of stress versus strain sensing. Indeed, force stresses and strains then covary, making it is impossible to decipher their individual influences. The demonstration in Coutand and Moulia (2000) is more than correlative. It stems from a mechanical (and hence mechanistic) dissection of the mechanosensing process, combining biomechanical modeling with experimental assessment.

The stress versus strain sensing debate has recently experienced a revival following the analysis of mechanosensitive microtubule reorientation in shoot apical meristems by Hamant et al. (2008) (see also Boudaoud 2010 for more insights). By combining mechanical modeling and imaging, the authors elegantly demonstrated that microtubule orientation in shoot apical meristems was driven by tissue mechanics. They also claimed that microtubule direction was better explained by principal tissue stresses than by principal tissue strains (especially at the hinge between the meristem and the primordial). This may result from the peculiar profile of apical meristematic cells, in which the cell wall is so thin and flexible that the stiffness of the cytoskeleton may be no longer negligible. However, this is at odds with the mechanical model developed to estimate stress and (auto)strain fields, which completely neglects the microtubules as a load-bearing cytoskeleton (see note S1 in Hamant et al. 2008, supplementary online data), and the phase-transition dynamics of these polymerizing–depolymerizing structures (Trepat et al. 2007;

Foethke et al. 2009). A better explanation may stem from the hypothesis that different sensing pathways may be involved, with different mechanical relations with the cell wall. Mechanosensing of external loads is thought to involve mechanosensitive ionic channels that sit in the soft cellular membranes, and are gated by membrane tensional stresses (Peyronnet, et al. 2008). They thus cannot sense wall stresses, but are just stretched following cell wall strains. On the contrary, direct adhesion domains exist where the cytoskeleton elements link to the cell wall (see Baluška et al. 2003; Coutand 2010). These linkers are partially embedded into the cell wall, and hence cell wall stresses are directly transmitted to them. At last, it should also be noted that tissue wall strains and stresses were computed by Hamant et al. (2008) through a complex model incorporating several other strong hypotheses on the rheology of cell wall and water fluxes and the processes governing anisotropic deposition of cellulose microfibrils and cell division. All these hypotheses have yet to be validated, and are all crucial for the definition of the rest shape of cell elements, and thus of their strain. The argument by Hamant et al. (2008) is thus more indirect than the one by Coutand and Moulia (2000) and should be confirmed by more studies. Anyhow, leaving the particular issue of the mechanosensing of growth-autostresses anisotropy by cytoskeleton in the very thin wall apical cells behind and coming back to thigmomorphogenetic responses to external loads, tissue strains have been found to be the relevant variable for measuring the internal mechanical stimulus applied to cells in the nonmeristematic plant tissues (also see Coutand 2010).

Once the mechanosensed local tissue variable has been established, the next step is to understand the mechanosensing integrative process. One approach has been to proceed by analogy with structural modeling in mechanics (as presented in Fig. 1). It was thus necessary to define (1) the local strain-sensing function of a cell (equivalent to the *constitutive equation* in mechanics), and (2) an integration mechanism within the perceptive structure of the plant that calculates the plant's total sensing activity, S_i. As the focus of interest includes the thigmomorphogenetic responses of the plant, plant response modules have been added relating S_i with primary growth (Coutand and Moulia 2000), secondary growth (Coutand et al. 2009), and with the quantitative expression of primary mechanosensitive genes (Coutand et al. 2009).

3.1.2 The Local Mechanosensing Function of a Cell

Among the mechanisms involved in mechanosensing, mechanosensitive ionic channels have attracted the most detailed quantitative studies through the patch–voltage–pressure–clamp technique on protoplasts (plant cells enzymatically stripped of their cell wall, e.g., Ding and Pickard 1993; Peyronnet et al. 2008). By altering turgor pressure, thereby inducing strains (and tensional stresses) in the plasma membrane, and monitoring ionic current after clamping the voltage, it becomes possible to quantitatively characterize the response of mechanosensitive channels. The general shape of these response curves is sigmoidal and can easily be linearized

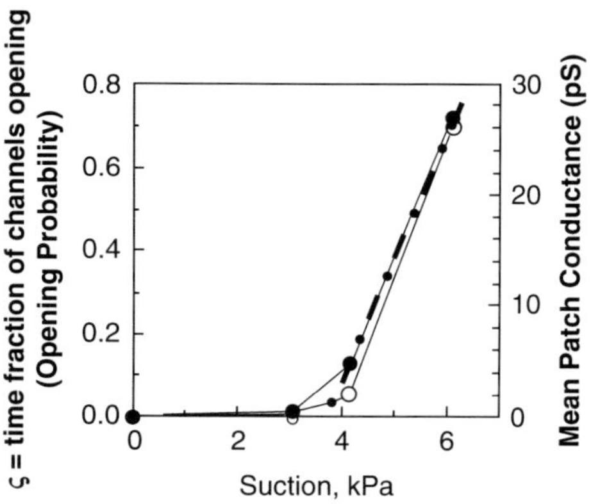

Fig. 2 Opening probability of mechanosensitive channels (MSCs) and mean patch conductance as a function of patch depression (and hence membrane tension and MSC strain) *open circle*, *filled circle* two replicates, *dashed dotted line* linear fit; modified from Ding and Prickard, 1993. Copyright © 1993, The Plant Journal, John Wiley and Sons

in the range of small strains (Fig. 2). Based on these data, Coutand and Moulia (2000) made the assumption that in *the range of small tissue strains*, the local mechanosensitive function of a tissue element could be approximated through a linear function over a threshold, i.e.,

$$dS_i = k_s(\varepsilon - \varepsilon_0)dV \quad \text{if } \varepsilon > \varepsilon_0, \text{ else } dS_i = 0, \tag{1}$$

where dS_i is the local signal in the cell (in Fig. 2, $dS_i = dI$, where I is the ionic current), k_s is a mechanosensitivity factor ($k_s = 0$ translates as insensitive tissue, while higher k_s equates to more sensing), ε is the local mechanical strain of the tissue element, ε_0 an eventual strain threshold or minimal effective strain ($\varepsilon_0 \geq 0$) (see Moulia et al. 2006 for a review), and dV is the volume of the tissue element.

Equation (1) assumes that only tensile strains are sensed ($\varepsilon > \varepsilon_0 \geq 0$), but also extends straightforwardly to the case where both tensile and compressive strains are sensed proportionally to their absolute value, as observed in animal bone tissues (Schrieffer et al. 2005).

Using a model for mechanosensing integration (to be discussed later), Moulia and coworkers have shown that this hypothesis could quantitatively explain the variation of thigmomorphogenetic responses for both primary growth (Coutand and Moulia 2000) and secondary growth (Coutand et al. 2009). Moreover, they were recently able to directly assess equation (1) experimentally (Coutand et al. 2009) by measuring the expression of a primary mechanosensitive gene coding for a zinc finger protein, ZFP2, with probable transcription factor function (Leblanc-Fournier et al. 2008; Martin et al. 2009). The expression pattern of *ZFP2* makes it a good marker for assessing the mechanosensing function. Indeed, it is very quickly and transiently overexpressed (detected as early as 5 min after tissue straining), and only in the strained tissues (probably in a cell-autonomous manner). The response of the cell mechanotransduction pathway (from primary reaction in the cytoplasm to primary gene expression in the nucleus) could thus be assessed by measuring the relative quantitative abundance of *ZFP2* transcripts, Q_r, using Quantitative Real-Time PCR (Coutand et al. 2009, in *Populus tremula*alba* (*Pta*)).

Q_r is the ratio between the content of *Pta ZFP2* transcripts in the strained tissue elements $n_t(\varepsilon)$ and the content of an unstrained control $n_t(\varepsilon = 0)$ (McMaugh and Lyon 2003; an eventual correction for similar volume in both samples is achieved through a multiplicative dilution prefactor, estimated through the assessment of the reference gene(s), not shown here). If the model in (1) holds for the entire mechanotransduction pathway, then it predicts that the increment of the content of *Pta-ZFP2* transcript in a strained tissue element should be:

$$n_{t(\varepsilon)} - n_{t(0)} = k_{mt}(\varepsilon - \varepsilon_0)\mathrm{d}V,$$

$$\text{i.e., } n_{t(\varepsilon)} = k_{mt}(\varepsilon - \varepsilon_0)\mathrm{d}V + C_0\,\mathrm{d}V, \tag{2}$$

where k_{mt} is the sensitivity of the local mechanotransduction pathway, and C_0 is the transcript concentration in the unstrained control (or baseline concentration), meaning the predicted Q_r, noted $\hat{Q}_r$, should lead to

$$\hat{Q}_r = \frac{n_{t(\varepsilon)}}{n_{t(0)}} = \frac{k_{mt}}{C_0}(\varepsilon) - \left(\frac{k_{mt}}{C_0}\varepsilon_0 - 1\right). \tag{3}$$

As the stems were strained though bending, the strains were not uniform across the cross-section. Moreover, the volume of stem collected for QPCR was small but not infinitesimal (typically 200 mm^3) meaning volume-averaged strains have to be considered. Thus, (3) cannot apply directly. If we call $N_t(\varepsilon)$ the total number of transcripts in the volume of strained tissue analyzed, and $N_t(\varepsilon)$ the content of an unstrained control $N_t(\varepsilon)$ over the same volume V, then both are the sum of the contents of all the tissue elements over the volume V

$$N_{t(\varepsilon)} = \iiint_V k_{mt}(\varepsilon - \varepsilon_0)\mathrm{d}V + \iiint_V C_0\,\mathrm{d}V. \tag{4}$$

If we assume that the mechanosensitivity of the cells, k_{mt}, and the baseline transcript concentration, C_0, are constant within the segment of organs under study (tissues of same age, and same history), then

$$Q_{r\,organ} = \frac{N_{t(\varepsilon)}}{N_{t(0)}} = \frac{k_{mt}}{C_0}\left(\frac{\iiint_V \varepsilon\,\mathrm{d}V}{\iiint_V \mathrm{d}V}\right) - \left(\frac{k_{mt}}{C_0}\frac{\iiint_V V\varepsilon_0\,\mathrm{d}V}{\iiint_V V\mathrm{d}V} - 1\right)$$

$$\text{i.e., } Q_{r\,organ} = \frac{k_{mt}}{C_0}\bar{\varepsilon} - \left(\frac{k_{mt}}{C_0}\bar{\varepsilon}_0 - 1\right). \tag{5}$$

Indeed, the experimental relation between measured Q_r and volume-averaged strain $\bar{\varepsilon}$ was found to be linear (Fig. 3), with (5) explaining 77% of the 1:500 variation of Q_r. The results in Fig. 3 thus validate the local model of (1). They also give the first *in planta* measurement of the mechanosensitivity of the mechanotransduction

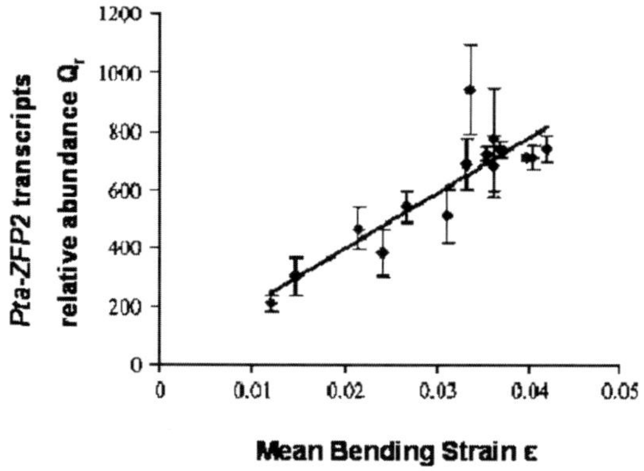

Fig. 3 Measured relative transcript abundance, Qr, of the primary mechanosensitive gene *Pta ZFP2* (assessed by Q-RT-PCR) and the volume-averaged strain in the bent stem-segment (i.e., Sum of the strain normalized by the volume of the bent tissue; Coutand et al. 2009). © 2009 Plant Physiology, American Society of Plant Biologists

pathway in a plant tissue. Under the conditions of this experiment (young poplars sitting in a controlled growth cabinet with minimal mechanical stimulation before the experiment), a 1% strain induces a (transient) 200-fold increment in the number of transcripts of the primary mechanosensitive gene *Pta-ZFP2*!

It was surprising that the bending strain range in which this linear mechanosensing model holds true goes up to (at least) 5%, i.e., quite large strains, way beyond the range of elastic strains in cell walls. This probably reflects the much greater flexibility of cell components compared to the cell wall (see Sato et al. 2005).

3.2 *Quantifying Global Thigmomorphogenetic Responses*

At this point, to properly set the problem of integrated mechanosensing, we now need to consider the global growth responses of the plant in greater depth. This was made possible by using a quantitatively controlled bending device while continuously monitoring primary elongation (Fig. 4a) or secondary thickening (Fig. 4b) using Linear Voltage Displacement Transducers (LVDT; Coutand et al. 2000). Figure 4 clearly demonstrates that an elastic bending restricted to the basal part of the stem induced a thigmomorphogenetic response in the distant primary growth zone, so that a long-range internal secondary signal S_i traveled from the bent tissues to the responding primary tissues (Coutand et al. 2000; also see Brenner et al. 2006). In contrast, the secondary growth response seems local to the bent zone (Coutand Martin et al. 2009), as previously argued by Mattheck and Bethge (1998). In both cases, the early response is that growth stops for one to a few hours, then restarts and eventually recovers the control rate. For primary growth, recovery time is highly dependent on the amount of bending strain, typically ranging from 100 to 1,000 min. From then on, no compensatory growth is observed, meaning that the final length of the plant is decreased (by -2 mm/bending in Fig. 4c). Secondary growth, though, shows clear and long-lasting growth stimulation after the initial inhibition, with growth rate increasing over 3 days then relaxing to the control rate for 3–4 more

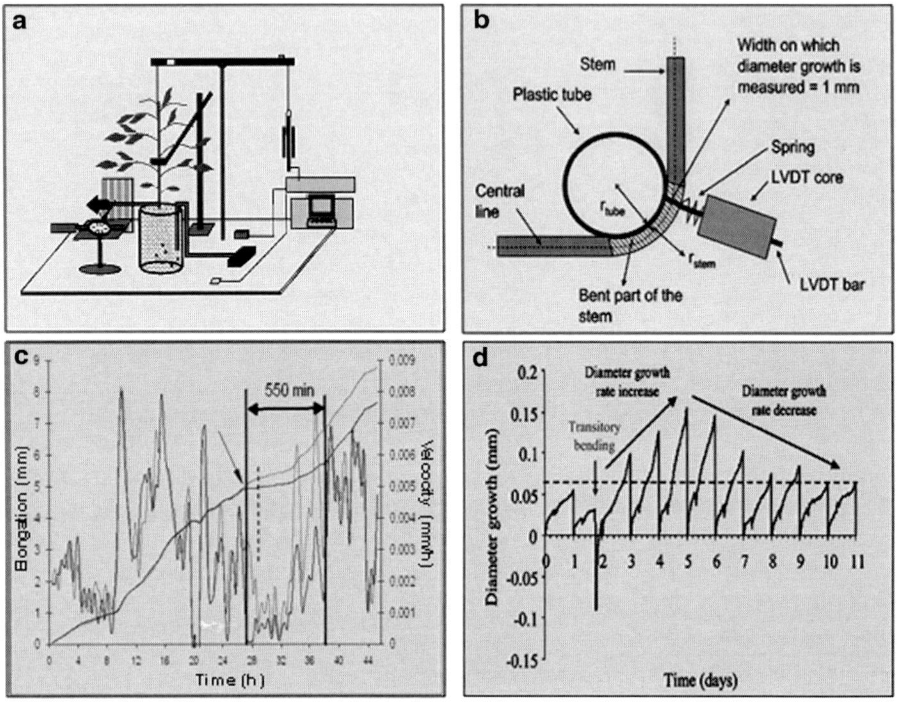

Fig. 4 Experimental setup for applying controlled, localized, and quantified bending strains while continuously monitoring growth using Linear Voltage Displacement Transducers (Coutand et al. 2000, 2009; Coutand and Moulia 2000 reproduced by kind permission of Journal of Experimental Botany and Oxford University Press). (**a**) Controlled elastic bending of the basal part of the stem (distant from the primary growth zone) while continuously monitoring primary elongation, r; (**b**) Controlled bending of the basal part of the stem, while continuously monitoring secondary radial expansion growth at the bend site; (**c**) Chart of primary elongation growth after one transient bending of the stem base *black*= bent plant *gray*= control plant; (**d**) chart of secondary radial growth after one transient bending of the stem base (as the responses lasted several days, LVDT was reset to zero daily to enable easy comparison of daily growth increments)

days. The effect of this stimulation of secondary growth (+0.35 mm/bending in Fig. 4d) was approximately 30 times higher than the effect of the initial inhibition, resulting in an overall stimulation of radial growth. In contrast with primary growth, the timing of the response seemed to be much less dependent on the level of bending strain than the peak (and total) increment in growth rate (Coutand et al. 2009, 2010).

3.3 Integrating Local Mechanotransduction into Plant Mechanosensing: The Sum of Strain Sensing Model (S^3m)

It is a huge task to integrate all the physiological factors and genetic regulations involved in thigmomorphogenesis. Transcriptomic studies in *Arabidopsis thaliana* revealed that some 760 genes (over 3% of the genome) had their expression regulated 30 min after a mechanical stimulation (Lee et al. 2005). Even considering leaves only, Fluch et al. (2008) found at least 192 genes under transcriptional regulation in poplar. Besides the size of the transcriptomic regulation network, the complexity of long-distance signaling is also a factor (e.g., Brenner et al. 2006).

When dealing with a structural mechanics model applied to the bending of a plant organ (e.g., Moulia and Fournier 1997; Sellier and Fourcaud 2009, see Fig. 1) there are billions of molecules and bounds involved. However, it takes just a handful of equations to (1) capture the change in scale between the mechanical behavior of a tissue element and of the whole organ, and (2) to understand the effect of the plant's mechanical structure on its mechanical function. Indeed, a useful mechanistic integrative model is not meant to integrate every aspect of real-world settings but only those necessary to explain a given phenomenology. It remains just a tool for testing a set of interacting hypotheses against experimental data.

Working from this idea, Moulia and coworkers intended to build a *minimal model* of mechanosensing integration, from the level of the strained tissue element up to the thigmomorphogenetic growth responses in the entire stem (Coutand and Moulia 2000). This model is designed to chart the effects on the global thigmomorphogenetic responses of both the mechanical structure and the *mechanoperceptive structure* (the anatomical distribution of mechanosensitive tissues within the plant) of the organ. It is built in analogy with the process of integrative modeling in structural mechanics (ISM, see Fig. 1).

The starting point was equation (1) which quantifies local mechanosensing in a tissue element. As mechanostimulation sparks signals to move out of the cell, it was assumed that the secondary signal output by each cell, dS_o, is directly proportional to the mechanotransduction signal over an eventual threshold – and hence to dS_i (hypothesis H1), and can thus be written as:

$$dS_o = k \cdot dSi = k_o(\varepsilon - \varepsilon_0)dV \tag{6}$$

(with $k_o = k \times k_i$). We have seen previously that the timeframe for long-distance signal propagation was very short compared with growth response. It was thus assumed that for our purposes, the details of the signal propagation, especially signal damping, could be neglected. The simplest model for the integration of the mechanical sensing is then that the output signals, dS_o, of all the mechanosensitive cells simply sum up into a global secondary internal signal S_i (hypothesis H2). In short, the more cells are strained, the higher the S_i.

However, the domains of mechanosensitive integration seem to differ for the responses of primary and secondary growth zones. Subapical primary growth responds to distant sensing throughout the stem volume V_s, whereas distributed cambial growth only seems to be affected by strain-induced signals propagating radially in the cell layer of the cross-section A_s. The internal signal propagated axially along the whole stem and controlling the response of primary growth $S_{i,1}$ can then be written as (Coutand and Moulia 2000):

$$S_{i,1(\varepsilon)} = \iiint\limits_{V_s} k_{o(\varsigma,y,z)}\left(\varepsilon_{(\varsigma,y,z)} - \varepsilon_0\right)\mathrm{d}V, \tag{7}$$

where ζ is the distance from the apex and (y, z) describes the position of the tissue elements across the cross-section of the stem. By analogy, the internal signal propagating along the stem radius and controlling secondary growth, $S_{i,2}$, in the cambium at a position ζ on the stem thus becomes (Coutand et al. 2009):

$$S_{i,2(\varepsilon,\varsigma)} = \iiint\limits_{\mathrm{lc}A_s(\varsigma)} k_{o(\varsigma,y,z)}\left(\varepsilon_{(y,z)} - \varepsilon_0\right)\mathrm{d}x\,\mathrm{d}y\,\mathrm{d}z = lc \iint\limits_{A_s(\varsigma)} k_{o(\varsigma,y,z)}\left(\varepsilon_{(y,z)} - \varepsilon_0\right)\mathrm{d}y\,\mathrm{d}z, \tag{8}$$

where l_c is the typical length of an initial cell in the cambium (or, in a more practical way, the longitudinal length over which radial growth is measured, i.e., the size of the LVDT pad sitting on the stem in Fig. 4b, which was 1 mm in that particular case; Coutand et al. 2009).

Note that the distribution of mechanosensitive tissues defining the mechanosensitive structure of the plant (of volume V_s and cross-sectional area $A_s(\zeta)$ at position ζ on the stem) does not span the whole stem volume but only the mechano-competent tissues. More precisely, the mechanosensitive structure of the plant (at a given time point) is given by the geometrical description of mechanosensitivity $k_o(\varsigma, y, z)$ and threshold $\varepsilon_0(\varsigma, y, z)$, just as the mechanical structure of the plant is given by the spatial distribution of the mechanical properties (e.g., the longitudinal Young's modulus $E_L(\varsigma, y, z)$ and a yield threshold $\sigma_0(\varsigma, y, z)$).

In a first approximation, Coutand and Moulia (2000) assumed that lignified tissues could be considered nonsensing, and that all living tissues had similar mechanosensitivity k_o. Using *Jr- ZFP2* in situ RNA hybridization as a marker of mechano-competence in walnut stems (*Juglans regia*), Leblanc-Fournier et al. (2008) found that the cortical and medullar parenchyma of stems (and to a lesser extent some phloem parenchyma cells) displayed marking. The stiffer epidermal cells, collenchyma, xylem, and sclerenchyma did not, nor did the meristematic cambium [this lack of marking in cambial and epidermal cells is another argument in favor of a nonautonomous mechanosensitive control of cell growth, as expressed in (7) and (8)]. However, comparative tests on the *Sum of Strain Sensing* model have shown that the model output only marginally depends on the detailed distribution of mechanosensitivity (Coutand and Moulia 2000), at least in the range of anatomical variability displayed by plants from the same cultivar at the same growth stage. The most determinant factor was the geometry of the stem. For simplicity purposes, more recent studies then took mechanosensitivity to be homogeneous over all tissues (e.g., Coutand Martin et al. 2009). If $k_o(\varsigma, y, z)$ and $\varepsilon_o(\varsigma, y, z)$ are constant, then they can be factorized in the spatial integrals, so that the model evolves to

$$S_{i,1(\varepsilon)} = k_o\left(\iiint\limits_{V_s} \varepsilon_{(\varsigma,y,z)}\,\mathrm{d}V\right) - k_o\varepsilon_0\,V_s = k_o S_{1\,\mathrm{strains}} - \Sigma_0, \tag{9}$$

$$S_{i,2_{(\varepsilon)}}(\varsigma) = k_o \left(l_c \cdot \iint\limits_{As(\varsigma)} \varepsilon_{(\varsigma,y,z)}\, dy\, dz \right) - k_o \varepsilon_0 l_c A_s = k_o S_{2\text{strains}}(\varsigma) - \Sigma_{0,2}(\varsigma). \quad (10)$$

This model thus predicts that the integrated signals are linearly dependent on integrals of the strain field over the domains of mechanosensitive integration for primary and secondary growth ($S_{1\text{strains}}$ and $S_{2\text{strains}}$, respectively). This is what prompted the original name for the "Sum of Strains" model. However, a more accurate name is the "*Sum of Strain Sensing*" *model* (S^3m), as it is not the strain that is summed but the output of strain sensing by cells.

This minimal model of mechanosensitive integration was tested quantitatively versus the corresponding growth responses described earlier (Fig. 4a, b) not only by varying the applied force but also by sampling the variability of plant structure (size, anatomy, distribution of stiff tissues, and mechanocompetent tissues). In this case, the prediction of the S^3model is not that the growth response should be linear with $S_{i\text{strains}}$ (as it was for the relative abundance of transcripts) but that the growth responses should give a dose–response curve on the Sum of Strains. Indeed, if collecting more cells necessarily entails adding RNA to the sample (linearity), the biological thigmomorphogenetic response of growing tissues to the (candidate) integrated signal S_i may not be additive.

As shown in Fig. 5, a tight logarithmic relation was found between primary growth response (recovery time τ_r) and $S_{1\text{strains}}$, explaining 75% of the 1:10 variation in the response (Coutand and Moulia 2000).

$$\tau_{\text{recovery}} = a_1 \cdot \ln\left(\frac{S_{1\,\text{strains}}}{S_{0_{1\text{strains}}}}\right) \quad \text{for } S_{1\,\text{strains}} > S_{0_{1\text{strains}}}. \quad (11)$$

Similarly, a relation was found between $S_{2\text{strains}}$ and the response of radial growth, again explaining 75% of the 1:5 variation generated by varying stem bending and size (Coutand et al. 2009). A first experiment on poplar seemed to reveal that a linear relationship between radial growth response and $S_{2\text{strains}}$ was statistically slightly more significant than a logarithmic relation. However, further analysis on a

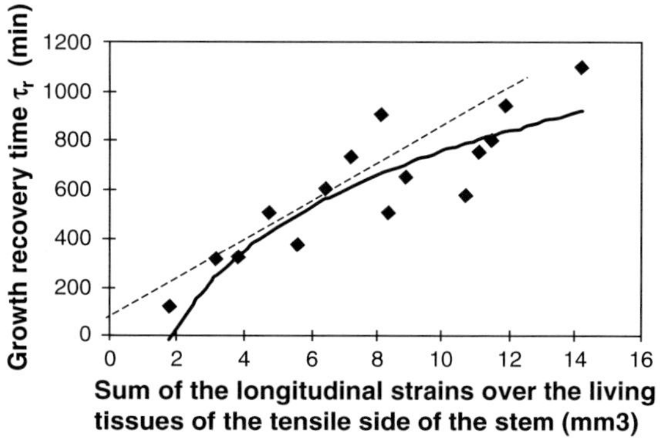

Fig. 5 Dose–response curve of the recovery time of the primary growth response to the Sum of Strains *dashed dotted line* linear fit, *block line* log fit, *filled diamond* experimental results (adapted from Coutand and Moulia 2000 reproduced by kind permission of Journal of Experimental Botany and Oxford University Press)

set of dicot tree species (Coutand et al. 2010) showed that the logarithmic relation is more generic, and thus to be preferred (also see Telewski 2006).

These logarithmic relations need to be discussed.

- First, they are in agreement with the "Weber–Fechner law" widely (though not always) observed in human and animal sensory physiology (Weber's Law 2010), and in plant gravisensing (see Moulia and Fournier 2009). The Weber–Fechner law states that "the change in a stimulus that will be just noticeable is a constant ratio of the original stimulus." Indeed, differentiating equation (11) and noting the growth response, G, gives

$$\mathrm{d}G = a\left(\frac{\mathrm{d}S_{\mathrm{i\,strains}}}{S_{\mathrm{i\,strains}}}\right) = \frac{a}{S_{\mathrm{i\,strains}}}\left(\mathrm{d}S_{\mathrm{i\,strains}}\right). \tag{12}$$

  The increment in growth response upon an increment in the Sum of Strains signal is inversely proportional to the prevailing level of the signal, so that the apparent sensitivity of the response decreases hyperbolically with the prevailing level of mechanical stimulus. This tuning of the sensitivity of the response is termed "accommodation" (Schrieffer et al. 2005; Moulia et al. 2006), and it is likely to be of major adaptive value by avoiding overreactions to noise. Indeed, in wind, the standard error of wind velocity fluctuations increases proportionally to mean wind velocity (Stull 2007). Using a protocol of 0–80 repeated back-and-forth manual basal bending sways in *Ulmus americana*, Telewski and Pruyn (1998) also found a logarithmic relationship between change in height and number of sways (the amount of bending was uncontrolled but the number of sways is likely to correlate with the total Sum of Strains due to the sways). The logarithmic dose–response curve is thus likely to apply to repeated bending during wind sways (swaying frequency was not measured either, but experience tells us it was probably close to the fundamental modal frequency of the plant, which is much easier to achieve manually).
- Second, the $S_{0_{1\mathrm{strains}}}$ threshold should not be confused with the Minimal Effective Strain threshold ε_0 of the local sensing function. $S_{0_{1\mathrm{strains}}}$ is the threshold for reception of the global systemic signal reaching the growth zone, expressed using its Sum of Strain component. A more complete specification of the model using (9) and (11) should be

$$G = a_1' \cdot \ln\left(\frac{S_{i,1\,(\varepsilon)}}{S_{0i,1}}\right) = a_1' \cdot \ln\left(\frac{k_o S_{1\,\mathrm{strains}} - \Sigma_{0,1}}{k_o S_{0_{1\,\mathrm{strains}}} - \Sigma_{0,1}}\right). \tag{13}$$

  Data analysis showed that the strain threshold ε_0 (and its integral Σ_0) could be neglected for both the thigmomorphogenetic responses of primary and secondary growth under the conditions of the experiments, whereas $S_{0_{\mathrm{strains}}}$ threshold could not (but see further discussion in Sect. 3.5). Thus, $S_{0_{\mathrm{strains}}}$ is likely to depend on the receptor pathway in the growing tissues (and possibly on propagation from the sender mechanosensitive tissues to the receiver growing tissues).

- Third, the global thigmomorphogenetic sensitivity of a plant can be described quantitatively using only two parameters for primary growth response (a_1, $So_{1strains}$) (Coutand and Moulia 2000) and two for secondary response (a_2, $So_{2strains}$) Coutand et al. 2010). These two quantities are integrative "macro-characters" (Tardieu 2003). They include the whole in-plant signaling process but through explicit and validated mechanical and mechanosensitive integrations of the interactions between the mechanical and mechanoperceptive structures of the plant and its mechanical environment (load). Varying load and/or plant size and anatomy affects the $S_{1strains}$ value along the *x*-axis in Fig. 5, and thus the value of the response, but the relation expressed in (11) and (12) (and the corresponding log response curve) remain invariant. This relation and the corresponding parameters in (11) are thus independent of both load intensity and plant size/ structure. They measure *intensive quantities*.
- Finally, (11) is not to be confused with a standard dose–response curve. Indeed, it involves an explicit integration of the effect of the mechanical and perceptive structures of the plant through the *Sum of Stain Sensing* model – a model that has been assessed experimentally. This is to be contrasted with purely correlative dose–response curves with an "arbitrarily chosen" measure of the stimulus (e.g., force; Jaffe et al. 1980).

3.4 S^3m, a Global Model of Thigmomorphogenesis?

The Sum of Strain Sensing model uses the distribution of tissue strains across the plant, $\varepsilon(x,y,z)$, called strain field, as an input. This distribution can be measured directly using kinematic methods (strain, contrary to stress, is an observable; see Moulia et al. 1994; Moulia and Fournier 2009). However, it is more informative to couple the S^3m model with a mechanical model which outputs the strain field as a function of the load, plant geometry, and plant tissue rheology (Fig. 6). Indeed, the filtering of environmental signals via the plant's mechanical structure and the consequences of multicellularity on the internal mechanical environment of the cells is then taken into account mechanistically. This one-way coupling (Mechanical model ⇨ $\varepsilon(x,y,z)$ ⇨ S^3m model) was realized by Coutand and Moulia (2000) to study primary growth response to controlled bending. This was actually the first attempt to experimentally assess the Sum of Strain Sensing model, and it also illustrates how an integrative model acts as an interpretative tool for an experiment.

An important step forward would be to achieve full coupling between a mechanic model and the S^3m model, namely to also implement the S^3m model(t) ⇨ growth response(t) ⇨ Mechanical model(t + 1) ⇨ $\varepsilon(x,y,z,t + 1)$ ⇨ S^3m model (t + 1) . . ., as described in Fig. 6. This would be the first attempt to time closure the model and explain thigmomorphogenetic acclimation over the plant growth and development time course (Moulia et al. 2006). Moreover, it represents an attempt to include mechanics and physics constraints into a very simple biological growth model (Fig. 6). This kind of fully coupled model could be characterized as a

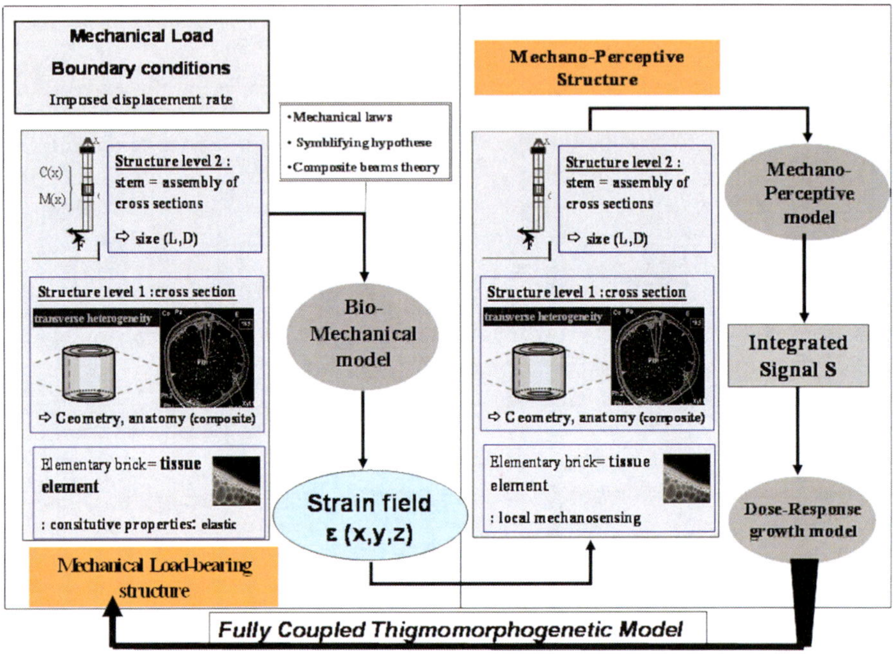

Fig. 6 Diagram of a fully coupled integrative dynamic model of thigmomorphogenesis. The ISM model of the mechanical load-bearing structure (*left*) was designed by Coutand and Moulia (2000) to analyze their bending experiments. It is based on a validated composite-beam model of plant organ flexion (Moulia and Fournier 1997). Its inputs are the curvature field, C, and the bending moment, M, along the stem (measured as in Moulia et al. 1994). Its parameters are (1) length, L, and diameters along the stem, $D(\zeta)$; (2) applied shear force, F; (3) anatomical cross-sectional images processed using the model by Moulia and Fournier (1997); and (4) estimates of tissue stiffness (longitudinal Young's moduli (Coutand and Moulia 2000). Like all models based on beam theory, this model defines two integration levels: the cross-section, and then the stem. The Mechanosensitive model is the model presented in Coutand and Moulia (2000). Its inputs are the strain field, $\varepsilon(x,y,z,t)$, in each stem, and the stem geometry factors L and $D(\zeta)$. Its output is the integrated signal S_i. It goes on to feed a model of thigmomorphogenetic growth responses (Coutand and Moulia 2000, Martin et al. 2009). In a fully coupled dynamic model of thigmomorphogenesis, the outputs of the thigmomorphogenetic growth response module can be used to update the size and geometry of the stem at the next time step, enabling time integration to be processed (or at least simulated)

dynamic model with dynamic structure (Prusinkiewicz and Rolland-Lagan 2006) performing the spatial and time integration of the thigmomorphogenetic control of growth, together with the integration of primary and secondary growth responses. Starting with initial conditions (the plant at an initial stage) and receiving external loading as an input, it may account for the complete thigmomorphogenetic syndrome of developmental acclimation over time. In particular, it could help explaining the phenotypical plasticity of stem allometries (slenderness and tapering) in response to different types of mechanical loads (artificial loads, isolated plant under wind, plants in canopy). This integration remains to be achieved and studied mathematically and experimentally. It does, however, raise the question of time integration over longer periods and over repeated loads.

3.5 Time Integration of Mechanosensing: The Problem of Slow Accommodation

Mechanical loads in nature do not usually occur as a single bending (although postbuckling bending is an exception). Meteorological variations in wind usually result in windy weather alternating with quiet weather, in time patterns that follow climatic trends but are usually at days scale (Stull 2007). The effect of daily recurring chronic mechanical loads on plant mechanosensing has been studied by Martin et al. (2010) on poplar. The crucial point here was that the amount of Sum of Strain was kept almost constant along the nine successive daily bendings. As stem diameter necessarily changed over time, the bending to be applied to achieve an almost constant Sum of Strains was recalculated daily.

Secondary growth after one single bending reproduced the time course expected from previous studies (Fig. 7a). The effects of repeated daily bending were additive for the three first bendings but they then clearly departed from being additive. This was tested by comparing experimental results against two hypothetical models of time integration. In model 9x1B, the nine successive bendings are assumed to induce additive effects, making it a time extension of the logic of the Sum of Strain Sensing model, with successive strains also summing up over time (see also Coutand et al. 2010). In model 3x1B, only the three first bendings were assumed to have additive effects and were followed by complete desensitization. The experimental results clearly departed from 9x1B but were statistically very similar to the 3x1B model throughout the loading period, thus demonstrating a clear and intense desensitization after 3 days (the alternative explanation of growth saturation was ruled out; see Martin et al. 2010). This desensitization could be reversed, but only after 7–10 days without stimulation (similar behavior has been observed in mammal bones with even longer characteristic times at around 3 weeks; Schrieffer et al. 2005). This demonstrates a long-term accommodation of mechanosensitivity to daily repeated bending (after an initial additive behavior), which contrasts with the almost instantaneous logarithmic "Weber–Fechner" accommodation of the response. It would be interesting to vary the delay between the successive loads to monitor the shift between the two modes of accommodation. A candidate mechanism for long-term accommodation could be a slow increase in the Minimal Effective Strain threshold ε_0 after a strain stimulation, as observed in bones by Schrieffer et al. (2005). ε_0 would then range from almost zero after a long state of mechanical protection (as in Coutand and Moulia 2000; Coutand et al. 2009, and at the beginning of the experiment by Martin et al. 2010) to a much higher state after long-term recurring loads (Martin et al. 2010) – a change likely to be under transcriptional and/or translational regulation (e.g., variations in the cell density of mechanoreceptors and/or of molecular actors of the local cellular mechanosensory pathway).

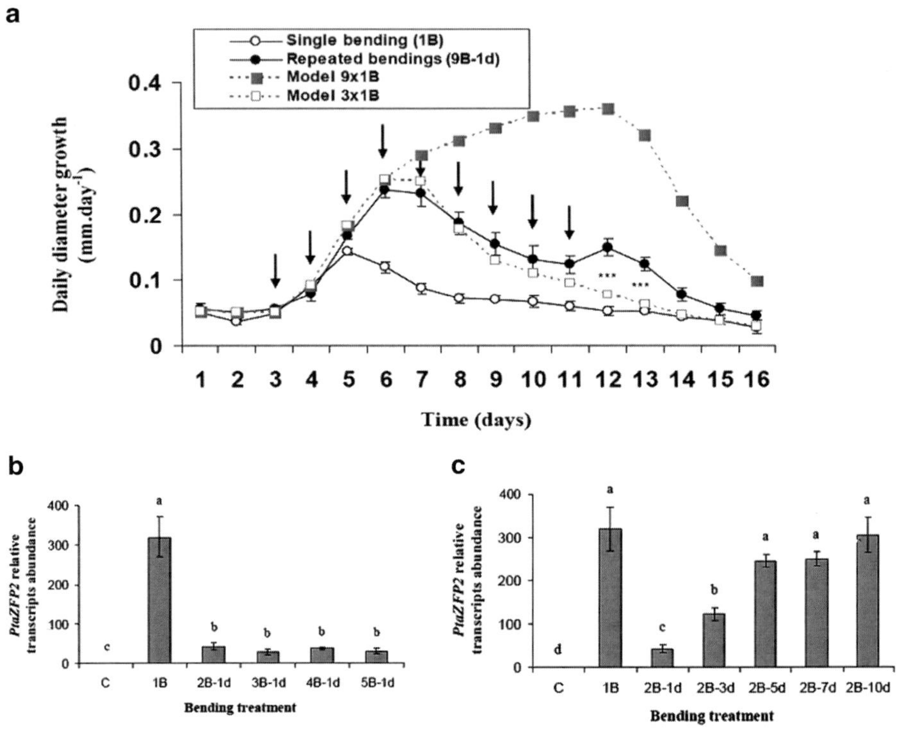

Fig. 7 Response of secondary growth (**a**) and *Pta-ZFP2* expression (**b**, **c**) to repeated daily bending (adapted from Martin et al. 2010 reproduced by kind permission of Journal of Experimental Botany and Oxford University Press). *Open circle* measured growth response to a single bending, *closed circle* measured growth response to nine successive bendings at 1-day intervals (9B-1d). *Dashed squared lines* model of additive effects (linear time integration) 9x1B, *open square* model with a sensitivity shift after 3 daily bendings (3x1B; accommodation). *C* control (no load); *1B* one single bending; *x*B–*y*d = *x* bendings each separated by *y* days; *Pta-ZFP2* Populus tremula*alba Zinc Finger Protein 2 gene

The putative adaptive significance of this slow accommodation process has not yet been studied in depth, but as windy weather involves repeated mechanostimulation, accommodation is visibly a major process governing plant response to wind loads in natural conditions. Slow accommodation also explains the quantitative discrepancy between the amount of response observed in the lab (where plants are usually left sitting for long periods under a very low levels of mechanical stimulation before experiments) and in natural conditions (as highlighted in the Introduction). Understanding and modeling the mechanisms and processes underpinning accommodation is therefore crucial to the analysis of wind acclimation in plants.

However, the detailed modeling of the time dynamics of mechanosensing from the analysis of growth response to repeated bending scenarios may well require a large series of many experiments. Understanding the gene regulation network underlying accommodation would be of great help. Moreover, mechanosensitive genes provide additional observable variables for analyzing the accommodation process. For example, *Pta-ZFP2* relative transcript abundance (Q_r) was shown to undergo major desensitization after the very first bending (Fig. 7b), and relaxation to a fully sensitive state occurred within 5 days (Fig. 7c). Similar patterns were found for genes within the local mechanotransduction pathway (e.g., *Touch2 calmodulin*, a gene encoding *ACC synthase* from the ethylene pathway, Arteca and Arteca 1999; Martin et al. 2010). Conversely, the cell wall remodeling xyloglucan endotransglycosylase/hydrolase *TCH4-XTH*, probably more related to the growth response setting in the receptor zones, displayed distinct relaxation kinetics (Martin et al. 2010).

3.6 *The Sum of Strain Sensing Model S^3m: A Template for the System Biology of Mechanosensing?*

Although the process of slow accommodation after a few bendings remains to be analyzed and coupled with the Sum of Strain Sensing model (S^3m), S^3m does, in its present state, represent an initial foray into the integrative "system biology" of mechanosensing (e.g. Telewski 2006; Moulia et al. 2006; Baskin 2006). System biology is usually defined as a field seeking to converge knowledge on the structure and the dynamics of biological systems, and integrating recent insights on gene and gene product regulation networks (e.g., Alvarez-Buylla et al. 2007; Traas and Moneger 2010). This task is only achievable using modular models (Tardieu 2003). Our aim here is to show how S^3m can be used as a template tool for further mechanistic and integrative analysis of mechanosensing and thigmomorphogenesis.

3.6.1 From Plant to Genes and Back

The challenge involved in the integrative biology of mechanosensing can be described as (Moulia et al. 2006; Hamant et al. 2008):

1. Going from a plant exposed to its mechanical environment to the internal mechanical state of its mechanocompetent tissues and on to the subsequent local mechanotransduction process, the expression of mechanosensitive genes, and finally the production of the corresponding proteins and molecular mechanosensing machinery.
2. Then integrating local signaling and gene regulation in tissue elements into the overall syndrome of thigmomorphogenetic growth responses, and its adaptive relevance in terms of mechanical acclimation.

As this loop involves several organizational levels and scales plus a host of interactions, it cannot be handled without the help of specific tools called structure–function models (Godin and Sinoquet 2005), placing this challenge within the realm of Systems Biology (Tardieu 2003; Moulia and Fournier 2009; Traas and Moneger 2010). The bottom line here is that System-Biology modeling and cross-comparison against the data produced through a suitable experiment makes it possible to assess a set of hypotheses. In cases where the combination of hypotheses cannot be worked out without calculus, modeling becomes an extension of the experimental method (Legay 1997).

3.6.2 A Template to Integrate More Detailed Modules

Although quite simple, the Sum of Strain Sensing model addresses the different process-based modules required for a systems biology approach (Fig. 6) (1) a local strain-sensing module, (2) a secondary internal signal integration module (possibly including the process of long-range propagation), (3) a growth response module (an accommodation module, yet to be formalized, has to be included whenever the response to chronic repeated load needs to be considered). Furthermore, it clearly defines and separates the scales involved (local modules vs. modules performing spatial integration). Any attempt to go into further mechanistic detail on a single module (making a "zoom sub-model") has to fit with the quantitative response of its minimal S^3m version. To illustrate, to study the mechanotransduction pathway upstream of *Pta-ZFP2* or the process of molecular accommodation of mechanosensitivity, a model of the gene regulation network underlying *Pta-ZFP2* net transcript accumulation could be produced. However, its output should still match equation (5) and Fig. 3. By the same token, any attempt to understand the growth response module (for instance by including the eco-physiological processes involved, Godin and Sinoquet 2005; Tardieu 2003) should be consistent with a logarithmic dose–response curve to an additive long-distance internal signal.

Wherever a more detailed zoom-model of a given module has been achieved, it can then be linked with the other existing modules so that the interplay of this module with the global behavior of the system can be studied. This may offer extended possibilities for testing a given zoom model and progressing more rapidly towards a proper zoom model. Moreover, it provides a template to start studying the interesting issue of long-range coupling between regulations or signaling networks in separated tissue elements (e.g., regulation of strain sensitivity in sensing tissues and of the sensitivity to secondary systemic signals underlying S_i in the responding growth zones).

3.6.3 A Tool for Genetic Dissection

Using the Sum of Strain Sensing model to analyze the response of six sympatric tree species with contrasted slendernesses and buckling risk factors, Coutand et al. (2010) found very significant interspecific differences in the mechanosensitivity of the thigmomorphogenetic control of secondary growth, with sensing threshold $S_{02strain}$ ranging from 1 to 1.8. This analysis would not have been possible without the Sum of Strain Sensing model, as the species studied differed in diameter and bending stiffness. This study also demonstrated that S^3m was robust to genetic variation. Another plus is that only one of the S^3m model parameters was undergoing genetic variation. S^3m thus appears a valuable tool for dissecting the genetic control of a highly composite and environmentally sensitive trait: stem slenderness. More generally, integrative biomechanical models offer a clear way for increasing the genetic heritability (h^2) of the variable by removing the effects of environment and plant size/structure (Sierra de Grado et al. 2008). This may also pave the way to improved detection of robust Quantitative Traits Loci and thereby help identify controlling genes via analysis of genetic $\times$ environment variability (Tardieu 2003). Similarly, S^3m may be also used to phenotype mutants and identify unknown control genes.

3.6.4 A Tool to Identify Missing Elements and Rewarding Molecular Studies

Another interesting heuristic property of integrative models like S^3m is that they can help identify modules that are (1) more influential than others in the global thigmomorphogenetic response, and (2) insufficiently understood. This will help better target molecular studies (such as gene regulation studies) that may be rewarding in the sense that (1) the genes may trigger very significant phenotypes, and (2) their identification improves our understanding of the system response significantly. The S^3 model probably needs to be developed further before it can be made completely suitable for this application, but even in its present stage, it already clearly pinpoints the accommodation process as a major deadlock that needs to be broken to improve our understanding of wind acclimation in plants. Moreover, it points toward different tissues involved in fast and slow accommodation, and provides variables quantifying the relevant local phenotypes (the sensing threshold of the long-range secondary signal $S0_{\text{strains}}$ in growing tissues for fast accommodation, and strain threshold ε_0 in slow accommodation) and methods to estimate them. Finally, the very clear shift of mechanosensitivity after three stimulations makes it a candidate for tentative modeling through a regulation network (Alvarez-Buylla et al. 2007).

3.6.5 A Platform to Ecological and Agricultural/Forestry Science?

Wind acclimation and storm hardening of crop and forest species are major challenges in agronomics and forestry research (and this may even increase with global changes; Quine and Gardiner 2002). While models are available for the mechanics of lodging, wind throw, and wind break (see de Langre 2008 and Gardiner et al. 2008 for reviews), the existing growth models disregard thigmomorphogenesis and therefore skip wind acclimation (Moulia et al. 2006). A major asset of the S^3 model is that it can be coupled with very different mechanical models. It can be coupled with mechanical models to analyze the effects of the static and dynamic strains produced by wind-induced vibrations in plants (e.g., Gardiner et al. 2008; Rodriguez et al. 2008; Sellier et al. 2008). As the S^3 model can also handle growth responses to wind, it can be coupled with structure–function growth models (see Moulia et al. 2006; Fourcaud et al. 2008 for general discussion). There is still a lot of work to be done before mechanosensing models can be coupled with (1) wind–plant interaction and wind climate models or (2) existing growth models, but integrative mechanosensing models are surely a key breakthrough paving the way to a better understanding of the ecological and economical relevance of the thigmomorphogenetic acclimation, and to exploring the consequences of global changes in terms of stand growth and resistance to wind hazards.

4 Conclusions: The Challenges of Integrative Mechanobiology

This review has shown that a picture is emerging in the field of integrative mechanobiology. This emergence is based on intensive, long-standing interdisciplinary collaborations harnessing solid mechanics, fluid mechanics, nonlinear physics of dynamic systems, biomechanics, and mechanobiology. Integrative modeling and continued confrontation with experiments is central to this movement. These models provide tools for (1) specifying the consequences on measurable variables of a set of hypotheses in interaction with plant structure and (2) designing and interpreting experiments to assess these hypotheses. This approach, employing systems biology as an extension of the experimental method, contrasts with (and complements) alternative setups in which experimentalists intensively collect data while bioinformaticians set up models and data mining programs. In the case of plant biomechanics and mechanobiology, the integrative models also rely (1) on a clear identification of what is purely physical (the ISM model in our case) and what is mechanobiological (the S^3m model), and (2) on an explicit coupling of the two (sub-)models. This makes it possible to integrate the constraints on plant phenotype related to the physics of plant structure into the coupled model. The mechanical model can then be adjusted whenever necessary, such as to cope with a different experimental setup or to natural conditions. In each submodel, the effects of the topology and geometry of the plant structure (or of the studied biological system, cell, meristem, etc.) have to be taken into account, while a clear modular design and explicit setting of organizational levels has to be implemented. However, models can only generate useful insight if they are kept simple enough. Multiplying the number of elements and degrees of freedom makes models very difficult to analyze and experimentally assess. Hence, a concomitant effort to simplify the models (as done in physics through dimensional analysis; see de Langre 2008; Rodriguez et al. 2008) and address them to specific questions for which their outputs and hypotheses can be experimentally falsified is extremely valuable.

This approach thus requires setting a common interdisciplinary culture (remember the stress vs. strain issue) and common projects.

This type of interdisciplinary research program is being applied to the integrative mechanobiology of (1) the interaction between plants and their mechanical environment (wind and gravity) and (2) the growth-induced autostresses in the apical meristem (that is not reviewed here; see chapter "Plants as Mechano-Osmotic Transducers" and Traas and Moneger 2010 for recent reviews). A first step in this program is to set a minimal model of mechanobiological integration over the considered structure, coupled with a structural mechanics model (Coutand and Moulia 2000; Hamant et al. 2008). Both models have to be validated on their output (e.g., Moulia and Fournier 1997; Coutand and Moulia 2000). But mechanistic models also have to be validated in terms of their basic mechanistic modules (e.g., Coutand et al. 2009) and their capacity to dissect natural genetic variation (e.g., Sierra de Grado et al. 2008; Almeras and Fournier 2009; Coutand et al. 2010). In the case of the thigmomorphogenetic responses to external bending loads, the Sum of Strain Sensing model, S^3m, has fulfilled these requirements.

However, evolution does not comply with minimal models. There are several reasons for this, such as the lack of pure optimality principles (apart from those governed by physics) (see Gould and Lewontin 1979; Moulia and Fournier-Djimbi 1997; Niklas 1998 for a discussion on a topic going far beyond the scope of this review). Selection for mutational robustness, for example, can induce a great deal of redundancy (Alvarez-Buylla et al. 2007).

The minimal model can, however, be used as a template to integrate more detailed and mechanistic zoom modules, and as a guide for pinpointing relevant and rewarding extensions.

We have illustrated this program in the example of the S^3m model. However, the field of integrative mechanobiology is still in its infancy, and we have a long way to go before achieving a satisfactory understanding of the processes involved. Hence, we shall end this review by highlighting some of the unresolved questions and challenges that, from our point of view, are key hurdles to the mechanobiology of how plants acclimate to their mechanical (wind and gravity) environment.

4.1 Wind Loading and Accommodation

The first challenge is to take into account the repetitive nature of wind loads. This entails integrating the vibrational mechanics of wind–plant interactions (Py et al. 2006; Sellier and Fourcaud 2009). Mechanical models of vibrational dynamics are already available but have not yet been re-engineering for coupling with mechanosensing models. Specific mechanosensitive responses to strain rate or strain frequency should also be assessed. Indeed, frequency dependence has been observed in other living systems (e.g., Ehrlich and Lanyon 2002), and there are clues to indicate a plant-led regulation of its modal frequencies (Rodriguez et al. 2008).

Besides this, a further major challenge is to understand the broad long-term accommodation of mechanosensitivity. This requires an explicit two-way coupling of the mechanosensing model and the mechanical model – a task that has yet to be undertaken. A biomechanical and mechanobiological model of cell (and tissue element) mechanosensing, and its possible regulation through gene regulation networks, is clearly lacking. Once developed, this model could be integrated into a plant model such as S^3m to handle crosstalk between distant tissue elements. This is a huge task, but one that may benefit from a similar program in the mechanobiology of meristems (Green 1999; Traas and Moneger 2010).

Finally, our understanding of the transmission and reception of secondary systemic signals remains far too coarse (Brenner et al. 2006). In particular, the mechanisms underlying the setting of the threshold for a growth response to the long-range signals $S_{01\mathrm{strain}}$ and $S_{02\mathrm{strain}}$ need to be studied, as changes in this parameter shape the thigmomorphogenetic response. This makes it necessary to understand and model the reception of mechanosensitive secondary signals, and the genetic regulation and accommodation potentially involved. This would in turn help define proper dimensionless variables for the internal signal and the Sum of Strain. Quantitative analysis of gene regulation in growing tissues does, however, remain a delicate task, but a proper formalism (set through interdisciplinary biomechanical research) has just been set (Merret et al. 2010).

4.2 In Search of a Function: Plant Mechanical Design and Biomechanical Strategies

Besides improving our understanding of mechanosensing, the adaptive value of this process should be addressed. To what extent does the thigmomorphogenetic response really acclimatize plants to wind, whatever the climate? Does thigmomorphogenesis control the risk of buckling? Is strain sensing involved in acclimation to water stresses? To tackle these questions, we need to compare the mechanosensitive responses of the plants and their behavior to environmental constraints such as height growth, water losses, and storms. To achieve a sufficient level of generality requires comparison (and possibly the coupling) between biomechanical models of risks (wind damage, buckling) and the mechanosensitive growth model. This is actually the most appropriate way to ascribe a real function to thigmomorphogenetic processes and to bring mechanosensitive models into the study of dynamic structure–function models (Godin and Sinoquet 2005).

By the same token, the combination of mechanical acclimation with the capacity to recover from mechanical hazards (Moulia and Fournier 2009) should be taken into account. Plant mechanical design has to be considered through a complete biomechanical strategy for ecological resilience to mechanical constraints (Fournier et al. 2006). This sets clear connections between integrative mechanobiology and biomechanical functional ecology. This issue is of major interest for assessing the potential resilience of plants (and of ecosystems) to the possible increase in storms.

The first successes of integrative mechanobiology have thus opened up a large set of questions for interdisciplinary research with a continuum of scientific challenges ranging from cell biology and gene regulation to functional ecology and more applied issues such as crop and forest management for better resilience to global changes in the mechanical environment. This is a lively, active, and attractive research community. We really believe the next version of a review like this, say in 5–10 years (?), would be very different from this one, and we really look forward to the kind of developments on the horizon!

References

Almeras T, Fournier M (2009) Biomechanical design and long-term stability of trees: morphological and wood traits involved in the balance between weight increase and the gravitropic reaction. J Theor Biol 256:370–381

Alvarez-Buylla ER, Benitez M, Balleza D, Chaos A, Espinosa-Soto C, Padilla-Longoria P (2007) Gene regulatory network models for plant development. Curr Opin Plant Biol 10:83–91

Ancelin P, Courbaud B, Fourcaud T (2004) Development of an individual tree-based mechanical model to predict wind damage within forest stands. For Ecol Manag 203:101–121

Arteca JM, Arteca RN (1999) A multi-responsive gene encoding 1-aminocyclopropane-1-carboxylate synthase (ACS6) in mature Arabidopsis leaves. Plant Mol Biol 39:209–219

Baluška F, Šamaj J, Wojtaszek P, Volkmann D, Menzel D (2003) Cytoskeleton-plasma membrane-cell wall continuum in plants. Emerging links revisited. Plant Physiol 133:482–491

Barbacci A, Constant T, Nepveu G (2009) Theoretical and experimental study of a mechanical model describing the trunk behaviour of mature beech trees (*Fagus sylvatica* L.) under the static loading of the crown. Trees Struct Funct 23:1137–1147

Baskin TI (2006) Ultradian growth oscillations in organs: physiological signal or noise? In: Mancuso S (ed) Rhythms in plants. Springer, Berlin

Boudaoud A (2010) An introduction to the mechanics of morphogenesis for plant biologists. Trends Plant Sci 15:353–360

Braam J (2005) In touch: plant responses to mechanical stimuli. New Phytol 165:373–389

Brenner ED, Stahlberg R, Mancuso S, Vivanco J, Baluška F, Van Volkenburgh E (2006) Plant neurobiology: an integrated view of plant signaling. Trends Plant Sci 11:513–519

Coutand C (2010) Mechanosensing and thigmomorphogenesis, a physiological and biomechanical point of view. Plant Sci 179:168–182. doi:10.1016/j.plantsci.2010.05.001

Coutand C, Moulia B (2000) Biomechanical study of the effect of a controlled bending on tomato stem elongation: local strain sensing and spatial integration of the signal. J Exp Bot 51:1825–1842

Coutand C, Julien JL, Moulia B, Mauget JC, Guitard D (2000) Biomechanical study of the effect of a controlled bending on tomato stem elongation: global mechanical analysis. J Exp Bot 51:1813–1824

Coutand C, Martin L, Leblanc-Fournier N, Decourteix M, Julien JL, Moulia B (2009) Strain mechanosensing quantitatively controls diameter growth and PtaZFP2 gene expression in poplar. Plant Physiol 151:1–10

Coutand C, Chevolot M, Rowe N, Lacointe A, Scotti I (2010) Mechanosensing of stem bending and its interspecific variability in five neotropical rainforest species. Ann Bot 105:341–347. doi:10.1093/aob/mcp286

de Langre E (2008) Effects of wind on plants. Annu Rev Fluid Mech 40:141–168

Delus V, Lesnino G, Wessolly L (2004) Diagnostic de tenue mécanique de l'arbre: Test de traction, premières applications en France. PHM Rev Hort 461:1–5

Ding JP, Pickard B (1993) Modulation of mechanosensitive calcium-selective cation channels by temperature. Plant J 3:713–720

Dupont S, Brunet Y (2008) Impact of forest edge shape on tree stability: a large-eddy simulation study. Forestry 81:299–315

Ehrlich PJ, Lanyon LE (2002) Mechanical strain and bone cell function: a review. Osteopor Int 13:688–700

Fluch S, Olmo CC, Tauber S, Stierschneider M, Kopecky D, Reichenauer TG, Matuníková I (2008) Transcriptomic changes in wind-exposed poplar leaves are dependent on developmental stage. Planta 228:757–764

Foethke D, Makushok T, Brunner D, Nédélec F (2009) Force- and length-dependent catastrophe activities explain interphase microtubule organization in fission yeast. Mol Syst Biol 5:241

Fourcaud T, Zhang X, Stokes A, Lambers H, Korner C (2008) Plant growth modelling and applications: the increasing importance of plant architecture in growth models. Ann Bot 101:1053–1063

Fournier M, Stokes A, Coutand C, Fourcaud T, Moulia B (2006) Tree biomechanics and growth strategies in the context of forest functional ecology. In: Herrel A, Speck T, Rowe N (eds) Ecology and biomechanics: a biomechanical approach to the ecology of animals and plants. CRC Taylor and Francis, Boca Raton, pp 1–33

Gardiner B, Byrne K, Hale S, Kamimura K, Mitchell S, Peltola H, Ruel JC (2008) A review of mechanistic modelling of wind damage risk to forests. Forestry 81:447–463

Godin C, Sinoquet H (2005) Structural–functional plant modelling. New Phytol 166:705–708

Gould SJ, Lewontin RC (1979) The spandrels of San Marco and the Panglossian paradigm: a critique of the adpatationist program. Proc R Soc Lond B 205:581–598

Green PB (1999) Expression of pattern in plants: combining molecular and calculus-based biophysical paradigms. Am J Bot 86:1059–1076

Hamant O, Heisler MG, Jonsson H, Krupinski P, Uyttewaal M, Bokov P, Corson F, Sahlin P, Boudaoud A, Meyerowitz EM, Couder Y, Traas J (2008) Developmental patterning by mechanical signals in *Arabidopsis*. Science 322:1650–1655

Haswell ES, Peyronnet R, Barbier-Brygoo H, Meyerowitz EM, Frachisse JM (2008) Two MscS homologs provide mechanosensitive channel activities in the *Arabidopsis* root. Curr Biol 18:730–734

Hejnowicz Z (1997) Graviresponses in herbs and trees: a major role for the redistribution of tissue and growth stresses. Planta 203:S136–S146

Hepworth DG, Vincent JFV (1999) The growth response of the stems of genetically modified tobacco plants (*Nicotiana tabacum* 'Samsun') to flexural stimulation. Ann Bot 83:39–43

Holbrook MN, Putz FE (1989) Influence of neighbors on tree form: effects of lateral shade and prevention of sway on the allometry of *Liquidambar styraciflua* (sweet gum). Am J Bot 76:1740–1749

Jaffe M, Biro R, Bridle K (1980) Thigmomorphogenesis: calibration of the parameters of the sensory function in beans. Physiol Plant 49:410–416

Jaffe MJ, Leopold AC, Staples RC (2002) Thigmo responses in plants and fungi. Am J Bot 89:375–382

Jaouen G, Alméras T, Coutand C, Fournier M (2007) How to determine sapling buckling risk with only a few measurements. Am J Bot 94:1583–1593

Leblanc-Fournier N, Coutand C, Crouzet J, Brunel N, Lenne C, Moulia B, Julien JL (2008) Jr-ZFP2, encoding a Cys2/His2-type transcription factor, is involved in the early stages of the mechano-perception pathway and specifically expressed in mechanically stimulated tissues in woody plants. Plant Cell Environ 31:715–726

Lee D, Polisensky DH, Braam J (2005) Genome-wide identification of touch- and darkness-regulated Arabidopsis genes: a focus on calmodulin-like and XTH genes. New Phytol 165:429–444

Legay JM (1997) L'experience et le modèle, un discours sur la méthode. Quae Editions, Paris, France

Martin L, Leblanc-Fournier N, Azri W, Lenne C, Henry C, Coutand C, Julien JL (2009) Characterization and expression analysis under bending and other abiotic factors of PtaZFP2, a poplar gene encoding a Cys2/His2 zinc finger protein. Tree Physiol 29:125–136

Martin L, Leblanc-Fournier N, Julien JL, Moulia B, Coutand C (2010) Acclimation kinetics of physiological and molecular responses of plants to multiple mechanical loadings. J Exp Bot 61:2403–2412. doi:10.1093/jxb/erq069

Mattheck C, Bethge K (1998) The structural optimization of trees. Naturwissenschaften 85:135–140

McMahon TA (1973) Size and shape in biology. Elastic criteria impose limits on biological proportions, and consequently on metabolic rates. Science 179:1201–1204

McMaugh SJ, Lyon BR (2003) Real-time quantitative RT-PCR assay of gene expression in plant roots during fungal pathogenesis. Biotechniques 34:982–986

Mechanics (2010) Britannica Student Library. Ultimate Reference Suite. Encycloaedia Britannica, Chicago

Meng SX, Lieffers VJ, Reid DEB, Rudnicki M, Silins U, Jin M (2006) Reducing stem bending increases the height growth of tall pines. J Exp Bot 57:3175–3182

Merret R, Moulia B, Hummel I, Cohen D, Dreyer E, Bogeat-Triboulot B (2010) Monitoring the regulation of gene expression in a growing organ using a fluid mechanics formalism. BMC Biol 8:18

Monshausen GB, Gilroy S (2009) Feeling green: mechanosensing in plants. Trends Cell Biol 19:228–235

Moulia B (2000) Leaves as shell structures: double curvature, autostresses and minimal mechanical energy constraints on leaf rolling in grasses. J Plant Growth Regul 19:19–30

Moulia B, Combes D (2004) Thigmomorphogenetic acclimation of plants to moderate winds greatly affects height structure in field-gown alfalfa (*Medicago sativa* L.), an indeterminate herb. Comp Biochem Physiol A Mol Integr Physiol 137:77

Moulia B, Fournier M (1997) Mechanics of the maize leaf: a composite beam model of the midrib. J Mater Sci 32:2771–2780

Moulia B, Fournier M (2009) The power and control of movements in plants: a plant biomechanics and system biology view. J Exp Bot 60:461–486

Moulia B, Fournier-Djimbi M (1997) Optimal mechanical design of plant stems: the models behind the allometric power laws. In: Jeronimidis G, Vincent JFV (eds) Plant biomechanics. University of Reading Press, Reading, England, pp 43–55

Moulia B, Fournier M, Guitard D (1994) Mechanics and form of the maize leaf: in vivo qualification of the flexural behaviour. J Mater Sci 29:2359–2366

Moulia B, Coutand C, Lenne C (2006) Posture control and skeletal mechanical acclimation in terrestrial plants: implications for mechanical modeling of plant architecture. Am J Bot 93:1477–1489

Niklas KJ (1992) Plant biomechanics: an engineering approach to plant form and function. University of Chicago Press, Chicago, Il

Niklas KJ (1998) The influence of gravity and wind on land plant evolution. Rev Paleobot Palynol 102:1–14

Niklas KJ, Spatz H.-C (2004) Growth and hydraulic (not mechanical) constraints govern the scaling of tree height and mass. Proceedings of the National Academy of Sciences, USA 101:15661–15663

Peyronnet R, Haswell ES, Barbier-Brygoo H, Frachisse JM (2008) AtMSL9 and AtMSL10: sensors of plasma membrane tension in Arabidopsis roots. Plant Signal Behav 3:726–729

Prusinkiewicz P, Rolland-Lagan AG (2006) Modeling plant morphogenesis. Curr Opin Plant Biol 9:83–88

Py C, de Langre E, Moulia B (2006) A frequency lock-in mechanism in the interaction between wind and crop canopies. J Fluid Mech 568:425–449

Quine C, Gardiner B (2002) Climate change impacts: storms. For Commun Bull 125:41–51

Rodriguez M, de Langre E, Moulia B (2008) A scaling law for the effects of architecture and allometry on tree vibration modes suggests a biological tuning to modal compartmentalization. Am J Bot 95:1523–1537

Sato K, Adachi T, Matsuoa M, Tomita Y (2005) Quantitative evaluation of threshold fiber strain that induces reorganization of cytoskeletal actin fiber structure in osteoblastic cells. J Biomech 38:1895–1901

Schrieffer JL, Warden SJ, Saxon LK, Robling AG, Turner CH (2005) Cellular accommodation and the response of bone to mechanical loading. J Biomech 38:1838–1845

Sellier D, Fourcaud T (2009) Crown structure and wood properties: influence on tree sway and response to high winds. Am J Bot 96:885–896

Sellier D, Brunet Y, Fourcaud T (2008) A numerical model of tree aerodynamic response to a turbulent airflow. Forestry 81:279–297

Sierra de Grado R, Pando V, Martinez-Zurimendi P, Penalvo A, Bascones E, Moulia B (2008) Biomechanical differences in the stem straightening process among *Pinus pinaster* provenances. A new approach for early selection of stem straightness. Tree Physiol 28:835–846

Spatz H-C, Speck O (2002) Oscillation frequencies of tapered plant stems. Am J Bot 89:1–11

Speck T, Spatz HC, Vogellener D (1990) Contribution to the biomechanics of plants I. Stabilities of plant stems with strengthening elements of different cross-sections against weight and wind forces. Bot Acta 103:111–122

Stress (2010) Encyclopaedia Britannica. Ultimate Reference Suite. Encyclopaedia Britannica, Chicago

Stull RB (2007) An introduction to boundary layer meteorology. Springer, Berlin

Tardieu F (2003) Virtual plants: modelling as a tool for the genomics of tolerance to water deficit. Trends Plant Sci 8:9–14

Telewski FW (2006) A unified hypothesis of mechanoperception in plants. Am J Bot 93:1466–1476

Telewski FW, Pruyn ML (1998) Thigmomorphogenesis: a dose response to flexing in *Ulmus americana* seedlings. Tree Physiol 18:65–68

Traas J, Moneger F (2010) Systems biology of organ initiation at the shoot apex. Plant Physiol 152:420–427

Trepat X, Deng L, An SS, Navajas D, Tschumperlin DJ, Gerthoffer WT, Butler JP, Fredberg JJ (2007) Universal physical responses to stretch in the living cell. Nature 447:592–595

Weber's Law (2010) Encyclopaedia Britannica. Ultimate Reference Suite. Encyclopaedia Britannica, Chicago

Hydraulics of Vascular Water Transport

John S. Sperry

Abstract The science of plant water transport is equal parts of physics and biology. Plants have evolved a complex wick system that harnesses the cohesive hydrogen bond energy of liquid water and suppresses the heterogeneous nucleation of cavitation. Trade-offs between making the wick safe against cavitation and implosion, yet efficient in moving water, result in the process being limiting to plant performance. Cavitation limits the range of negative pressures that can be harnessed to move water, and the hydraulic conductance of the wick limits the flow rate that can be moved at a given negative pressure gradient. Both limits constrain CO_2 uptake via the water-for-carbon trade-off at the stomatal interface. Research in the area concerns the mechanisms of cavitation, its reversal by embolism repair, consequences for plant ecology and evolution, and the coupling of water transport to plant productivity. Very little is known of the molecular biology underlying xylem physiology.

1 Introduction

Few transport processes generate more controversy and misunderstanding than the long-distance movement of water in the xylem of plants. The transport mechanism is very simple and very remarkable – remarkable enough to be contentious. In its essentials, it is a physical process. Yet the physics impose major constraints on the biology of plants. This chapter explains water transport and identifies issues where research is active or needed. It is written to be accessible for the nonspecialist.

Of all biological transport systems, plant xylem transports the most fluid over the longest distances (Vogel 1994). A tree with a trunk the size of a man's waist transports on the order of 100 l of water on a clear day (Enquist et al. 2000). Summing over a watershed, plant water use matches the runoff in rivers, and constitutes just under half the annual precipitation (Schlesinger 1997). This conspicuous consumption makes water a major limiting resource for plant life. What is

J.S. Sperry
Department of Biology, University of Utah, 257S 1400E, Salt Lake City, UT 84112, USA
e-mail: j.sperry@utah.edu

P. Wojtaszek (ed.), *Mechanical Integration of Plant Cells and Plants*,
Signaling and Communication in Plants 9, DOI 10.1007/978-3-642-19091-9_12,
© Springer-Verlag Berlin Heidelberg 2011

all this water used for? A miniscule percentage, less than 1%, is split in the photosynthetic reaction to release O_2. A few more percentage points are pulled into cells by osmosis in the growth process (see chapter "Plants as Mechano-osmotic Transducers"). The remaining bulk, usually over 95%, is evaporated to the atmosphere through open stomata: the process of transpiration (Sutcliffe 1968). If vascular plants could exist without stomata, plants would not be water-limited and arid and semi-arid landscapes would be far more productive.

But, plants need stomata to let in sufficient CO_2 for photosynthesis, and so they must transpire. The exchange rate of water for carbon is very poor, with hundreds of molecules of water lost per molecule of CO_2 fixed under typical conditions. The transpiration stream represents a river of water flowing in exchange for a relative trickle of carbon.

2 The Cohesion–Tension Transport Mechanism

With such a poor exchange rate of water moved per unit carbon acquired, plants had to evolve an energetically cheap transport process to avoid starvation. It is hard to imagine a more economical process than the prevailing "cohesion–tension" explanation (Pickard 1981). According to this mechanism, water is wicked up the plant. Like a wick, the plant sucks fluid to a porous evaporating surface by capillary force. It is a physical process that operates without the need for metabolic energy once the anatomically complex "wick" system has been grown.

The essential physics of the wick are illustrated in Fig. 1a. Water evaporates from a meniscus that is anchored by capillary forces in a tube. These forces are the adhesion of water to the hydrophilic tube and the surface tension of water. Because the meniscus is held in a relatively stationary position as water evaporates from its surface, the pressure in the liquid water drops, and the evaporated molecules are replaced by bulk flow towards the meniscus (Rand 1978). The energy source is the vapor pressure deficit that initiates evaporation. The energy for evaporation is transferred to the hydrogen bonds that generate the capillary forces anchoring the meniscus. Hydrogen bond energy creates the drop in water pressure which drives flow up the tube.

Although energetically efficient for the plant, the cohesion–tension mechanism is limited in the subatmospheric water pressure that it can generate. This pressure, P, is usually expressed relative to atmospheric and is a negative value. One theoretical limit to P is "capillary failure" of the forces holding the meniscus in place. The most negative pressure the meniscus can generate without giving way, P^{CAP}, is given by the capillary equation:

$$P^{\mathrm{CAP}} = -4T\cos(a)/D_{\mathrm{c}}, \qquad (1)$$

where T is the surface tension of water, a is the contact angle between water and the tube wall (a measure of wettability or adhesion), and D_{c} is the diameter of the

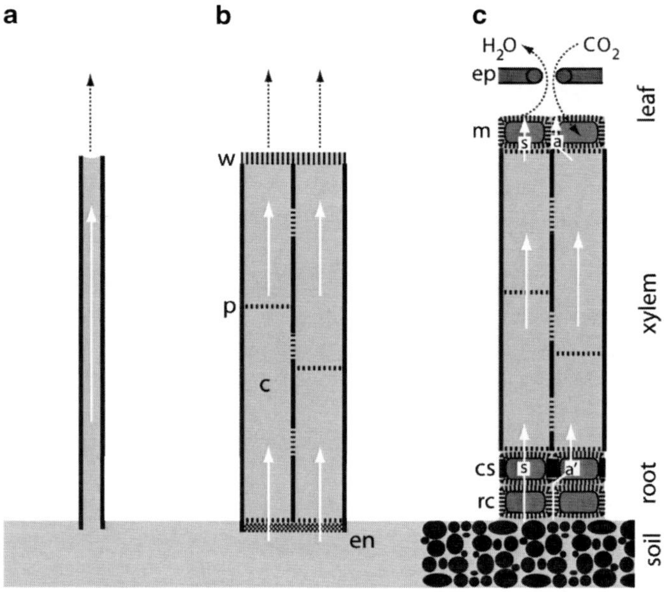

Fig. 1 The cohesion–tension mechanism for transpiration-driven xylem flow. (**a**) The basic wicking process: evaporation from a meniscus coupled to bulk flow by capillary forces. (**b**) In plants, large capillary suction is generated in narrow pores of cell walls (w) and long-distance bulk flow occurs in a low-resistance network of dead xylem conduits (c). Conduits are connected by pits (p), which also protect against air-entry and embolism in the inevitable event of damage to individual conduits. Water is filtered through living cell membranes at the root endodermis (en) by reverse osmosis. (**c**) Living cells (rounded rectangles enclosed in hatched cell walls) do not generate transpirational flow (*ep* leaf epidermis; *m* mesophyll; *cs* Casparian strip of the endodermis; *rc* root cortex). Water flows through them to the site of evaporation via symplastic (*s* arrows) and apoplastic routes (*a* arrows). In the root, the apoplastic route (a') is interrupted by the Casparian strip. Transpiration is actively regulated by stomatal opening through which the water-for-carbon gas exchange takes place

circular tube holding the meniscus. If P drops to P^{CAP}, the meniscus is pulled inward, emptying the tube by capillary failure.

A second limit on P concerns the boiling point of water. The vapor pressure of water at 20°C is about 2.3 kPa in absolute pressure. At sea level, that yields a potential minimum (most negative) P of −99 kPa at the boiling point. Boiling is generally heterogeneously nucleated by gas bubbles, hydrophobic surfaces, or other materials that destabilize the cohesion of water molecules via hydrogen bonding (Debenedetti 1996). If heterogeneous nucleation is suppressed, boiling will not occur, and liquid water will sustain a P below the boiling point. The water is in a metastable "super-sucked" state: a state equivalent to being super-heated (Debenedetti 1996), except the water pressure is being lowered past the boiling point at ambient temperature. As explained in Sect. 4, P in plant xylem must drop below −99 kPa for water to be extracted from soil and pulled up the xylem against gravity and friction. Therefore, the cohesion–tension mechanism requires the xylem sap to be in a metastable liquid state.

Water under negative P is often referred to as being under "tension" because the liquid exhibits tensile strength as if it were a solid. In Lyman Briggs' classic "Z" tube experiments, liquid water placed under tension by centrifugal force sustained a P of below -25 MPa ($-25{,}000$ kPa) before the abrupt appearance of the vapor phase (Briggs 1950). This abrupt "boiling" is termed cavitation. In Brigg's experiments, cavitation was probably nucleated by irregularities in the contact between the water and the Z-shaped capillaries. In the absence of any heterogeneous nucleation, theoretical models predict that pure water can develop a P of at least -30 MPa at room temperature before the kinetic energy of the water molecules is sufficient to overcome their cohesion and spontaneously nucleate cavitation (Kwak and Panton 1985).

The biggest threat to metastable water in plant xylem is "bubble nucleation": cavitation triggered by gas bubbles in the sap (Sect. 6). The most negative P that can be sustained in the presence of an air bubble, P^{BUB}, can be derived from La Place's law for spheres (Vogel 1988):

$$P^{\mathrm{BUB}} = -4T/D_{\mathrm{b}}, \tag{2}$$

where D_{b} is the diameter of curvature for the bubble. If P drops below P^{BUB}, the surface tension forces that tend to collapse the bubble are exceeded by the negative P that tends to expand the bubble, and the gas phase grows (Oertli 1971). The similarity of (1) and (2) indicate the central importance of surface tension in limiting the magnitude of P generated by the wicking process.

A third limit to P concerns the strength of the capillary tube itself. The internal fluid is under negative pressure, which can lead to the collapse of a weak tube. The P causing failure by tube implosion, P^{IMP}, is difficult to express in simple equation because it depends on the shape of the tube, the relative thickness of the wall, and the internal structure of the wall, among other things (Young 1989). In the simple case of an isolated cylindrical tube where the wall is isotropic and less than a tenth of the tube radius (e.g., Fig. 1a), La Place's law for cylindrical hoop stress can be used to estimate P^{IMP}:

$$P^{\mathrm{IMP}} = -2\sigma(t/D_{\mathrm{c}}), \tag{3}$$

where σ is the strength (force per cross-sectional wall area) of the wall against implosion and t is the thickness of the tube wall. While (3) does not apply to actual xylem conduits, it is useful for showing the importance of the "thickness-to-span" ratio (t/D_{c}) for determining P^{IMP}. The greater this ratio, the stronger the conduit wall, regardless of additional structural complexities in actual conduits (Hacke et al. 2001).

The role of these three potential limits: P^{CAP}, P^{BUB}, and P^{IMP}, in constraining the range of P achievable in plant xylem will be considered after looking at the anatomy of the "wick" grown by plants.

3 The Anatomy of the Flow Path

The anatomy of the flow path couples narrow pores at the site of evaporation, which maximize capillary suction, to much wider water-filled channels in the vascular tissue for maximizing ease of transport. The physical concept is illustrated in Fig. 1b. Living cells of leaves and roots are added in Fig. 1c.

3.1 *The Evaporating Surface*

The evaporating surface is the cell walls abutting internal air spaces in the leaf or other transpiring tissue. The meniscus of Fig. 1a is located in the nanometer-scale pores of the cell wall, represented by the surface of narrow channels marked "w" in Fig. 1b, and by the leaf mesophyll cell walls labeled "m" in Fig. 1c. Cell wall capillary forces generate the negative *P* that pulls water up the xylem and out of the soil. Being narrow and hydrophilic, the wall channels can theoretically generate extremely negative *P*: a wall pore 5 nm in diameter would not exhibit capillary failure until $P^{CAP} = -58.4$ MPa, assuming a contact angle of zero and a surface tension of 0.073 N m^{-1} (1).

The negative water pressure in the cell wall channels pulls water from the xylem conduits either directly via an apoplastic route of contiguous walls (Fig. 1c, arrow labeled "a"), or indirectly through protoplasts by a symplastic route (Fig. 1c arrow labeled "s"; Canny 1990). In the symplastic route, water is ultimately pulled from the protoplast by reverse osmosis to the site of evaporation at the cell wall surface. Osmosis itself does not power the transpiration stream; it keeps cells hydrated against the negative *P* of the apoplast. In the absence of transpiration, however, osmotic uptake of water by growing cells will initiate xylem transport (Boyer et al. 1985). Osmosis into the expanding cell exerts an inward pull on the wall menisci, which hold fast and generate negative *P* for water uptake from the xylem.

3.2 *The Xylem Conduits*

The xylem conduits (tubes labeled "c" in Fig. 1c) provide a high-conductance pipeline for delivering the transpiration stream to the site of evaporation. Without them, the negative *P* required to supply transpiration in a plant taller than a few centimeters would stress the plant beyond physiological tolerance (Raven 1987). Xylem conduits are many microns in diameter as opposed to the nanometer-scale cell wall channels. Poiseuille's law for laminar flow (which prevails in xylem) predicts that hydraulic conductivity (flow rate per pressure gradient) is proportional to tube diameter to the fourth power. Consequently, the conductivity through a xylem conduit is many orders of magnitude greater than through cell wall channels.

High conductivity means the pressure gradient of the transpiration stream can be relatively low: ca. 0.01 MPa m^{-1} in trunks and major branches (Zimmermann 1983). Pressure gradients increase as conduits become narrower moving from trunks to minor branches and twigs. The highest gradients appear to be in petioles and leaf veins where they can exceed 10 MPa m^{-1} (Zwieniecki et al. 2002). Modest pressure gradients in the major stems mean that more water can be carried longer distances without requiring extremely negative *P*. The total pressure drop required to pull the transpiration stream against friction is often around 1 MPa (1,000 kPa) in angiosperms, and somewhat greater in conifers (Mencuccini 2003).

In keeping with the physical basis of transport, the xylem conduits function after they die and lose their protoplasts. They are cell wall skeletons, with thick secondary walls that are rigidified by lignin to minimize the risk of implosion by negative sap pressure (3). The thick walls help maximize the "thickness to span" ratio (t/D_c). The lignification and layering of the secondary wall maximizes wall strength (σ). Increases in both of these variables make P^{IMP} more negative (3), and provide greater protection from conduit implosion.

The ancestral type of conduit, the tracheid, is a single elongated cell several millimeters long. Most seedless vascular plants and gymnosperms function with tracheids only. Angiosperms and gnetophytes (and some ferns) have the derived vessel type of conduit. Vessels consist of many unicellular vessel members joined by perforation plates to form a multicellular tube several centimeters long. Tracheids evolved into vessel members to give rise to vessels (Bailey and Tupper 1918).

Importantly, as the conduits mature and lose their protoplasts to become functional, they start out full of fluid. If not, their lumens would be unable to generate sufficient capillary suction to fill themselves with water in most conditions. Because the lumen diameter is many microns, the P^{CAP} of the conduit lumen is only a few kPa negative (1) which is much less negative than the mid-day *P* of the transpiration stream. Although capillary forces are responsible for driving the transpiration stream, capillary *rise* is not involved. The conceptual wick in Fig. 1c cannot function unless it is primed by being water-filled. In the event that conduits do become gas-filled, many species have mechanisms for actively pumping water into empty conduits (see Sect. 8).

All vascular systems have to minimize impacts of the inevitable leaks that occur. Leaks in the transpiration stream are inward. If a xylem conduit is damaged, the transpiration stream does not pour out because it is under subatmospheric pressure. Instead, air leaks into the conduit. The impact of leaks is minimized by the presence of multiple conduits in parallel and in series. The conduits are coupled by interconduit pits that act as "check valves" allowing passage of water but resisting passage of an air–water interface (dashed openings labeled "p" in Fig. 1b). Air is thereby inhibited from spreading from a gas-filled ("embolized") conduit (Zimmermann 1983). Whether the network is plumbed by tracheids or vessels, the basic compartmentalization represented in Fig. 1b is preserved; the only difference is in the size of the compartments and details of the interconduit pitting.

3.3 *Soil Water Uptake*

The network of overlapping conduits extends to the roots where the negative *P* pulls water from the soil (Fig. 1c). As in the leaves, there is a symplastic (Fig. 1c, "s" arrow) and apoplastic (Fig. 1c, "a′" arrow) route for water to move across the root cortex (cell layer "rc" in Fig. 1c; Steudle 1994). At the endodermis, however, (and in some plants also at the exodermis) most water is forced through cell membranes because of the Casparian strip which forms an apoplastic barrier (Fig. 1c, "cs"). Under transpirational conditions, the water is pulled across the membranes by reverse osmosis. The endodermal filter (hatched zone labeled "en" in Fig. 1b, cell layer labeled "cs" in Fig. 1c) allows the plant to control the content of the transpiration stream. When transpiration is negligible, and the soil is wet, the endodermis can allow water to move into the xylem by osmosis, creating positive "root pressure" in the xylem (Steudle 1994).

Soil water is also usually under a negative *P*, and for the same reason as in the plant: capillary forces holding water in pore spaces (Fig. 1c, "soil"). The cohesion–tension mechanism can be summarized as a tug-of-war on a rope of water between capillary forces in cell walls versus soil.

4 The Magnitude of Xylem Water Pressure (*P*)

Most textbook descriptions of water transport in plants emphasize the need to pull water against the force of gravity. The hydrostatic slope is -9.8 kPa m^{-1} (20°C, 45° latitude, sea level). Hence, a 10-m tall tree must have xylem pressure of -98 kPa just to counteract gravity (Vogel 1988). This is near the boiling point of water (-99 kPa at sea level), a fact which excites controversy, because trees can be much taller than 10 m. Obviously, for the cohesion–tension mechanism to work, heterogeneous nucleation of boiling (cavitation) must be suppressed in the xylem conduits and cell wall pores.

Gravity actually exerts a relatively trivial influence on xylem *P*. As mentioned above (Sect. 3.2), about 1,000 kPa of pressure is required to move water from soil to leaves against friction, a value that does not appear to be markedly size-dependent. Overcoming friction and gravity requires a 10-m tall tree at mid-day to have a xylem pressure of about $-1{,}100$ kPa, or more conveniently, -1.1 MPa. The other significant determinant of xylem pressure is the soil water potential. Because of the reverse osmosis at the root, the xylem must extract water not only from the soil pore space, but also against the osmotic strength of the soil solution. A 10 m mangrove rooted in seawater must pull against ca. -2.4 MPa of osmotic potential, which added to -1.1 for gravity and friction, comes to -3.5 MPa. In deserts, plants have to pull against soil water potentials of -5 MPa or lower. The lowest xylem pressure the author is aware of is -13.1 MPa (mean value) in a shrub from the California chaparral (*Ceanothus cuneatus*) during the dry season (Jacobson et al. 2007).

5 The Limits to Negative Pressure (*P*) in Xylem: "Vulnerability Curves"

How low can xylem pressures drop before xylem transport fails? This question strikes at the heart of the cohesion–tension mechanism, because for it to work, water pressures must be able to become sufficiently negative to extract water from soil and move it against gravity and friction.

"Vulnerability curves" provide the answer for how low the sap pressure can drop before water transport fails (Tyree and Sperry 1989). These curves show the hydraulic conductivity of the xylem as it is exposed to progressively more negative pressures. Curves from stem xylem of three species are shown in Fig. 2. The conductivity is expressed as the percentage loss from the initial value (PLC). The range of pressure causing vascular dysfunction is from where PLC first becomes significant to where it reaches 100. At 100 PLC, enough xylem conduits have become nonfunctional to stop transport completely. The loss of conductivity is generally gradual, indicating that some conduits in the stem are more vulnerable than others.

The species in Fig. 2 differ in their vulnerability to failure, but all can generate sufficient negative pressure to function by the cohesion–tension mechanism in their environment. Cottonwood (*Populus fremontii*), the most vulnerable species, begins to fail between −1.0 and −1.25 MPa. But it grows in riparian environments where soil water potentials are always near zero. It needs less than −1.25 MPa of negative pressure to move water against friction and gravity (Pockman and Sperry 2000). In contrast, hoary-leaf ceanothus (*Ceanothus crassifolius*) is the most resistant to failure. It grows in seasonally very dry chaparral of southern California where it

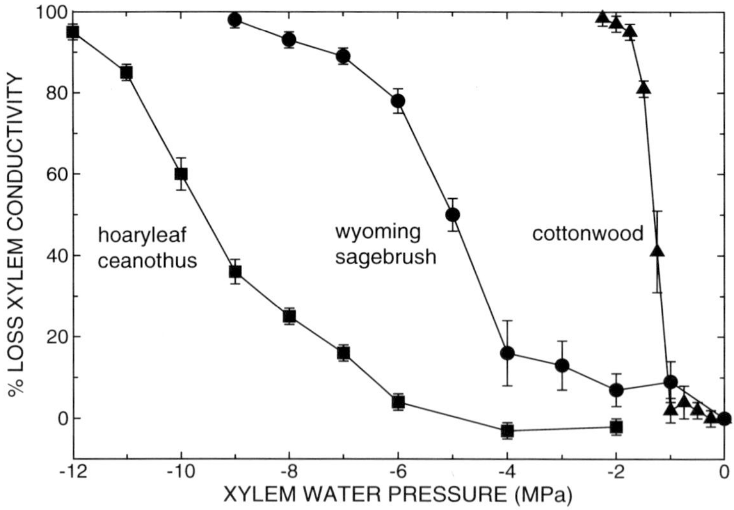

Fig. 2 Vulnerability curves of stems from hoaryleaf ceanothus (*Ceanothus crassifolius*), wyoming sagebrush (*Artemisia tridentata* var. *wyomingensis*), and cottonwood (*Populus fremontii*). From Sperry (2000)

needs very negative pressure (to −11 MPa or lower) to extract water from dry soils (Davis et al. 2002). The sagebrush (*Artemisia tridentata*) is intermediate, and grows in environments intermediate in stress (Kolb and Sperry 1999). It appears to be a general rule that vulnerability to vascular failure correlates with the physiological range of P required by a plant in its habitat (Maherali et al. 2003).

Vulnerability curves can be measured in a variety of ways. The ones in Fig. 2 come from the centrifugal force method (Pockman et al. 1995). This technique adapts Lyman Brigg's "Z-tube" experiments on capillary tubes to plant xylem. Stems are spun in custom rotors to put the xylem water under tension, and the conductivity of the stem is measured either during spinning (Cochard 2002) or between bouts of spinning using a hydraulic head (Alder et al. 1997). The two approaches appear to give similar results (Li et al. 2008). A more tedious, but more natural method is to dehydrate plants or plant parts to different pressures measured using either a pressure bomb or psychrometer (Sperry 1986). Conductivity is measured on stem pieces cut under-water (to avoid inducing embolism) and expressed relative to either the nonstressed value or to a maximum value obtained after flushing the stem to reverse in situ embolism. This "dehydration" method generally matches centrifugal techniques (Pockman et al. 1995).

5.1 Freeze-Related Dysfunction

Freeze–thaw cycles can also induce xylem dysfunction as illustrated in Fig. 3a for water birch (*Betula occidentalis*). The dashed line is the room temperature "water

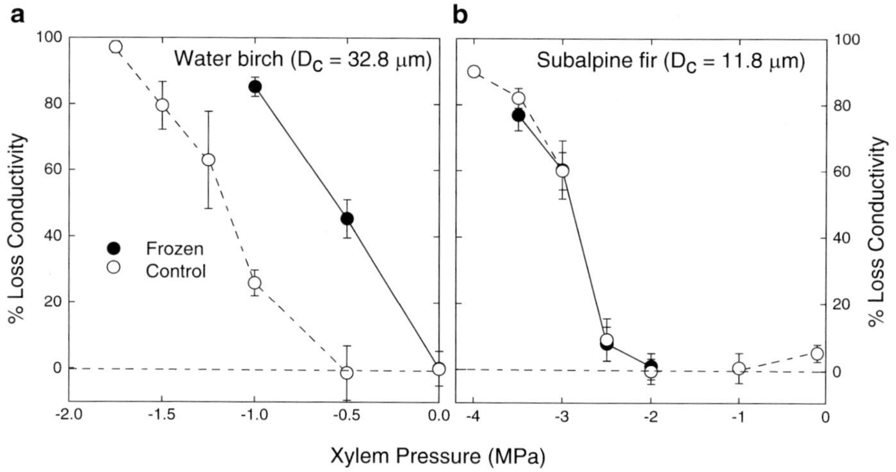

Fig. 3 Vulnerability curves with (*solid symbols, lines*) versus without (*open symbols, dashed lines*) a freeze–thaw cycle at each xylem pressure. Freeze–thaw cycles induce additional dysfunction in some species (**a**, Water birch, *Betula occidentalis*) but not in others (**b**, subalpine fir, *Abies lasiocarpa*, Dc = mean xylem conduit diameter). From Sperry and Sullivan (1992)

stress" vulnerability curve ("control", open symbols), indicating a threshold of between -0.5 and -1.0 MPa for the onset of failure. The curve was produced using the centrifugal force method. The solid line is a "freeze–thaw" curve ("frozen," solid symbols), representing stems that were frozen and thawed in the centrifuge at the indicated xylem pressure. The freeze–thaw treatment caused much more dysfunction in this species than water stress alone. Species differ in their response to freeze–thaw cycles. In contrast to birch, stems of sub-alpine fir (*Abies lasiocarpa*) shown in Fig. 3b show no additional failure caused by the freezing treatment (Sperry and Sullivan 1992). Numerous studies have documented naturally occurring loss of xylem conductance associated with freeze–thaw events in nature (Cavender-Bares 2005).

6 Causes of Limits to Negative Xylem Pressure (*P*)

Vulnerability curves support the cohesion–tension mechanism. They indicate that nucleation of cavitation can be suppressed and metastable water can exist. Negative pressures in xylem are sufficient to pull water from soil, against friction and gravity, to the site of transpiration. But the curves also show that P is limited. What limits P?

Research has documented xylem dysfunction by all three modes: capillary failure ($P < P^{CAP}$), bubble-nucleation ($P < P^{BUB}$), and implosion ($P \leq P^{IMP}$). Implosion appears to be the rarest cause of dysfunction, having been observed only in the axial tracheids of pine (*Pinus*) needles (Cochard et al. 2004), transfusion tracheids of *Podocarpus* (Brodribb and Holbrook 2005), vessels of *Eucalyptus* species (Cuevas 1969), and vessels of cell-wall mutants in tobacco (Piquemal et al. 1998). In *Pinus* and *Podocarpus*, the implosion is known to be reversible, with buckled walls at or below P^{IMP} recovering their original shape with rehydration as P rises back above P^{IMP}. In most species, implosion has not been observed, and dysfunction is caused by cavitation and gas-phase embolism before P can reach P^{IMP}. Estimated safety factors from implosion (P^{IMP}/P at failure) range from 1.9 for angiosperm vessels to 6.8 for the tracheids of conifer stem wood (Hacke et al. 2001).

Much research points to gas bubbles as the culprits that trigger cavitation and break the water column in nonimploding conduits. Under water-stress conditions, bubbles are pulled into the functional conduits by the "air-seeding" process (Zimmermann 1983). As sap pressures become increasingly negative in a water-filled conduit, air is pulled through the interconduit pits from previously embolized conduits (Fig. 4d; Crombie et al. 1985; Sperry and Tyree 1988). In many cases, the air enters by capillary failure at the pit membrane [$P < P^{CAP}$, (1)]. Once inside the water-filled conduit, the bubble can seed the phase change to vapor [$P < P^{BUB}$, (2)] and the water pressure rises abruptly to near atmospheric (Shen et al. 2002). The gas void expands to fill the entire conduit as water is drained by the transpiration stream flowing through adjacent conduits. Air-seeding could also occur through microfractures in the conduit wall created by incipient implosion (Jacobsen et al. 2005),

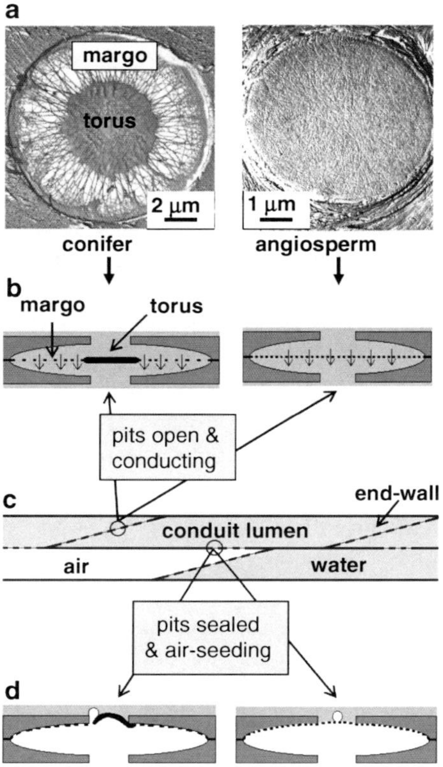

Fig. 4 Structure and function of interconduit pits in conifer tracheids (*left*) and angiosperm vessels (*right*). (**a**) Pit membranes, *face view*. (**b**) *Side view* schematic of membranes within the pit chamber formed by the secondary walls. Pits open and functioning in water transport. (**c**) Schematic of pit location within conduit network. (**d**) *Side view* of pits in sealed position showing proposed air-seeding process. From Pittermann et al. (2005)

or from microbubbles trapped in wall crevices (Pickard 1981; Shen et al. 2002), but there is no evidence for either mechanism.

The air-seeding process described above requires an embolized conduit to begin with, but embolism by damage is inevitable. Conduits become air-filled from leaf abscission, branch breakage, insect feeding, or even protoxylem rupture during tissue growth. Although the pits evolved to prevent leakage from embolized conduits, they also evolved to pass water freely. These two functions are in conflict, and consequently these pit valves appear to be the Achilles' heel of the pipeline.

The fact that cavitation can be caused by air entry into the xylem conduits makes it possible to measure vulnerability curves using positive air pressure rather than negative sap pressure (Cochard et al. 1992). In the "air-injection" method, a hydrated stem with xylem pressures at atmospheric is injected with air at progressively greater pressure, and the effect on hydraulic conductivity is measured. In this experiment, air is being pushed into the xylem rather than being pulled by negative sap pressure. As expected, air-injection curves generally agree with other vulnerability curve methods that use negative xylem pressure (Sperry et al. 1996).

Air-seeding is not the only source of bubbles in the xylem sap. According to the "thaw-expansion" hypothesis, freeze–thaw cycles create air bubbles that potentially trigger cavitation (Sucoff 1969). Gases are practically insoluble in ice, and tend to freeze out in bubbles as the sap freezes. If the sap pressure is negative enough, the bubbles will nucleate cavitation. "Negative enough" means P is less than $P^{\mathrm{BUB}} = -4T/D_{\mathrm{b}}$, where D_{b} is the bubble diameter [(2); Yang and Tyree 1992]. In support of the hypothesis, experiments under controlled conditions where transpiration is negligible have shown that it is indeed the thawing phase that creates the cavitation and not the freezing phase (Mayr and Sperry 2010). Furthermore, water that is super-saturated with air creates much more cavitation after a freeze–thaw cycle than water at an equilibrium gas concentration (Sperry and Robson 2001). Bubble nucleation according to the thaw-expansion mechanism is assumed to cause transport failure associated with freeze–thaw cycles (Fig. 3a).

7 Anatomy of Cavitation

Textbooks often remark that larger conduits are more vulnerable to cavitation, but the reality is much more complex, and in many cases not well resolved. Although a size-by-vulnerability trend can be found for cavitation by water stress (Fig. 5a), it is generally statistically weak and often not significant depending on the sample size and species composition (Tyree et al. 1994; Pockman and Sperry 2000; Hacke et al. 2006). An ambiguous dependency on conduit size is not surprising, because the vulnerability to air-seeding should depend on the structure of the interconduit pits rather than the dimensions of the conduit.

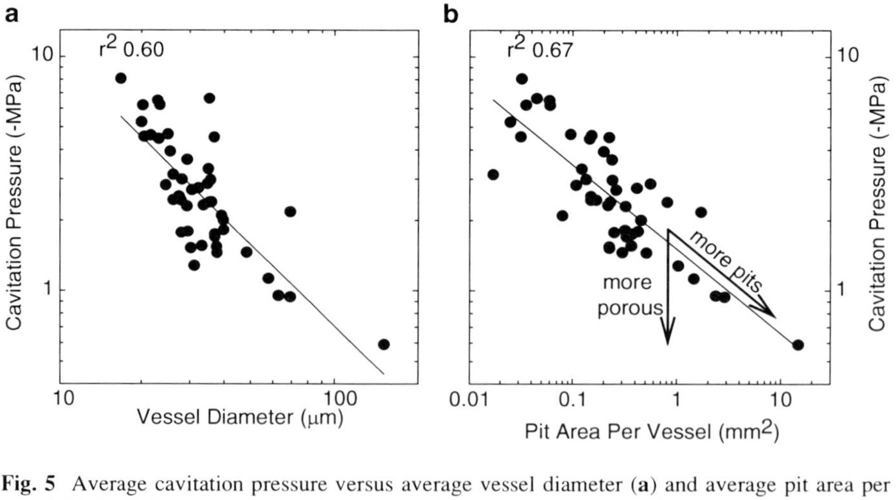

Fig. 5 Average cavitation pressure versus average vessel diameter (**a**) and average pit area per vessel (**b**) in angiosperms. Data points are species averages, compiled from Hacke et al. (2006), Christman et al. (2009), Christman and Sperry (2010)

7.1 Pit Valve Structure and Function

Interconduit pits are thin areas where secondary wall has not been deposited (Fig. 4). Only the primary walls divide the adjacent conduits, forming a thin sheet of cellulosic meshwork termed as pit membrane. In most plants (excepting most gymnosperms), the pit membrane is homogenous: uniformly thin and microporous (Fig. 4a, right). The pores in this membrane form a capillary seal and inhibit air entry. Air-seeding appears to occur by capillary failure rather than membrane rupture (Fig. 4d, right). If the air bubble is large enough to nucleate cavitation [(2); Shen et al. 2002], the pressure will relax to atmospheric and the gas void will grow to fill the entire conduit as water is drained out by the transpiration stream. According to (1), larger pores (bigger D_c = pore diameter) will make for leakier pit membranes and more vulnerable xylem.

What complicates this simple conclusion is that most membrane pores are much smaller than those indicated by the cavitation pressure of a stem (Choat et al. 2008). Pores of air-seeding size are apparently quite rare, which suggests a role for probability. A single angiosperm vessel can have 10,000 or more interconduit pits (Christman et al. 2009). No matter how faithfully the developmental program is performed, not all of these pits will be the same. Of the thousands of pits that might be holding out the air from an adjacent embolized vessel, it only takes one leaky membrane to compromise the seal. The more pits there are, the leakier will be the seal by chance. Indeed, there is a tendency for more vulnerable angiosperm xylem to have a greater extent of pitting per vessel (Fig. 5b, "more pits" arrow; Wheeler et al. 2005). This may also underlie the tendency for bigger vessels to be more vulnerable to cavitation (Fig. 5a): because they tend to have more pits.

Pit structure may also influence the probability of leaky membranes in a vessel. Species with thinner membranes tend to be more vulnerable to air-seeding (Choat et al. 2008). Fewer layers of cellulose microfibrils may increase the chances of creating thin spots or large holes during pit membrane development. For a given number of pits, thinner and more porous membranes would tend to increase vulnerability (Fig. 5b, "more porous" arrow). The consensus is that it is the combination of individual pit structure and pit number that determines vulnerability to cavitation by air-seeding (Fig. 5b, "more pits" and "more porous" arrows).

A related problem is the difficulty of measuring the vulnerability of large-vesseled species like ring-porous trees, many tropical trees, and lianas. Centrifuge and air-injection methods generally indicate that many large vessels are exceptionally vulnerable, with some cavitating at a negative P of only a few tenths of an MPa (Li et al. 2008). Some field studies also show many large vessels are not functioning at mid-day (Taneda and Sperry 2008). However, long vessels may cause artifacts in at least some centrifugal force techniques. Long vessels allow microbubbles or other nucleators to enter the center of the stem that is placed under tension, causing premature cavitation (Cochard et al. 2005). More techniques need to be brought to bear on cavitation including Magnetic Resonance Imaging (MRI) and single-vessel measures of air-seeding pressures (Holbrook et al. 2001; Choat et al. 2005).

Conifers and most gnetophytes have a different kind of interconduit pit (Fig. 4, left). The pits are uniformly circular, and rather than having a homogenous membrane, the center of the membrane is greatly thickened to form a "torus" (Fig. 4a, left). Surrounding and holding the torus is a very porous "margo." To seal against air entry, capillary forces at the large margo pores, though relatively weak, are strong enough to aspirate the membrane, sealing the thick torus over the pit aperture (Fig. 5d, left). Air-seeding through torus-margo pits seems to occur by slippage of the torus from sealing position, exposing (and perhaps even rupturing) the margo whose pores are too large to prevent air-seeding (Sperry and Tyree 1990). It is possible that air-seeding could also occur through pores in the torus, but they are usually too small to correspond to the measured air-seeding pressures.

Why did the torus–margo membrane evolve within the gymnosperms? These gymnosperms exhibit the same wide range of vulnerability to cavitation as other plants, so the torus–margo system is not necessarily more air-tight than homogenous pits. The reason appears to be that the torus–margo membrane is about 60 times more conductive to water flow than the microporous homogenous membrane (Pittermann et al. 2005). This is a major benefit when water must cross these pits every few millimeters through the short unicellular tracheids of these species. The tracheids are particularly short in the trunks and branches of conifers and vesselless gnetophytes because they do double duty in providing mechanical strength.

7.2 Conduit Structure and Freeze–Thaw Cavitation

The link between conduit size and vulnerability to cavitation by freezing and thawing is relatively straightforward, both experimentally and theoretically. Experimentally, whether in the field or in the lab, species with wider conduits exhibit a greater loss of conductivity from freeze–thaw cycles (Hammel 1967; Cochard and Tyree 1990; Cavender-Bares 2005). In controlled centrifuge experiments, conifer tracheids and angiosperm vessels exhibit a similar relationship between conduit diameter and vulnerability (Fig. 6; Davis et al. 1999a; Pittermann and Sperry 2003). The reason that birch is more vulnerable to freezing than the fir in Fig. 3 is because it had larger diameter conduits. There is apparently nothing special about fir tracheids that make them resistant, other than their tendency to be narrow (Pittermann and Sperry 2003). Narrow angiosperm vessels are just as resistant to cavitation by freezing and thawing as narrow conifer tracheids (Fig. 6).

Theoretically, the greater vulnerability of wide conduits is explained by their predicted tendency to form larger diameter bubbles during freezing (Sperry and Robson 2001). If freezing starts at the conduit wall and moves towards the center of the lumen, air will outgas in the lumen center. The greater the cross-sectional area of the lumen, the larger the bubble diameters (D_b), and the more likely the sap P on thawing will be more negative than P^{BUB} (2) and cause cavitation and embolism. The bubbles dissolve rapidly during the thaw because the surrounding water has

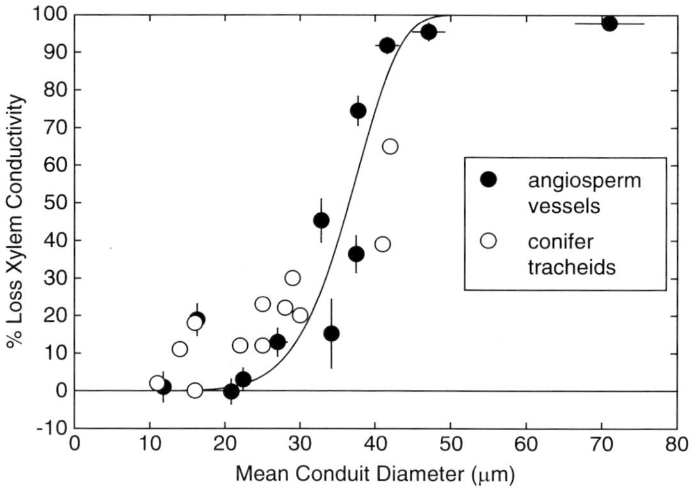

Fig. 6 Loss of hydraulic conductivity caused by a single freeze–thaw cycle at −0.5 MPa versus average conduit diameter. Data are species averages for angiosperm vessels (*open symbols*) and conifer tracheids (*closed symbols*). From Pittermann and Sperry (2003)

been degassed, so D_b is decreasing as liquid P is restored during melting (Sucoff 1969). This explains why some narrow-conduit species show no freeze–thaw effect even if exposed to very negative P (Fig. 3b): the bubbles shrink faster than negative P can be restored. It also explains the observation that more rapid thaws and also more negative P tend to increase the amount of cavitation caused by freeze–thaw cycles (Davis et al. 1999b).

Some freezing-related observations are more difficult to explain. In some species, the embolism caused by freezing increases with decreasing minimum ice temperature (Pittermann and Sperry 2006). This is not predicted by the thaw-expansion hypothesis, which depends only on the bubble size during the thaw period. A possible explanation is that the low temperature causes tissue damage which makes the vascular system more vulnerable to cavitation by water stress during the thaw (Pockman and Sperry 1997). More research is needed on this phenomenon, which can be an important factor in range limits of plants (Sect. 9.3).

8 Reversal of Cavitation

Cavitated and embolized conduits can be refilled under some circumstances. From a physical standpoint, for a conduit to be refilled, the water pressure, P, needs to be less negative than the fairly weak capillary suction the conduit lumen (diameter D_c) is capable of generating [$P^{CAP} = -4T\cos(a)/D_c$, (1)]. For a conduit of $D_c = 20$ μm, P must rise above −14.6 kPa to allow water to be pulled into the conduit by capillary action (assuming $a = 0$). The surface tension of the meniscus puts the air under pressure, eventually dissolving away the bubble (Yang and Tyree 1992; Grace 1993). Surface sculpturing and contact angle (a) could modify the process, but little is known of such effects.

The P of the bulk xylem sap carried around the embolized conduit is rarely above the refilling threshold. Exceptions are in short plants rooted in wet soil under nontranspiring conditions (rain, night), or when P is elevated by osmotically driven root pressure (Sect. 3.3). Root pressure is particularly important in overnight refilling of herbs, and in annual spring refilling in many temperate trees and vines (Milburn and McLaughlin 1974; Sperry et al. 1987; Cochard et al. 1994; Hacke and Sauter 1996). Spring root pressures in many temperate woody species typically peak prior to leaf expansion and in a few days refill xylem conduits embolized by freeze–thaw cycles over the winter. Stem pressures generated in a few temperate trees in early spring (*Acer* species, most famously) also act to reverse winter embolism (Sperry et al. 1988).

Mysteriously, refilling has been observed in many species despite the fact that the P of the bulk xylem sap is far below the refilling threshold (Salleo et al. 1996; Tyree et al. 1999; Zwieniecki et al. 2000; Bucci et al. 2003; Stiller et al. 2005). These species apparently have a pump, no doubt powered directly or indirectly by osmosis, which delivers water to the embolized vessel at high-enough pressure to refill it. The process is metabolically dependent, because it is prevented by poisons and light starvation (Zwieniecki et al. 2000; Stiller et al. 2005). It also appears to be linked to phloem transport, because it can be inhibited by stopping phloem transport to the area by girdling (Salleo et al. 2006). The refilling has been coupled to starch hydrolysis in xylem parenchyma (Bucci et al. 2003; Salleo et al. 2009). Aside from these observations, nothing is known of how the water is pressurized and delivered to the conduit, and how it is kept pressurized in the conduit while neighboring water-filled conduits are at negative P. There are many hypotheses (Holbrook and Zwieniecki 1999; Hacke and Sperry 2003; Vesala et al. 2003), and solving the mystery is attracting much research effort.

9 Ecological and Physiological Consequences of Cavitation

Cavitation is an important factor in the adaptation of plants to their environment. Vulnerability curves reveal the physical limits to xylem pressure. Vulnerable plants like the cottonwood in Fig. 2 cannot live in the California chaparral like the hoaryleaf ceanothus because they would cavitate to death. Conversely, the ceanothus does not flourish in riparian zones presumably because it is out-competed by vulnerable species. Traits that confer tolerance to cavitation appear to come at the cost of reduced competitive ability (Sperry et al. 2006). One cost of cavitation resistance is the tendency towards narrow conduits (Fig. 5a) which reduce transport capacity. A second cost is the tendency towards dense wood tissue in arid-adapted plants that has been related to the need to reinforce the xylem conduits against implosion under more negative pressures [Sect. 3.2; (3)]. Greater thickness-to-span ratio [t/D_c; (3)] required to resist implosion also increases wood density (Hacke et al. 2001). Denser wood is more expensive, potentially reducing growth rates.

9.1 Stomatal Regulation

Vulnerability curves have implications for stomatal responses to both soil and atmospheric humidity. Given the large effect of friction on *P* (Sect. 4), the regulation of transpiration is also the regulation of *P*. Vulnerable species, like cottonwood, have little safety margin from cavitation (Rood et al. 2000). They require extremely good stomatal control in response to soil moisture and humidity to keep *P* from falling critically negative. They can also be phreatophytic, tapping into a stable water table. Such plants have been termed "isohydric" in that they maintain a nearly constant mid-day *P* under transpirational conditions regardless of soil or humidity (Tardieu and Davies 1993). Isohydric plants are "drought avoiders" because they prevent overly negative *P* by stomatal control and sometimes with reliance on ground water.

Some vulnerable plants may survive without good stomatal regulation if they live in extremely benign habitats where soil is always wet and the evaporative gradient low. Such plants are "drought susceptible" because in the absence of stomatal regulation their vulnerable xylem would quickly cavitate during soil drying and increasing evaporative gradients.

Very cavitation-resistant plants, like sagebrush or ceanothus in Fig. 2, do not require tight stomatal regulation unless soils dry out to the point where frictional gradients threaten to cause excessive cavitation. These plants typically allow their *P* to become more negative during drought, and have been termed "anisohydric" (Franks et al. 2007). They often do not tap into permanent ground water. Because they exhibit large negative *P* during droughts, they are "drought tolerators." Despite being very resistant to cavitation, they also experience significant cavitation during drought because their pressures can become very negative (Davis et al. 2002).

9.2 Drought Responses

Plant hydraulics can inform predictions of how vegetation will respond to projected conditions of rapid climate change (McDowell et al. 2008). In areas of increased drought, drought-susceptible species should be quickly eliminated. Drought avoiders would initially do well, being protected by stomatal closure. But if droughts were long or frequent enough, chronic stomatal closure would induce physiological dormancy. In such an energy-deprived state, they would be more susceptible to pathogens, competition, and starvation. This appears to explain the massive drought-related dieback of the isohydric pinyon pine (*Pinus edulis*) in the southwestern US (McDowell et al. 2008). Drought tolerators would initially show more cavitation than the drought avoiders, but not as much as the drought-susceptible species. Tolerators might show some dieback, but by maintaining at least some physiological activity, they would survive all but the most severe droughts. *Juniperus* species are drought tolerators (Willson and Jackson 2006). Where junipers co-occur with the afore-mentioned pinyon pine in extensive woodlands, they survive much better, although suffering more cavitation. Very severe droughts, however, will kill even junipers (West et al. 2008).

Physiological responses to environment can be quantified by the incorporation of vulnerability curves into models of the soil–plant–atmosphere continuum (Sperry et al. 1998). Such models are capable of predicting species' specific responses of plant water use to a changing environment (Jackson et al. 2000). The analysis and prediction of drought responses based on the performance of the plant's hydraulic system is receiving increased interest (Rice et al. 2004; Brodribb and Cochard 2009).

9.3 Low Temperature Responses

Vulnerability to freezing-induced cavitation also influences plant ecophysiology. Among diffuse-porous angiosperm trees, there is a pronounced trend toward narrower conduits at higher latitude (Baas 1986). Presumably, this is adaptive in reducing the amount of winter embolism in temperate habitats. Within a given region, trees with narrower vessels tend to leaf out earlier and have longer growing seasons than trees with wider vessels, a phenology consistent with the greater resistance of narrow vessels to freeze–thaw cavitation (Wang et al. 1992). Diffuse-porous trees generally rely on multiple growth rings to supply their water, so embolism during winter would reduce growing season transport capacity, particularly in those species that do not exhibit spring refilling. Those that do rely on refilling are vulnerable to events that impair the refilling process. Large-scale dieback of paper birch (*Betula papyrifera*) in northeastern North America has been linked to winter-freezing damage of roots and the inhibition of spring root pressures for restoring transport capacity (Cox and Malcolm 1997).

Ring-porosity, which is largely a north-temperate phenomenon, may have evolved in response to the freezing problem. Ring-porous trees rely on a single growth ring for their transport, producing a ring of very large early-wood vessels to carry their water (Ellmore and Ewers 1986). These large (typically $>80\ \mu m$) vessels are exceptionally vulnerable to cavitation by freezing and thawing (Cochard and Tyree 1990). They are produced late in the spring when the risk of freezing is low, and consequently, ring-porous trees are among the last to leaf out and have the shortest growing seasons (Lopez et al. 2008). The advantage, however, is that these plants can rely on a consistent transport capacity that is less influenced by the vagaries of winter freezes and refilling mechanisms. The first freezes of fall cavitate the earlywood vessels, and they typically never function again (Sperry and Sullivan

1992). Ring-porous trees occur in the lower temperate latitudes where the frost-free period is apparently long enough for them to compete with more narrow-vesseled trees that have longer growing seasons.

In some species, for reasons that are not well understood, freezing-related cavitation is sensitive to minimum temperature (Sect. 7.2). Diploid *Larrea tridentata*, the dominant shrub of the Mojave desert in southwestern North America, becomes completely cavitated at minimum temperatures of −20°C or below. The northern limit of the Mojave diploid race corresponds closely with the −20°C minimum temperature isotherm (Pockman and Sperry 1997). The implication is that the cavitation response is an important factor limiting the northern range of the population. More research is necessary to understand why cavitation can be sensitive to the freezing temperature (Cavender-Bares 2005).

9.4 Evolutionary Responses

Vascular evolution has produced innumerable variations on the basic wick theme. Much can be learned by studying how structural modifications influence vascular function in different lineages. These "natural experiments" can be used to test a number of structure–function hypotheses.

The transition from tracheids to vessels has happened three or more times during evolution, including at least once in the flowering plants (Carlquist 1988). The classic explanation is that the multicellular vessels reduce flow resistance relative to the unicellular tracheid. Data from early-diverging angiosperm lineages contradict this explanation, however, because vessel-bearing species with "primitive" vessel morphology have similar sapwood-specific flow resistivity as their vesselless relatives. Instead, the initial advantage of vessel evolution may be the origin of separate, specialized tissues for transport, storage, and mechanical support (Sperry et al. 2007; Hudson et al. 2010). The poor hydraulic performance of primitive vessels can be attributed in part to their scalariform perforation plates. The obstructing bars of these plates are vestiges of the pitted intertracheid walls (Carlquist 1992), and they approximately double the flow resistivity of the vessel lumen. Hydraulically efficient vessels apparently evolved with the origin of obstruction-free simple perforation plates (Christman and Sperry 2010).

Vessel evolution may also have been associated with a temporary increase in vulnerability to cavitation by water stress (Sperry et al. 2007). As tracheids linked together to form long vessels, the number of pits per conduit necessarily increased, greatly increasing the probability of failure (Fig. 5b, "more pits" arrow). This trend was hypothetically compensated for by the evolution of less porous pit membrane structure (Fig. 5b, opposite the "more porous" direction).

Recent work has focused on the importance of leaf hydraulics in limiting plant productivity (Brodribb et al. 2005). Low vein densities (vein length per leaf area) of nonflowering plants are associated with low maximum photosynthetic rates. A causal explanation is that low vein density means longer distance from the xylem conduits to the site of evaporation (Brodribb et al. 2007). The consequently greater pressure drop is compensated for by lower diffusive conductance through stomata. The evolution of higher vein densities in later-diverging angiosperm lineages, perhaps associated with vessel evolution, may have been critical for the rise of high-productivity plants, including all major food crops (Brodribb and Feild 2010).

10 Conclusions

The science of plant water transport is equal parts of physics and biology. Plants have evolved a complex wick system that harnesses the cohesive hydrogen bond energy of liquid water and suppresses the heterogeneous nucleation of cavitation. Trade-offs between making the wick safe against failure, yet efficient in moving water, result in the process being limiting to plant performance. Failure in most cases arises from air-seeding and bubble nucleation which leads to cavitated and embolized conduits. Cavitation limits the range of negative pressures that can be harnessed to move water, and the hydraulic conductance of the wick limits the flow rate of water that can be moved at a given pressure gradient. Both limits constrain CO_2 uptake via the water-for-carbon trade-off at the stomatal interface.

The cohesion–tension mechanism has withstood numerous challenges over the years, and has considerable experimental support (summarized in Tyree 1997; Cochard et al. 2001). But there is more to be learned. Cavitation by air-seeding at interconduit pits appears to be a stochastic process that is influenced by pit quality and quantity, but the details are poorly understood. The function of the pit valves is a topic of much research because these valves are crucial to both the safety and efficiency of water transport. Other cavitation mechanisms are possible, but have not been documented. Freezing-related cavitation appears to be largely explained by the thaw-expansion mechanism, but low-temperature effects and the possibility of cavitation during the freezing phase need further research (Mayr and Sperry 2010). Perhaps the biggest mystery is how refilling can occur by an active pump that works against the negative pressure of the transpiration stream. Finally, next to nothing is known about the molecular and developmental biology underlying xylem function.

At a larger scale, the largely physical constraints on water transport translate into biological constraints on plant metabolism, growth, and survival. A quantitative grasp of plant hydraulics has considerable explanatory power. Stomatal regulation takes on new significance when viewed in the light of the limitations on the delivery of the transpiration stream (Brodribb and Holbrook 2004). Mortality events and range limits often have important links to hydraulic limitations. The productivity of plants, and how it scales with plant size, have theoretical links to vascular architecture as captured by the influential and controversial Metabolic Scaling Theory (MST; Enquist et al. 2000). The fact that the growth in the hydraulic conductance of a tree falls increasingly short of its growth in volume ultimately may play a role in limiting tree size (Ryan et al. 2006). At the ecosystem level, the physical limits on the transpiration stream can constrain the CO_2 uptake and hence the carbon sink strength of the vegetation. The wicking of water through plant vasculature is not only a remarkable biophysical process, but it is one with far-reaching biological implications.

References

Alder NN, Pockman WT, Sperry JS, Nuismer S (1997) Use of centrifugal force in the study of xylem cavitation. J Exp Bot 48:665–674

Baas P (1986) Ecological patterns of xylem anatomy. In: Givnish TJ (ed) On the economy of plant form and function. Cambridge University Press, Cambridge, pp 327–351

Bailey IW, Tupper WW (1918) Size variation in tracheary cells. I. A comparison between the secondary xylems of vascular cryptograms, gymnosperms and angiosperms. Proc Am Acad Arts Sci 54:149–204

Boyer JS, Cavalieri AJ, Schulze ED (1985) Control of the rate of cell enlargement: excision, wall relaxation, and growth-induced water potentials. Planta 163:527–543

Briggs LJ (1950) Limiting negative pressure of water. J Appl Phys 21:721–722

Brodribb TJ, Cochard H (2009) Hydraulic failure defines the recovery and point of death in water-stressed conifers. Plant Physiol 149:575–584

Brodribb TJ, Feild TS (2010) A surge in leaf photosynthetic capacity during early angiosperm diversification. Ecol Lett. doi:10.1111/j.1461-0248.2009.01410.x

Brodribb TJ, Holbrook NM (2004) Stomatal protection against hydraulic failure: a comparison of coexisting ferns and angiosperms. New Phytol 162:663–670

Brodribb TJ, Holbrook NM (2005) Water stress deforms tracheids peripheral to the leaf vein of a tropical conifer. Plant Physiol 137:1139–1146

Brodribb TJ, Holbrook NM, Zwieniecki MA, Palma B (2005) Leaf hydraulic capacity in ferns, conifers, and angiosperms: impacts on photosynthetic maxima. New Phytol 165:839–846

Brodribb TJ, Feild TS, Jordan GJ (2007) Leaf maximum photosynthetic rate and venation are linked by hydraulics. Plant Physiol 144:1890–1898

Bucci SJ, Scholz FG, Goldstein G, Meinzer FC, Sternberg LDSL (2003) Dynamic changes in hydraulic conductivity in petioles of two savanna species: factors and mechanisms contributing to the refilling of embolized vessels. Plant Cell Environ 26:1633–1645

Canny MJ (1990) What becomes of the transpiration stream? New Phytol 114:341–368

Carlquist S (1988) Comparative wood anatomy. Springer, Berlin

Carlquist S (1992) Pit membrane remnants in perforation plates of primitive dicotyledons and their significance. Am J Bot 79:660–672

Cavender-Bares J (2005) Impacts of freezing on long distance transport in woody plants. In: Holbrook NM, Zwieniecki MA (eds) Vascular transport in plants. Academic, Amsterdam, pp 401–424

Choat B, Lahr EC, Melcher PJ, Zwieniecki MA, Holbrook NM (2005) The spatial pattern of air-seeding thresholds in mature sugar maple trees. Plant Cell Environ 28:1082–1089

Choat B, Cobb AR, Jansen S (2008) Structure and function of bordered pits: new discoveries and impacts on whole-plant hydraulic function. New Phytol 177:608–626

Christman MA, Sperry JS (2010) Single vessel flow measurements indicate scalariform perforation plates confer higher resistance to flow than previously estimated. Plant Cell Environ 33:431–433

Christman MA, Sperry JS, Adler FR (2009) Testing the rare pit hypothesis in three species of *Acer*. New Phytol 182:664–674

Cochard H (2002) A technique for measuring xylem hydraulic conductance under high negative pressures. Plant Cell Environ 25:815–819

Cochard H, Tyree MT (1990) Xylem dysfunction in Quercus: vessel sizes, tyloses, cavitation and seasonal changes in embolism. Tree Physiol 6:393–408
Cochard H, Cruiziat P, Tyree MT (1992) Use of positive pressures to establish vulnerability curves: further support for the air-seeding hypothesis and implications for pressure-volume analysis. Plant Physiol 100:205–209
Cochard H, Ewers FW, Tyree MT (1994) Water relations of a tropical vine-like bamboo (*Rhipidocladum racemiflorum*): root pressures, vulnerability to cavitation and seasonal changes in embolism. J Exp Bot 45:1085–1089
Cochard H, Ameglio T, Cruiziat P (2001) The cohesion theory debate continues. Trends Plant Sci 6:456
Cochard H, Froux F, Mayr S, Coutand C (2004) Xylem wall collapse in water-stressed pine needles. Plant Physiol 134:401–408
Cochard H, Gaelle D, Bodet C, Tharwat I, Poirier M, Ameglio T (2005) Evaluation of a new centrifuge technique for rapid generation of xylem vulnerability curves. Physiol Plant 124:410–418
Cox RM, Malcolm JM (1997) Effects of duration of a simulated winter thaw on dieback and xylem conductivity of *Betula papyrifera*. Tree Physiol 17:389–396
Crombie DS, Hipkins MF, Milburn JA (1985) Gas penetration of pit membranes in the xylem of *Rhododendron* as the cause of acoustically detectable sap cavitation. Aust J Plant Physiol 12:445–454
Cuevas LE (1969) Shrinkage and collapse studies on *Eucalyptus viminalis*. J Inst Wood Sci 4:29–38
Davis SD, Sperry JS, Hacke UG (1999a) The relationship between xylem conduit diameter and cavitation caused by freezing. Am J Bot 86:1367
Davis SD, Ewers FW, Wood J, Reeves JJ, Kolb KJ (1999b) Differential susceptibility to xylem cavitation among three pairs of Ceanothus species in the Transverse Mountain ranges of Southern California. Ecoscience 6:180–186
Davis SD, Ewers FW, Portwood KA, Sperry JS, Crocker MC, Adams GC (2002) Shoot dieback during prolonged drought in *Ceanothus* chaparral in California: a possible case of hydraulic failure. Am J Bot 89:820–828
Debenedetti PG (1996) Metastable liquids. Princeton University Press, Princeton, NJ
Ellmore GS, Ewers FW (1986) Fluid flow in the outermost xylem increment of a ring-porous tree, *Ulmus americana*. Am J Bot 73:1771–1774
Enquist BJ, West GB, Brown JH (2000) Quarter-power allometric scaling in vascular plants: Functional basis and ecological consequences. In: West GB, Brown JH (eds) Scaling in biology. Oxford University Press, Oxford, pp 167–198
Franks PJ, Drake PL, Froend RH (2007) Anisohydric but isohydrodynamic: seasonally constant plant water potential gradient explained by a stomatal control mechanism incorporating variable plant hydraulic conductance. Plant Cell Environ 30:19–30
Grace J (1993) Refilling of embolized xylem. In: Grace J, Raschi A, Borghetti M (eds) Water transport in plants under climatic stress. Cambridge University Press, Cambridge, pp 52–62
Hacke U, Sauter JJ (1996) Xylem dysfunction during winter and recovery of hydraulic conductivity in diffuse-porous and ring-porous trees. Oecologia 105:435–439
Hacke UG, Sperry JS (2003) Limits to xylem refilling under negative pressure in *Laurus nobilis* and *Acer negundo*. Plant Cell Environ 26:303–311
Hacke UG, Sperry JS, Pockman WP, Davis SD, McCulloh KA (2001) Trends in wood density and structure are linked to prevention of xylem implosion by negative pressure. Oecologia 126:457–461
Hacke UG, Sperry JS, Wheeler JK, Castro L (2006) Scaling of angiosperm xylem structure with safety and efficiency. Tree Physiol 26:689–701
Hammel HT (1967) Freezing of xylem sap without cavitation. Plant Physiol 42:55–66
Holbrook NM, Zwieniecki MA (1999) Embolism repair and xylem tension: do we need a miracle? Plant Physiol 120:7–10

Holbrook NM, Ahrens ET, Burns MJ, Zwieniecki MA (2001) In vivo observation of cavitation and embolism repair using magnetic resonance imaging. Plant Physiol 126:27–31

Hudson PJ, Razanatsoa J, Feild TS (2010) Early vessel evolution and the diversification of wood function – insights from Malagasy Canellales. Am J Bot 97:80

Jackson RB, Sperry JS, Dawson TE (2000) Root water uptake and transport: using physiological processes in global predictions. Trends Plant Sci 5:482–488

Jacobsen AL, Ewers FW, Pratt RB, Paddock WA, Davis SD (2005) Do xylem fibers affect vessel cavitation resistance? Plant Physiol 139:546–556

Jacobson AL, Pratt RB, Ewers FW, Davis SD (2007) Cavitation resistance among 26 chaparral species of southern California. Ecol Monogr 77:99–115

Kolb KJ, Sperry JS (1999) Differences in drought adaptation between subspecies of sagebrush (*Artemisia tridentata*). Ecology 80:2373–2384

Kwak H, Panton RL (1985) Tensile strength of simple liquids predicted by a model of molecular interactions. J Phys Appl Phys 18:647–659

Li Y, Sperry JS, Bush SE, Hacke UG (2008) Evaluation of centrifugal methods for measuring xylem cavitation in conifers, diffuse- and ring-porous angiosperms. New Phytol 177:558–568

Lopez OR, Farris-Lopez K, Montgomery RA, Givnish TJ (2008) Leaf phenology in relation to canopy closure in southern Appalachian trees. Am J Bot 95:1395–1407

Maherali H, Pockman WT, Jackson RB (2003) Adaptive variation in the vulnerability of woody plants to xylem cavitation. Ecology 85:2184–2199

Mayr S, Sperry JS (2010) Freeze-thaw induced embolism in Pinus contorta: centrifuge experiments validate the "thaw-expansion" hypothesis but conflict with ultrasonic data. New Phytol 185:1016–1024

McDowell N, Pockman WT, Allen CD, Breshears DD, Cobb N, Kolb T, Plaut J, Sperry JS, West A, Williams DG, Yepez EA (2008) Mechanisms of plant survival and mortality during drought. Why do some plants survive while others succumb to drought? New Phytol 178:719–739

Mencuccini M (2003) The ecological significance of long-distance water transport: short-term regulation, long-term acclimation and the hydraulic costs of stature across plant life forms. Plant Cell Environ 26:163–182

Milburn JA, McLaughlin ME (1974) Studies of cavitation in isolated vascular bundles and whole leaves of *Plantago majo*r L. New Phytol 73:861–871

Oertli JJ (1971) The stability of water under tension in the xylem. Z Pflanzenphysiol 65:195–209

Pickard WF (1981) The ascent of sap in plants. Progr Biophys Mol Biol 37:181–229

Piquemal J, LaPierre C, Myton K, O'Connel A, Schuch W, Grima-Pettenati J, Boudet AM (1998) Down-regulation of cinnamoyl-CoA reductase induces significant changes of lignin profiles in transgenic tobacco plants. Plant J 13:71–83

Pittermann J, Sperry JS (2003) Tracheid diameter is the key trait determining extent of freezing-induced cavitation in conifers. Tree Physiol 23:907–914

Pittermann J, Sperry JS (2006) Analysis of freeze-thaw embolism in conifers: the interaction between cavitation pressure and tracheid size. Plant Physiol 140:374–382

Pittermann J, Sperry JS, Hacke UG, Wheeler JK, Sikkema EH (2005) Torus-margo pits help conifers compete with angiosperms. Science 310:1924

Pockman WT, Sperry JS (1997) Freezing-induced xylem cavitation and the northern limit of Larrea tridentata. Oecologia 109:19–27

Pockman WT, Sperry JS (2000) Vulnerability to cavitaiton and the distribution of Sonoran desert vegetation. Am J Bot 87:1287–1299

Pockman WT, Sperry JS, O'Leary JW (1995) Sustained and significant negative water pressure in xylem. Nature 378:715–716

Rand RH (1978) The dynamics of and evaporating meniscus. Acta Mechanica 29:135–146

Raven JA (1987) The evolution of vascular land plants in relation to supracellular transport processes. Adv Bot Res 5:153–219

Rice KJ, Matzner SL, Byer W, Brown JR (2004) Patterns of tree dieback in Queensland, Australia: the importance of drought stress and the role of resistance to cavitation. Oecologia 139:190–198

Rood SB, Patino S, Coombs K, Tyree MT (2000) Branch sacrifice: cavitation-associated drought adaptation of riparian cottonwoods. Trees 14:248–257

Ryan MG, Phillips N, Bond BJ (2006) The hydraulic limitation hypothesis revisited. Plant Cell Environ 29:367–381

Salleo S, Lo Gullo MA, De Paoli D, Zippo M (1996) Xylem recovery from cavitation-induced embolism in young plants of *Laurus nobilis*: a possible mechanism. New Phytol 132:47–56

Salleo S, Trifilo P, LoGullo MA (2006) Phloem as a possible major determinant of rapid cavitation reversal in *Laurus nobilis* (laurel). Funct Plant Biol 33:1063–1074

Salleo S, Trifilo P, Esposito S, Nardini A, LoGullo MA (2009) Starch-to-sugar conversion in wood parenchyma of field-growing *Laurus nobilis* plants: a component of the signal pathway for embolism repair? Funct Plant Biol 36:815–825

Schlesinger WH (1997) Biogeochemistry. Academic, San Diego

Shen F, Rongfu G, Liu W, Zhang W (2002) Physical analysis of the process of cavitation in xylem sap. Tree Physiol 22:655–659

Sperry JS (1986) Relationship of xylem embolism to xylem pressure potential, stomatal closure, and shoot morphology in the palm *Rhapis excelsa*. Plant Physiol 80:110–116

Sperry JS (2000) Hydraulic constraints on plant gas exchange. Agric For Meteorol 2831:1–11

Sperry JS, Robson DJ (2001) Xylem cavitation and freezing in conifers. In: Colombo SJ, Bigras FJ (eds) Conifer cold hardiness. Kluwer, Dordrecht, pp 121–136

Sperry JS, Sullivan JEM (1992) Xylem embolism in response to freeze-thaw cycles and water stress in ring-porous, diffuse-porous, and conifer species. Plant Physiol 100:605–613

Sperry JS, Tyree MT (1988) Mechanism of water stress-induced xylem embolism. Plant Physiol 88:581–587

Sperry JS, Tyree MT (1990) Water-stress-induced xylem embolism in three species of conifers. Plant Cell Environ 13:427–436

Sperry JS, Holbrook NM, Zimmermann MH, Tyree MT (1987) Spring filling of xylem vessels in wild grapevine. Plant Physiol 83:414–417

Sperry JS, Donnelly JR, Tyree MT (1988) Seasonal occurrence of xylem embolism in sugar maple (Acer saccharum). Am J Bot 75:1212–1218

Sperry JS, Saliendra NZ, Pockman WT, Cochard H, Cruiziat P, Davis SD, Ewers FW, Tyree MT (1996) New evidence for large negative xylem pressures and their measurement by the pressure chamber method. Plant Cell Environ 19:427–436

Sperry JS, Adler FR, Campbell GS, Comstock JP (1998) Limitation of plant water use by rhizosphere and xylem conductance: results from a model. Plant Cell Environ 21:347–359

Sperry JS, Hacke UG, Pittermann J (2006) Size and function in conifer tracheids and angiosperm vessels. Am J Bot 93:1490–1500

Sperry JS, Hacke UG, Feild TS, Sano Y, Sikkema EH (2007) Hydraulic consequences of vessel evolution in angiosperms. Int J Plant Sci 168:1127–1139

Steudle E (1994) Water transport across roots. Plant Soil 167:79–90

Stiller V, Lafitte HR, Sperry JS (2005) Embolized conduits of rice (*Oryza sativa* L.) refill despite negative xylem pressure. Am J Bot 92:1970–1974

Sucoff E (1969) Freezing of conifer xylem and the cohesion-tension theory. Physiol Plant 22:424–431

Sutcliffe JF (1968) Plants and water. Edward Arnold, London

Taneda H, Sperry JS (2008) A case-study of water transport in co-occurring ring- versus diffuse-porous tresses: contrasts in water-status, conducting capacity, cavitation and vessel refilling. Tree Physiol 28:1641–1652

Tardieu F, Davies W (1993) Integration of hydraulic and chemical signalling in the control of stomatal conductance and water status of droughted plants. Plant Cell Environ 16:341–349

Tyree MT (1997) The cohesion-tension theory of sap ascent: current controversies. J Exp Bot 48:1753–1765
Tyree MT, Sperry JS (1989) Vulnerability of xylem to cavitation and embolism. Annu Rev Plant Physiol Plant Mol Biol 40:19–38
Tyree M, Davis S, Cochard H (1994) Biophysical perspectives of xylem evolution – is there a tradeoff of hydraulic efficiency for vulnerability to dysfunction? IAWA J 15:335–360
Tyree MT, Salleo S, Nardini A, Lo Gullo MA, Mosca R (1999) Refilling of embolized vessels in young stems of laurel. Do we need a new paradigm? Plant Physiol 102:11–21
Vesala T, Holtta T, Peramaki M, Nikinmaa E (2003) Refilling of a hydraulically isolated embolized xylem vessel: model calculations. Ann Bot 91:419–428
Vogel S (1988) Life's devices. Princeton University Press, Princeton
Vogel S (1994) Life in moving fluids: the physical biology of flow, 2nd edn. Princeton University Press, Princeton, NJ
Wang J, Ives NE, Lechowicz MJ (1992) The relation of foliar phenology to xylem embolism in trees. Funct Ecol 6:469–475
West AG, Hultine KR, Sperry JS, Bush SE, Ehleringer JR (2008) Transpiration and hydraulic strategies in a pinyon-juniper woodland. Ecol Appl 18:911–927
Wheeler JK, Sperry JS, Hacke UG, Hoang N (2005) Inter-vessel pitting and cavitation in woody Rosaceae and other vesselled plants: a basis for a safety vs. efficiency trade-off in xylem transport. Plant Cell Environ 28:800–812
Willson C, Jackson RB (2006) Xylem cavitation caused by drought and freezing stress in four co-occurring *Juniperus* species. Physiol Plant 127:374–382
Yang S, Tyree MT (1992) A theoretical model of hydraulic conductivity recovery from embolism with comparison to experimental data on *Acer saccharum*. Plant Cell Environ 15:633–643
Young WC (1989) Roark's formulas for stress and strain. McGraw Hill, New York
Zimmermann MH (1983) Xylem structure and the ascent of sap. Springer, Berlin
Zwieniecki MA, Hutyra L, Thompson MV, Holbrook NM (2000) Dynamic changes in petiole specific conductivity in red maple (*Acer rubrum* L.), tulip tree (*Liriodendron tulipifera* L.) and northern fox grape (*Vitis labrusca* L.). Plant Cell Environ 23:407–414
Zwieniecki MA, Melcher PJ, Boyce KC, Sack L, Holbrook NM (2002) Hydraulic architecture of leaf venation in *Laurus nobilis* L. Plant Cell Environ 25:1445–1450

microRNAs and Mechanical Stress

Shanfa Lu

Abstract Mechanical stress, one of the major abiotic stresses, significantly affects plant growth, survival, and reproduction. Recent studies showed that cytoplasmic calcium, reactive oxygen species, and phytohormones play important roles in mechanosensing and subsequent anatomical and morphological changes in plants exposed to mechanical stress. A large number of mechanical stress-responsive genes, particularly those involved in reaction wood formation in tree species, have been isolated and characterized. Importantly, a group of mechanical stress-responsive miRNAs was identified in *Populus trichocarpa* plants. They were predicted or experimentally validated to regulate genes encoding transcription factors, metabolism and cellular process-related proteins, and many function-unknown proteins. Analyses of miRNA expression patterns and target gene functions indicate that miRNAs play crucial regulatory roles in reaction wood formation. Recent results provide novel and useful information for fully elucidating the genetic and molecular mechanisms of plant responses to mechanical stress.

1 Introduction

Recent studies have made great progress in understanding the molecular mechanism of plant responses to one of the major abiotic stresses, mechanical stress. It includes the identification of a group of mechanical stress-responsive miRNAs (Lu et al. 2005, 2008b). This chapter begins with a brief overview of our up-to-date knowledge about miRNAs, followed by a more detailed review on recent progress in plant responses to mechanical stress. The identification of mechanical stress-responsive miRNAs and the characterization of their target functions in tension

S. Lu
Medicinal Plant Cultivation Research Center, Institute of Medicinal Plant Development, Chinese Academy of Medical Sciences and Peking Union Medical College, No. 151 Malianwa North Road, Haidian District, Beijing 100193, People's Republic of China
e-mail: sflu@implad.ac.cn

P. Wojtaszek (ed.), *Mechanical Integration of Plant Cells and Plants*,
Signaling and Communication in Plants 9, DOI 10.1007/978-3-642-19091-9_13,
© Springer-Verlag Berlin Heidelberg 2011

wood formation are highlighted. Conclusions and perspectives of mechanical stress-responsive miRNA studies are also included.

2 microRNAs

microRNAs (miRNAs) are a class of small noncoding RNA molecules with size about 21 nucleotides in length. They were first discovered in an animal species, *Caenorthabditis elegans*, through analysis of *lin-4* and *lin-14* mutants that showed abnormal larval development (Lee et al. 1993; Wightman et al. 1993). In recent studies, miRNAs have been found to exist not only in animals but also in plants and viruses (Llave et al. 2002; Park et al. 2002; Reinhart et al. 2002; Mette et al. 2002; Pfeffer et al. 2004). According to the current miRBase database (release 14), a total of 10,581 miRNAs in 115 species of plants and animals and viruses have been identified through direct cloning methods or computational approaches (Griffiths-Jones et al. 2008; http://www.mirbase.org/). These miRNAs play very important regulatory roles in organisms by targeting other RNA molecules for cleavage or for translational repression. In plants, miRNA targets are usually mRNAs that are involved in transcriptional regulation, cell metabolism, signal transduction, and stress response. However, some noncoding RNAs have also been found to be cleaved by miRNAs in *Arabidopsis* and *Pinus* (Allen et al. 2005; Lu et al. 2007). The cleavage of noncoding RNAs by plant miRNAs may activate the process of the other class of small RNAs, known as ta-siRNAs, which regulate plant development and defense responses (Allen et al. 2005; Lu et al. 2007).

miRNAs are derived from primary miRNAs (pri-miRNAs) that are transcribed from miRNA loci by RNA polymerase II (Lee et al. 2002). In plants, primary miRNAs are cleaved by Dicer-like 1 protein (DCL1) in the nucleus through interaction with several other proteins (Voinnet 2009). The produced hairpin structures, known as miRNA precursors (pre-miRNAs), are then processed by DCL1 into miRNA:miRNA* duplexes that comprise mature miRNAs of about 21 nucleotides and similar-sized miRNA* fragments on the opposing arm of miRNA precursors. Duplexes are further unwound by a helicase to release the single-stranded mature miRNAs that are loaded onto AGO1 protein complexes for identification and negative regulation of target RNAs (Chen 2005; Jones-Rhoades et al. 2006; Mallory and Vaucheret 2006; Voinnet 2009).

miRNAs have been found to regulate the development of all organs, such as root, stem, leaf, flower, and seed in plants. For instance, miR164 and miR166 are root-associated miRNAs. They regulate *NAC* and *HD-ZIP* transcription factor genes, respectively. Downregulation of miR164 in *Arabidopsis* by T-DNA insertion causes an increase in lateral root number (Guo et al. 2005), while *Medicago* plants overexpressing the *MIR166* gene show a decrease (Boualem et al. 2008). miR156, which targets a group of squamosa promoter binding protein-like (SPL) transcription factor genes, is involved in flowering time modulation. Overexpression of a *MIR156* gene causes late flowering in *Arabidopsis* (Schwab et al. 2005). Consistently, plants overexpressing a miR156-resistant version of the target gene, *SPL3*, are early flowering (Wu and Poethig 2006). Other miRNAs involving in flower development include miR159, miR160, miR164, miR165/166, miR172, miR319, and so forth (Chen 2005; Jones-Rhoades et al. 2006; Mallory and Vaucheret 2006). All of them are transcription factor-regulating miRNAs. Some of these flower development-associated miRNAs also regulate the development of leaf. Overexpression of miR319 causes leaf crinkled in addition to flowering delayed in *Arabidopsis* (Palatnik et al. 2003). Overexpression of a miRNA-resistant version of miR160 target gene, *ARF17*, causes dramatic developmental defects, including rosette and cauline leaf margin serration, upward curling of the leaf margins, early flowering, altered floral morphology, and reduced fertility (Mallory et al. 2005). Additional leaf development-associated miRNA is miR824. It targets the *MADS*-box gene *Agamous-like 16* (*AGL16*) for cleavage. Expression of a miR825-resistant *AGL16* mRNA in *Arabidopsis* increases the incidence of stomata in higher order complexes. By contrast, plants with overexpressed miR824 show a decreased incidence of stomata (Kutter et al. 2007). In addition, miR165/166 targets a group of class III *HD-ZIP* transcription factor genes that control the development of various organs, including stem, leaf, and root, through affecting vascular, apical meristem, and organ polarity (Kim et al. 2005; Williams et al. 2005; Zhou et al. 2007a; Boualem et al. 2008).

Increasing evidence indicates that miRNAs are also involved in plant responses to biotic and abiotic stresses (Sunkar et al. 2007; Lu et al. 2008c). In galled loblolly pine (*Pinus taeda*) stems infected with the fungus *Cronartium quercuum* f. sp. *fusiforme*, expression of 10 of the 11 analyzed miRNA families was found to be significantly repressed (Lu et al. 2007). Eighty-two putative miRNA targets that are likely to be associated with disease responses have been revealed in loblolly pine through a comparative genomic analysis (Lu et al. 2007). In *Arabidopsis*, miR393 mediates antibacterial resistance through repressing auxin signaling (Navarro et al. 2006). During plants response to various abiotic stresses, including nutrient starvation, oxidative stress, cold, dehydration, salinity, and UV-B radiation, the expression of many miRNAs is altered (Jones-Rhoades and Bartel 2004; Sunkar and Zhu 2004; Fujii et al. 2005; Chiou et al. 2006; Sunkar et al. 2006; Zhao et al. 2007a; Zhou et al. 2007b; Lu et al. 2008b). In our recent studies, miRNAs were found to be involved in mechanical stress responses in *Populus trichocarpa*, suggesting a novel role of miRNAs in mechanical fitness of plants (Lu et al. 2005, 2008b).

3 Mechanical Stress and Plant Responses

Plants are frequently exposed to numerous environmental stresses throughout their life span. One of these stresses, mechanical stress, is critical to plant growth, survival, and reproduction. In nature, mechanical stress is generated by various environmental factors, such as wind, snow, touching, rubbing, and bending. Plants that are continuously subjected to these mechanical strains show significant morphological changes, including growth retardation, increase of radial growth and root allocation, and reduction of dry mass and seed production (Jaffe 1980; Niklas 1998). The generation of altered morphologies in response to mechanical stress is known as thigmomorphogenesis and is thought to be significant for making plants more resistant to the mechanically stressful conditions (Jaffe 1973, 1980).

Up to date, the mechanism and signal transduction pathway of thigmomorphogenesis is still not well understood. Cytoplasmic calcium may play a role in transducing the extra-mechanical signal into an intracellular signal. In plants subjected to mechanical stimuli, the cytoplasmic calcium level is increased rapidly (Knight et al. 1991; Haley et al. 1995). Changes of calcium concentration are coincident with the increases of reactive oxygen species (ROS) (Mori and Schroeder 2004). In a recent study, Monshausen et al. (2009) showed that calcium-regulated ROS production during mechanosensing in *Arabidopsis* roots, strongly suggesting a functional link between calcium and ROS in plant response to mechanical stress. In addition to calcium and ROS, phytohormones play vital roles in thigmomorphogenesis and a crosstalk among various kinds of phytohormone, such as ethylene, auxin, and ABA, has been shown to be very important for the subsequent anatomical and morphological changes (Jaffe 1980).

During thigmomorphogenesis, expression of a variety of genes is altered in plants. It includes touch-inducible (*TCH*) genes (Braam and Davis 1990; Lee et al. 2005), protein kinase genes (Botella et al. 1996; Mizoguchi et al. 1996), extension genes (Shirsat et al. 1996; Hirsinger et al. 1999), an H^+-ATPase gene (Oufattole et al. 2000), a cytosolic ascorbate peroxidase gene (Gadea et al. 1999), and many other protein-coding genes (Braam 2005). A genome-wide search for *TCH* genes using microarray analysis revealed that over 2.5% of total *Arabidopsis* genes were touch-inducible (Lee et al. 2005). These *TCH* genes encode calcium-binding proteins, cell wall modifying proteins, disease resistance proteins, kinases, and transcription factors, suggesting the complexity of gene network in response to mechanical stimuli in plants.

Because of the long life span and perennial growth habit, responses of trees to mechanical stress are more significant for growth and development as compared to herbaceous species. In order to support the mechanical loads generated by wind, snow, bending, and other environmental factors, trees constantly develop specialized woody tissues, known as reaction wood, to correct inclined branch and stem growth (Sinnott 1952; Wardrop and Davies 1964; Scurfield 1971, 1973). Reaction wood in angiosperm trees is known as tension wood (TW) that is developed on the upper part of leaning stems or branches. Accordingly, wood located on the lower part is called opposite wood (OW). TW is unique to angiosperm trees. Although no substantial difference in the lignin and neutral sugar composition is observed between OW and normal wood, TW generated in angiosperm trees contains a decreased vessel number and an increased fiber cell number. Typical TW fibers differ from normal fibers in containing a cell wall layer, known as gelatinous layer (G-layer) (Dadswell and Wardrop 1955). Consistently, the deposition of lignin and hemicellulose is reduced in TW while the biosynthesis of cellulose is increased (Fujii et al. 1982). Differing from that in angiosperm trees, reaction wood developed in gymnosperm trees locates on the lower part of leaning stems or branches and is called compression wood (CW). OW in gymnosperm trees is referred to the wood developed on the upper part. CW is unique to gymnosperm trees. Compared to normal wood, CW contains more lignin and galactose and xylose and less glucose and mannose and arabinose. Similar to angiosperm OW, conifer OW has the same content of lignin and cellulose and hemicellulose as normal wood (Timell 1973).

Through the analysis of gene expression, numerous genes were found to be involved in reaction wood formation in pine (Whetten et al. 2001), cypress (Yamashita et al. 2008), poplar (Wu et al. 2000; Joshi 2003; Déjardin et al. 2004; Sterky et al. 2004; Andersson-Gunnerås et al. 2006; Bhandari et al. 2006; Nishikubo et al. 2007; Tiimonen et al. 2007), and *Eucalyptus* (Paux et al. 2005; Lu et al. 2008a). Analyzing 67 cDNA fragments cloned from a cypress species *Chamaecyparis obtusa* (Siebold & Zucc.) Endl., Yamashita et al. (2008) identified 24 genes with changed expression patterns during CW formation. The identified genes encode methionine synthase, laccase, caffeic acid *O*-methyltransferase, ascorbate oxidase, pinoresinol–lariciresinol reductase, import intermediate-associated protein 100, and other functional proteins. In *Populus tremula* × *P. alba* trees, five arabinogalactan proteins, a sucrose synthase, and a fructokinase were found to be specific or overexpressed in TW (Déjardin et al. 2004). Analyzing the formation of G-layers of TW fibers, Nishikubo et al. (2007) identified a set of xyloglucan *endo*-transglycosylase/hydrolase (*XTH16*) and xyloglucan *endo*-transglycosylase (*XET*) genes that were responsible to mechanical stress in the hybrid aspen *P. tremula* L. × *P. tremuloides* Michx. TW formation-associated genes identified in bent stems of quaking aspen trees include three cellulose synthase genes (*CesA*s) and a *KORRIGAN* endoglucanase gene (*KOR*) (Wu et al. 2000; Joshi 2003; Bhandari et al. 2006). Consistently, tension stress-responsive *CesA*s have also been identified in *Eucalyptus* (Lu et al. 2008a). From the xylem cells of *Eucalyptus* stems undergoing secondary cell wall biosynthesis, we cloned three *CesA* genes. Two of them, including *EgraCesA2* and *EgraCesA3*, are significantly upregulated in tension-stressed *Eucalyptus* xylem cells. Accordingly, *GUS* expression directed by the *EgraCesA2* and *EgraCesA3* promoter is also upregulated by tension stress in tobacco stems (Lu et al. 2008a). Thus, upregulation of *CesA*s could be important for the accumulation of crystalline cellulose in G-layer generated in TW. Microarray analysis of 231 genes expressed in *Eucalyptus* xylem showed that 196 of them were regulated by tension stress (Paux et al. 2005). It includes genes encoding Aux/IAA proteins, a cellulose synthase, lignin biosynthetic proteins, and an expansin-related protein, suggesting significant roles of auxin and cell wall-related proteins in reaction wood formation.

4 Mechanical Stress-Responsive miRNAs

4.1 Identification of Mechanical Stress-Responsive miRNAs

Although a large number of miRNAs have been found in more than 100 plant species, mechanical stress-responsive miRNAs are characterized only in *P. trichocarpa* (Griffiths-Jones et al. 2008; Zhang et al. 2010; Lu et al. 2005, 2008b). Examining the expression of miRNAs in stems bent into an arch for 4 days and comparing them with those in stems without bending, we identified a total of 22 mechanical stress-responsive miRNAs (Table 1) (Lu et al. 2005, 2008b). Among them, miR156, miR162, miR164, miR475, miR478, miR480, miR481, and miR482 are suppressed by tension stress that induces TW generation and by compression stress that causes OW formation in angiosperm plants. By contrast, miR408 and miR1444 are upregulated in both TW and OW. These miRNAs may play important regulatory roles in counteracting overall mechanical stimuli since no significantly different expression is exhibited in TW and OW. Among the twelve mechanical stress-responsive miRNAs showing differential expression in TW and OW, five (miR159, miR476, miR479, miR1446, and miR1447) exhibit preferentially upregulated expression in OW, five (miR160, miR172, miR530, miR827, and miR1450) are significantly suppressed only in OW, whereas the other two, miR168 and miR171, seems to be downregulated more significantly in TW than in OW. These

Table 1 Mechanical stress-responsive miRNAs in *Populus* (Lu et al. 2005, 2008b)

miRNA	Expression of miRNAs in respose to mechanical stress	
	In tension wood (TW)	In opposite wood (OW)
miR156	Suppressed	Suppressed
miR159	No significant change	Upregulated
miR160	No significant change	Suppressed
miR162	Suppressed	Suppressed
miR164	Suppressed	Suppressed
miR168	Suppressed	No significant change
miR171	Suppressed	No significant change
miR172	No significant change	Suppressed
miR408	Upregulated	Upregulated
miR475	Suppressed	Suppressed
miR476	No significant change	Upregulated
miR478	Suppressed	Suppressed
miR479	No significant change	Upregulated
miR480	Suppressed	Suppressed
miR481	Suppressed	Suppressed
miR482	Suppressed	Suppressed
miR530	No significant change	Suppressed
miR827	No significant change	Suppressed
miR1444	Upregulated	Upregulated
miR1446	No significant change	Upregulated
miR1447	No significant change	Upregulated
miR1450	No significant change	Suppressed

TW and OW differentially expressed miRNAs are possibly associated with more specialized regulations that lead to a preferential development of either TW or OW (Lu et al. 2005, 2008b).

4.2 Roles of miRNAs in Plant Responses to Mechanical Stress

The main mode of plant miRNA-mediated gene regulation is the direct cleavage of other RNA molecules by binding to sequence regions with perfect or near-perfect complementarity (Rhoades et al. 2002). Searching miRNA-complementary regions by computational approaches and mapping cleavage sites using the modified 5′ RNA ligase-mediated (RLM)-RACE method have been successfully applied to miRNA target prediction and experimental validation, respectively (Rhoades et al. 2002; Vazquez et al. 2004; Lu et al. 2005). Using a sensitive procedure that includes a penalty scoring system developed by Jones-Rhoades and Bartel (2004), we predicted a set of target genes for the mechanical stress-responsive miRNAs (Lu et al. 2005, 2008b). They encode transcription factors and proteins involved in miRNA biogenesis, metabolism, and cellular processes. Many genes with unknown functions were also predicted to be targets of mechanical stress-responsive miRNAs.

More than forty transcription factors have been predicted to be targets of eight mechanical stress-responsive miRNAs, indicating miRNAs may serve as master regulators in TW and OW formation (Table 2). miR156 is involved in mechanical stress by regulating the levels of squamosa-promoter binding protein-likes (SPLs), which is a family of SPB domain-containing transcription factors involving in vegetative phase change, floral transition, pollen sac development, gibberellin (GA) biosynthesis and signaling, programmed cell death, and plant architecture development in *Arabidopsis* (Cardon et al. 1997; Unte et al. 2003; Wu and Poethig 2006; Gandikota et al. 2007; Zhang et al. 2007; Stone et al. 2005).

Both miR159 and miR164 were predicted to target members of MYB transcription factor family proteins. Functions of miRNA-regulated MYB genes are mostly unknown, except for two *Arabidopsis* miR159 targets, MYB33 and MYB65, which work redundantly in anther development (Millar and Gubler 2005). The other transcription factors predictively regulated by miR164 are NAC domain proteins that control auxiliary meristem formation (Raman et al. 2008). Suppression of miR164 in response to mechanical stress may cause upregulation of NAC domain proteins that enhance the activity of stem meristem cells and subsequently promote TW and OW formation.

miR171 targets seven GRAS domain-containing SCARECROW-like (SCL) transcription factor family members. GRAS proteins were previously found to function in the response of plants to mechanical stress (Mayrose et al. 2006). A founding member of GRAS protein family, known as *SCR* or *SHOOT GRAVITROPISM 1*, was shown to play a role in regulating cell divisions and elongations in response to gravitational force, a major mechanical stress for reaction

Table 2 Transcription factors predicted to be targets of mechanical stress-responsive miRNAs in *Populus trichocarpa* (Lu et al. 2005, 2008b)

miRNA	Function of targets	Gene model name[a]
miR156	Squamosa-promoter binding protein-like (SPL) transcription factor	eugene3.00160416, eugene3.01640028[b], fgenesh1_pg.C_LG_X001404, grail3.0010026801, eugene3.00400033[b], fgenesh4_pg.C_LG_II001303, estExt_Genewise1_v1.C_1240186, estExt_Genewise1_v1.C_LG_XV2187
miR159	MYB transcription factor	fgenesh4_pm.C_scaffold_40000020, eugene3.00091462, gw1.I.6885.1, gw1.III.41.1
miR160	Auxin responsive factor (ARF)	eugene3.00660262, eugene3.00640087, estExt_fgenesh4_pm.C_LG_X0888, fgenesh4_pg.C_LG_IX001411, fgenesh4_pg.C_LG_VIII000301, estExt_fgenesh4_pm.C_LG_XVI0323, fgenesh4_pg.C_LG_II000830, gw1.28.631.1, gw1.28.632.1, estExt_fgenesh4_pg.C_LG_V0901
miR164	NAC domain protein	gw1.V.3536.1, gw1.VII.2722.1, eugene3.00150202, fgenesh4_pm.C_LG_XII000069
	MYB transcription factor	estExt_Genewise1_v1.C_LG_VI0902
miR171	SCARECROW-like (SCL) transcription factor	gw1.127.243.1, gw1.II.1043.1, gw1.III.2060.1, gw1.VII.3405.1, estExt_Genewise1_v1.C_LG_II3184, eugene3.44860001, gw1.57.294.1, fgenesh4_pg.C_LG_II000787
miR172	AP2 domian-containing transcription factor	grail3.0019003502, eugene3.00050501, fgenesh4_pg.C_LG_X001965, eugene3.00280122, eugene3.00160775
miR530	Homeobox transcription factor KN3	estExt_Genewise1_v1.C_LG_II1820
miR1446	Gibberellin response modulator-like protein	fgenesh4_pg.C_LG_XII000915
	Homeobox transcription factor	gw1.I.9208.1

[a]Gene model names are present in *Populus trichocarpa* v1.1 except those indicated by the superscript letter b

[b]The gene model is present in *Populus trichocarpa* v1.0, but not in *Populus trichocarpa* v1.1

wood induction (Di Laurenzio et al. 1996). These results indicate the importance of GRAS domain-containing targets of miR171 in reaction wood formation (Lu et al. 2005). Additionally, one of the validated miR1446 targets, gibberellin (GA) response modulator-like protein, sharing a 66% amino acid identity with a rice GA response modulator-like protein (LOC_Os01g67650), is a GA-related GRAS domain-containing transcription factor that could be a repressor of GA signaling (Tables 2 and 3) (Peng et al. 1997; Lu et al. 2008b). Overexpression of a GA-related GRAS protein has been shown to diminish stem elongation and induce a dwarf phenotype (Itoh et al. 2005). Thus, miR1446 could be a master regulator in GA

Table 3 Experimentally validated targets of mechanical stress-responsive miRNAs in *Populus trichocarpa* (Lu et al. 2005, 2008b)

miRNA	Function of targets	Gene model name[a]
miR408	Plantacyanin	e_gw1.40.264.1, estExt_fgenesh4_pm.C_LG_II1118
	Early responsive to dehydration-related protein	estExt_fgenesh4_pm.C_LG_XI0023
miR475	Pentatricopeptide repeat protein (PPR)	eugene3.00061747, eugene3.00190210, eugene3.24000001, eugene3.00062011, fgenesh1_pg.C_scaffold_1064000001[b]
miR476	PPR	eugene3.00131192[b]
miR478	Organic anion transporter-like protein	estExt_fgenesh4_pg.C_LG_XI1100
miR479	Unknown protein	eugene3.00031416
miR480	Proton-dependent oligopeptide transport family protein	e_gw1.XIX.1979.1
miR482	Putative disease resistance protein	grail3.0140004801, eugene3.00102261, fgenesh4_pg.C_LG_XIX000056, eugene3.02400005[b]
miR1444	Polyphenol oxidase	gw1.182.27.1, eugene3.00110271
miR1446	GCN5-related *N*-acetyltransferase (GNAT) family protein	eugene3.00060403
	Gibberellin response modulator-like protein	fgenesh4_pg.C_LG_XII000915

[a]Gene model names are present in *Populus trichocarpa* v1.1 except those indicated by the superscript letter b

[b]The gene model is present in *Populus trichocarpa* v1.0, but not in *Populus trichocarpa* v1.1

signaling in plant response to mechanical stress. In addition to the gibberellin response modulator-like protein, miR1446 also regulates a predicted target gene encoding a homeodomain transcription factor and an experimentally validated target gene coding for a GCN5-related *N*-acetyltransferase (GNAT) family protein, suggesting different roles of miR171 and miR1446 played in TW and OW formation (Tables 2 and 3). Consistently, the expression pattern of miR171 is different from miR1446 in response to mechanical stress. miR171 is downregulated more significantly in TW than in OW, whereas miR1446 exhibits preferentially upregulated expression in OW (Table 1).

The other mechanical stress-responsive miRNA, miR172, regulates five AP2 domain-containing transcription factors that have been shown to mediate floral patterning, seed development, and stem cell fate (Jofuku et al. 1994; Würschum et al. 2006; Zhao et al. 2007b). miR172 could be involved in the regulation of stem meristem cell activities by controlling AP2 domain-containing transcription factors during the development of specialized tissues in TW and OW. miR530 was also predicted to regulate a transcription factor, the homeobox transcription factor KN3. Functions of this protein are currently unknown. Interestingly, the targets of miR160 include ten auxin-responsive factors (ARF) that regulate the expression of auxin-responsive genes, indicating that miR160 plays a master regulatory role in auxin signaling in plant stems receiving mechanical stress.

Twenty of the predicted targets have been experimentally validated by the modified RLM-RACE method (Table 3). Most of them are involved in cell metabolic processes. The validated targets of miR408 include two plantacyanins and an early responsive to dehydration-related protein. Plantacyanins have been shown to play a role in reproduction in *Arabidopsis* and are also believed to be involved in redox reactions occurring during primary defense responses in plants and/or in lignin formation (Nersissian et al., 1998; Dong et al. 2005). Downregulation of plantacyanins resulted by the accumulation of miR408 may be important in lignin deposition during the formation of reaction wood that has low lignin content. Thus, miR408 appears a key regulator of cell wall composition in plant responses to mechanical stress.

Both miR475 and miR476 are predicted to target pentatricopeptide repeat proteins (PPRs) for cleavage (Lu et al. 2005). Six of them were experimentally validated (Table 3). PPR is a large plant protein family and is believed to be constitutively involved in the posttranscriptional regulation of organelle gene expression and RNA processing, suggesting the regulatory roles of miR475 and miR476 in RNA editing (Small and Peeters 2000; Lurin et al. 2004). The involvement of miR475 and miR476 in plant responses to mechanical stress indicates the importance of RNA processing in specialized tissue generation. In bent stems of *P. trichocarpa*, miR475 and miR476 exhibit distinct expression patterns. miR475 is suppressed in both TW and OW, whereas miR476 exhibits preferentially upregulated expression in OW, suggesting the different roles of miR475 and miR476 in response to tension stress and compression stress and the involvement of different members of PPR protein family in TW and OW formation (Table 1).

miR482, a miRNA downregulated by mechanical stress, was predicted to regulate a group of disease resistance proteins, of which four were validated by the modified 5′ RNA ligase-mediated-RACE method (Tables 1–3) (Lu et al. 2005, 2008b). Although the actual functions of these disease resistance proteins are currently unknown, the involvement of a miRNA that regulates disease resistance proteins in plant responses to mechanical stress suggests a cross-talk existed between biotic and abiotic stress responses and reveals possible roles for miRNAs in complex defense pathways (Lu et al. 2008b).

miR1444 is one of the two miRNAs that are strongly upregulated in both TW and OW (Table 1). It was predicted to target at least 13 polyphenol oxidase (PPO) genes for cleavage (Lu et al. 2008b). Among them, two were experimentally validated (Table 3). PPOs are copper-containing enzymes. They oxidize mono or dihydroxy phenols to quinines in plants and have been shown to be involved in plant resistance to biotic and abiotic stresses (Mayer, 2006). The role of PPOs in plant responses to mechanical stress has not been characterized. The other validated targets of mechanical stress-responsive miRNAs include an organic anion transporter-like protein gene targeted by miR478, a proton-dependent oligopeptide transport family protein gene targeted by miR480, and a unknown gene targeted by miR479 (Table 3). Similar to *PPOs*, the functions of these genes remain to be elucidated.

5 Conclusions and Perspectives

Mechanical stress is one of the major abiotic stresses affecting plant growth, survival, and reproduction. In plants continuously subjected to mechanically stressful conditions, significant anatomical and morphological changes were observed and were believed to be important for making plants more resistant to mechanical stress (Jaffe 1973, 1980). The mechanisms driving these changes and the signal transduction pathways of mechanosensing have been intensely studied. Cytoplasmic calcium, ROS, and phytohormones have been shown to play important roles in plant responses to mechanical stress. A large number of mechanical stress-responsive genes, particularly those involving in reaction wood formation in tree species, have been identified. Importantly, a group of recently discovered small noncoding RNAs, known as miRNAs, were found to play regulatory roles in the development of specialized woody tissues in *P. trichocarpa* plants (Lu et al. 2005, 2008b).

miRNAs are derived from primary miRNAs through the interaction of various proteins (Voinnet 2009). They play regulatory roles in the development of plant organs and in plant responses to biotic and abiotic stresses mainly by targeting other RNAs for cleavage (Chen 2005; Jones-Rhoades et al. 2006; Mallory and Vaucheret 2006; Sunkar et al. 2007; Lu et al. 2008c). Studies on the bent stems of *P. trichocarpa* plants result in identifying a group of mechanical stress-responsive miRNAs (Lu et al. 2005, 2008b). These miRNAs were predicted or experimentally validated to target genes encoding transcription factors, metabolism and cellular process-related proteins, and many function-unknown proteins. Analysis of miRNA expression and target gene function suggested that miRNAs could be master regulators in phytohormone signaling and played very important roles in reaction wood formation. The results provide novel and useful information for fully understanding the genetic and molecular mechanisms of plant responses to mechanical stress.

Further genome-wide identification of miRNAs and other small regulatory RNAs involved in mechanical stress responses in *P. trichocarpa* and other plant species will help discover new regulators in plants. Additionally, functional characterization of mechanical stress-responsive miRNAs and their target genes through genetic manipulation will certainly shed light on the elucidation of miRNA-mediated regulatory network of thigmomorphogenesis (Jaffe 1973, 1980).

Acknowledgments This work was supported by the National High-Tech Research and Development Program (863 Program) of China (2009AA10Z101) and the National Natural Science Foundation of China (31070534).

References

Allen E, Xie Z, Gustafson AM, Carrington JC (2005) microRNA-directed phasing during trans-acting siRNA biogenesis in plants. Cell 121:207–221

Andersson-Gunnerås S, Mellerowicz EJ, Love J, Segerman B, Ohmiya Y, Coutinho PM, Nilsson P, Henrissat B, Moritz T, Sundberg B (2006) Biosynthesis of cellulose-enriched tension wood in *Populus*: global analysis of transcripts and metabolites identifies biochemical and developmental regulators in secondary wall biosynthesis. Plant J 45:144–165

Bhandari S, Fujino T, Thammanagowda S, Zhang D, Xu F, Joshi CP (2006) Xylem-specific and tension stress-responsive coexpression of KORRIGAN endoglucanase and three secondary wall-associated cellulose synthase genes in aspen trees. Planta 224:828–837

Botella JR, Arteca JM, Somodevilla M, Arteca RN (1996) Calcium-dependent protein kinase gene expression in response to physical and chemical stimuli in mungbean (*Vigna radiata*). Plant Mol Biol 30:1129–1137

Boualem A, Laporte P, Jovanovic M, Laffont C, Plet J, Combier JP, Niebel A, Crespi M, Frugier F (2008) MicroRNA166 controls root and nodule development in *Medicago truncatula*. Plant J 54:876–887

Braam J (2005) In touch: plant responses to mechanical stimuli. New Phytol 165:373–389

Braam J, Davis RW (1990) Rain-, wind- and touch-induced expression of calmodulin and calmodulin-related genes in *Arabidopsis*. Cell 60:357–364

Cardon GH, Höhmann S, Nettesheim K, Saedler H, Huijser P (1997) Functional analysis of the *Arabidopsis thaliana* SBP-box gene *SPL3*: a novel gene involved in the floral transition. Plant J 12:367–377

Chen X (2005) MicroRNA biogenesis and function in plants. FEBS Lett 579:5923–5931

Chiou T-J, Aung K, Lin S-I, Wu C-C, Chiang S-F, Su C-L (2006) Regulation of phosphate homeostasis by microRNA in *Arabidopsis*. Plant Cell 18:412–421

Dadswell HE, Wardrop AB (1955) The structure and properties of tension wood. Holzforschung 9:97–104

Déjardin A, Leplé J-C, Lesage-Descauses M-C, Costa G, Pilate G (2004) Expressed sequence tags from poplar wood tissues–a comparative analysis from multiple libraries. Plant Biol (Stuttg) 6:55–64

Di Laurenzio L, Wysocka-Diller J, Malamy JE, Pysh L, Helariutta Y, Freshour G, Hahn MG, Feldmann KA, Benfey PN (1996) The *SCARECROW* gene regulates an asymmetric cell division that is essential for generating the radial organization of the *Arabidopsis* root. Cell 86:423–433

Dong J, Kim ST, Lord EM (2005) Plantacyanin plays a role in reproduction in *Arabidopsis*. Plant Physiol 138:778–789

Fujii H, Chiou TJ, Lin SI, Aung K, Zhu JK (2005) A miRNA involved in phosphate-starvation response in *Arabidopsis*. Curr Biol 15:2038–2043

Fujii M, Azuma J, Tanaka F, Kato A, Koshijima T (1982) Studies on hemicelluloses in tension wood. I. Chemical composition of tension, opposite and side woods of Japanese beech (*Fagus crenata Blume*). Wood Res 68:8–21

Gadea J, Conejero V, Vera P (1999) Developmental regulation of a cytosolic ascorbate peroxidase gene from tomato plants. Mol Gen Genet 262:212–219

Gandikota M, Birkenbihl RP, Hohmann S, Cardon GH, Saedler H, Huijser P (2007) The miRNA156/157 recognition element in the 3′ UTR of the Arabidopsis SBP box gene *SPL3* prevents early flowering by translational inhibition in seedlings. Plant J 49:683–693

Griffiths-Jones S, Saini HK, van Dongen S, Enright AJ (2008) miRBase: tools for microRNA genomics. Nucleic Acids Res 36(Database Issue):D154–D158

Guo HS, Xie Q, Fei JF, Chua NH (2005) MicroRNA directs mRNA cleavage of the transcription factor NAC1 to downregulate auxin signals for *Arabidopsis* lateral root development. Plant Cell 17:1376–1386

Haley A, Russell AJ, Wood N, Allan AC, Knight M, Campbell AK, Trewavas AJ (1995) Effects of mechanical signaling on plant cell cytosolic calcium. Proc Natl Acad Sci USA 92:4124–4128

Hirsinger C, Salva I, Marbach J, Durr A, Fleck J, Jamet E (1999) The tobacco extensin gene *Ext 1.4* is expressed in cells submitted to mechanical constraints and in cells proliferating under hormone control. J Exp Bot 50:343–355

Itoh H, Shimada A, Ueguchi-Tanaka M, Kamiya N, Hasegawa Y, Ashikari M, Matsuoka M (2005) Overexpression of a GRAS protein lacking the DELLA domain confers altered gibberellin responses in rice. Plant J 44:669–679

Jaffe MJ (1973) Thigmomorphogenesis: the response of plant growth and development to mechanical stimulation. With special reference to Bryonia dioica. Planta 114:143–157

Jaffe MJ (1980) Morphogenetic responses of plants to mechanical stimuli or stress. BioScience 30:239–243

Jofuku KD, den Boer BGW, Montagu MV, Okamuro JK (1994) Control of *Arabidopsis* flower and seed development by the homeotic gene *APETALA2*. Plant Cell 6:1211–1225

Jones-Rhoades MW, Bartel DP (2004) Computational identification of plant microRNAs and their targets, including a stressinduced miRNA. Mol Cell 14:787–799

Jones-Rhoades MW, Bartel DP, Bartel B (2006) MicroRNAs and their regulatory roles in plants. Annu Rev Plant Biol 57:19–53

Joshi CP (2003) Xylem-specific and tension stress responsive expression of cellulose synthase genes from aspen trees. Appl Biochem Biotechnol 105:17–26

Kim J, Jung JH, Reyes JL, Kim YS, Kim SY, Chung KS, Kim JA, Lee M, Lee Y, Narry Kim V, Chua NH, Park CM (2005) microRNA-directed cleavage of *ATHB15* mRNA regulates vascular development in *Arabidopsis* inflorescence stems. Plant J 42:84–94

Knight MR, Campbell AK, Smith SM, Trewavas AJ (1991) Transgenic plant aequorin reports the effects of touch and cold-shock and elicitors on cytoplasmic calcium. Nature 352:524–526

Kutter C, Schob H, Stadler M, Meins F, Si-Ammour A (2007) MicroRNA-mediated regulation of stomatal development in *Arabidopsis*. Plant Cell 19:2417–2429

Lee D, Polisensky DH, Braam J (2005) Genome wide identification of touch and darkness-regulated Arabidopsis genes: a focus on calmodulin-like and *XTH* genes. New Phytol 165: 429–444

Lee RC, Feinbaum RL, Ambros V (1993) The *C. elegans* heterochronic gene *lin-4* encodes small RNAs with antisense complementarity to *lin-14*. Cell 75:843–854

Lee Y, Jeon K, Lee JT, Kim S, Kim VN (2002) MicroRNA maturation: Stepwise processing and subcellular localization. EMBO J 21:4663–4670

Llave C, Kasschau KD, Rector MA, Carrington JC (2002) Endogenous and silencing-associated small RNAs in plants. Plant Cell 14:1605–1619

Lu S, Li L, Yi X, Joshi CP, Chiang VL (2008a) Differential expression of three eucalyptus secondary cell wall-related cellulose synthase genes in response to tension stress. J Exp Bot 59:681–695

Lu S, Sun Y-H, Amerson H, Chiang VL (2007) MicroRNAs in loblolly pine (*Pinus taeda* L.) and their association with fusiform rust gall development. Plant J 51:1077–1098

Lu S, Sun Y-H, Chiang VL (2008b) Stress-responsive microRNAs in *Populus trichocarpa*. Plant J 55:131–151

Lu S, Sun Y-H, Shi R, Clark C, Li L, Chiang VL (2005) Novel and mechanical stress-responsive microRNAs in *Populus trichocarpa* that are absent from Arabidopsis. Plant Cell 17:2186–2205

Lu Y-D, Gan QH, Chi XY, Qin S (2008c) Roles of microRNA in plant defense and virus offense interaction. Plant Cell Rep 27:1571–1579

Lurin C, Andrés C, Aubourg S, Bellaoui M, Bitton F, Bruyère C, Caboche M, Debast C, Gualberto J, Hoffmann B, Lecharny A, Le Ret M, Martin-Magniette ML, Mireau H, Peeters N, Renou JP, Szurek B, Taconnat L, Small I (2004) Genome-wide analysis of *Arabidopsis* pentatricopeptide repeat proteins reveals their essential role in organelle biogenesis. Plant Cell 16:2089–2103

Mallory AC, Bartel DP, Bartel B (2005) MicroRNA-directed regulation of *Arabidopsis AUXIN RESPONSE FACTOR17* is essential for proper development and modulates expression of early auxin response genes. Plant Cell 17:1360–1375

Mallory AC, Vaucheret H (2006) Functions of microRNAs and related small RNAs in plants. Nat Genet 38:S31–S36

Mayer AM (2006) Polyphenol oxidases in plants and fungi: going places? A review. Phytochemistry 67:2318–2331

Mayrose M, Ekengren SK, Melech-Bonfil S, Martin GB, Sessa G (2006) A novel link between tomato *GRAS* genes, plant disease resistance and mechanical stress response. Mol Plant Phathol 7:593–604

Mette MF, van der Winden J, Matzke M, Matzke AJ (2002) Short RNAs can identify new candidate transposable element families in *Arabidopsis*. Plant Physiol 130:6–9

Millar AA, Gubler F (2005) The Arabidopsis *GAMYB*-like genes, *MYB33* and *MYB65*, are microRNA-regulated genes that redundantly facilitate anther development. Plant Cell 17:705–721

Mizoguchi T, Irie K, Hirayama T, Hayashida N, Yamaguchi-Shinozaki K, Matsumoto K, Shinozaki K (1996) A gene encoding a mitogen-activated protein kinase kinase kinase is induced simultaneously with genes for a mitogen-activated protein kinase and an S6 ribosomal protein kinase by touch, cold, and water stress in *Arabidopsis thaliana*. Proc Natl Acad Sci USA 93:765–769

Monshausen GB, Bibikova TN, Weisenseel MH, Gilroy S (2009) Ca^{2+} regulates reactive oxygen species production and pH during mechanosensing in *Arabidopsis* roots. Plant Cell 21:2341–2356

Mori IC, Schroeder JI (2004) Reactive oxygen species activation of plant Ca^{2+} channels. A signaling mechanism in polar growth, hormone transduction, stress signaling, and hypothetically mechanotransduction. Plant Physiol 135:702–708

Navarro L, Dunoyer P, Jay F, Arnold B, Dharmasiri N, Estelle M, Voinnet O, Jones JD (2006) A plant miRNA contributes to antibacterial resistance by repressing auxin signaling. Science 21:436–439

Nersissian AM, Immoos C, Hill MG, Hart PJ, Williams G, Herrmann RG, Valentine JS (1998) Uclacyanins, stellacyanins, and plantacyanins are distinct subfamilies of phytocyanins: plant-specific mononuclear blue copper proteins. Protein Sci 7:1915–1929

Niklas KJ (1998) Effects of vibration on mechanical properties and biomass allocation pattern of *Capsella bursapastoris* (*Cruciferae*). Ann Bot 82:147–156

Nishikubo N, Awano T, Banasiak A, Bourquin V, Ibatullin F, Funada R, Brumer H, Teeri TT, Hayashi T, Sundberg B, Mellerowicz EJ (2007) Xyloglucan endo-transglycosylase (XET) functions in gelatinous layers of tension wood fibers in poplar: a glimpse into the mechanism of the balancing act of trees. Plant Cell Physiol 48:843–855

Oufattole M, Arango M, Boutry M (2000) Identification and expression of three new *Nicotiana plumbaginifolia* genes which encode isoforms of a plasma-membrane H^+-ATPase, and one of which is induced by mechanical stress. Planta 210:715–722

Palatnik JF, Allen E, Wu X, Schommer C, Schwab R, Carrington JC, Weigel D (2003) Control of leaf morphogenesis by microRNAs. Nature 425:257–263

Park W, Li J, Song R, Messing J, Chen X (2002) CARPEL FACTORY, a Dicer homolog, and HEN1, a novel protein, act in microRNA metabolism in *Arabidopsis thaliana*. Curr Biol 12:1484–1495

Paux E, Carocha V, Marques C, Mendes de Sousa A, Borralho N, Sivadon P, Grima-Pettenati J (2005) Transcript profiling of *Eucalyptus* xylem genes during tension wood formation. New Phytol 167:89–100

Peng J, Carol P, Richards DE, King KE, Cowling RJ, Murphy GP, Harberd NP (1997) The Arabidopsis *GAI* gene defines a signaling pathway that negatively regulates gibberellin responses. Genes Dev 11:3194–3205

Pfeffer S, Zavolan M, Grasser FA, Chien M, Russo JJ, Ju J, John B, Enright AJ, Marks D, Sander C, Tuschl T (2004) Identification of virus-encoded microRNAs. Science 304:734–736

Raman S, Greb T, Peaucelle A, Blein T, Laufs P, Theres K (2008) Interplay of miR164, *CUP-SHAPED COTYLEDON* genes and *LATERAL SUPPRESSOR* controls axillary meristem formation in *Arabidopsis thaliana*. Plant J 55:65–76

Reinhart BJ, Weinstein EG, Rhoades MW, Bartel B, Bartel DP (2002) MicroRNAs in plants. Genes Dev 16:1616–1626

Rhoades MW, Reinhart BJ, Lim LP, Burge CB, Bartel B, Bartel DP (2002) Prediction of plant microRNA targets. Cell 110:513–520

Schwab R, Palatnik JF, Riester M, Schommer C, Schmid M, Weigel D (2005) Specific effects of microRNAs on the plant transcriptome. Dev Cell 8:517–527

Scurfield G (1971) Histochemistry of reaction wood cell walls in tow species of *Eucalyptus* and in *Tristania conferta* R. BR Aust J Bot 20:9–26

Scurfield G (1973) Reaction wood: its structure and function. Science 179:647–655

Shirsat AH, Bell A, Spence J, Harris JN (1996) The *Brassica napus* extA extensin gene is expressed in regions of the plant subject to tensile stresses. Planta 199:618–624

Sinnott EW (1952) Reaction wood and the regulation of tree form. Am J Bot 39:69–78

Small ID, Peeters N (2000) The PPR motif: a TPR-related motif prevalent in plant organellar proteins. Trends Biochem Sci 25:46–47

Sterky F, Bhalerao RR, Unneberg P, Segerman B, Nilsson P, Brunner AM, Charbonnel-Campaa L, Lindvall JJ, Tandre K, Strauss SH, Sundberg B, Gustafsson P, Uhlen M, Bhalerao RP, Nilsson O, Sandberg G, Karlsson J, Lundeberg J, Jansson S (2004) A *Populus* EST resource for plant functional genomics. Proc Natl Acad Sci USA 101:13951–13956

Stone JM, Liang X, Nekl ER, Stiers JJ (2005) *Arabidopsis AtSPL14*, a plant-specific SBP-domain transcription factor, participates in plant development and sensitivity to fumonisin B1. Plant J 41:744–754

Sunkar R, Chinnusamy V, Zhu J, Zhu JK (2007) Small RNAs as big players in plant abiotic stress responses and nutrient deprivation. Trends Plant Sci 12:301–309

Sunkar R, Kapoor A, Zhu JK (2006) Posttranscriptional induction of two Cu/Zn superoxide dismutase genes in *Arabidopsis* is mediated by downregulation of miR398 and important for oxidative stress tolerance. Plant Cell 18:2051–2065

Sunkar R, Zhu J-K (2004) Novel and stress-regulated microRNAs and other small RNAs from *Arabdopsis*. Plant Cell 16:2001–2019

Tiimonen H, Häggman H, Tsai CJ, Chiang V, Aronen T (2007) The seasonal activity and the effect of mechanical bending and wounding on the *PtCOMT* promoter in *Betula pendula* Roth. Plant Cell Rep 26:1205–1214

Timell TE (1973) Studies on opposite wood in conifers part I: chemical composition. Wood Sci Technol 7:1–5

Unte US, Sorensen AM, Pesaresi P, Gandikota M, Leister D, Saedler H, Huijser P (2003) *SPL8*, an SBP-Box gene that affects pollen sac development in *Arabidopsis*. Plant Cell 15:1009–1019

Vazquez F, Gasciolli V, Crété P, Vaucheret H (2004) The nuclear dsRNA binding protein HYL1 is required for microRNA accumulation and plant development, but not posttranscriptional transgene silencing. Curr Biol 14:346–351

Voinnet O (2009) Origin, biogenesis, and activity of plant microRNAs. Cell 136:669–687

Wardrop AB, Davies GW (1964) The structure of reaction wood: the structure and differentiation of compression wood. Aust J Bot 12:24–38

Whetten R, Sun YH, Zhang Y, Sederoff R (2001) Functional genomics and cell wall biosynthesis in loblolly pine. Plant Mol Biol 47:275–291

Wightman B, Ha I, Ruvkun G (1993) Posttranscriptional regulation of the heterochronic gene *lin-14* by *lin-4* mediates temporal pattern formation in *C. elegans*. Cell 75:855–862

Williams L, Grigg SP, Xie M, Christensen S, Fletcher JC (2005) Regulation of *Arabidopsis* shoot apical meristem and lateral organ formation by microRNA miR166g and its AtHD-ZIP target genes. Development 132:3657–3668

Wu G, Poethig RS (2006) Temporal regulation of shoot development in *Arabidopsis thaliana* by *MIR156* and its target *SPL3*. Development 133:3539–3547

Wu L, Joshi CP, Chiang VL (2000) A xylem-specific cellulose synthase gene from aspen (*Populus tremuloides*) is responsive to mechanical stress. Plant J 22:495–502

Würschum T, Gross-Hardt R, Laux T (2006) *APETALA2* regulates the stem cell niche in the *Arabidopsis* shoot meristem. Plant Cell 18:295–307

Yamashita S, Yoshida M, Yamamoto H, Okuyama T (2008) Screening genes that change expression during compression wood formation in *Chamaecyparis obtusa*. Tree Physiol 28:1331–1340

Zhang Y, Schwarz S, Saedler H, Huijser P (2007) SPL8, a local regulator in a subset of gibberellins-mediated developmental processes in *Arabidopsis*. Plant Mol Biol 63:429–439

Zhang Z, Yu J, Li D, Zhang Z, Liu F, Zhou X, Wang T, Ling Y, Su Z (2010) PMRD: plant microRNA database. Nucleic Acids Res 38(Database Issue): D806–D813

Zhao B, Liang R, Ge L, Li W, Xiao H, Lin H, Ruan K, Jin Y (2007a) Identification of drought-induced microRNAs in rice. Biochem Biophys Res Commun 354:585–590

Zhao L, Kim Y, Dinh TT, Chen X (2007b) miR172 regulates stem cell fate and defines the inner boundary of APETALA3 and PISTILLATA expression domain in *Arabidopsis* floral meristems. Plant J 51:840–849

Zhou GK, Kubo M, Zhong R, Demura T, Ye ZH (2007a) Overexpression of *MIR165* affects apical meristem formation, organ polarity establishment and vascular development in *Arabidopsis*. Plant Cell Physiol 48:391–404

Zhou X, Wang G, Zhang W (2007b) UV-B responsive micro-RNA genes in *Arabidopsis thaliana*. Mol Syst Biol 3:103

Index

P. Wojtaszek (ed.), *Mechanical Integration of Plant Cells and Plants*, Signaling and Communication in Plants 9, DOI 10.1007/978-3-642-19091-9,

© Springer-Verlag Berlin Heidelberg 2011

MIX
Papier aus verantwortungsvollen Quellen
Paper from responsible sources
FSC® C105338

Printed by Books on Demand, Germany